Bacteriophage T4 and its relatives

A series of critical reviews

Editors:

Jim D Karam

Department of Biochemistry, Tulane University Health Sciences Center, 1430 Tulane Avenue, New Orleans, LA, USA

Eric S Miller

Department of Microbiology, Campus Box 7615, North Carolina State University, Raleigh, NC 27695, USA

VIROLOGY JOURNAL

Editor-in-Chief: Robert Garry

Published by

BioMed Central

Published by BioMed Central Ltd.

British Library Cataloguing-in-Publication Data.
A catalogue record for this book is available from the British Library.

ISBN-13: 978-0-9540278-7-2
ISBN-10: 0-9540278-7-6

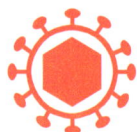

VIROLOGY JOURNAL

Virology Journal (ISSN 1743-422X) is published online by :

BioMed Central Ltd
Floor 6, 236 Gray's Inn Road
London WC1X 8HB, UK
T: +44 (0) 20 3192 2000
F: +44 (0) 20 3192 2010
E: info@biomedcentral.com
E: rfgarry@tulane.edu (editorial enquiries)

Virology Journal can be found on the web under the following address:
http://www.virologyj.com

Open Access

For further information about the journal, please see the description available on the website (http://www.virologyj.com)

Disclaimer

Indexing services

Virology Journal is indexed by CABI, CAS, Citebase, EmBase, EmBiology, Google, Google Scholar, Index Copernicus, MEDLINE, OAIster, PubMed, PubMed Central, Scirus, Scopus, SOCOLAR, Thomson Reuters, Zetoc.

Cover images

Front cover: In vitro display of antigens on bacteriophage T4 capsid. See Rao and Black, http://www.virologyj.com/content/7/1/356
Back cover: A cartoon model of leading and lagging strand DNA synthesis by the Bacteriophage T4 Replisome. See Mueser *et al.,* http://www.virologyj.com/content/7/1/359

Bacteriophage T4 and its relatives

A series of critical reviews

Contents

VIROLOGY JOURNAL

PREFACE

Bacteriophage T4 and its Relatives

The articles appearing in this compilation were published online by BioMed Central's *Virology Journal* in late 2010 and are now being made available in hard copy through the highly appreciated efforts of the journal's editorial staff. Our motivation to plan this series of reviews arose from the success of the last book entirely dedicated to T4, *Molecular Biology of Bacteriophage T4* (*ASM Press*, 1994), which went out of print a few years ago but remains of interest to many researchers and educators. Since 1994, the emergence and growth of the field of microbial genomics has had a tremendous impact on our understanding of the molecular bases of genetic processes and their diversity. It is no coincidence that T4 and its phylogenetic relatives have been very much part of the progress that has been made in studies of the mechanisms of genome replication and maintenance, gene expression and macromolecular assembly. We are deeply grateful to the corresponding and other contributing authors who agreed to join us in making this series possible.

Initially, we explored several publishing avenues, but ultimately recognized that online publishing is the most desirable economically and in terms of the effectiveness of reaching a broad readership. The relationship between BioMed Central's *Virology Journal* and one of us (Jim D Karam; an Editor) facilitated the process. The commitment from *Virology Journal*'s Editor-in-Chief, Robert Garry, to the project and the work of BioMed Central's editorial staff was invaluable.

All of the included monographs underwent external evaluation and review, often by experts in the field recommended by the authors. It was always a pleasure to work with the authors, who conscientiously and patiently worked with the editors to develop up-to-date monographs on their topics. Through this process, many memories were awakened of colleagues who contributed to the 1994 T4 book, including experts and mentors who have since passed (i.e., Gisela Mosig, Nancy Nossal, Ulf Henning, Eduard Kellenberger) and who stand as visionaries in the field. We hope that new approaches to genomics, and molecular and structural biochemistry will establish a new generation of scientists who find phage T4 and its relatives rewarding systems of biological study.

The front and rear covers of the print version were chosen from two of the most widely accessed online articles of the series.

The Editors,
Jim D Karam and Eric S Miller

Karam and Miller *Virology Journal* 2010, **7**:293
http://www.virologyj.com/content/7/1/293

VIROLOGY JOURNAL

EDITORIAL

Bacteriophage T4 and its relatives

Jim D Karam[1]*, Eric S Miller[2]

Bacteriophage T4 and its relatives (A series of critical reviews)
Jim Karam & Eric Miller

In the coming months Virology Journal will publish a number of authoritative reviews about the biochemistry, structural biology and genomics of the bacteriophage T4 and the T4-related phages. Phage T4 is one of the most extensively investigated viruses and has been the central focus of several monographs and reviews over the last 25 years. Its popularity among experimental biologists is related to the ease with which this phage and some of its relatives can be propagated in widely available non-pathogenic laboratory strains of *Escherichia coli* and the diversity of experimental approaches that can be used to analyze its DNA genome and the RNA and protein products it encodes. The T4 biological system is amenable to investigation by genetic, phylogenetic, biochemical, biophysical, structural, computational and other tools.

Advances in T4 science have paralleled advances in Molecular Biology since the birth of this interdisciplinary field around the middle of the 20th Century [1,2]. Such seminal discoveries as the chemical nature of the gene, the existence of messenger RNA, how the genetic code is read, how genes determine protein structure, how DNA is replicated by multicomponent protein machines and many other findings that have become integral to our current understanding of basic molecular mechanism in biology have typically involved important contributions from the T4 and T4-related experimental systems. The last monograph to comprehensively review all aspects of the molecular biosciences of the T4 virus was published in 1994 [3]. Since that time, the field of Molecular Biology has undergone considerable transformation, particularly as a consequence of advancements in the methods for sequencing microbial and eukaryotic genomes and using DNA sequence data for novel experimental designs that have yielded numerous rewards in resolving biological mysteries and stimulating

the growth of biotechnology. The review series to be published in Virology Journal will emphasize advances and seminal discoveries in four major areas of T4 research: Genomics, Gene Expression, DNA Replication and Phage Morphogenesis.

Genomics

Phages that share an evolutionary history with T4 are highly abundant in nature and can be detected by simple plating techniques using a diversity of bacterial genera or species as hosts. Over the last several years, advances in DNA sequencing technologies have made it possible to analyze the genomes of a large number of these phages, including both close and distant phylogenetics relatives of T4. The sequence database for these T4-like phages is a rich source of insights into the mechanisms of genome replication, expression, packaging and diversification in evolution. In many cases, the experimental systems and genetic tools to test these insights are available. In a review entitled ***Genomes of the T4-related phages as windows on microbial evolution,*** V. Petrov, S. Ratnayaka, J. Nolan, E. Miller and J. Karam summarize the genome sequence information currently available in databases for more than 40 T4 relatives that represent a wide array of specificities to bacterial hosts. Genomes have been sequenced from T4-related phages that infect strains of *Enterobacteria, Aeromonas, Acinetobacter, Klebsiella, Pseudomonas, Vibrio* and marine *Synechococcus* and *Prochlorococcus.* Comparisons between these genomes reveal a high degree of genetic diversity around a conserved core of genes that determine the replication, temporal expression and packaging (phage morphogenesis) of a specifically designed dsDNA viral chromosome. The review draws parallels between the diversity of this large and mosaically organized group of phage genomes and the type and extent of diversity that is being observed within groups of prokaryotic and eukaryotic genomes in general. The broad natural distribution of the T4-related phages includes the largest ecosystem of our planet, the marine environment. A review by M. Clokie, A. Millard and N. Mann (***T4-related phages of the marine***

* Correspondence: karamoff@tulane.edu
[1]Department of Biochemistry, Tulane University Health Sciences Center, 1430 Tulane Avenue, New Orleans, LA, USA
Full list of author information is available at the end of the article

ecosystem) focuses on the comparative genomics and other studies of T4 relatives that infect cyanobacteria, particularly the genera *Synechococcus* and *Prochlorococcus*. The results of these studies have implications about the possible roles of the T4-related cyanophages as traffickers of bacterial genes, including genes involved in photosynthesis, and the potential impact of host-phage interactions on control of the marine ecology. A remarkable feature of T4-related genomes, irrespective of the host range or geographical origin, is the abundance and diversity of mobile DNA elements that have colonized this group of phages in evolution. Studies with phage T4 were among the first to show the natural existence of mobile introns in the prokaryotic world and to elucidate the mechanisms of intron mobility through the action of homing endonucleases [4]. Almost every category of homing endonuclease genes has been detected in the group of T4-related genomes sequenced so far. These genes can exist inside as well as outside of introns and the homing enzymes they produce can mobilize a diversity of DNA sequences laterally between genomes of the T4 family of phages[5]. In the article entitled *Mobile DNA elements of the T4-related phages*, D. Edgell, E. Gibb and M. Belfort review the major progress that has been made over the last 15 years in our understanding of the structures, mechanisms of action and physiological roles of these mobile elements and their potential impact on phage and microbial evolution.

Gene expression

Temporal gene expression constitutes an important part of the strategies used by all viruses to coordinate the different biochemical processes involved in viral genome replication, genome packaging and the release of new generations of virus. In general, the control of gene expression, which is highly permissive to diversification in the evolution of organisms [6], contributes significantly to the adaptation of viruses to new physiological conditions such as the encounter of these infectious genetic entities with new potential hosts. The T4-related phages exhibit many examples of such diversification. In three reviews that deal with the control of T4 gene expression, the authors discuss advances in research on the structures and functions of the phage-encoded proteins that determine the temporal utilization of the phage genome during the different stages of phage development in the bacterial host. The review by D. Hinton (*Control of transcription in the prereplicative phase of T4 development*) discusses the current state of knowledge about the structures and functions of the protein factors and DNA sites that control phage genome transcription shortly after the entry of the phage DNA into the bacterial host cell. The protein-DNA

interactions during the early phase of the phage developmental program set the stage for diverting the host RNA polymerase from transcription of the bacterial genome to transcription of the T4 genes required for phage DNA replication, repair and the other replication-related processes that ultimately ensure the coordination between phage DNA metabolism and phage morphogenesis. Some of the phage-encoded proteins made during the prereplicative phase introduce modifications onto subunits of the host RNA polymerase while others associate with this enzyme and alter the specificity of the transcription apparatus so that expression of the phage genes that determine the structure and assembly of infectious virions is maximized. The review by E. P. Geiduscheck and G. A. Kassavetis (*Transcriptional control during the late phase of T4 development*) discusses recent progress in the analysis of the structures and biochemical functions of the key proteins, especially gp33, gp45 and gp55, involved in this transition in RNA polymerase specificity from early to late transcription. These proteins play roles in coordinating late transcription with genome replication during the late phase of phage development. The third review, *Posttranscriptional control of T4 development* by M. Uzan and E. S. Miller discusses the several biochemical strategies used by phage T4 to control gene expression beyond the level of transcription. This phage encodes a number of proteins that exert differential effects on the utilization of the mRNA for specific phage induced proteins. These strategies include controls over mRNA activation (RNA processing), inactivation (RNA decay and repression of translation) and host ribosome function. All 3 reviews highlight the insights gained from the sequence polymorphism that has been observed among allelic proteins in the databases for T4 relatives.

DNA replication

T4 encodes all of the proteins required for replication of the phage DNA genome, including the components of a complete DNA replisome and several other proteins that perform important auxiliary functions in the replication, repair and recombination of the genome. The ease with which the T4 system replication system can be analyzed by a wide range of experimental tools and the many similarities this system exhibits to eukaryotic DNA replication machines have made it a widely recognized model for investigators in the DNA replication field. Genetic and biochemical studies of the multi-protein complexes that carry out T4 DNA replication have brought to light the important role that genetic recombination plays in replication [7], elucidated several of the enzyme mechanisms involved at DNA replication forks and provided the generally accepted model for coordination of DNA synthesis between the leading and lagging strands (i.e., the trombone model; [8]). Three reviews will highlight the recent advances in research on the mechanisms of the initiation and DNA chain elongation stages of T4 DNA replication and the structures of the proteins that carry out these processes. A review by T.C. Mueser, J.M. Devos, J.M. Hinerman and K.J. Williams *(Structural analysis of T4 DNA replication)* discusses these structures with emphasis on the determinants of biochemical function and by drawing parallels to available structural information about replication proteins from other biological systems. In a review entitled *Initiation of T4 DNA replication and replication fork dynamics,* K. N. Kreuzer and R. J. Brister describe recent advances in understanding the interplay between two modes of intiation of T4 DNA replication, one based on the recognition of specific origins on the T4 dsDNA chromosome and one based on the use of the enzymes of homologous recombination to create initiation sites through the invasion of the circularly permuted and terminally redundant phage dsDNA chromosomes by the ends of homologous molecules. Remarkably, like the linear dsDNA chromosomes of eukaryotes, the T4 chromosome harbors multiple sites for intiation (origins) of replication and the review discusses the evidence for differential use of these origins and the take over by recombination-driven initiation during the course of the phage developmental program. Over the last 15 years, the structures of several of the proteins of the T4 DNA replisome and/or homologous proteins from the T4-related phage RB69 have been analyzed at the atomic level by X-ray crystallography. A review by J. Liu and S.W. Morrical (**Assembly and dynamics of the T4 homologous recombination machinery**) emphasizes advances in research on the structures and biochemical mechanisms of the T4 encoded proteins that support genetic recombination and the initiation of phage DNA replication. As an integral part of the biochemical strategy for generating hundreds of phage genomes per infected cell, the T4-encoded proteins for genetic recombination have evolved to be abundant and highly active and as a consequence, have been accessible for detailed biochemical analysis. They are serving as models for evolutionarily related counterparts in eukaryotes and bacteria.

Phage morphogenesis

Two reviews in this thematic series focus on the synthesis, structures and assembly of the two major components of the T4 virion, the capsid in which the phage DNA is packaged (T4 heads) and the phage tail and tail fibers, which make it possible for this bacterial virus to recognize its bacterial host and deliver its DNA into the cell. Far from being a hindrance to the experimental biologist, the complexity of the structure of the T4 virion has proven to be a great asset in the elucidation of many biochemical mechanisms that are broadly represented in other systems of viral assembly in cellular hosts. The structure of T4 phage particles, or what is sometimes referred to as the "T4 morphotype", exhibits several features that are conserved among phylogenetic relatives of this phage and that appear to be mimicked by a large number of viruses in nature. A review entitled *Morphogenesis of T4 heads* (by V. Rao and L. Black) discusses the new insights that have been gained over the last few years about T4 head assembly from the direct structural analysis of a protein (gp24) related to the major component of the phage capsid (gp23), solid NMR analysis of T4 particles, other biophysical studies and refinements in in vitro assays of DNA packaging. A second review (*Morphogenesis of the T4 tails and tail fibers* by P. G. Leiman, F. Arisaka, M.J. Van Raiij, A. A. Aksyuk, V. A. Kostychendo, S. Kanamaru and M. G. Rossmann further underscores the impact of recent advances in the structural sciences on understanding of the biochemical processes that underlie the assembly of multi-component nucleoprotein biological structures. The studies reviewed here have led to vivid images of T4 phage particles and the dynamics of phage infection. This review discusses the application of a variety of approaches that determined the structure of the contractile tail of T4 and the broad implications of the findings to the structural organization of other phages with contractile tails.

Some of the reviews in this series will be supplemented by web-based information to be updated as new developments in the field come to light. These supplements and updates will include summaries in the form of PowerPoint charts (including simple animations), Tables or videos that can be used by research scientists and educators alike.

Virology Journal is taking a leading role in facilitating the dissemination of new information in fast-growing areas of phage biology and the series on T4 and its relatives constitutes a first example of the journal's efforts in this regard. We are grateful to Robert F. Garry, Ph.D., Editor in Chief of Virology Journal and Professor at Tulane University for his guidance during the preparation of manuscripts for this thematic series.

Author details
[1]Department of Biochemistry, Tulane University Health Sciences Center, 1430 Tulane Avenue, New Orleans, LA, USA. [2]Department of Microbiology, Campus Box 7615, North Carolina State University, Raleigh, NC 27695, USA.

Received: 3 August 2010 Accepted: 28 October 2010
Published: 28 October 2010

References

1. Cairns J, Stent GS, Watson JD, (eds.): **Phage and the Origins of Molecular Biology.** New York: Cold Spring Harbor Laboratory Press; 1992.
2. Holmes FL: **Seymor Benzer and the definition of the gene.** In *The concept of the gene in development and evolution: Historical and episemological perspectives.* Edited by: Beurton P, Falk R, Rheinberger H-J. Cambridge, UK: Cambridge University Press; 2000:115-158.
3. Karam JD, (ed.), *et al*: **Molecular Biology of Bacteriophage T4.** Washington, DC: American Society for Microbiology; 1994.
4. Belfort M: **Scientific serendipity initiates an intron odyssey.** *J Biol Chem* 2009, **284(44)**:29997-30003.
5. Belle A, Landthaler M, Shub DA: **Intronless homing: site-specific endonuclease SegF of bacteriophage T4 mediates localized marker exclusion analogous to homing endonucleases of group I introns.** *Genes Dev* 2002, **16(3)**:351-362.
6. Barton NH, Briggs DEG, Eisen JA, Goldstein DB, Nipam HP: **Evolution.** New York: Cold Spring Harbor Laboratory Press; 2007.
7. Mosig G, Eiserling F: **T4 and related phages: structure and development.** *The Bacteriophages* Oxford University Press; 2006, 225-267.
8. Alberts B: **DNA replication and recombination.** *Nature* 2003, **421(6921)**:431-435.

doi:10.1186/1743-422X-7-293
Cite this article as: Karam and Miller: Bacteriophage T4 and its relatives. *Virology Journal* 2010 **7**:293.

Petrov et al. Virology Journal 2010, **7**:292
http://www.virologyj.com/content/7/1/292

VIROLOGY JOURNAL

REVIEW

Genomes of the T4-related bacteriophages as windows on microbial genome evolution

Vasiliy M Petrov[1], Swarnamala Ratnayaka[1], James M Nolan[2], Eric S Miller[3], Jim D Karam[1*]

Abstract

The T4-related bacteriophages are a group of bacterial viruses that share morphological similarities and genetic homologies with the well-studied *Escherichia coli* phage T4, but that diverge from T4 and each other by a number of genetically determined characteristics including the bacterial hosts they infect, the sizes of their linear double-stranded (ds) DNA genomes and the predicted compositions of their proteomes. The genomes of about 40 of these phages have been sequenced and annotated over the last several years and are compared here in the context of the factors that have determined their diversity and the diversity of other microbial genomes in evolution. The genomes of the T4 relatives analyzed so far range in size between ~160,000 and ~250,000 base pairs (bp) and are mosaics of one another, consisting of clusters of homology between them that are interspersed with segments that vary considerably in genetic composition between the different phage lineages. Based on the known biological and biochemical properties of phage T4 and the proteins encoded by the T4 genome, the T4 relatives reviewed here are predicted to share a genetic core, or "Core Genome" that determines the structural design of their dsDNA chromosomes, their distinctive morphology and the process of their assembly into infectious agents (phage morphogenesis). The Core Genome appears to be the most ancient genetic component of this phage group and constitutes a mere 12-15% of the total protein encoding potential of the typical T4-related phage genome. The high degree of genetic heterogeneity that exists outside of this shared core suggests that horizontal DNA transfer involving many genetic sources has played a major role in diversification of the T4-related phages and their spread to a wide spectrum of bacterial species domains in evolution. We discuss some of the factors and pathways that might have shaped the evolution of these phages and point out several parallels between their diversity and the diversity generally observed within all groups of interrelated dsDNA microbial genomes in nature.

Background

Discovery of the three T-even phages (T2, T4 and T6) and their subsequent use as model systems to explore the nature of the gene and genetic mechanisms had a profound impact on the proliferation of interdisciplinary biological research. Indeed, work with these bacterial viruses during the period between 1920 and 1960 laid down several important foundations for the birth of Molecular Biology as a field of research that freely integrates the tools of almost every discipline of the life and physical sciences [1,2]. Phage T2, the first of the T-even phages to be isolated (see [3] for a historical perspective) occupied center stage in most of the early studies, although the underlying genetic closeness of this phage to T4 and T6

gave reason to treat all three phages as the same biological entity in discussions of what was being learned from each of them. The switch in attention from T2 to T4 came about largely as a response to two major studies in which T4 rather than T2 was chosen as the experimental system. These were the studies initiated by Seymour Benzer in the mid-1950s on the fine-structure of the phage *rIIA* and *rIIB* genes (see [4] for an overview) and the collaborative studies by Richard Epstein and Robert Edgar [5] through which an extensive collection of T4 conditional lethal (temperature-sensitive and amber) mutants was generated [6] and then freely shared with the scientific community. Use of the Epstein-Edgar collection of T4 mutants, as well as comparative studies with T2 and T6 and other T4 relatives isolated from the wild, ultimately led to detailed descriptions of the structure, replication and expression of the T4 genome and the morphogenetic pathways that underlie phage assembly and the release of

* Correspondence: karamoff@tulane.edu
[1]Department of Biochemistry, Tulane University Health Sciences Center, 1430 Tulane Avenue, New Orleans, LA, USA
Full list of author information is available at the end of the article

phage progeny from infected *Escherichia coli* hosts (see [2,7,8] for comprehensive reviews). As the best-studied member of this group of phages, T4 has become the reference or prototype for its relatives.

Over the last 50 years, hundreds of T4-related phages have been isolated from a variety of environmental locations and for a number of different bacterial genera or species [9,10]. The majority of these wild-type phages were isolated by plating raw sewage or mammalian fecal samples on the same *E. coli* strains that are commonly used in laboratories for growing T4 phage stocks or enumerating T4 plaques on bacterial lawns. The archived *E. coli* phages include both close and highly diverged relatives of the canonical T-even phages, as originally surmised from their serological properties and relative compatibilities with each other in pair-wise genetic crosses [11] and later confirmed through partial or complete sequencing of representative phage genomes [12-16]. In addition to the large number of archived T-even-related phages that grow in *E. coli*, there are several (<25) archived relatives of these phages that do not use *E. coli* as a host, but instead grow in other bacterial genera, including species of *Acinetobacter, Aeromonas, Klebsiella, Pseudomonas, Shigella, Vibrio* or photosynthesizing marine cyanobacteria ([9,10] and recent GenBank submissions, also see below). The sequencing of the genomes of a number of these phages has shown that they are all highly diverged from the T-even phages and that in general, there is a higher degree of genetic diversity among T4 relatives that are presumably genetically or reproductively separated from one another in nature because of their differences in the range of bacterial hosts they can infect [14-17]. The list of sequenced T4-related phage genomes has more than doubled during the last 3-4 years, further reinforcing the evidence for extensive genetic diversity within this group of phages. A major goal of the current review is to provide updated information about the sequence database for T4-related genomes and to summarize their commonalities and differences in the context of what is also being learned from the comparative genomics of other microbial organisms in nature. Ecologically, the lytic T4-related phages occupy the same environmental niches as their bacterial hosts and together with their hosts probably exercise major control over these environments.

What is a T4-related or T4-like phage?

The International Committee for the Taxonomy of Viruses (ICTV) has assigned the T-even phages and their relatives to the "T4-like Viruses" genus, which is one of six genera of the Myoviridae Family http://www.ncbi.nlm.nih.gov/ICTVdb/index.htm. Broadly, the Myoviridae are tailed phages (order Caudovirales) with icosahedral head symmetry and contractile tail structures. Phages listed under the "T4-like Viruses" genus exhibit

morphological features similar to those of the well-characterized structure of phage T4, as visualized by electron microscopy, and encode alleles of many of the T4 genes that determine the T4 morphotype [8]. The diversity of morphotypes among the bacterial viruses is staggering and to the untrained eye, subtle differences between different Myoviridae or different T4 relatives can be difficult to discern under the electron microscope [9,10]. In recent years there has been an increased reliance on information from phage genome sequencing to distinguish between different groups of Myoviridae and between different phages that can be assigned to the same group. The hallmark of the T4-like Viruses is their genetic diversity, which can blur their commonalities with each other, especially for taxonomists and other biologists who wish to understand how these and other groups of dsDNA phages evolve in their natural settings. As is the case for many other dsDNA phages, the genomes of T4 and its analyzed relatives are mosaics of one another, consisting of long and short stretches of homology that intersperse with stretches that lack homology between relatives [14-18]. Much of this mosaicism is thought to have resulted from DNA rearrangements, including genetic gains and losses ("indels"), replacements, translocations, inversions and other types of events similar to those that have shaped the evolution of all microbial genomes in nature. It appears that for the T4-like Viruses, DNA rearrangements have occurred rampantly around a core of conserved (but mutable) gene functions that all members of this group of Myoviridae encode. Sequence divergence or polymorphism within this functionally conserved core is often used to gain insights into the evolutionary history of these phages [16,19,20]. As the genome sequence database for T4 relatives has grown over the last several years, it has also become increasingly evident that the T4-like Viruses exist as different clusters that can be distinguished from one another by the higher levels of predicted genetic and biological commonalities between phages belonging to the same cluster as compared to phages in different clusters. Clusters of closely interrelated genomes have also been observed with other groups of dsDNA phages and microbial genomes in general, e.g., [21,22]. Many of the distinguishing features between clusters of T4-related phages are predicted to be the result of an evolutionary history of isolation within distinct hosts and extensive lateral gene transfer (LGT), i.e., the importation of genes or exchanges with a diversity of biological entities in nature. Genomic mosaicism, which appears to be a common feature of many groups of interrelated dsDNA phages [23,24], underscores the discontinuities that can be created by LGT between different lineages of the same group of interrelated phage genomes.

The inventory of sequenced T4-related genomes

In Table 1, we have listed 41 T4-related phages for which substantive genome sequence information is currently available in public databases, particularly GenBank and http://phage.bioc.tulane.edu (or http://phage.ggc.edu). This listing highlights the bacterial genera and species for which such phages are known to exist [10] and includes recent entries in GenBank for three phages that grow in *Klebsiella, Pseudomonas* and *Shigella* species, respectively. The largest number of archived T4 relatives have originated from raw sewage or mammalian fecal matter and detected as plaque formers on lawns of laboratory strains of *E. coli* B and by using plating conditions that are particularly favorable for clear plaque formation by T4. *E. coli* K-12 strains have also been used in some cases (Table 1). The RB phages listed in Table 1 are part of the largest number of T4 relatives to have been collected around the same time from approximately the same environmental source. This collection consists of ~60 phages (not all T4-related) that were isolated by Rosina Berry (an undergraduate intern) from various sewage treatment plants in Long Island, New York during the summer of 1964 for Richard Russell's PhD project on speciation of the T-even phages [25]. The RB phages, which were isolated by using *E. coli* B as a host, include both close and distant relatives of the T-even phages and have received broad attention in comparative studies of the biochemistry and genetics of the T4 biological system [2,7,8]. The genomes of most of the distant relatives of T4 from this collection were sequenced and annotated several years ago [14-16]. More recently, draft or polished sequences have also become available for several close relatives of T4 from this collection as well as for phages T2 and T6 (see http://phage.ggc.edu for updates). The other phages listed in Table 1 are from smaller collections that originated through studies by various laboratories, as noted in the references cited in Table 1.

Each of the genomes we discuss in this review has a unique nucleotide sequence and a genetic composition that unambiguously distinguish it from the others. Yet, all of these genomes can be assigned to a single umbrella group based on shared homologies for a number of genes

Table 1 An overview of sequenced T4-related phage genomes [1]

Bacteria	Phages [2]	Bacterial strain used in phage isolation
Proteobacteria		
Enterobacteria	T2, T4, T6	*E. coli* B (see [3] for references)
	RB3, RB14, RB15, RB16, RB18, RB26, RB32, RB43, RB49, RB51, RB70, RB69	*E. coli* B/5 [25]
	LZ2	*E . coli* B strain NapIV [62]
	JS8, JS10, JSE	*E. coli* K-12 strain K802 [69,74]
	CC31	*E. coli* B strain S/6/4 (Karam lab; New Orleans sewage, unpublished)
	phi1	*E. coli* K-12 F+ (I. Andriashvili, 1971, unpublished); Tbilisi sewage; (M. Kutateladze pers. commun.)
Acinetobacter	133	*Ac. johnsonii* (see [14] for references)
	Acj9, Acj61	*Ac. johnsonii* (Karam lab; New Orleans sewage, unpublished)
	42 (=Ac42)	*Acinetobacter sp.* (H. Ackermann, D'Herelle Center, Canada; pers. commun.)
Aeromonas	44RR, 31, 25, 65	Various *Ae. salmonicida* strains (see [14] for references)
	Aeh1	*Ae. hydrophila* C-1 (see [14] for references)
	PX29	*Ae. salmonicida* strain 95-65 (Karam lab; New Orleans sewage, unpublished)
Klebsiella	KP15	*Klebsiella pneumoniae* (Z. Drulis-Kawa, pers. commun.; Warsaw, Poland sewage).
Pseudomonads	phiW-14	*Delftia acidovorance* (see GenBank Accession no. NC_013697)
Shigella	phiSboM-AG3	*Shigella boydii* (see GenBank Accession no. NC_013693)
Vibriobacteria	KVP40, nt-1	See [14] for references
Cyanobacteria		
Synechococcus	SPM2	*S. marinus* [27]
	S-RSM4	*S. marinus* [31]
	Syn9	*S. marinus.* Also grows in *Prochlorococcus* [75]
Prochlorococcus	P-SSM2, P-SSM4	P-SSM2, P-SSM4: [42]

[1] The phages are listed under the major divisions (phyla) and genera of the bacterial hosts used for their isolation.

that we refer to here as the "Core Genome" of the T4-related phages, or T4-like Viruses. The genetic background for the Core Genome can vary considerably between T4 relatives and constitutes an important criterion for distinguishing between close and distant relatives among the ~40 phage genomes sequenced so far. The three T-even phages have traditionally been considered to be closely interrelated on the basis that they share ~85% genome-wide homology, similar genetic maps and certain biological properties in common with each other [8,26]. By using comparable criteria for phage genome organization and assortment of putative genes, i.e., predicted open-reading frames (ORFs) and tRNA encoding sequences, we could group the phages listed in Table 1 into 23 different *types* of T4 relatives, with the *T-even type* phages representing the largest group or cluster of closely interrelated phage genomes sequenced so far. These 23 *types* and their distinguishing features are listed in Table 2. The abundance of sequence data for the *T-even type* phages is largely the result of an effort by J. Nolan (in preparation) to analyze the genomes of RB phages that had been predicted by Russell [25] to be closely related to the T4 genome. We presume that in nature, each *type* of T4-related phage listed in Table 2 is representative of a naturally existing cluster or pool of closely interrelated phages that contains a record of evolutionary continuities between members of the pool. A pool of closely interrelated phages would be expected to exhibit low levels of sequence divergence between pool members, but might also show evidence of sporadic deletions, acquisitions, exchanges or other DNA rearrangements in the otherwise highly conserved genetic composition.

The listing shown in Table 2 should be regarded as somewhat arbitrary since setting the homology standard to a higher or lower value than ~85% can result in different groupings. In fact, as will be explained below for the *T-even type* phages, small differences in the genetic composition can have major biological consequences, which might merit further subdivisions within this cluster. In addition, as evidenced by information from the recently analyzed T4 relatives listed in Tables 1 and 2, the isolation of new T4-related phages for known and newly recognized bacterial hosts is likely to reveal a greater diversity of phage genome *types* and virion morphologies than the listing in Table 2 provides.

Genetic commonalities between T4 relatives
A few years ago, a comparative analysis of ~15 completely or almost completely sequenced T4-related genomes showed that they share two important characteristics [14]:

1. Their genes are contained in a circularly permuted order within linear dsDNA chromosomes. In most cases, this characteristic became evident during the assembly and annotation of DNA sequence data into single contiguous sequences (contigs) and in some cases, the ends of the single contigs were further confirmed to be contiguous with each other by use of the PCR [14,17,27]
2. The genomes were each predicted to encode a set of 31-33 genes that in T4 have been implicated in the ability of the phage to exercise autonomous control over its own reproduction. This control includes the biochemical strategies that determine the circularly permuted chromosomal design, which is generated through the integration of the protein networks for DNA replication, genome packaging and viral assembly in the phage developmental program [8]. This set of genes amounts to a mere ~12% of the T4 genome.

Expansion of the sequence database to >20 different *types* of T4-related genome configurations (Table 2) has reinforced the observation that a core set of 31-33 genes is a unifying feature of all T4 relatives. However, it has also become increasingly evident that other phage genes enjoy a very wide distribution among these genomes, suggesting that the minimum number of genes required to generate a plaque-forming phage with generally similar morphology to T4 is greater than the number of the universally distributed genes and might vary with specific adaptations of different clusters of closely interrelated phages in nature. As is the case with other host-dependent, but partially autonomously replicating genetic entities in the microbial world, particularly the bacterial endosymbionts [28-30], there is usually a dependence on auxiliary functions from the entity and this dependence can vary with the host in which the entity propagates. In T4, it is already known that some phage-encoded functions are essential for phage growth in some *E. coli* strains but not others and that in many instances mutations in one gene can result in decreased dependence on the function of another gene. Many such examples of intergenic suppression have been published and referenced in comprehensive reviews about the T4 genome [2,7,8]. The analysis of the genomes of some T4 relatives has also yielded observations suggesting that ordinarily indispensable biochemical activities might be circumvented or substituted in certain genetic backgrounds of the phage or host genome. Examples include two separate instances where the need for the recombination and packaging Endonuclease VII (gp49; encoded by gene *49*), which is essential in T4, appears to have been circumvented by the evolution of putative alternative nucleases (through replacements or new acquisitions) in the *E. coli* phage RB16 (*RB16ORF270c*) and the *Aeromonas* phage 65 (*65ORF061w*) [14]. Another example is the possible substitution of the essential dUTPase function provided by gp56 in T4 by host-like dUTPase genes in the

Table 2 T4-related phages with sequenced genomes

Phage or genome type	Phage	Genome size (bp)	Database reference [1]	ORFs (T4-like/ Total)	tRNA genes	Shared or unique properties of the genomes[2]
T-even	E. coli phage T4	168,903	NC_000866	278	8	The *T-even type* genomes share 85-95% ORF homology with one another and >90% nucleotide sequence identity between most of their shared alleles. Also, these genomes encode glucosyl transferases and dCMP hydroxymethylases, but their DNA modification patterns vary (see text and Table 3). Some members of this cluster are known to be partially compatible with each other in genetic crosses (see text). Phage RB70 (Table 1) might be identical to phage RB51.
	E. coli phage T4T	168920	HM137666	280	8	
	E. coli phage T2	163,793	Tulane	232/269	9	
	E. coli phage T6	168,974	Tulane	228/270	7	
	E. coli phage RB3	~168,000	Tulane	~240/ ~270	10	
	E. coli phage RB14	165,429	NC_012638	235/274	10	
	E. coli phage RB15	~167,000	Tulane	~236/ ~269	7	
	E. coli phage RB18	166,677	Tulane	237/268	10	
	E. coli phage RB26	163,036	Tulane	232/~269	10	
	E. coli phage RB32	165,890	NC_008515	237/270	8	
	E. coli phage RB51	168,394	NC_012635	242/273	9	
	E. coli phage LZ2	>159,664	Tulane	~240/ >260	10	
RB69	E. coli phage RB69	167,560	NC_004928	212/273	2	~20% of the ORFs in this genome are unique to RB69; this phage excludes T4 in RB69 × T4 crosses [25].
RB49	E. coli phage RB49	164,018	NC_005066	120/279	0	The 3 genomes of this *type* share 96-99% ORF homology with one another
	E. coli phage phi1	164,270	NC_009821	115/276	0	
	E. coli phage JSE	166,418	NC_012740	122/277	0	
JS98	E. coli phage JS98	170,523	NC_010105	202/266	3	JS98 and JS10 share ~98% ORF homology with each other.
	E. coli phage JS10	171,451	NC_012741	197/265	3	
CC31	E. coli phage CC31	165,540	GU323318	156/279	8	~43% of the CC31 ORFs are unique to this phage. Also, CC31 is the only known non*T-even type* phage predicted to encode *glucosyl transferase* genes (see Table 3)
RB43	E. coli phage RB16	176,789	HM134276	115/260	2	The genomes of RB16 and RB43are similarly organized and share >85% ORF homology with each other [14]
	E. coli phage RB43	180,500	NC_007023	118/292	1	
133	Acinetobacter phage133	159,897	HM114315	110/257	14	Each of these *Acinetobacter* phages has a unique set of ORFs that occupy ~35% of the genome. That is, each represents a different *type* of T4-related phage genome.
Acj9	Acinetobacter phage Acj9	169,953	HM004124	97/253	16	
Acj61	Acinetobacter phage Acj61	164,093	GU911519	101/241	13	
Ac42	Acinetobacter phage Ac42	167,718	HM032710	117/257	3	
44RR	Aeromonas phage 44RR	173,591	NC_005135	118/252	17	Phages 44RR and 31 share ~98% ORF homology (and ~97% sequence identity) with each other. Also, they exhibit ~80% ORF homology with phage 25
	Aeromonas phage 31	172,963	NC_007022	117/247	15	
25	Aeromonas phage 25	161,475	NC_008208	116/242	13	The phage 25 genome is 11-12 kb shorter than the genome of 44RR (or 31). Also, ~14% of the phage 25 ORFs are unique to this phage.
Aeh1	Aeromonas phage Aeh1	233,234	NC_005260	106/352	23	Phages Aeh1 and PX29 share ~95% ORF homology with each other and partially overlap in host-range properties
	Aeromonas phage PX29	222,006	GU396103	109/342	25	

Table 2 T4-related phages with sequenced genomes (Continued)

65	Aeromonas phage 65	235,289	GU459069	102/439	17	~55% of the ORFs in this genome are unique to phage 65
KVP40	Vibrio phage KVP40	244,834	NC_005083	99/381	29	Phages KVP40 and nt-1 share ~85% ORF homology with each other and partially overlap in host range properties
	Vibrio phage nt-1	247,144	Tulane	95/400	26	
S-PM2	Marine Synechococcus phage S-PM2	196,280	NC_006820	40/236	1	See [31] for comparisons between the marine cyano phages. Based on their diversity, each represents a different type of T4-related phage genome.
S-RSM4	Marine Synechococcus phage S-RSM4	194,454	NC_013085	41/237	12	
Syn9	Marine Synechococcus phage Syn9	177,300	NC_008296	43/226	6	
P-SSM2	Marine Prochlorococcus phage P-SSM2	252,401	NC_006883	47/329	1	
P-SSM4	Marine Prochlorococcus phage P-SSM4	178,249	NC_006884	46/198	0	
KP15	Klebsiella pneumoniae phage KP15	174,436	GU295964	116/239	1	~80% of KP15 ORFs are homologous and similarly organized to ORFs in RB43
W14	Delftia acidovorance phage phiW-14	157,486	NC_013697	60/236	0	
AG3	Shigella boydii phage phiSboM-AG3	158,006	NC_013693	64/260	4	

[1] In this column, numbers with the prefixes NC, GU and HM refer to GenBank accession numbers and the designation "Tulane" refers to the database at http://phage.bioc.tulane.edu (soon to be transferred to http://phage.ggc.edu). The NC_000866 accession is for the T4 genome sequence that was compiled from data contributed by many laboratories [2,7]. The HM137666 accession is for the T4 genome sequence determined on DNA from a single source, termed T4T, which is the wild-type T4D strain maintained by the Karam laboratory at Tulane University, New Orleans.

[2] In this column, "% ORF homology" refers to the percentage of ORFs that are alleles between the compared genomes.

Aeromonas phages 65 and Aeh1 and the vibriophages KVP40 and nt-1 [14,17].

Taking into consideration the distribution of T4-like genes in the >20 different *types* of phage genome configurations listed in Table 2 and the examples of putative genetic substitutions/acquisitions mentioned above, we estimate that the Core Genome of the T4-related phages consists of two genetic components, one highly resistant and one somewhat permissive to attrition in evolution. We refer to the genes that are essential under all known conditions as "Core genes" and those that can be substituted or circumvented in certain genetic backgrounds of the phage and/or bacterial host as "Quasicore genes". In Table 3 and Figure 1 we list the two sets of genes and highlight their functional interrelationships and some of the conditions under which some Quasicore genes might not be required. Interestingly, the absence of members of the Quasicore set is most often observed in the T4-related marine cyanophages, which also exhibit the smallest numbers of T4-like genes and the greatest sequence divergence in Core genes from any of the other host-specificity groups of T4 relatives listed in Tables 1 and 2. Possibly, the marine cyanobacteria represent a natural environment that has favored the evolution of a specific streamlining of the genetic background for the Core Genome of T4-related phages. This streamlining might have been driven through a combination of what the

cyanobacterial hosts could provide as substitutes for physiologically important, but occasionally dispensable functions of these phages and what the phage genomes themselves might have acquired as alternatives to lost genes by LGT from other biological entities. We view each *type* of phage genomic framework listed in Table 2 as a specific adaptation of the Core Genome in the evolution of these phages in the different bacterial genera or species where T4 relatives have been detected.

An overview of how the sequenced T4-like Viruses differ from each other

The T4-related genomes sequenced so far exhibit divergence from one another in several respects including; (a) the range of bacterial host species that the respective phages infect, (b) the sizes of these genomes and the capsids (phage heads) in which they are packaged, (c) the types of modifications, if any, that the genomic DNA undergoes in vivo, (d) their assortment of protein- and tRNA-encoding genes, (e) their assortment of T4-like genes (alleles of T4 genes), (f) the sequence divergence (mutational drift) and in some cases, the intragenic mosaicism between alleles and (e) the topological arrangement of alleles and their regulatory signals in the different genomes. Divergence between genomes within some of these categories appears to have occurred independently of other categories. For example, phages that share a bacterial

Table 3 Genes of the Core Genome of T4-like Viruses

T4 genes[1]	Gene products and/or activities[1]	Comments[2]
DNA replication, repair and recombination		
43; **45**; **44** and **62**; **41** &*61*; *59*;**32**; **46** &**47**; **uvsW**; *uvsX, uvsY*; *30*; *rnh*; *39*+*60* &*52*; *dda*; *49*	**gp43** (DNA polymerase); **gp45** (trimeric sliding clamp); **gp44/gp62** sliding clamp loader complex (gp44 tetramer+gp62 monomer); **gp41**/*gp61*helicase-primase complex (hexamers of both proteins); **gp59** (helicase-primase loader & gp43 regulator); **gp32** (single-strand binding protein); **gp46-gp47** (subunits of a recombination nuclease complex required for initiation of DNA replication); **UvsW protein** (recombination DNA-RNA helicase, DNA-dependent ATPase); *uvsX* (RecA-like recombination protein); *uvsY* (uvsX helper protein); *gp30* (DNA ligase); *Rnh* (Ribonuclease H); *gp39* *+60* & *gp52* (subunits of a Type II DNA topoisomerase); *Dda* protein (short-range DNA helicase); *gp49* (Endonuclease VII, required for recombination & DNA packaging).	Many of the Quasicore genes in this group are absent in one or more T4-related marine cyanophages. In T4, some these genes are not required in certain *E. coli* hosts or become dispensable in the presence of mutations in specific other genes (intergenic suppression).
Auxiliary metabolism		
nrdA &**nrdB**; *nrdC*; *nrdG*; *nrdH*; *56*; *cd*; *frd*; *td*; *tk*; *1*; *denA*; *dexA*	**NrdA-NrdB** (subunits of an aerobic ribonucleotide reductase complex); *NrdG & NrdH* (subunits of an anaerobic ribonucleotide reductase complex); *NrdC* (thioredoxin); *gp56* (dCTPase-dUTPase); *Cd* (dCMP deaminase); *Frd* (DHFR; (dihydrofolate reductase); *Td* (thymidylate synthetase), *Tk* (thymidine kinase); *gp1* (dNMP kinase); *DenA* (Endonuclease II); *DexA* (Exonuclease A).	A combination of at least some of these genes is required to supplement the intracellular pool of nucleotides for phage DNA and RNA synthesis.
Gene expression		
33; **55**; *regA*	**gp33** (essential protein that mediates gp55-gp45-RNA polymerase interactions in late transcription); **gp55** (sigma factor for late transcription); **RegA** (mRNA-binding translational repressor; also involved in host nucleoid unfolding)	In T4, *regA* mutations are not lethal, yet all the T4 relatives examined so far encode homologues of this gene.
Phage morphogenesis		
2; **3**; **4**; **5**; **6**; **8**; **13**; **14**; **15**; **16 17**; **18**; **19**; **20**; **21**; **22**; **23**; **25**; **26**; **34**; **35**; **36**; "**37**"; **49**; **53**	*gp2* (protects ends of packaged DNA against RecBCD nuclease); **gp3** (sheath terminator); **gp4** (Head completion protein); **gp5** (baseplate lysozyme hub component); **gp6** (baseplate wedge component); **gp8** (baseplate wedge), **gp13** (head completion protein); **gp14** (head completion protein); **gp15** (tail completion protein); **gp16** &**gp17** (subunits of the terminase for DNA packaging); **gp18** (tail sheath subunit), **gp19** (tail tube subunit); **gp20** (head portal vertex protein); **gp21** (prohead core protein and protease); **gp22** (prohead core protein); **gp23** (precursor of major head protein); **gp25** (base plate wedge subunit); **gp26** (base plate hub subunit); **gp34** (proximal tail fiber protein subunit); **gp35** (tail fiber hinge protein); **gp36** (small distal tail fiber protein subunit); **gp37** (large distal tail fiber protein subunit; heterogeneous among T4 relatives); gp49 (Endo VII; required for DNA packaging); **gp53** (baseplate wedge component)	T4 gp2 is not required in recBCD mutant hosts and no gene 2 homologues are detected in some marine cyanophages. Also, the "**37**" designation means that in some T4 relatives (e.g. the marine cyanophages and the vibriophages), the identification of gene 37 and other tail fiber genes can be difficult or impossible to make by bioinformatic tools because of extensive mosaicism or putative substitutions with non-homologous tail-fiber genes.
Other		
rIIA &*rIIB*	The precise functions of the *rIIA* and *rIIB* gene products are not known. In T4, *rIIA* or *rIIB* mutations exhibit multiple effects on phage physiology, but are only lethal in the presence of a lambda prophage.	Like many other Quasicore genes, the *rIIA* and *rIIB* genes are found in all T4 relatives, except the marine cyanophages. The wide natural distribution of these 2 genes might be a reflection of the distribution of prophages that restrict T4 relatives in various bacterial hosts.

[1]Core genes and their products are shown in bold font and Quasicore genes and their products in unbolded italic.

[2]See text for additional explanations.

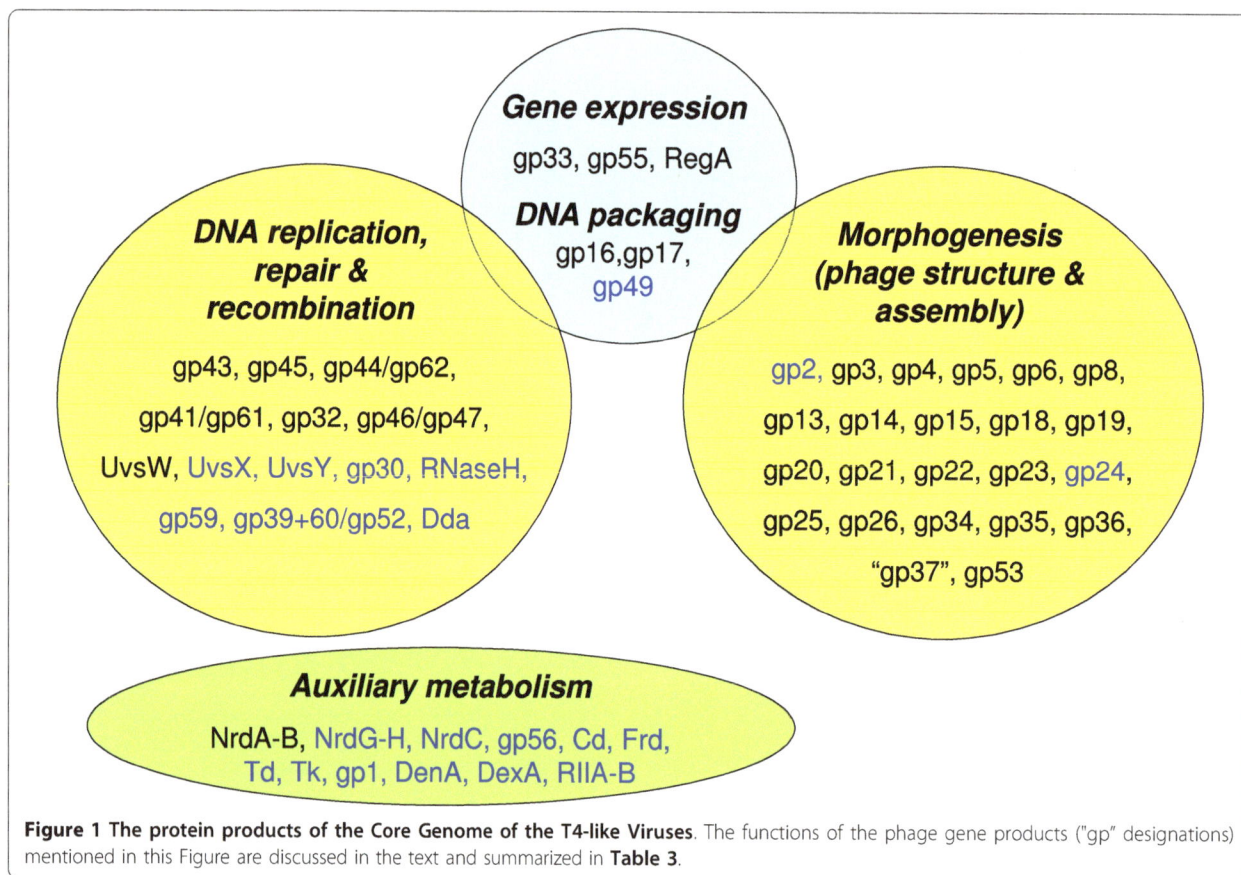

Figure 1 The protein products of the Core Genome of the T4-like Viruses. The functions of the phage gene products ("gp" designations) mentioned in this Figure are discussed in the text and summarized in **Table 3**.

host do not necessarily share similar genome sizes, similar genetic compositions at a global level, similar DNA modifications or similar genome topologies. On the other hand, phages that infect different bacterial host species seem to exhibit the highest degree of divergence from each other in most or all categories. The assignment of T4 relatives to the different groups or *types* listed in Table 2 takes into account shared similarities in most categories, the implication being that members of a phage/genome *type* are probably more closely related to each other than they are to members of other clusters of interrelated phages. For example, in pair-wise comparisons, the *T-even type* phages listed in Table 2 exhibited 85-95% genome-wide homology (shared alleles) as well as high levels of nucleotide sequence identity with each other. Most of the dissimilarities between members of this cluster of phages map to genomic segments that have long been known to be variable between T2, T4 and T6, based on electron microscopic analysis of annealed DNA mixtures from these phages [26]. Phage genome sequencing has shown that the hypervariability of these segments among all *types* of T4 relatives involves: (a) an often-observed mosaicism in tail fiber genes, (b) unequal distribution of ORFs for putative homing endonucleases, even between the closest of relatives and (c) a clustering of novel ORFs in the phage

chromosomal segment corresponding to the ~40-75 kb region of the T4 genome [14-16]. The biological consequences of these genetic differences are significant [2,7,8]. Although distant relatives of the three T-even phages have been isolated that also use *E. coli* as a bacterial host (e.g. phages RB43, RB49, RB69 and others; Table 2), no close relatives of these canonical members of the T4-like Viruses genus have yet been found among the phages that infect bacterial hosts other than *E. coli*. By using the ORF composition of the T4 genome as a criterion, we estimate that the range of homology to this genome (i.e., percentage of T4-like genes) among the coliphage relatives analyzed so far is between ~40% (for phage RB43) and ~78% (for phage RB69). Among the T4 relatives that grow in bacterial hosts other than the Enterobacteria, the homology to the T4 genome ranges between ~15% T4-like genes in the genomes of some marine cyanophages and ~40% T4-like genes in the genomes of some *Aeromonas* and *Acinetobacter* phages (Table 2). These homology values reflect the extent of the heterogeneity that exists in the genetic backgrounds of the two components of the Core Genome (Figure 1, Table 3) among the different phages or phage clusters listed in Table 2. The five *types* of genome configurations currently catalogued among the T4-related marine cyanophages (Table 2) range in size between ~177 kb

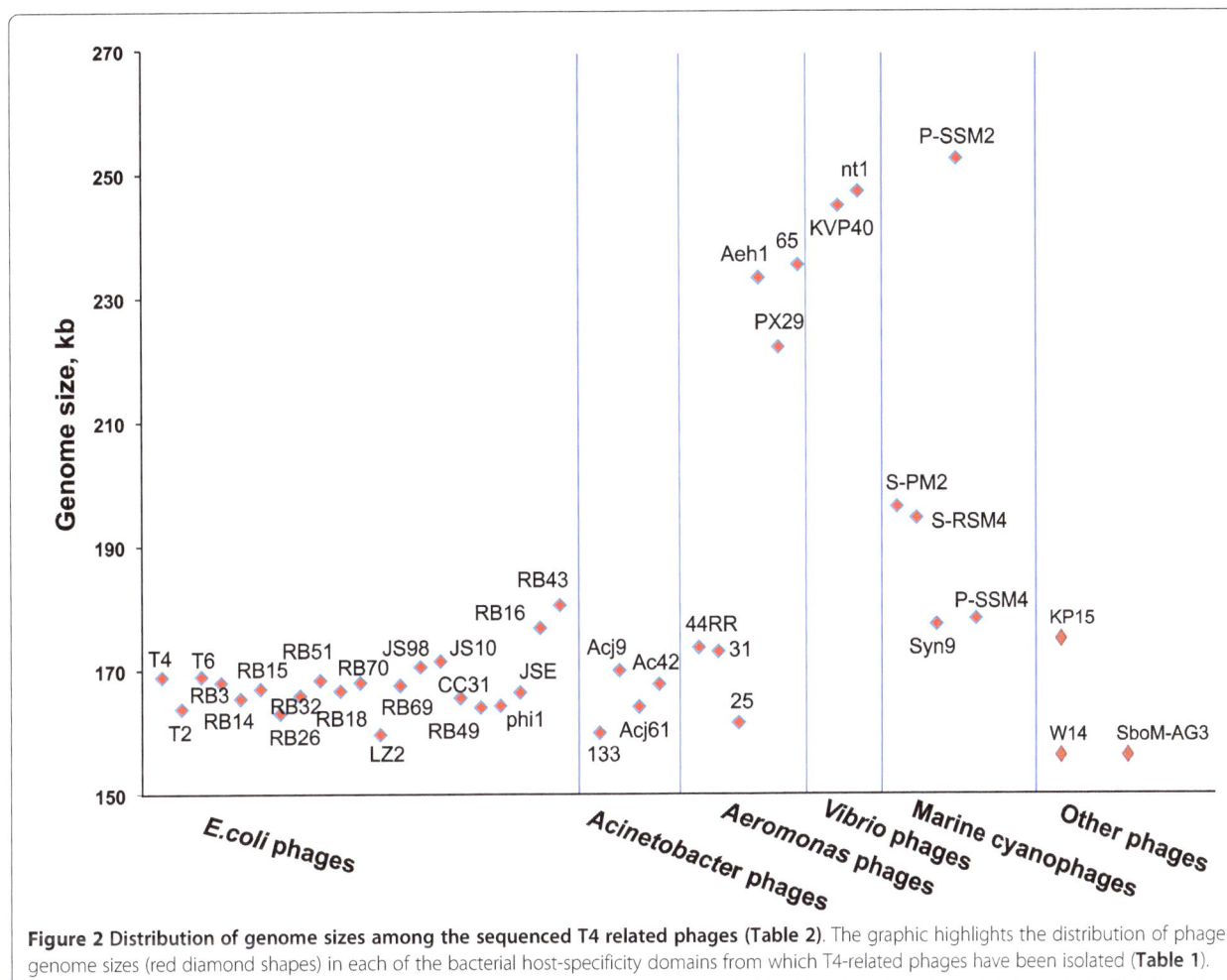

Figure 2 Distribution of genome sizes among the sequenced T4 related phages (Table 2). The graphic highlights the distribution of phage genome sizes (red diamond shapes) in each of the bacterial host-specificity domains from which T4-related phages have been isolated (**Table 1**).

(for phage Syn9) and ~252 kb (for phage P-SSM2) and carry the smallest number of T4-like genes among all currently recognized *types* of T4 relatives. The range here is between 40 (for S-PM2) and 47 (for P-SSM2) T4-like genes per genome [31]. A comprehensive listing of T4 alleles in most of the phages listed in Tables 1 and 2 can be found in Additional file 1 or online at http://phage.bioc.tulane.edu and http://phage.ggc.edu. The recent genome entries in GenBank mentioned earlier for phiSboM-AG3 and phiW-14 predict ~60 T4-like genes, mostly Core and Quasicore genes, for each. Taken together, these observations are consistent with the notion that components of the Core Genome have been somewhat resistant to dispersal in evolution, but that the host environment must also play an important role by determining the most appropriate genetic background of this unifying feature of T4-related genomes.

Genome size heterogeneity among T4 relatives

In Figure 2 we show a graphic representation of the heterogeneity in genome sizes for the phages listed in Table 2.

The size range observed so far for genomes of the T4-like Viruses is between ~160,000 and ~250,000 bp (or ~160-250 kb). Relatives of T4 with genomes near or larger than 200 kb also exhibit larger and more elongated heads than phages with genomes in the ~170 kb size range [9,10]. These extraordinarily large T4 relatives have sometimes been referred to as "Schizo T-even" phages [32] and rank among the largest known viruses, i.e., the so-called "giant" or "jumbo" viruses [33]. T4-related giants have been isolated for *Aeromonas, Vibrio* and marine cyanobacterial host species, but no such giants have yet been isolated for T4 relatives that grow in *E. coli* or the other host species listed in Table 1. For the *Vibrio* bacterial hosts, only giant T4 relatives have been isolated so far, whereas a wide range of phage genome sizes has been observed among the *Aeromonas* and cyanobacterial phages. Comparative genomics has not yet revealed any genetic commonalities between the T4-related giant phages of *Aeromonas, Vibrio* and marine bacteria (Fgure 1) that might explain the cross-species similarities in head morphology. So, it remains unclear what might have determined the evolution of different stable genome sizes in different phage lineages or clusters. It is equally possible that giant genomes can evolve from smaller precursors or can themselves serve as progenitors of smaller genomes. Detailed studies of the comparative genomics of the functional linkage between DNA replication, packaging and morphogenesis for the different genome size categories shown in Figure 2 might be needed to provide explanations for what determines the evolution of different genome sizes in different phage clusters or lineages. Also, fine-structure morphological differences do exist among T4 relatives that are of similar size and share homologies for structural genes, indicating that the determination of head size and shape can vary with different combinations of these genes.

Some observations in the T4 biological system further underscore the plasticity of head-size determination and the dependence of this plasticity on multiple genetic factors in phage development [8]. Based on mutational analyses, the interplay of at least four T4 genes can generate larger (more elongated) phage heads containing DNA chromosomes that are larger than the ~169 kb size of wild-type T4 DNA. These are the genes for the major capsid protein (gene *23*), portal protein (gene *20*), scaffold protein (gene *22*) and vertex protein (gene *24*). In addition, the recombination endonuclease Endo VII (gp49) and the terminase (gp16 and gp17) play important roles in determining the size of the packaged DNA in coordination with head morphogenesis (headful packaging). Possibly, it is the regulation of these conserved gene functions that can diverge coordinately with increased genetic acquisitions that lead to larger genomes and larger heads in certain cellular environments. The T4-related *Aeromonas* phages would be particularly attractive as experimental systems to explore the evolutionary basis for head-genome size determination because this subgroup of phages is easy to grow and contains representatives of the entire range of phage genome and head sizes observed so far (Figure 2 and Table 2).

Lateral mobility and the Core Genome of the T4-like Viruses

It is clear that the Core Genome of the T4-related phages has spread to the biological domains of a diversity of bacterial genera (Table 1), although it is unclear how this spread might have occurred and to what degree genetic exchange is still possible between T4 relatives that are separated by bacterial species barriers and high sequence divergence between alleles of the Core and Quasicore genes listed in Table 3 and Figure 1. Such exchange would require the availability of mechanisms for transferring Core Genome components from one bacterial species domain into another. In addition, shuffled genes would have to be compatible with new partners. Experimentally,

there is some evidence indicating that the products of some Core genes, e.g., the DNA polymerase (gp43) and its accessory proteins (gp45 and gp44/62), can substitute for their diverged homologues *in vivo* [12,34-36]. Such observations suggest that the shuffling of Core Genome components between diverged T4 relatives can in some cases yield viable combinations. However, for the most part there appear to be major barriers to the shuffling of Core Genome components between distantly related T4-likeViruses in nature. In some respects, the mutational drift within this common core should provide valuable insights into its evolutionary history since the last common ancestor of the T4 related genomes examined so far [19,20]. On the other hand, it should be recognized that the evolutionary history of the Core Genome is not necessarily a good predictor of whole phage genome phylogeny because the majority of the genetic background of this common core varies considerably between the different *types* of T4 relatives (Table 2) and is probably derived from different multiple sources for different phage lineages or clusters.

Although the Core Genome of the T4-related phages might resist fragmentation in evolution, it is unclear if there could have been one or more than one universal common phage ancestor for all of the genes of this unifying feature of the analyzed T4 relatives. Some answers about the origins of the different multi-gene clusters that constitute the Core Genome of these phages might come from further exploration of diverse environmental niches for additional plaque-forming phages and other types of genetic entities that might bear homologies to the Core and Quasicore genes (Table 3 and Figure 1). For example, it remains to be seen if there are autonomously replicating phages or plasmids in nature that utilize homologues of the T4 DNA replication genes, but lack homologues of the DNA packaging and morphogenetic genes of this phage. Conversely, are there phages in nature with alleles of the genes that determine the T4 morphotype, but no alleles of the T4 DNA replication genes? The natural existence of such biological entities could be revealed through the use of the currently available sequence database for T4-related genomes to design appropriate probes for metagenomic searches of a broader range of ecological niches than has been examined so far. Such searches could be directed at specific Core or Quasicore genes [37] or specific features of the different *types* of phage genomes listed in Table 2. It is worth noting that putative homologues of a few T4 genes have already been detected in other genera of the Myoviridae, e.g. the *Salmonella* phage Felix 01 (NC_005282) and the archaeal *Rhodothermus* phage RM378 (NC_004735). Both of these phages bear putative homologues of the T4 gene for the major capsid protein gp23. So, it appears that at least some of the

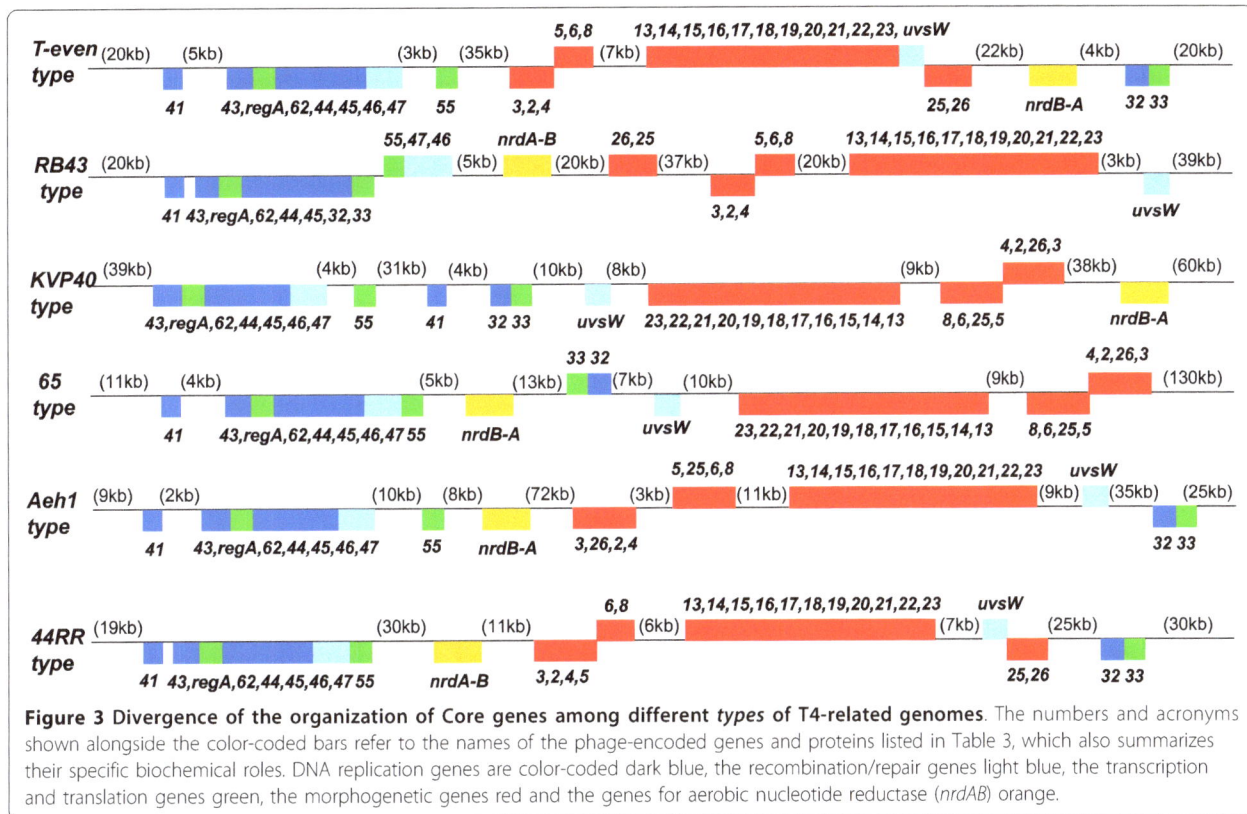

Figure 3 Divergence of the organization of Core genes among different *types* of T4-related genomes. The numbers and acronyms shown alongside the color-coded bars refer to the names of the phage-encoded genes and proteins listed in Table 3, which also summarizes their specific biochemical roles. DNA replication genes are color-coded dark blue, the recombination/repair genes light blue, the transcription and translation genes green, the morphogenetic genes red and the genes for aerobic nucleotide reductase (*nrdAB*) orange.

Core and Quasicore genes of the T4-related phages (Figure 1, Table 3) can survive lateral transfer and function in genetic backgrounds that lack homologies to their presumed ancestral partner genes. In addition, a very recent report [38] describes two *Campylobacter* phages (CPt10 and CP220) that appear to be related to T4, based on the large number of putative T4-like genes that they bear (see GenBank Accession nos. FN667788 and FN667789). Other recent submissions to GenBank that deserve attention and further analysis include the genomes of *Salmonella* phage Vi01 (FQ312032), and *E. coli* phage IME08 (NC_014260; an apparent close relative of phage JS98). Clearly, the sequence database for T4-related genomes requires further enhancements and detailed EM characterization of all of the sequenced phages is needed before a clear picture can emerge about the contributions of the host or host ecology to evolution of the genetic framework and morphological fine-structure within the extended family of T4 relatives.

Additional evidence suggesting that some Core Genome components of T4 relatives can be subjected to lateral transfer in natural settings comes from the variety of topologies (different genetic arrangements) that have been observed for the Core genes in the phages analyzed so far. In Figure 3, we show six examples of naturally existing topologies for the set of Core genes listed in Table 3. The topology exhibited by the *T-even type*

phages is shared by the majority of the other T4-related *E. coli* phages and by all 4 of the T4-related *Acinetobacter* phages listed in Table 2. Interestingly, the two *E. coli* phages RB16 and RB43 exhibit a unique genome topology that has most of the DNA replication genes clustered together in one genomic sector. This *RB43 type* topology is also observed in the recently annotated genome of *Klebsiella* phage KP15 (as we surmise from by our own examination of GenBank Accession no. GU295964). Interestingly, the RB16 and RB43 genomes are rich in a class of putative homing endonuclease genes (HEGs) that bear sequence similarities to the genes for a class of DNA-binding proteins that mediate genetic rearrangements in the developmental programs of plants [14,39-41]. The other unique genome topologies shown in Figure 3 have been observed for the *Vibrio* phage KVP40 (and its close relative nt-1) and several *Aeromonas* phages, including the giant phages 65 and Aeh1 (and its close relative phage PX29) and the smaller phages 25 and 44RR (and its close relative phage 31), respectively. The marine cyanophages exhibit yet other topologies for Core Genome components [31,42]. The diversity of Core Genome topologies underscores the ability of Core and Quasicore genes to function in different orientations and in a variety of genetic backgrounds and regulatory frameworks [14]. The genetic regulatory sequences for a number of Core genes, like phage replication genes *43* (DNA polymerase) and *32* (Ssb protein), are highly diverged between representatives of the different *types* of T4 relatives listed in Table 2[14], further reflecting the adaptive potential of the T4-related Core Genome. Another indication that this genetic core can be prone to lateral transfer is the observed colonization of some of the Core or Quasicore genes or their vicinities by mobile DNA elements, especially intron-encoded and freestanding HEGs [14,43,44]. We will discuss the possible roles of these elements in the evolution of T4-related genomes later in this review.

The Pangenome of the T4-like Viruses

Collectively, the genetic backgrounds for the Core Genome of the T4 relatives examined for the current report are predicted to encode a total of ~3000 proteins that do not exhibit statistically significant sequence matches to any other proteins outside of the databases for the T4-related phages. This number of ORFs is ~1.5 orders of magnitude larger than our estimate of the number of Core plus Quasicore genes in the Core Genome of these phages (Figure 1, Table 3), and might be several orders of magnitude smaller than the union of all the different ORFs that exist in T4-related phages in nature. We refer to this union as the "Pangenome" of the T4-like Viruses, in analogy to the pan genomes of other known groups of autonomously replicating organisms [30]. Based on results from the recent isolation and analysis of the T4-related coliphage CC31 and the *Acinetobacter* phages Acj9 and Acj61 listed in Table 2 , novel and highly divergent members of the T4-like Viruses might be easily detected in environmental samples by taking advantage of the bacterial host diversity of these phages, the uniqueness of certain sequences in specific phage genomes or lineages and other characteristics that distinguish between the different clusters or *types* of phage genomes listed in Table 2. The analysis of the genomes of phages CC31, Acj9 and Acj61, predicted that each encodes ~120 newly recognized ORFs that can be added to the growing count of the Pangenome of the T4-like Viruses (unpublished observations). Such observations suggest that additional diversity is likely to be uncovered through the isolation and analysis of larger numbers of T4 relatives for the known as well as previously unexplored potential bacterial hosts of these phages [38,45].

Despite their plasticity in genome size and their increasing inventory of new ORFs, there are indications that natural diversity of the T4-related phages is not unlimited. We already know of pairs and triplets of nearly identical (yet distinct) genomes that have been isolated years apart from each other and from different geographical areas (Tables 1 and 2). The natural existence of such nearly identical phage genomes might

mean that there are limits to the number of genetic backgrounds that can evolve around a certain Core Genome composition. The limitations might be imposed by the specific partnership that an evolving phage ultimately establishes with its bacterial host(s). More examples of nearly identical genomes in nature would be desirable to find since they might provide clues to the incremental changes by which progenitor genomes can begin to branch into different lineages through additions, deletions and exchanges in the genetic background of the Core Genome.

Genetic isolation between T4 relatives

Genetic separation between interrelated phages can evolve within a shared bacterial host range, as for example might have occurred for the *E. coli* phages T4 and RB69 [25] or come about as a consequence of the transfer of the capacity for whole genome propagation from one host species to another, as might be represented by the different host-specificities of the phages listed in Tables 1 and 2. Insights into the biochemical processes that might lead to the genetic isolation of a T4-related genome from close relatives can be drawn from the number of studies that have been carried out on phage-phage exclusion and host-mediated restriction of the T-even phages [8,46,47]. As explained below, the three T-even phages and their close relatives (*T-even type* phages, Table 2) represent a scenario in which small changes in a genome might result in major effects on its compatibility with a parental genotype.

Phages T2, T4 and T6 can undergo genetic recombination and phenotypic mixing with each other *in vivo* (in pair-wise co-infections of their shared *E. coli* hosts), but they are also partially incompatible with each other under these conditions [11]. The genomes of these phages encode similar, but distinct enzyme networks that modify their genomes and prevent their restriction by gene products encoded by the bacterial hosts and/or certain prophages or defective prophages that can reside in some of these hosts [46,47]. In addition, a few genetic differences between these otherwise closely interrelated phages cause them to be partially incompatible. The genes known to be involved in T-even phage genome modification and restriction are listed in Table 4. Some of these genes specify the modification of phage genomic DNA with glucosylated hydroxymethyl (gluc-Hm) groups at dCMP residues, whereby the DNA becomes resistant to host restriction activities, particularly the *E. coli* Mcr (Rgl) enzyme system. Other phage genes are responsible for commandeering the host transcription system for expression of the modified phage DNA and away from the expression of any DNA (including the host genome) that does not carry the phage-induced modifications [8,48,49]. Subtle differences in phage

Table 4 Distribution of alleles of the T4 DNA modification, restriction and antirestriction genes in T4-related phages[1]

T4 Gene	Product	Role	Phages with alleles of the T4 gene
42	dCMP - hydroxymethylase	Hm-dCMP synthesis	All *T-even type* and *JS98 type* phages. Also phages RB69 and CC31, all 4 *Acinetobacter* phages (133, Acj9, Acj61 and Ac42) and the *Aeromonas* phages 44RR, 31 and 25.
56	dCTPase - dUTPase	Increases dCMP pool, decreases dCTP pool; provides dUMP for dTMP synthesis	All phages listed in Table 2, except the giant *Aeromonas* phages Aeh1 and 65 and the giant *Vibrio* phages KVP40 and nt-1.
α-gt	α glucosyl transferase	α glucosylation of Hm-dCMP DNA	*T-even type* phages only
β-gt	β glucosyl transferase	β-glucosylation of Hm-dCMP DNA	Phages T4 and CC31 only
βα-gt	β-1, 6-glucosyl-α-glucose transferase	β glucosylation of α-glucosylated Hm-dCMP DNA	All T-even type phages, except T4; also present in CC31
denA	Endonuclease II (Endo II)	Limited cleavage of unmodified (dCMP-containing) DNA	All T4 relatives, except the *Acinetobacter* phages and marine cyanophages listed in Table 2
denB	Endonuclease IV (Endo IV)	Extensive cleavage of unmodified (dCMP containing) DNA	Same distribution as gene *42*
alc	Alc protein	Disallows transcription of unmodified (dCMP containing) DNA	Same distribution as gene *42*
arn	Arn protein	Counters the restriction effects of the host McR (Rgl) system	All *T-even type* phages; and phage CC31 only.

[1] The information in this Table is for the phages listed in Table 2.

DNA modification and the interplay between phage- and host-encoded proteins can limit the opportunities for genetic recombination between the very similar phage genomes.

T2, T4 and T6 encode homologous dCTPase-dUTPase (gp56; gene 56), dCMP-hydroxymethylase (gp42; gene 42) and dNMP kinase (gp1; gene 1) enzymes that together create a pool of hydroxymethylated-dCTP (Hm-dCTP) for phage DNA synthesis. The Hm-dCMP of the synthesized DNA is further modified by the addition of glucose molecules to the Hm groups. The glucosylation is carried out differently and to different extents between the three phage relatives. They all encode homologues of an α-glucosyltransferase (αgt gene) that adds glucose molecules to the Hm groups in the α-configuration; however, the T2 and T4 enzymes glucosylate 70% whereas the T6 enzyme glucosylates only 3% of these groups in the respective genomes. The three phages also differ in a second wave of glucosylations of the genomic Hm-dCMP. T4 encodes a β-glucosyltransferase (βgt gene) that adds glucose (in the β-configuration) to the rest of the unglucosylated Hm-dCMP residues in the phage DNA, whereas T2 and T6 lack a βgt gene and instead encode a β-1,6-glucosyl-α-glucose transeferase (βαgt gene) that adds glucose to the glucose moieties of some of the preexisting α-glucosylated Hm-dCMP residues, thus resulting in modification of the respective Hm-dCMP residues with gentobiose. This second glucosylation occurs at 70% of the α-glucosylated residues in T2 as compared to only ~3% of these residues in T6. That is, ~25% of the Hm-dCMP residues in T2 and T6 remain unglucosylated. Enzymes of the bacterial host synthesize the UDP-glucose (UDPG) used for the glucosylation reactions by the phage-induced enzymes. Interestingly, all of the close relatives of the T-even phages listed in Table 2 (*T-even type* phages) are predicted to encode αgt and βαgt genes, i.e., they are similar to T2 and T6 in their glucosylation genes. However, the glucosylation patterns of these relatives have not been analyzed. Also, it is worth noting that currently, T4 is the only member of the T4-like Viruses genus known to encode α- and β-glucosyltransferases. A distant relative of the *T-even type* phages, the coliphage CC31 (GU323318), is predicted to encode the unique combination of βgt and βαgt genes and currently, is the only other phage besides T4 in which a βgt gene has been detected by bioinformatic analyses.

Differences in DNA modification patterns, such as those that exist between the three T-even phages might open windows for phage-encoded nucleases that are able to distinguish between their own genomes and the genomes of dissimilarly modified close relatives. Also, as has been observed in T4, a lack of Hm-dCMP glucosylation can render the Hm-dCMP-containing phage DNA susceptible to the host-encoded Mcr (Rgl) restriction system, as well as the restriction systems of some prophages that can reside in *E. coli* or other potential Enterobacterial hosts [46,47]. Possibly, the unglucosylated Hm-dCMP sites in the T2 and T6 genomes escape restriction activities originating from the host through protection by the DNA modifications in their vicinity or through evolutionary adjustments in the expression of phage genes that control the susceptibility of phage DNA to the host-encoded restriction activities. In T4, the gene *2* protein (gp2), which attaches to DNA ends, protects against degradation by the host RecBCD exonuclease (Exo V) and the *arn* gene product (Arn protein) protects unglucosylated Hm-dCMP DNA against the host Mcr system [50-52] (Table 4). It would be interesting to find out if the *arn* gene and gene *2* are controlled differently in the different *T-even type* phages. All the phages in this cluster are predicted to encode homologues of T4 genes *56*, *42*, *2* and *arn* (Table 4) and at least some of them exhibit partial mutual exclusion with the T-even phages [25]. Elucidating the molecular basis for the partial incompatibilities within this cluster of closely interrelated phages might shed light on some subtle differences in phage genome adaptation that can begin to transition close relatives towards total genetic isolation from each other.

Additional factors that can potentially contribute to phage-phage exclusion between relatives that share the same bacterial host are the products of phage-specific nuclease genes, some of which might be imported into evolving phage genomes through lateral DNA transfer. Among these are genes for homing enzymes (HEGs), which exist as different types and in variable numbers among T4-related phage genomes. At least three HEG-encoded nucleases have been implicated in the partial exclusion of T2 by T4 [53-55]. Other types of inhibition of one T4-related phage by another are also possible and might potentially be discovered among the predicted products of the numerous novel ORFs in the Pangenome of the T4-like Viruses. The distribution of HEGs in the genomes of the phages listed in Tables 1 and 2 is discussed later in this review.

There are some distant relatives of the T-even phages that encode homologues of genes *42* and *56*, but that lack homologues of the glucosyltransferase genes. Examples are the coliphages RB69 and JS98 and the *Aeromonas salmonicida* phages 44RR, 31 and 25 (see Table 2 for GenBank Accession nos.). These gene *42*-encoding phages also encode homologues of the T4 genes that have been implicated in phage-induced degradation or inhibition of the expression of unmodified (dCMP-containing) DNA, i.e., the *alc*, *denA* and *denB* genes (Table 4). It is not yet known if phages like RB69 and JS98 are adapted to having Hm-dCMP instead of glucosylated Hm-dCMP in their DNA (e.g., through effective inhibition of the host restriction systems) or if they encode other types of modifications to the Hm-dCMP residues that provide similar protection from restriction by the host as does the glucosylation in *T-even type* phages. In addition, there are many T4 relatives that lack homologies to the entire gene network that controls DNA modification and expression of glucosylated DNA in phage T4, including genes *42* and *56*, the glucosyl-transferase genes and the *arn*, *alc* and *denB* genes. The dCMP of

the genomes of these phages probably lacks major modifications, as suggested by studies that have demonstrated a sensitivity of some of these genomes to certain Type II restriction endonucleases that fail to digest wild-type (modified) T4 genomic DNA [56]. Elucidation of the host-phage interactions that allow these seemingly unmodified phage genomes to propagate without being restricted by their hosts would be important for developing a better understanding of how the Core Genome of the T4-related phages has succeeded in spreading across bacterial species barriers in nature.

One example of a total incompatibility between phage T4 and a relative that also grows in *E. coli* is the exclusion of T4 by phage RB69 [25]. The T4 and RB69 genomes are >75% homologous over very long stretches of their genomes, but when introduced into the same host cells they generate no viable phage recombinants between them and only RB69 phage progeny are made. The sequencing of the RB69 genome has revealed considerable divergence in the nucleotide sequences of most of its alleles of T4 genes. So, it is not surprising that the T4 and RB69 have not been observed to exchange DNA through homologous recombination [12,35]. However, the sequence divergence between the two genomes does not explain why RB69 completely excludes T4 [25]. Interestingly, the RB69 genome is predicted to lack HEGs whereas T4 is predicted to encode many such nuclease genes. Yet, it is T4 rather than RB69 that suffers exclusion by its relative. The six *types* of T4-related phages that can grow in *E. coli* (Table 2) could potentially serve as excellent sources of material for studies of the multiple factors that can transition T4-related genomes from partial to total genetic isolation from each other despite access to the same bacterial host domain. Technological developments in DNA and genome analysis since the early studies on T4-related phage-phage exclusion should make it possible to develop PCR-based high-throughput methodologies for examining large populations of phage progeny from crosses between compatible, partially compatible or incompatible phages.

Agents of lateral DNA transfer in T4-related genomes

Although horizontal DNA transfer is suspected to play a major role in the evolution of the T4-related phages, particularly in diversification of the Pangenome of these phages, there are few clues about the agents that might mediate such transfer. Typically, the junctions between Core Genome components and adjacent DNA presumed to be imported by lateral transfer show no similarities to the familiar sequence signatures of known bacterial mobile elements that insert through site-specific and transpositional recombination [57]. Ectopic insertions (DNA additions) and illegitimate reciprocal or nonreciprocal recombination (DNA replacements) in the natural pools of evolving T4-related phages are possible causes for diversification of phage genomes through DNA rearrangements [58,59]; however, it is unclear if such events are more likely to occur in dsDNA phage evolution (or the evolution of the T4-like Viruses in particular) than in the evolution of bacterial and other cellular genomes in the microbial world. The diversity observed among the T4-related genomes examined so far appears to be of a similar magnitude to the diversity seen between distantly interrelated bacterial genera [60]. For example, in Aeh1, KVP40 and the cyanobacterial phages (Table 2), >85% of the genetic composition is unique to the *type* of T4-related phage genome and presumed to have originated through DNA rearrangements that assembled these genomes from core and variable components. The plasticity of genome size and the ability of modules of Core genes to function in a variety of orientations and genetic neighborhoods (Figure 3) suggest that genomes of the T4-like Viruses are particularly receptive to genetic gains and losses that might improve their adaptation to new environments. In addition, based on studies with T4 [8,61], these genomes are predicted to encode a highly active enzyme system for homologous recombination that has evolved to be an integral part of the machinery for genome replication, maintenance and packaging. It is known that the enzymes for homologous recombination can also mediate non-homologous (or "illegitimate") exchanges between marginally similar or even dissimilar genetic sequences in all DNA-based biological systems. An evolving T4-related genome might incorporate foreign DNA through at least two pathways that involve illegitimate recombination; (a) traditional reciprocal exchanges with foreign genetic entities (genetic replacements) and (b) initiation of DNA replication through the invasion of intracellular phage DNA pools by free 3' ends of foreign DNA (genetic additions; see also [8]). The production of viable phage recombinants by way of such events might be rare, but the observed mosaicism between the known T4-related phages is clear evidence that genetic shuffling has been rampant in the evolution of these phages.

Homing endonucleases as possible mediators of T4-related genome diversification

Other agents that might facilitate the acquisition of novel DNA into evolving T4-related genomes are the DNA endonucleases, especially homing endonucleases. Homing enzymes have been experimentally shown to mediate the unidirectional transfer of DNA between closely related T4-like genomes in two types of scenarios, intron homing [43,44] and intronless homing [53,54]. Both types of homing utilize homologous recombination between phages co-infecting the same bacterial host to complete the transfer of genetic information from the endonuclease-encoding genome to a recipient genome

that lacks the gene for the endonuclease. In Table 5, we summarize the distribution of putative HEGs among the T4-related genomes sequenced so far. The abundance and variable distributions of these genes in this pool of interrelated phage genomes suggests that T4 and its relatives are attractive natural homes for this category of transposable elements. Also, as indicated in Table 5, most of the known or predicted HEGs in these phages exist as freestanding ORFs in the phage genomes. There are only three HEGs known that reside inside self-splicing group I introns and that have been experimentally implicated in intron homing [62]. All three reside in the cluster of *T-even type* phages [63] and have probably spread within this cluster in natural settings. In contrast, there is no convincing evidence that these elements have moved across the bacterial species and genera that separate the different clusters or phage/genome *types* listed in Table 2. Nevertheless, recently observed novel activities of HEGs suggest that this category of transposable genes might be capable of generalized transposition without leaving traces of their involvement in the lateral transfer.

In both intron-homing and intronless-homing the primary role of the homing endonuclease is to introduce a dsDNA break in the genome destined to receive the HEG-containing intron or freestanding HEG. It is the repair process for the dsDNA break that ultimately provides a copy of the donor DNA for recombination into the recipient through a gene conversion event. In this regard, any endonuclease that creates dsDNA breaks might be a potential mediator of lateral DNA transfer [64,65]. Since the enzymes for homologous recombination can mediate exchanges between marginally similar or even dissimilar sequences, it is possible that a variety of endonucleases can initiate illegitimate genetic exchanges.

There are at least three examples of freestanding HEGs in T4-related phages that are suspected to encode the homing enzymes for introns lacking HEGs of their own [36,55,65]. The natural existence of such HEGs raises the possibility that some homing enzymes can mediate the transposition of DNA that is distantly located from their own structural genes without necessarily co-transferring the HEG itself. Such a role for HEGs would be consistent with the observation that much of the mosaicism between T4-related genomes is usually not associated with closely linked HEGs; however, no experimental evidence is currently available in support of the notion that HEGs can create mosaicism at distant genetic loci. Considering the wide distribution of HEGs in what is probably only a small sampling of the diversity of T4-related genomes in nature, this class of genomes might ultimately prove to be a rich repository of other as yet unidentified families of HEGs.

Table 5 Distribution of HEGs or putative HEGs in sequenced T4-related genomes

Phage genome analyzed	seg-like GIY-YIG	other GIY-YIG	mob-like HNH	HNH-AP2	other HNH	hef-like	Intron encoded GIY-YIG	HNH	Total
			Category and number of HEGs found						
T4	7		5				2	relic	15
T2					1				1
T6	6		3		1		1		11
RB3	4		2				1	1	8
RB14	2								2
RB15	1		1						2
RB18	2								2
RB26									None
RB32	2		2						4
RB51 or RB70	2		1		1				4
LZ2	3						1		4
RB69									None
RB49				1	1				2
phi-1	1				1				2
JSE	1		2	2	2			relic?	7
JS98									None
JS10									None
CC31									None
RB43			1	3					4
RB16	2	1	1	6					10
133	2	1	1			1			5
Acj9	2		1		1	1			5
Acj61	3								3
Acc42	4	1							5
44RR									None
31									None
25	2		1		1				4
Aeh1			1						1
PX29	2								2
65			1						1
KVP40		1							1
nt-1	1								1
S-PM2									None
S-RSM4									None
Syn9		1							1
P-SSM2		1							1
P-SSM4									None
KP15	1			1					
W-14					1				1
Sbo-AG3			2						2

It is perhaps not surprising that introns appear to be much less abundant than HEGs in T4-related genomes. To persist in evolution, introns must be able to guarantee the survival of their host by maintaining their self-splicing activities. Introns depend on homing enzymes for their spread, although they can integrate less frequently through reverse splicing [66,67]. In contrast, untranslated intercistronic regions offer a much larger selection of potential targets for the insertion of HEGs, which might also enter genomes through rare ectopic insertion [68]. The three group I introns that have been described for the *T-even type* phages all encode their own HEGs , i.e., the introns in the *td* (I-TevI), *nrdB* (I-TevII) and *nrdB* (I-TevII) genes (Table 5). A fourth group I intron was recently described for the DNA polymerase gene (gene *43*) of the *Aeromonas salmonicida* phage 25 (Intron *25.g43B*) [36]. This intron lacks its own HEG, but is predicted to use a freestanding HEG for mobility. Another putative group I intron can be detected in gene *43* of the recently published genome sequence of phage JSE, a close relative of phage RB49 [69]. Our own examination of this sequence suggests that the JSE intron contains a truncated derivative of a former HEG, i.e., much like the existence of a truncated HEG in the intron of the T4 *nrdB* gene [70]. Such HEG truncations might add to the difficulties in detecting traces of these mobile elements in contemporary phage genomes.

In summary, the observations cited above suggest that the self-mobilizing freestanding HEGs are potential agents of lateral transfer that might contribute to genomic mosaicism by mobilizing a variety of genetic sequences in phage genomes, including introns and flanking as well as distant DNA and genes or gene clusters.

Concluding remarks

Genomes of the T4-like Viruses are repositories of a diversity of genes for which no biological roles have been assigned or can be predicted on the basis of comparisons to other sequences in databases. The reference for these phages, phage T4, has been extensively studied [2,7,8] and provides a rational basis for suspecting that the diversity among its relatives is a reflection of adaptations of a core phage genome to a variety of challenges in evolution, including encounters with new host environments. Experimentally, many T4 genes that are not essential for phage propagation in some bacterial hosts or genetic backgrounds are nevertheless essential in others (see [8] for examples). Bacterial genomes are themselves dynamic entities that are subject to the trafficking of prophages, plasmids and possibly other entities that can restrict or complement the propagation of other invaders of bacteria. There are at least three examples in the T4 biological system where prophages or defective prophages can restrict T4 phage growth. These are the restriction of T4 *rII* mutants by lambda lysogens, the restriction of unglucosylated HMC-DNA by P1 lysogens and the restriction of late phage gene expression by the e14 element [8]. Such examples underscore the important role that the host (and its resident prophages) must play in determining the T4-related genotype required for survival in the host environment. The range of natural bacterial hosts for any of the phages listed in Tables 1 and 2 might be much broader than what is available or has been used in laboratories to propagate these phages and evaluate their physiology. The isolation of new T4 relatives for

known bacterial hosts as well as the identification of new bacterial hosts for known and new types of T4-related phages would be important for bridging the many gaps in our understanding of how the T4-like Viruses have managed to spread across bacterial species barriers. At the very least, the current sequence database for these Myoviridae should prove to be a rich source of genetic markers for bioprospecting as well as being a mine of reagents for basic research and biotechnology.

In regard to studies of the basic mechanisms of molecular evolution, the T4-like Viruses constitute a large pool of interrelated autonomously replicating entities that are highly accessible to analysis of broadly applicable concepts in biology. The genomes of these viruses are large by viral standards and exhibit many parallels to the mosaicism and diversity of prokaryotic cellular genomes. The phage genomes analyzed so far (Table 2) could be used as reference points for the analysis, especially through metagenomic tools, of large populations of closely interrelated phages within specific ecological domains without having to isolate these phages as plaque-forming units. This would be particularly important for the detection of commonalities between T4-related genomes and other types of genomes in the microbial world. In addition, such metagenomic approaches would be useful for detecting the continuities and abrupt discontinuities that occur at the branch points between phage lineages.

As potential sources of interesting gene products for studies of biological structure and function, one needs only to scan the literature for the numerous examples where T4-encoded proteins have been used to elucidate the mechanisms of processes common to most organisms, such as DNA replication, transcription, translation, genetic recombination, mutation, homing and others. One of the most important paths to biological diversification is the path to changes in the specificities of proteins and nucleic acids that retain their essential biochemical activities. The collection of sequenced T4-related phages is already a rich source of such examples of diversification of protein specificity.

Finally, we should mention the resurgence of interest in bacterial viruses as sources of toxins [71] and as potential therapeutic agents against bacterial pathogens [72,73]. T4 and its known relatives are classical examples of how virulent a virus can be against one bacterial host and ineffective against many other bacteria. These phages have no other lifestyle but the one leading to cell death and they use multiple targets in their attacks on hosts. The different specificities with which the T4-like Viruses recognize and inhibit different bacterial host species raise hopes that phage-induced gene products can be found that are highly specific to targets in specific bacterial pathogens. By using combinations of these gene products to attack multiple targets the development of bacterial resistance against these biological drugs would become highly unlikely. Bacteriophage genomics and particularly the genomics of T4-related phages are opening windows to many new frontiers of basic and applied biology.

List of Abbreviations

contigs: Contiguous sequences; dsDNA: Double-stranded DNA; HEG: Homing endonuclease gene; Hm: Hydroxymethyl; ICTV: International Committee for the Taxonomy of Viruses; LGT: Lateral gene transfer; ORF: Open-reading frame; PCR: Polymerase chain reactions; UDPG: Uridine diphosphate-glucose

Additional material

Additional file 1: Table S1. A comprehensive listing of T4 alleles in most of the phages listed in Tables 1 and 2 can be found in Additional file 1

Acknowledgements

We thank David Edgell for many helpful comments on the manuscript and Hans Ackermann for enlightening us about the importance of diligence when comparing the morphologies of different Myoviridae by electron microscopy. Ultimately, a rational nomenclature for viruses belonging to the T4 family will require the use of both genetic and morphological criteria. We are also grateful Martha Clokie, Andy Millard and Nick Mann for contributing information about the cyanophages for Additional file 1 and to the many other colleagues who have discussed phage genomics with us. Jill Barbay and Marlene Jones provided excellent clerical and other assistance during the preparation and submission of the manuscript.

Author details

[1]Department of Biochemistry, Tulane University Health Sciences Center, 1430 Tulane Avenue, New Orleans, LA, USA. [2]School of Science and Technology, Georgia Gwinnett College, 1000 University Center Lane, Lawrenceville, GA 30043, USA. [3]Department of Microbiology, Campus Box 7615, North Carolina State University, Raleigh, NC 27695, USA.

Authors' contributions

J. D. Karam wrote the first draft of the manuscript with considerable help from V. Petrov and S. Ratnayaka, who prepared summaries for the Tables and Figures. Also, V. Petrov had participated heavily in the analysis of a large number of the genomes reviewed here and prepared most of the genomes sequenced in the Karam laboratory for submission to GenBank. S. Ratnayaka assisted V. Petrov in these efforts. J. Nolan created and manages the websites http://phage.bioc.tulane.edu (more recently http://phage.ggc.edu), which was used extensively in the preparation of the summaries presented in this review. J. Nolan also contributed unpublished information about the sequences of several close relatives of T4 and he and E. Miller contributed numerous suggestions for improvement of the manuscript. In addition E. Miller facilitated the sequence analysis of a number of the phage genomes discussed here. All authors read and approved the final manuscript.

Competing interests

The authors declare that they have no competing interests.

Received: 21 May 2010 Accepted: 28 October 2010
Published: 28 October 2010

References

1. Cairns J, Stent GS, Watson JD: **Phage and the Origins of Molecular Biology.** New York: Cold Spring Harbor Laboratory Press; 1992.
2. Karam JD, et al: **Molecular Biology of Bacteriophage T4.** Washington, DC: American Society for Microbiology; 1994.
3. Abedon ST: **The murky origin of Snow White and her T-even dwarfs.** Genetics 2000, **155(2)**:481-486.
4. Benzer S: **The fine structure of the gene.** Sci Am 1962, **206**:70-84.
5. Edgar B: **The genome of bacteriophage T4: an archeological dig.** Genetics 2004, **168(2)**:575-582.
6. Epstein RH, Bolle A, Steinberg CM, Kellenberger E, Boy De La Tour E, Chevalley R, Edgar RS, Susman M, Denhardt GH, Lielausis A: **Physiological studies of conditional lethal mutants of bacteriophage T4D.** Symposia on Quantitative Biology: 1963; Cold Spring Harbor Laboratory of Quantitative Biology Cold Spring Harbor Press, New York; 1963, 375-394.
7. Miller ES, Kutter E, Mosig G, Arisaka F, Kunisawa T, Ruger W: **Bacteriophage T4 genome.** Microbiol Mol Biol Rev 2003, **67(1)**:86-156.
8. Mosig G, Eiserling F: **T4 and related phages: structure and development.** The Bacteriophages Oxford University Press; 2006, 225-267.
9. Ackermann HW, Krisch HM: **A catalogue of T4-type bacteriophages.** Arch Virol 1997, **142(12)**:2329-2345.
10. Ackermann HW: **5500 Phages examined in the electron microscope.** Arch Virol 2006, **152(2)**:227-243.
11. Russell RL: **Comparative genetics of the T-even bacteriophages.** Genetics 1974, **78(4)**:967-988.
12. Wang CC, Yeh LS, Karam JD: **Modular organization of T4 DNA polymerase. Evidence from phylogenetics.** J Biol Chem 1995, **270(44)**:26558-26564.
13. Desplats C, Dez C, Tetart F, Eleaume H, Krisch HM: **Snapshot of the genome of the pseudo-T-even bacteriophage RB49.** J Bacteriol 2002, **184(10)**:2789-2804.
14. Petrov VM, Nolan JM, Bertrand C, Levy D, Desplats C, Krisch HM, Karam JD: **Plasticity of the gene functions for DNA replication in the T4-like phages.** J Mol Biol 2006, **361(1)**:46-68.
15. Nolan JM, Petrov V, Bertrand C, Krisch HM, Karam JD: **Genetic diversity among five T4-like bacteriophages.** Virol J 2006, **3**:30.
16. Comeau AM, Bertrand C, Letarov A, Tetart F, Krisch HM: **Modular architecture of the T4 phage superfamily: a conserved core genome and a plastic periphery.** Virology 2007, **362(2)**:384-396.
17. Miller ES, Heidelberg JF, Eisen JA, Nelson WC, Durkin AS, Ciecko A, Feldblyum TV, White O, Paulsen IT, Nierman WC, et al: **Complete genome sequence of the broad-host-range vibriophage KVP40: comparative genomics of a T4-related bacteriophage.** J Bacteriol 2003, **185(17)**:5220-5233.
18. Zuber S, Ngom-Bru C, C B: **Genome analysis of phage JS98 defines a fourth major subgroup of T4-like phages in Escherichia coli.** J Bacteriol 2007, **189(22)**:8206-8214.
19. Lavigne R, Darius P, Summer EJ, Seto D, Mahadevan P, Nilsson AS, Ackermann HW, Kropinski AM: **Classification of Myoviridae bacteriophages using protein sequence similarity.** BMC Microbiol 2009, **9**:224.
20. Filee J, Bapteste E, Susko E, Krisch HM: **A selective barrier to horizontal gene transfer in the T4-type bacteriophages that has preserved a core genome with the viral replication and structural genes.** Mol Biol Evol 2006, **23(9)**:1688-1696.
21. Bull A, (ed): **Microbial Diversity and Bioprospecting.** Washington; ASM Press; 2004.
22. Hatfull GF, Cresawn SG, Hendrix RW: **Comparative genomics of the mycobacteriophages: insights into bacteriophage evolution.** Res Microbiol 2008, **159(5)**:332-339.
23. Rohwer F, Edwards R: **The phage proteomic tree: a genome-based taxonomy for phage.** J Bacteriol 2002, **184(16)**:4529-4535.
24. Lawrence JG, Hatfull GF, Hendrix RW: **Imbroglios of viral taxonomy: genetic exchange and failings of phenetic approaches.** J Bacteriol 2002, **184(17)**:4891-4905.
25. Russell RL: **Speciation among the T-even bacteriophages.** Ph.D. Dissertation Pasadena: California Institute of Technology; 1967.

26. Kim JS, Davidson N: Electron microscope heteroduplex study of sequence relations of T2, T4, and T6 bacteriophage DNAs. *Virology* 1974, **57**(1):93-111.

27. Mann NH, Clokie MR, Millard A, Cook A, Wilson WH, Wheatley PJ, Letarov A, Krisch HM: The genome of S-PM2, a "photosynthetic" T4-type bacteriophage that infects marine Synechococcus strains. *J Bacteriol* 2005, **187**(9):3188-3200.

28. Klasson L, Andersson SGE: Evolution of minimal-gene-sets in host-dependent bacteria. *Trends Microbiol* 2004, **12**(1):37-43.

29. Gil R, Silva FJ, Pereto J, Moya A: Determination of the core of a minimal bacterial gene set. *Microbiol Mol Biol Rev* 2004, **68**(3):518-537.

30. Abby S, Daubin V: Comparative genomics and the evolution of prokaryotes. *Trends Microbiol* 2007, **15**(3):135-141.

31. Millard AD, Zwirglmaier K, Downey MJ, Mann NH, Scanlan DJ: Comparative genomics of marine cyanomyoviruses reveals the widespread occurrence of Synechococcus host genes localized to a hyperplastic region: implications for mechanisms of cyanophage evolution. *Environ Microbiol* 2009, **11**(9):2370-2387.

32. Tetart F, Desplats C, Kutateladze M, Monod C, Ackermann HW, Krisch HM: Phylogeny of the major head and tail genes of the wide-ranging T4-type bacteriophages. *J Bacteriol* 2001, **183**(1):358-366.

33. Hendrix RW: Jumbo bacteriophages. In *Lesser Known Large dsDNA Viruses. Volume 328.* Heidelberg, Germany: Springer-Verlag Berlin; 2009:229-240.

34. Wang CC, Pavlov A, Karam JD: Evolution of RNA-binding specificity in T4 DNA polymerase. *J Biol Chem* 1997, **272**(28):17703-17710.

35. Yeh LS, Hsu T, Karam JD: Divergence of a DNA replication gene cluster in the T4-related bacteriophage RB69. *J Bacteriol* 1998, **180**(8):2005-2013.

36. Petrov VM, Ratnayaka S, Karam JD: Genetic insertions and diversification of the PolB-type DNA polymerase (gp43) of T4-related phages. *J Mol Biol* 2010, **395**(3):457-474.

37. Filee J, Tetart F, Suttle CA, Krisch HM: Marine T4-type bacteriophages, a ubiquitous component of the dark matter of the biosphere. *Proc Natl Acad Sci USA* 2005, **102**(35):12471-12476.

38. Timms AR, Cambray-Young J, Scott AE, Petty NK, Connerton PL, Clarke L, Seeger K, Quail M, Cummings N, Maskell DJ, *et al*: Evidence for a lineage of virulent bacteriophages that target Campylobacter. *BMC Genomics* 11:214.

39. Wuitschick JD, Lindstrom PR, Meyer AE, Karrer KM: Homing endonucleases encoded by germ line-limited genes in Tetrahymena thermophila have APETELA2 DNA binding domains. *Eukaryot Cell* 2004, **3**(3):685-694.

40. Magnani E, Sjolander K, Hake S: From endonucleases to transcription factors: evolution of the AP2 DNA binding domain in plants. *Plant Cell* 2004, **16**(9):2265-2277.

41. Wessler SR: Homing into the origin of the AP2 DNA binding domain. *Trends Plant Sci* 2005, **10**(2):54-56.

42. Sullivan MB, Coleman ML, Weigele P, Rohwer F, Chisholm SW: Three Prochlorococcus cyanophage genomes: signature features and ecological interpretations. *PLoS Biol* 2005, **3**(5):e144.

43. Edgell DR, Belfort M, Shub DA: Barriers to intron promiscuity in bacteria. *J Bacteriol* 2000, **182**(19):5281-5289.

44. Belfort M: Scientific serendipity initiates an intron odyssey. *J Biol Chem* 2009, **284**(44):29997-30003.

45. Wu LT, Chang SY, Yen MR, Yang TC, Tseng YH: Characterization of extended-host-range pseudo-T-even bacteriophage Kpp95 isolated on Klebsiella pneumoniae. *Appl Environ Microbiol* 2007, **73**(8):2532-2540.

46. Carlson K, Raleigh EA, Hattman S: Restriction and modification. In *Molecular Biology of Bacteriophage T4.* Edited by: Karam J. Washington, D. C.: American Society for Microbiology Press; 1994:369-381.

47. Snyder L, Kaufman G: T4 phage exclusion mechanism. In *Molecular Biology of Bacteriophage T4.* Edited by: Karam J. Washington, D. C.: American Society for Microbiology; 1994:391-396.

48. Drivdahl RH, Kutter EM: Inhibition of transcription of cytosine-containing DNA in vitro by the alc gene product of bacteriophage T4. *J Bacteriol* 1990, **172**(5):2716-2727.

49. Severinov K, Kashlev M, Severinova E, Bass I, McWilliams K, Kutter E, Nikiforov V, Snyder L, Goldfarb A: A non-essential domain of Escherichia coli RNA polymerase required for the action of the termination factor Alc. *J Biol Chem* 1994, **269**(19):14254-14259.

50. Dharmalingam K, Goldberg EB: Phage-coded protein prevents restriction of unmodified progeny T4 DNA. *Nature* 1976, **260**(5550):454-456.

51. Dharmalingam K, Revel HR, Goldberg EB: Physical mapping and cloning of bacteriophage T4 anti-restriction endonuclease gene. *J Bacteriol* 1982, **149**(2):694-699.

52. Kim BC, Kim K, Park EH, Lim CJ: Nucleotide sequence and revised map location of the arn gene from bacteriophage T4. *Mol Cells* 1997, **7**(5):694-696.

53. Belle A, Landthaler M, Shub DA: Intronless homing: site-specific endonuclease SegF of bacteriophage T4 mediates localized marker exclusion analogous to homing endonucleases of group I introns. *Genes Dev* 2002, **16**(3):351-362.

54. Liu Q, Belle A, Shub DA, Belfort M, Edgell DR: SegG endonuclease promotes marker exclusion and mediates co-conversion from a distant cleavage site. *J Mol Biol* 2003, **334**(1):13-23.

55. Wilson GW, Edgell DR: Phage T4 mobE promotes trans homing of the defunct homing endonuclease I-TevIII. *Nucleic Acids Res* 2009, **37**(21):7110-7123.

56. Monod C, Repoila F, Kutateladze M, Tetart F, Krisch HM: The genome of the pseudo T-even bacteriophages, a diverse group that resembles T4. *J Mol Biol* 1997, **267**(2):237-249.

57. Craig NL: Mobile DNA II. Washington, DC: American Society for Microbiology Press; 2002.

58. Brussow H, Hendrix RW: Phage genomics: small is beautiful. *Cell* 2002, **108**(1):13-16.

59. Hendrix RW, Hatfull GF, Smith MC: Bacteriophages with tails: chasing their origins and evolution. *Res Microbiol* 2003, **154**(4):253-257.

60. Bull A, Stach J: An overview of biodiversity-estimating the scale. *Microbial Diversity and Bioprospecting* Washington, DC: American Society for Microbiology Press; 2004, 15-28.

61. Mosig G: Recombination and recombination-dependent DNA replication in bacteriophage T4. *Annu Rev Genet* 1998, **32**:379-413.

62. Eddy SR: Introns in the T-even bacteriophages. *Ph.D. Dissertation* Boulder: University of Colorado at Boulder; 1991.

63. Sandegren L, Sjoberg BM: Distribution, sequence homology, and homing of group I introns among T-even-like bacteriophages: evidence for recent transfer of old introns. *J Biol Chem* 2004, **279**(21):22218-22227.

64. Eddy SR, Gold L: Artificial mobile DNA element constructed from the EcoRI endonuclease gene. *Proc Natl Acad Sci USA* 1992, **89**(5):1544-1547.

65. Zeng Q, Bonocora RP, Shub DA: A free-standing homing endonuclease targets an intron insertion site in the psbA gene of cyanophages. *Curr Biol* 2009, **19**(3):218-222.

66. Roman J, Rubin MN, Woodson SA: Sequence specificity of in vivo reverse splicing of the Tetrahymena group I intron. *RNA* 1999, **5**(1):1-13.

67. Roy SW, Irimia M: Mystery of intron gain: new data and new models. *Trends Genet* 2009, **25**(2):67-73.

68. Gibb EA, Edgell DR: An RNA hairpin sequesters the ribosome binding site of the homing endonuclease mobE gene. *J Bacteriol* 2009, **191**(7):2409-2413.

69. Denou E, Bruttin A, Barretto C, Ngom-Bru C, Brussow H, Zuber S: T4 phages against Escherichia coli diarrhea: potential and problems. *Virology* 2009, **388**(1):21-30.

70. Eddy SR, Gold L: The phage T4 nrdB intron: a deletion mutant of a version found in the wild. *Genes Dev* 1991, **5**(6):1032-1041.

71. Waldor M, Friedman D, Adhya S: Phages: Their Role in Bacterial Pathogenesis and Biotechnology. Washington, D.C: American Society for Microbiology Press; 2005.

72. Fischetti VA, Nelson D, Schuch R: Reinventing phage therapy: are the parts greater than the sum? *Nat Biotechnol* 2006, **24**(12):1508-1511.

73. Fischetti VA: Bacteriophage lysins as effective antibacterials. *Curr Opin Microbiol* 2008, **11**(5):393-400.

74. Chibani-Chennoufi S, Sidoti J, Bruttin A, Dillmann ML, Kutter E, Qadri F, Sarker SA, Brussow H: Isolation of Escherichia coli bacteriophages from the stool of pediatric diarrhea patients in Bangladesh. *J Bacteriol* 2004, **186**(24):8287-8294.

75. Weigele PR, Pope WH, Pedulla ML, Houtz JM, Smith AL, Conway JF, King J, Hatfull GF, Lawrence JG, Hendrix RW: Genomic and structural analysis of Syn9, a cyanophage infecting marine Prochlorococcus and Synechococcus. *Environ Microbiol* 2007, **9**(7):1675-1695.

doi:10.1186/1743-422X-7-292
Cite this article as: Petrov *et al.*: Genomes of the T4-related bacteriophages as windows on microbial genome evolution. *Virology Journal* 2010 **7**:292.

Clokie *et al. Virology Journal* 2010, **7**:291
http://www.virologyj.com/content/7/1/291

VIROLOGY JOURNAL

REVIEW

T4 genes in the marine ecosystem: studies of the T4-like cyanophages and their role in marine ecology

Martha RJ Clokie[1], Andrew D Millard[2*], Nicholas H Mann[2]

Abstract

From genomic sequencing it has become apparent that the marine cyanomyoviruses capable of infecting strains of unicellular cyanobacteria assigned to the genera *Synechococcus* and *Prochlorococcus* are not only morphologically similar to T4, but are also genetically related, typically sharing some 40-48 genes. The large majority of these common genes are the same in all marine cyanomyoviruses so far characterized. Given the fundamental physiological differences between marine unicellular cyanobacteria and heterotrophic hosts of T4-like phages it is not surprising that the study of cyanomyoviruses has revealed novel and fascinating facets of the phage-host relationship. One of the most interesting features of the marine cyanomyoviruses is their possession of a number of genes that are clearly of host origin such as those involved in photosynthesis, like the *psbA* gene that encodes a core component of the photosystem II reaction centre. Other host-derived genes encode enzymes involved in carbon metabolism, phosphate acquisition and ppGpp metabolism. The impact of these host-derived genes on phage fitness has still largely to be assessed and represents one of the most important topics in the study of this group of T4-like phages in the laboratory. However, these phages are also of considerable environmental significance by virtue of their impact on key contributors to oceanic primary production and the true extent and nature of this impact has still to be accurately assessed.

Background

The cyanomyoviruses and their hosts

In their review on the interplay between bacterial host and T4 phage physiology, Kutter et al [1] stated that "efforts to understand the infection process and evolutionary pressures in the natural habitat(s) of T-even phages need to take into account bacterial metabolism and intracellular environments under such conditions". This statement was made around the time that the first cyanophages infecting marine cyanobacteria were being isolated and characterized and the majority of which exhibited a T4-like morphology (Figure 1) and [2-4]. Obviously, the metabolic properties and intracellular environments of obligately photoautotrophic marine cyanobacteria are very different to those of the heterotrophic bacteria that had been studied as the experimental hosts of T4-like phages and no less significant are the differences between the

environments in which they are naturally found. It is not surprising, therefore, that the study of these phages has led to the recognition of remarkable new features of the phage-host relationship and this is reflected by the fact that they have been referred to as "photosynthetic phages" [5,6]. These T4-like phages of cyanobacteria have extensively been referred to as cyanomyoviruses and this is the term we have used throughout this review. Without doubt the most exciting advances have been associated with an analysis of their ecological significance, particularly with respect to their role in determining the structure of marine cyanobacterial populations and diverting fixed carbon away from higher trophic levels and into the microbial loop. Associated with this have been the extraordinary developments in our understanding of marine viral communities obtained through metagenomic approaches e.g. [7-9] and these are inextricably linked to the revelations from genomic analyses that these phages carry a significant number of genes of clearly host origin such as those involved in photosynthesis, which raises important questions regarding the metabolic function of these genes and

* Correspondence: a.d.millard@warwick.ac.uk
[2]Department of Biological Sciences, University of Warwick, Gibbet Hill Road, Coventry, CV4 7AL, UK
Full list of author information is available at the end of the article

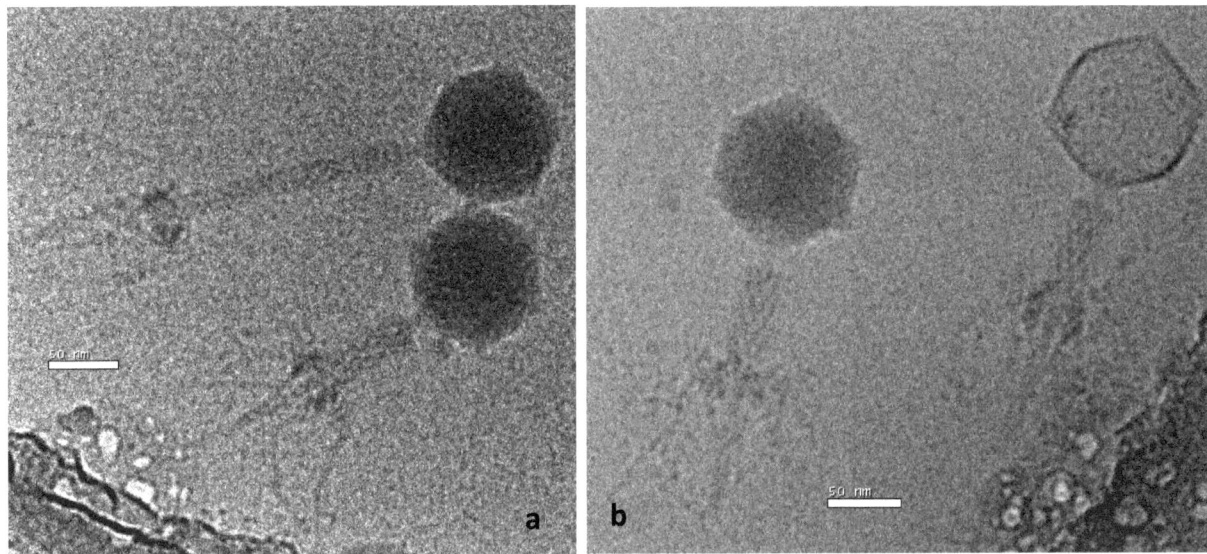

Tail: l = 62 , 63 nm, w (top) = 28 , 27 nm

Figure 1 Cryoelectron micrographs of purified S-PM2 phage particles. (A) Showing one phage particle in the extended form and one in the contracted form both still have DNA in their heads and (B) Two phage particles with contracted tail sheaths, the particle on the left has ejected its DNA. The lack of collar structure is particularly visible in (B). The diameter of the head is 65 nm. Pictures were taken at the University of Warwick with the kind assistance of Dr Svetla Stoilova-McPhie.

their contribution to phage fitness. Obviously, this has major implications for horizontal gene transfer between phages, but also between hosts. Finally, from genomic sequencing it has also become apparent that the cyanomyoviruses are not only morphologically similar to T4, but are also genetically interrelated. It is still too early for these key areas, which form the major substance of this review, to have been extensively reviewed, but aspects of these topics have been covered [10-12].

Central to discussing these key aspects of cyanomyoviruses is a consideration of their hosts and the environment in which they exist. Our knowledge of marine cyanomyovirus hosts is almost exclusively confined to unicellular cyanobacteria of the genera *Synechococcus* and *Prochlorococcus*. These organisms are highly abundant in the world's oceans, and together they are thought to be responsible for 32-89% of the total primary production in oligotrophic regions of the oceans [13-15]. Although members of the two genera are very closely related to each other they exhibit major differences in their light-harvesting apparatus. Typically cyanobacteria possess macromolecular structures, phycobilisomes, that act as light-harvesting antennae composed of phycobilin-bearing phycobiliproteins (PBPs) and non-pigmented linker polypeptides. They are responsible for absorbing and transferring excitation energy to the protein-chlorophyll reaction centre complexes of PSII and PSI. Cyanobacterial PBSs are generally organised as a hemidiscoidal complex with a core structure, composed of a PBP allophycocyanin (APC), surrounded by six peripheral rods, each composed of the PBP phycocyanin (PC) closest to the core and phycoerythrin (PE) distal to the core. These PBPs, together with Chl *a*, give cyanobacteria their characteristic colouration; the blue-green colour occurs when PC is the major PBP. In marine *Synechococcus* strains, classified as sub-cluster 5.1 (previously known as marine cluster A) [16], the major light-harvesting PCB is phycoerythrin giving them a characteristic orange-red colouration. Other marine *Synechococcus* strains, more commonly isolated from coastal or estuarine waters, have phycocyanin as their major PCB and classified as sub-cluster 5.2 (previously known as marine cluster B) [16].

In contrast marine *Prochlorococcus* strains do not possess phycobilisomes and instead utilize a chlorophyll a_2/b_2 light-harvesting antenna complex [17]. The genetic diversity within each genus represented by a wide variety of ecotypes is thought to be an important reason for their successful colonization of the world's oceans and there is now clear evidence of spatial partitioning of individual cyanobacterial lineages at the basin and global scales [18,19]. There is also a clear partitioning of ecotypes on a vertical basis within the water column, particularly when stratification is strong e.g. [20], which at least in part may be attributable to differences in their ability to repair damage to PSII [21]. This diversity of ecotypes obviously raises questions regarding the host ranges of the cyanomyoviruses.

Diversity

The T4-like phages are a diverse group, but are unified by their genetic and morphological similarities to T4. The cyanomyoviruses are currently the most divergent members of this group and despite clear genetic relatedness exhibit only a modest morphological similarity to the T-evens, with smaller isometric heads and tails of up to ~180 nm in length Figure 1 and [22-24], and so have been termed the ExoT-evens [22]. It has been suggested that the isometric icosahedral capsid structures of the cyanomyoviruses may reflect the fact that they only possess two (gp23 and gp20) of the five T4 capsid shell proteins with consequent effects on the lattice composition. Despite forming a discrete sub-group of the T4-like phages they exhibit considerable diversity. One study on phages isolated from the Red Sea using a *Synechococcus* host revealed a genome size range of 151-204 kb. However, the *Prochlorococcus* phage P-SSM2 is larger at 252 kb [25] and a study of uncultured viruses from Norwegian coastal waters revealed the presence of phages as large as 380 kb that could be assumed to be cyanoviruses, by virtue of their possession of the *psbA* and *psbD* genes [26].

Attempts to investigate the diversity of cyanomyoviruses began with the development of primers to detect the conserved *g20* encoding the portal vertex protein [27] and other primer sets based on *g20* were subsequently developed [28,29]. Diversity was found to vary both temporally and spatially in a variety of marine and freshwater environments, was as great within a sample as between oceans and was related to *Synechococcus* abundance [30-34]. With the accumulation of *g20* sequence information from both cultured isolates and natural populations phylogenetic analysis became possible and it became apparent that were nine distinct marine clades with freshwater sequences defining a tenth [28,29,32,34-36]. Only three of the nine marine clades contained cultured representatives. Most recently a large scale survey confirmed the three marine clades with cultured representatives, but cast doubt on the other six marine clades, while at the same time identifying two novel clades [37]. The key observation from this study was that *g20* sequences are not good predictors of a phage's host or the habitat. A substantial caveat that must be applied to these molecular diversity studies is that although the primers were designed to be specific for cyanomyoviruses there is no way of knowing whether they also target other groups of myoviruses e.g. [29].

A study employing degenerate primers against *g23*, which encodes the major capsid protein in the T4-type phages, to amplify *g23*-related sequences from a diverse range of marine environments revealed a remarkable degree of molecular variation [38]. However, sequences clearly derived from cyanomyoviruses of the Exo-Teven subgroup were only found in significant numbers from surface waters. Most recently Comeau and Krisch [39] examined *g23* sequences obtained by PCR of marine samples coupled with those in the Global Ocean Sampling (GOS) data set. One of their key findings was that the GOS metagenome is dominated by cyanophage-like T4 phages. It is also clear from phylogenetic analysis that there is an extremely high micro-diversity of cyanomyoviruses with many closely related sequence subgroups with short branch lengths.

Host ranges

Studies on the host range of marine cyanomyoviruses have shown wide variations. Waterbury and Valois [3] found that some of their isolates would infect as many as 10 of their 13 *Synechococcus* strains, whereas one would infect only the strain used for isolation. One myovirus isolated on a phycocyanin-rich *Synechococcus* strain, would also infect phycoerythrin-rich strains. None of the phages would infect the freshwater strain tested. Similar observations were made by Suttle and Chan [4]. A study by Millard et al., which investigated host ranges of 82 cyanomyovirus isolates showed that the host ranges were strongly influenced by the host used in the isolation process [40]. 65% of phages isolates on *Synechococcus* sp. WH7803 could infect *Synechococcus* sp. WH8103, whereas of the phages isolated on WH8103 ~91% could also infect WH7803. This may reflect a restriction-modification phenomenon. The ability to infect multiple hosts was widespread with ~77% of isolates infecting at least two distinct host strains. Another large scale study using 33 myoviruses and 25 *Synechococcus* hosts revealed a wide spread of host ranges from infection only of the host used for isolation to 17/25 hosts [41]. There was also a statistical correlation of host range with depth of isolation; cyanophage from surface stations tended to exhibited broader host ranges. A study on the host ranges of cyanophages infecting *Prochlorococcus* strains found similar wide variations in the host ranges of cyanomyoviruses, but also identified myoviruses that were capable of infecting both *Prochlorococcus* and *Synechococcus* hosts [42].

Genetic commonalities and differences between T4-like phages from different environmental niches

The first reported genetic similarity between a cyanomyovirus and T4 was by Fuller *et al*, 1998 who discovered a gene homologous to *g20* in the cyanomyovirus S-PM2 [27]. In 2001 Hambly *et al*, then reported that it was not a single gene that was shared between S-PM2 and T4, but remarkably a 10 Kb fragment of S-PM2 contained the genes *g18-g23*, in a similar order to those found in T4 [22]. With the subsequent sequencing of the complete genomes of the cyanomyoviruses S-PM2 [5], P-SSM4 [25],

P-SSM2 [25], Syn9 [23] and S-RSM4 [43], it has become apparent that cyanomyoviruses share a significant number of genes that are found in other T4-like phages.

General properties of cyanophage genomes

The genomes of all sequenced cyanomyovirus are all at least 10 Kb larger than the 168 Kb of T4, with P-SMM2 the largest at 252 Kb. Genomes of cyanomyovirus have some of the largest genomes of the T4-like phages with only Aeh1 and KVP40 [44] of other T4-like phage having genomes of comparable size. The general properties of cyanophage genomes such as mol G+C content and % of genome that is coding are all very similar to that of T4 (Table 1). The number of tRNAs found within is variable, with the 2 cyanomyoviruses P-SMM2 and P-SMM4 isolated on *Prochlorococcus* having none and one respectively. In contrast the two cyanophages S-PM2 and S-RSM4 that to date are only known to infect *Synechococcus* have 12 and 25 tRNAs respectively. Previously it has been suggested a large number of tRNAs in a T4-like phage may be an adaptation to infect multiple hosts [44], this does not seem fit with the known data for cyanomyoviruses with Syn9 which is known to infect cyanobacteria from two different genera has 9 tRNAs, significantly fewer than the 25 found in S-PM2 that only infects cyanobacteria of the genus *Synechococcus*.

Common T4-like genes

A core genome of 75 genes has previously been identified from the available T4-like genomes, excluding the cyanomyovirus genomes [25]. The cyanomyoviruses S-PM2, P-SSM4, P-SSM2 and Syn9 have been found to share 40, 45, 48 and 43, genes with T4 [5,23,25]. The majority of these genes that are common to a cyanophage and T4 are the same in all cyanomyoviruses (Figure 2).

Table 1 General properties of cyanomyoviruses genomes in comparison to T4 and KVP40

Phage	No of Genes	tRNAs	% Coding	Genome Size (Kb)	% mol G+C
T4	288	10	93	168.9	35
KVP40	386	30	92	244.8	42
S-PM2	236	25	92	196.2	37
P-SSM4	198	0	92	178.2	36
P-SSM2	330	1	94	252.4	35
Syn9	232	6	97	177.3	40
S-RSM4	238	12	94	194.4	41

Data was extracted from the genbadnk submission of each genome sequence in May 2009. T4 (accession NC_000866), KVP40 (accession number NC_005083), S-PM2 (accession number NC_006820), P-SSM4 (accession number NC006884), P-SSM2 (accession number NC006883), Syn9 (accession number NC_005083), S-RSM4 (accession number FM207411).

Transcription

Only four genes involved in transcription have been identified as core gene in T4-like phages [25]. The cyanomyoviruses are found to have three of these genes *g33*, *g55* and *regA*. A trait common to all cyanomyoviruses is the lack of homologues to *alt*, *modA* and *modB*, that are essential in moderating the specificity of the host RNA polymerase in T4 to recognize early T4 promoters [45]. As cyanomyoviruses do not contain these genes it is thought that the expression of early phage genes may be driven by an unmodified host RNA polymerase that recognizes a σ^{-70} factor [5]. In S-PM2 and Syn9 homologues of early T4 genes have an upstream motif that is similar to that of the σ^{-70} promoter recognition sequence [5,23], however these have not been found in S-RSM4 (this lab, unpublished data). Cyanomyoviruses are similar to the T4-like phage RB49 in that they do not contain homologues of *motA* and *asi* which are responsible for production of a transcription factor that replaces the host σ^{-70} factor that has been deactivated by Asi. In RB49 the middle mode of transcription is thought to be controlled by overlapping both early and late promoters [46], this is thought to be the case in S-PM2 with all homologues of T4 genes that are controlled by MotA in T4 having both an early and late promoter [5]. This also seems to be the case in Syn9 which has a number of genes that contain a number of both early and late promoters upstream [23]. However, Q-PCR was used to demonstrate that a small number of genes from S-PM2 that had middle transcription in T4, did not have a middle transcription profile in S-PM2 [46]. Subsequent global transcript profiling of S-PM2 using microarrays has suggested a pattern of transcription that is clearly different to the identified early and late patterns [Millard et al unpublished data]. Whether this pattern of transcription is comparable to the middle mode of transcription in T4 is still unknown. Furthermore, a putative promoter of middle transcription has been identified upstream of T4 middle homologues in the phage P-SMM4 and Syn9, but not in P-SSM2, S-PM2 [23] or S-RSM4 (this lab, unpublished data). Therefore, the exact mechanism of how early and middle transcription may occur in cyanomyoviruses and if there is variation in the control mechanism between cyanophage as well as difference compared to other T4-like phages is still unclear.

The control of late transcription in cyanomyoviruses and other T4 like phages seems to be far more conserved than early or middle transcription with all cyanophages sequenced to date having a homologue of *g55*, which encodes for an alternative transcription factor in T4 and is involved in the transcription of structural proteins [45]. Homologues of the T4-genes *g33* and *g45* which are also involved in late transcription in T4 are all found in

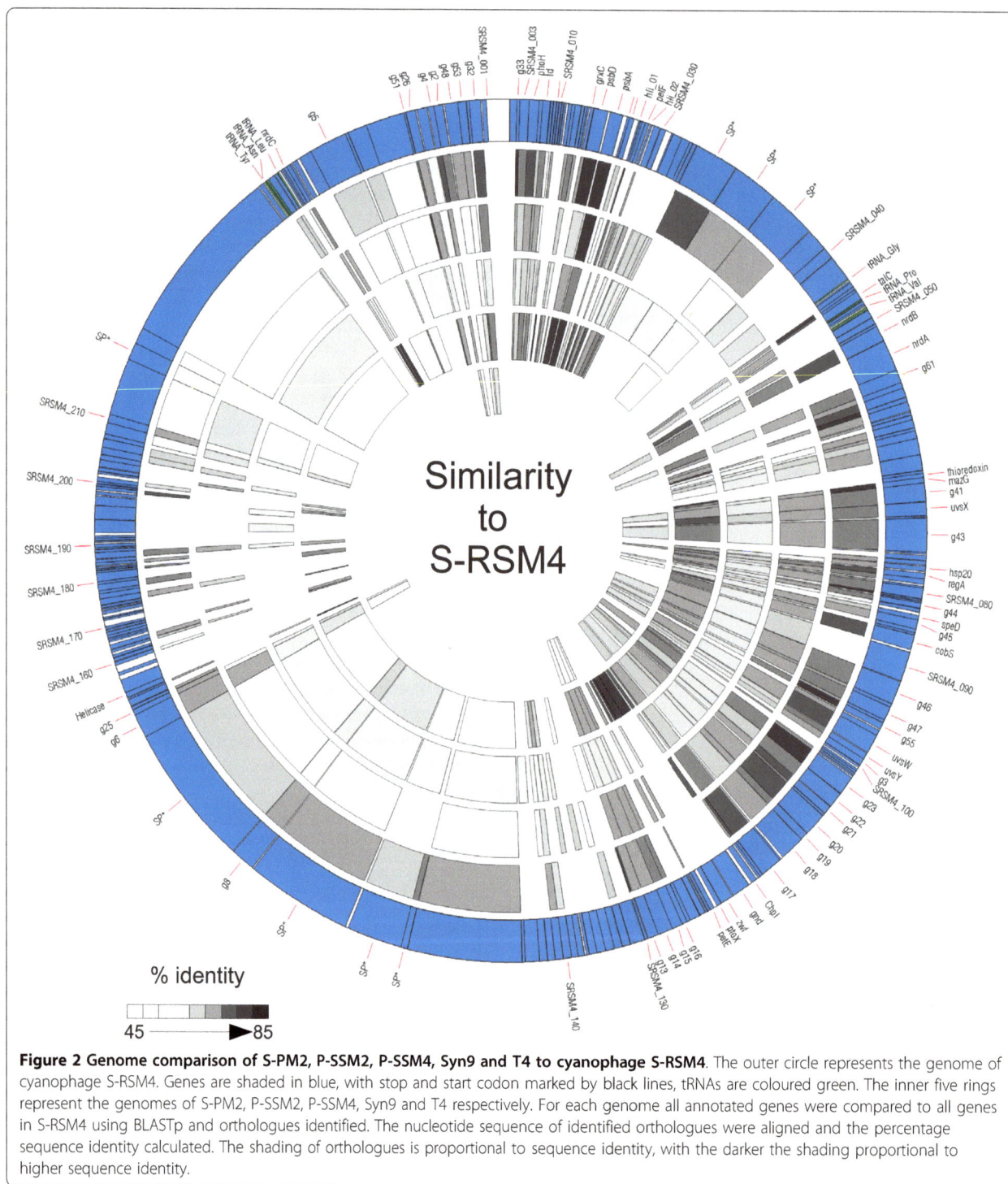

Figure 2 Genome comparison of S-PM2, P-SSM2, P-SSM4, Syn9 and T4 to cyanophage S-RSM4. The outer circle represents the genome of cyanophage S-RSM4. Genes are shaded in blue, with stop and start codon marked by black lines, tRNAs are coloured green. The inner five rings represent the genomes of S-PM2, P-SSM2, P-SSM4, Syn9 and T4 respectively. For each genome all annotated genes were compared to all genes in S-RSM4 using BLASTp and orthologues identified. The nucleotide sequence of identified orthologues were aligned and the percentage sequence identity calculated. The shading of orthologues is proportional to sequence identity, with the darker the shading proportional to higher sequence identity.

cyanomyoviruses, but no homologues of *dsbA* (RNA polymerase binding protein) have been found. A late promoter sequence of NATAAATA has been identified in S-PM2 [5], which is very similar to the late promoter of TATAAATA that is found in T4 and KVP40 [44,45]. The motif was found upstream of a number of homologues of known T4 late genes in S-PM2 [5] and Syn9 [23]. It has

since been found upstream of a number of genes in all cyanophage genomes in positions consistent of a promoter sequence [43].

Nucleotide metabolism

Six genes involved in nucleotide metabolism are found in all cyanomyoviruses and also in the core of 75 genes found in T4-like phages [25]. The genes lacking in cyanomyoviruses from this identified core of T4-like genes are *nrdD*, *nrdG* and *nrdH*, which are involved in anaerobic nucleotide biosynthesis [45]. This is presumably as a reflection of the marine environment that cyanomyoviruses are found in, the oxygenated ocean open, where anaerobic nucleotide synthesis will not be needed. A further group of genes that are noticeable by their absence is *denA*, *ndd* and *denB*, the products of these genes are all involved in the degradation of host DNA at the start of infection [45]. The lack of homologues of these genes is not limited to cyanomyoviruses, with the marine phage KVP40 also lacking these genes [45], thus suggesting cyanomyoviruses either are less efficient at host DNA degradation [23] or that they utilise another as yet un-described method of DNA degradation.

Replication and Repair

The replisome complex of T4 consists of the genes: *g43*, *g44*, *g62*, *g45*, *g41*, *g61* and *g32* are found within all cyanomyovirus genomes [5,23,25], suggesting that this part of the replisome complex is conserved between cyanomyoviruses and T4. Additionally, in T4 the genes *rnh* (RNase H) and *g30* (DNA ligase) are also associated with the replisome complex and are involved in sealing Ozaki fragments [45] However, homologues of these genes are not found in cyanomyoviruses, with the exception of an RNase H that has been identified in S-PM2. Therefore, either the other cyanomyoviruses have distant homologues of these proteins that have not yet been identified or they do not contain them. The latter is more probable as it is known for T4 and *E. coli* that host DNA I polymerase and host ligase can substitute for RNase H and DNA ligase activity [45].

The core proteins involved in join-copy recombination in T4 are gp32, UvsX, UvsY, gp46 and gp47 [45], homologues of all of these proteins have been identified in all cyanomyovirus genomes [5,23,25], suggesting the method of replication is conserved between cyanomyoviruses and other T4-like phages. In the cyanomyovirus Syn9 a single theta origin of replication has been predicted [23], thus contrasting with the multiple origins of replication found in T4 [45]. The theta replication in Syn9 has been suggested to be as result of the less complex environment it inhabits compared to T4 [23]. However, as already stated it does contain all the necessary genes for recombination-dependent replication, and it is not known if other sequenced cyanomyoviruses have single theta predicted method of replication.

With cyanomyoviruses inhabiting a environment that is exposed to high-light conditions it could be assumed that the damage to DNA caused by UV would have to be continuously repaired, in T4 *denV* encodes for endonuclease V that repairs pyrimidine dimers [45], a homologue of

this gene is found in the marine phage KVP40 [44], but not in any of the cyanophage genomes [5,23,25]. Given the environment in which cyanomyoviruses are found in it is likely that there is an alternative mechanism of repair, and a possible alternative has been identified in Syn9 [23]. Three genes were identified that have a conserved prolyl 4-hyroxylase domain that is a feature of the super family of 2-oxoglutarate-dependent dioxygenases, with the *E. coli* DNA repair protein AlkB part of this 2-oxoglutarate-dependent dioxygenase superfamily [23]. In Syn9 the genes 141, and 176 which contain the conserved domain were found to be located next adjacent to other repair enzymes UvsY and UvsX [23], this localization of these genes with other repair enzymes is not limited to Syn9 with putative homologues of these genes found adjacent to the same genes in P-SSM4. Interestingly, although putative homologues to these genes can be identified in the other cyanomyoviruses genomes they do not show the same conserved gene order.

Unlike other T4-like phages there is no evidence that any cyanomyoviruses utilize modified nucleotides such as hyroxymethyl cytosine or that they glycosylate their DNA. In addition all of the *r* genes in T4 that are known to be involved in superinfection and lysis inhibition [45] are missing in cyanophage genomes, as is the case in KVP40 [45].

Structural Proteins

Fifteen genes have previously been identified to be conserved among T4-like phages, excluding the cyanomyoviruses, that are associated with the capsid [25] Only 9 of these genes are present within all cyanomyoviruses and other T4-like phages, whilst some of them can be found in 1 or more cyanomyoviruses. The portal vertex protein (*g24*) is absent from all cyanomyoviruses, it has been suggest that cyanomyoviruses may have an analog of the vertex protein that provides a similar function [23]. Alternatively it has been proposed that cyanomyoviruses have done away with the need for gp24 due to the slight structural alteration in gp23 subunits [39]. The proteins gp67 and gp68 are also missing from all cyanophage genomes [5,23,25], it is possible that analogs of these proteins do not occur in cyanomyoviruses as mutations in these genes in T4 have been shown to alter the structure of the T4 head from a prolate structure to that of isometric head [47,48], which is the observed morphology of cyanomyovirus heads [5,23,25]. The protein gp2, has been identified in S-PM2 [5] and S-RSM4 [43], but not any other cyanophage genomes, similarly the *hoc* gene is present only in P-SSM2, whether the other cyanomyoviruses have homologues of these genes remains unknown.

In keeping with the conservation of capsid proteins in T4-like phages, 19 proteins associated with the tail have previously been identified in T4-like phages [25], again not all these genes are present in cyanomyoviruses, those that are not include *wac, g10, g11, g12, g35, g34* and *g37*. It would seem unlikely that cyanomyoviruses do not have proteins that will provide an analogous function to some of these proteins, indeed proteomic studies of S-PM2 [24] and Syn9 [23] has revealed structural proteins that have no known function yet have homologues in other cyanomyovirus genomes and therefore may account for some of these "missing" tail fiber proteins. Furthermore as new cyanomyoviruses are being isolated and characterised some of these genes may change category, for example a cyanomyovirus recently isolated from St. Kilda was shown to have distinct whiskers which we would anticipate would be encoded by a *wac* gene (Clokie unpublished observation).

Unique cyanomyovirus genome features

The sequence of the first cyanomyovirus S-PM2 revealed an "ORFanage" region that runs from ORF 002 to ORF 078 where nearly all ORFs are all database orphans [5]. Despite the massive increase in sequence data since the publication of the genome, this observation still holds true with the vast majority of these sequences still having no similarity to sequences in the nr database. Sequences similar to some of these unique S-PM2 genes can now be found in the GOS environmental data set. The large region of database orphans in S-PM2 is similar to a large region in KVP40 that also contains its own set of ORFs that encode database orphans [44].

All cyanomyovirus genomes contain genes that are unique, with at least 65 genes identified in each cyanomyovirus that are not present in other cyanomyoviruses [43]. However, it does not appear to be a general feature of cyanomyoviruses genomes to have an "ORFanage" region as found in S-PM2. Another feature unique to one cyanomyovirus genome is the presence of 24 genes thought to be involved in LPS biosynthesis split into two clusters in the genome of P-SSM2 [49].

It has been observed for T4-like phages that there is conservation in both the content and synteny of a core T4-like genome; conserved modules such as that for the structural genes *g1-g24* are separated by hyperplastic regions which are thought to allow phage to adapt to their host [50]. Recent analysis of the structural module in cyanomyoviruses has identified a specific region between *g15* and *g18* that is hyper-variable with the insertion of between 4 and 14 genes [43]. The genes within this region may allow cyanomyoviruses to adapt to their host as predicted function of these genes includes alternative plastoquinones and enzymes that may alter carbon metabolism such glucose 6-phosphate dehydrogenase and 6-phosphoglunate dehydrogenase. Whilst hyperplastic regions are found within T4-like

phages the position of this hyperplastic region is unique to cyanophages.

Finally, recent work has identified CfrI, an ~225 nt antisense RNA that is expressed by S-PM2 during its infection of *Synechococcus* [51]. CfrI runs antisense to an homing endonuclease encoding gene and *psbA*, connecting these two distinct genetic elements. The function of CfrI is still unknown, however it is co-expressed with *psbA* and the homing endonuclease encoding gene and therefore thought to be involved in regulation of their expression [51]. This is the first report of an antisense RNA in T4-like phages, which is surprising given antisense transcription is well documented in eukaryotic and increasingly so in prokaryotic organisms. Although an antisense RNA has only been experimentally confirmed in S-PM2, bioinformatic predictions suggest they are present in other cyanomyovirus genomes [51].

Signature cyanomyovirus genes

Whilst there are a large number of similarities between cyanomyoviruses and other T4-like phages as described above, and some features unique to each cyanomyovirus genome, there still remains a third category of genes that are common to cyanomyovirus but not other T4-like phages. These have previously been described as "signature cyanomyovirus genes" [25]. What constitutes a signature cyanomyovirus gene will constantly be redefined as the number of complete cyanomyovirus genomes sequenced increases. There are a number of genes common to cyanomyoviruses but not widespread or present in the T4-like super group (Table 2). Although the function of most signature cyanomyovirus genes is not known, some can be predicted as they are homologues of host genes.

The most obvious of these is the collection of genes that are involved in altering or maintaining photosynthetic function of the host. The most well studied and first discovered gene is the photosynthetic gene *psbA* which was found in S-PM2 [52], since then this gene has be found in all complete cyanomyovirus genomes [5,23,25]. The closely associated gene *psbD*, is found in all completely sequenced cyanomyovirus genomes with the exception of P-SSM2 [25]. However this is not a universal signature as although one study using PCR has found *psbA* to present in all cyanomyovirus isolates tested [49] or a different study showed that it was only present in 54% cyanomyoviruses [53]. The presence of *psbD* in cyanomyoviruses appears to be linked to the host of the cyanomyovirus with 25% of 12 phage isolated on *Prochlorococcus* and 85% of 20 phage isolated on *Synechococcus* having *psbD* [53]. With the most recent study using a microarray for comparative genomic hybridisations, found 14 cyanomyoviruses, known to infect only *Synechococcus*, contained both *psbA* and

Table 2 Shared genes in cyanomyoviruses

Functional Category	Gene	S-RSM4	S-PM2	P-SSM2	P-SSM4	Syn9	Product/Function
	denV	✗	✗	✗	✓	✗	Pryrimidine dimer repair
	59[S]	✗	✗	✓	✓	✓	ssDNA binding protein
	rnh[S]	✗	✓	✗	✗	✗	RNaseH
	49[S]	✗	✓	✓	✗	✗	Recombination endonuclease VII
	2[S]	✗	✓	✗	✗	✗	Protein protecting DNA ends
	hoc	✗	✗	✓	✗	✗	Capsid protein
	9[S]	✗	✗	✓	✓	✓	Baseplate socket
Structural Proteins	S-PM2_043	✓	✓	✓	✓	✓	structural*
	S-PM2_163	✓	✓	✓	✓	✓	structural*
	S-PM2_165	✓	✓	✓	✓	✓	structural*
	S-PM2_251	✓	✓	✓	✓	✓	structural*
Photosynthesis	psbA	✓	✓	✓	✓	✓	D2: core PSII protein
	psbD	✓	✓	✗	✓	✓	D1: core PSII protein
	petE	✓	✗	✓	✗	✓	Plastocyanin
	petF	✓	✗	✓	✗	✗	Ferredoxin
	cepT	✓	✓	✗	✗	✓	PE regulatory protein
	ptoX	✓	✗	✗	✓	✓	Plastoquinol terminal oxidase
	speD	✓	✓	✗	✓	✗	Polyamine biosynthesis
	hli	✓$^{\times 2}$	✓$^{\times 2}$	✓$^{\times 6}$	✓$^{\times 4}$	✓$^{\times 2}$	High light inducible protein
Carbon/phosphate metabolism	gnd	✓	✗	✗	✗	✓	6-phosphogluconate dehydrogenase
	zwf	✓	✗	✗	✗	✓	Glucose 6-phoshate dehydrogenase
	talC	✓	✗	✓	✓	✓	Transaldolase
	trx	✓	✗	✗	✗	✓	Thioredoxin
	phoH	✓	✓	✓	✓	✓	Phosphate -induced stress protein
	pstS	✗	✗	✓	✓	✗	Phosphate -induced stress protein
Conserved Cyanophages genes	mazG	✓	✓	✓	✓	✓	Nucleoside triphosphate pyrophosphohydrolase
	cobS	✓	✓	✓	✓	✓	Cobalamin biosynthesis
	prnA	✓	✓	✓	✗	✓$^{\times 2}$	Trpytophan halogenase
	S-PM2_225	✓	✓	✓	✓	✓	Oxygenase superfamily-like protein
	S-PM2_232	✓	✓	✓	✓	✓	Putative Helicase
	S-PM2_113	✓	✓	✓	✓	✓	Unknown
	S-PM2_117	✓	✓	✓	✓	✓	Unknown
	S-PM2_119	✓	✓	✓	✓	✓	Unknown
	S-PM2_138	✓	✓	✓	✓	✓	Unknown
	S-PM2_141	✓	✓	✓	✓	✓	Unknown
	S-PM2_164	✓	✓	✓	✓	✓	Unknown
	S-PM2_056	✓	✓	✓	✓	✓	Unknown
	S-PM2_186	✓	✓	✓	✓	✓	Unknown
	S-PM2_187	✓	✓	✓	✓	✓	Unknown
	S-PM2_194	✓	✓	✓	✓	✓	Unknown
	S-PM2_198	✓	✓	✓	✓	✓	Unknown

The table was modified from [25,45]. Genes were called present (#10003;) or absent (#10007) using previous annotations [5,23,25] and BLASTp with a cut off value of <10^{-5}.

*Genes previously identified as structural proteins by mass spectrometry [23,24].

[S]Genes previously identified as core to T4-like phages [25].

psbD [43]. *psbA* and *psbD* have also been detected in a large number of environmental samples from subtropical gyres to Norwegian coastal waters [26,54,55]. With cyanomyovirus derived *psbA* transcripts being detected during infection in both culture [56] and in the environment [57].

In summary, both *psbA* and *psbD* are widespread in cyanomyovirus isolates and that *psbD* is only present if *psbA* is also present [49,53] and cyanomyovirus are thought to have gained these genes on multiple occasions independently of each other [46,49,53].

In addition to *psbA* and *psbD*, other genes not normally found in phage genomes have been identified, these include *hli, cobS, hsp* that are found in all complete cyanomyovirus genomes. Additionally the genes *petE, petF, pebA, speD, pcyA, prnA, talC, mazG, pstS, ptoX, cepT,* and *phoH* have all been found in at least one or more cyanomyovirus genomes. In addition to being found in complete phage genomes these accessory genes have been identified in metagenomic libraries [54,55]. Not only are these genes present in the metagenomic libraries they are extremely abundant; e.g. there were 600 sequences homologous to *talC* in the GOS data set, in comparison there were 2172 sequences homologous to a major capsid protein [55]. The metabolic implications of these genes are discussed in the next section.

Cyanomyovirus-like sequences in metagenomes

In the last few years there has been a massive increase in the sequence data from metagenomic studies. The Sorcerer II Global Ocean Expedition (GOS) alone has produced 6.3 billion bp of metagenomic data from various Ocean sites [58], with the viral fraction of the metagenome dominated by phage like sequences [55]. Subsequent analysis by comparison of these single reads against complete genomes allows, recruitment analysis, allows identification of genomes that are common in the environment. In the GOS data set, only the reference genome of P-SSM4 was dominant [55].

A further study that examined 68 sampling sites, representative of the four major marine regions, showed the wide spread distribution of T4-like cyanomyovirus sequences in all four major biomes [7]. With increased cyanomyovirus sequences in the Sargasso Sea biome compared to the other regions examined [7]. In a metagenomic study of the viral population in the Chesapeake Bay the viral population was dominated by the *Caudovirales*, with 92% of the sequences that could be classified falling within this broad group [8]. A finer examination of this huge data set revealed that 13.6% and 11.2% of all homologues identified were against genes in the cyanomyovirus P-SSM2 and P-SSM4 respectively [8].

Even in metagenomic studies that have not specifically focused on viruses, cyanomyovirus sequences have been found. For example, in a metagenomic study of a subtropical gyre in the Pacific, up to 10% of fosmid clones contained cyanophages-like sequences, with a peak in cyanophages-like sequences at a depth of 70 m, which correlated with the maximal virus:host ratio [54]. All of the metagenomic studies to date have demonstrated the widespread distribution of cyanomyovirus like sequences in the ocean and provided a huge reservoir of sequence from the putative cyanomyovirus pan-genome. However, with only five sequenced cyanomyovirus it is not known how large the pan-genome of cyanomyoviruses really is. With every newly sequenced cyanomyovirus genome there has been ~25% of total genes in an individual phage that are not found in other cyanomyoviruses. Even for core T4-like genes their full diversity has probably not been discovered. By examining the diversity of ~1,400 gp23 sequences from the GOS data set it was observed that the cyanomyovirus-like sequences are extremely divergent and deep branching [39]. It was further concluded that diversity of T4-like phages in the world's Oceans is still to be fully delimited [39].

Metabolic Implications of unique cyanomyovirus genes
Cyanomyoviruses and Photosynthesis

Cyanomyoviruses are unique among T4-like phages in that their hosts utilize light as their primary energy source; therefore it is not to surprising cyanomyoviruses carry genes that may alter the photosynthetic capability of their hosts. The most well studied of the photosynthetic phage genes are *psbA* and *psbD*, which encode for the proteins D1 and D2 respectively. The D1 and D2 proteins form a hetero-dimer at the core of photosystem II (PSII) where they bind pigments and other cofactors that ultimately result in the production of an oxidant that is strong enough to remove electrons from water. As an unavoidable consequence of photosynthesis there is photo-damage to D1 and to a lesser extent the D2 protein, therefore all oxygenic photosynthetic organisms have evolved a repair cycle for PSII [59]. The repair cycle involves the degradation and removal of damage D1 peptides, and replacement with newly synthesized D1 peptides [59]. If the rate of removal and repair is exceeded by the rate of damage then photoinhibiton occurs with a loss of photochemical efficiency in PSII [60]. A common strategy of T4-like phages is to shutdown the expression of host genes after infection, but if this was to occur in cyanomyoviruses then there would be a reduction in the reduction efficiency of the PSII repair cycle and thus reduced photosynthetic efficiency of the host. This would be detrimental to the replication of phage and it has therefore been proposed that cyanomyoviruses carry their own copies of *psbA* to maintain the D1 repair cycle [52]. There is strong evidence to suggest that this is the case with Q-PCR data proving the *psbA* gene is expressed during the infection cycle for the phage S-PM2 and that there is no loss in photosynthetic efficiency during the infection cycle [56]. Further evidence for the function of these genes can be gained from P-SSP7 a podovirus that also express *psbA* during infection with phage derived D1 peptides also being detected in infected cells [61]. Although as yet phage mutants lacking these genes have yet to be constructed the results of modelling with in silico mutants suggests that *psbA* is a non essential gene [62] and that its fitness advantage is greater under higher irradiance levels [62,63].

The carriage of *psbD* is assumed to be for the same reason in the maintenance of photosynthetic efficiency during infection, indeed it has been shown that *psbD* is also expressed during the infection cycle [Millard et al unpublished data]. However, not all phage are known to carry both *psbD* and psbA, in general that the broader the host range of the phage the more likely it is to carry both genes [40,49]. It has therefore been suggested that by carrying both of these genes that phage can ensure the formation of a fully functional phage D1:D2 heterodimer [49].

Cyanomyoviruses may maintain the reaction centres of their host in additional and/or alternative ways to the replacement of D1 and D2 peptides. The reaction centre of PSII may also be stabilized by *speD* a gene that has been found in S-PM2, P-SSM4 and S-RMS4. *speD* encodes S-adenosylmethionine decarboxylase a key enzyme in the synthesis of the polyamines spermidine and spermine. With polyamines implicated in the stabilising the *psbA* mRNA in the cyanobacterium *Synechocystis* [64], altering structure of PSII [65] and restoring photosynthetic efficiency [66], it has been proposed they also act to maintain the function of the host photosystem during infection [11].

Whilst *psbA* and *psbD* are the most studied genes that may alter photosynthetic ability, they are certainly not the only genes. The carriage of *hli* genes that encode high light inducible proteins (HLIP) are also thought to allow the phages host to maintain photosynthetic efficiency under different environmental conditions. HLIP proteins are related to the chlorophyll *a/b*-binding proteins of plants and are known to be critical for allowing a freshwater cyanobacteria *Synechocystis* to adapt to high-light conditions [67]. The exact function in cyanomyoviruses is still unknown, they probably provide the same function of as HLIPs in their hosts, although this function is still to be fully determined. It is apparent that the number of *hli* genes in phage genome is linked to the host of the cyanomyovirus with phage that were isolated on *Prochlorococcus* (P-SSM2 & P-SSM4) having double the number of *hli* genes found on the those phage isolated on *Synechococcus* (S-RSM4, Syn9, S-PM2) (Table 2). The phylogeny of these genes suggest that some of these *hli* genes are *Prochlorococcus* specific [68], probably allowing adaptation to a specific host.

A further photosynthetic gene that may be advantageous to infection of a specific host is *cepT*. S-PM2 was the first phage found to carry a *cepT* gene [5], it is also now found in Syn9 [23], S-RSM4 and 10 other phages infecting *Synechococcus* [43], but is not found in the phage P-SSM2 and P-SSM4 which were isolated on *Prochlorococcus* [49]. *cepT* is thought to be involved in regulating the expression of phycoerythrin (PE) biosynthesis [69], PE is a phycobiliprotein that forms part of the phycobilisome that is responsible for light-harvesting in cyanobacteria [70], the phycobilisome complex allows adaptation to variable light conditions such as increased UV stress [70]. Recently it has been shown that amount of PE and chlorophyll increases per cell when the phage S-PM2 infects its host *Synechococcus* WH7803, with this increases in light harvesting capacity thought to be driven by the phage to provide enough energy for replication [6] with phage *cpeT* gene responsible for regulation of this increase [71]. As *Prochlorococcus* do not contain a phycobilisome complex that contains PE, which the *cpeT* regulates expression of, it is possibly a gene advantageous to cyanomyoviruses infecting *Synechococcus*.

Phage genes involved in bilin synthesis are not limited to *cepT*, within P-SSM2 the bilin reductase genes *pebA* and *pcyA* have been found and are expressed during infection [72]. The *pebA* gene is functional *in vitro* and catalyses a reaction that normally requires two host genes (*pebA* &*pebB*) and has since being renamed *pebS*, this single gene has been suggested to provide the phage with short tern efficiency over long term flexibility of the two host genes [72]. Despite evidence of expression and that the products are functional it is unclear how these genes are advantageous to cyanomyoviruses infecting *Prochlorococcus* which do not contain standard phycobilisome complexes.

Alteration of host photosynthetic machinery appears to be of prime importance to cyanomyoviruses with a number of genes that may alter photosynthetic function. In addition to maintaining PSII centres and altering bilin synthesis, a further mechanism for diverting the flow of electrons during photosynthesis may occur. A plastoquinol terminal oxidase (PTOX)-encoding gene was first discovered in P-SMM4 [25] and then in Syn9 [23] and more recently has been found to be widespread in cyanomyoviruses infecting *Synechococcus*. The role of PTOX in cyanobacteria, let alone cyanomyoviruses, is not completely understood, but it is thought to play a role in photo-protection. In *Synechococcus* it has been found that under iron-limited conditions CO_2 fixation is saturated at low light intensities, yet the reaction centres of PSII remain open at far higher light intensities. This suggests an alternative flow of electrons to receptors other than CO_2 and the most likely candidate acceptor is PTOX [73]. The alternative electron flow eases the excitation pressure on PSII by the reduction of oxygen and thus prevents damage by allowing an alternative flow of electrons from PSII [73]. Further intrigue to this story in that PTOX encoding genes are not present in all cyanobacterial genomes and are far more common in *Prochlorococcus* genomes than in *Synechococcus* genomes. Therefore, phage may not only maintain the current *status quo* of the cell as in the same manner *psbA* is thought to, but may offer an alternative pathway of electron flow if its host does not carry its own PTOX genes. Although this is speculative it is already known that cyanomyoviruses that carry PTOX genes can infect and replicate in *Synechococcus* WH7803 that does not have PTOX-encoding gene of its own.

Carbon Metabolism

All sequenced cyanomyoviruses have genes that may alter carbon metabolism in their hosts, although not all cyanomyoviruses have the same complement of genes [5,23,25]. Syn9 [23] and S-RSM4 have *zwf* and *gnd* genes encoding the enzymes glucose 6-phosphate dehydrogenase (G6PD) and 6-phosphogluconate dehydrogenase which are enzymes utilised in the oxidative stage of the pentose phosphate pathway (PPP). The rate-limiting step in the PPP is the conversion of glucose-6-phosphate, which is catalysed by G6PD. It could be advantageous for a phage to remove this rate-limiting step in order to increase the amount of NADPH or ribulose 5-phosphate it requires for replication. Whether the phage removes this rate limitation by encoding a G6PD that is more efficient than the host G6PD or simply producing more, is not known. Without experimental data the proposed advantages of these genes are speculative.

There are at least 5 modes in which the PPP can operate depending on the requirements of the cell [74]. It might be assumed that for a phage the priority might be to produce enough DNA and protein for replication, thus use the mode of PPP that produces more ribulose 5- phosphate at the expense of NAPH. The production of ribulose 5-phosphate could then be used as the precursors for nucleotide synthesis. This mode of flux would result in the majority of glucose-6-phosphate being converted to fructose-6-phosphate and glyceraldehyde 3-phosphate. These molecules could then be converted to ribulose 5-phosphate by a transaldolase and transketolase.

Therefore, it is not surprising that *talC* has been detected in four of the five sequenced cyanomyovirus genomes, in viral metagenomic libraries [54], and in fragments of cyanomyovirus genomes S-BM4 [53] and SWHM1 (this lab unpublished data). *talC* encodes a transaldolase, an important enzyme in linking PPP and glycolysis, that if functional would catalyze the transfer of dihydroxyacetone from fructose 6-phospate to erythrose 4-phosphate, giving sedoheptulose 7-phosphate and glyceraldehyde 3-phosphate. However, currently this alteration of the PPP is speculation as other modes of flux are just as possible depending on the circumstances

the phage find it self within its host with alternative modes leading to an increase in the production ATP and NADPH [23].

It does appear that maintaining or altering carbon metabolism is important to cyanomyoviruses as the genes *trx* is also found Syn9 and S-RSM4. The product of *trx* is thioredoxin, an important regulatory protein that is essential in the co-ordination of the light-dark reactions of photosynthesis by the activation of a number of enzymes, one of the few enzymes that it suppresses is glucose-6-phosphate dehydrogenase [75]. The reduced form of thioredoxin controls enzyme activity, with thioredoxin itself reduced by ferredoxin in a process catalysed by ferredoxin-thioredoxin reductase [76]. Whilst no cyanomyovirus have been found to have ferredoxin-thioredoxin reductase, the cyanomyovirus S-RSM4 and P-SSM4 do have *petF*, that encodes ferredoxin,. Ferredoxin acts as an electron transporter which is associated with PSI, whether the phage petF replaces host petF function is not known.

The function of another electron transporter is also unclear, some cyanophages (S-RSM4, Syn9, P-SSM2) have a homologue of *petE*. Host *petE* encodes plastocyanin, which transfers electrons from the cytochrome b_6f complex of photosystem II to P700$^+$ of photosystem I. It is known cyanobacterial *petE* mutants show both a reduced photosynthetic capacity for electron transport and slower growth rate [77]. Thus, it is possible that the phage *petE* is beneficial by means of maintaining photosynthetic function.

Whilst there are a number of genes, *trx*, *zwf*, *gnd*, *petE*, *petF* that may alter host carbon metabolism, unravelling their function is not a trivial task, this is exemplified genes such as *trx* that can regulate enzymes in the Calvin cycle, PPP, and gluconeogenesis. This is further complicated by the fact that to date no two cyanomyovirus to date have exactly the same complement of genes that may alter carbon metabolism, with S-PM2 having none of the above mentioned and at the opposite end of the spectrum S-RSM4 has the full complement. However, the widespread distribution of these genes in cyanomyoviruses suggests their presence is not coincidental and they may be advantageous to cyanomyovirus under certain environmental conditions.

Phosphate Metabolism

The gene *phoH* has been found in all sequenced cyanomyovirus genomes, and in KVP40 [44]. The function of the gene in cyanomyovirus is not known; in *E. coli* it is known that *phoH* forms part of the pho regulon, with *phoH* regulated by *phoB* with increased expression under phosphate-limited conditions [78]. A further protein implicated in adaptation to phosphate limitation is PstS that shows increased expression in *Synechococcus* under phosphate limitation [79]. Both P-SSM2 and

P-SSM4 have the gene *pstS* [25]. It is thought that cyanomyoviruses maintain *phoH* and *pstS* to allow their host to allow increased phosphate uptake during infection, although the mechanism of how this occurs is unknown.

Non-cyanobacterial genes with unknown function in cyanomyoviruses

There are many genes in cyanomyovirus genomes that are similar to hypothetical genes in their hosts, where the host function is not known. Additionally all phage contain bacterial genes that are not found in their cyanobacterial hosts, but appear to have been acquired from other bacterial hosts, this includes the genes *prnA* and *cobS* which encode tryptophan halogenase and an enzyme that catalyses the final step in cobalamin synthesis respectively. Tryptophan halogenase is not found in any known host of cyanomyoviruses, however it is known to catalyse the first step in the biosynthesis of the fungicide pyrrolnitrin in *Pseudomonas fluorescens* [80]. It has been suggested that it may function to provide antibiotic protection to its host, however as stated by the authors this idea is speculative [23]. It has been suggested that *cobS* may boost the production of cobalamin during phage infection [25], the resulting effect of increased cobalamin levels is not known. Potentially it may increase the activity of ribonucleotide reductases, although if it did the process would be unique to cyanophages [25].

Metabolic coup d'etat

Cyanomyoviruses may also affect host metabolism on a far grander scale than simply expressing genes to replace the function of host genes such as *psbA* or *talC*. The gene *mazG* has been found in all cyanomyovirus genomes sequenced to data and has also been found to be widespread in cyanomyovirus isolates [81]. MazG has recently been shown to hydrolyse ppGpp in *E. coli* [82]. ppGpp is known as a global regulator of gene expression in bacteria, it also shows increased expression in cyanobacteria under high-light conditions [83]. It has been proposed that the phage fools its host cell into believing it is in nutrient replete conditions, rather than the nutrient deplete conditions of an oligotrophic environment where *Synechococcus* and *Prochlorococcus* dominate [11]. It is thought to do this by the reducing the pool of ppGpp in the host which regulates global gene expression causing the host to modify its physiological state for optimal macromolecular synthesis thus most favourable conditions for production of progeny phage [84].

Gene transfer between the T4-likes and their hosts (impact on host genome evolution in the microbial world)

As discussed in the preceding sections there is clear evidence that cyanophages have acquired a plethora of genes from their bacterial hosts. These are recognisable either by being highly conserved such as *psbA* which is conserved the amino acid level, or by the presence of a shared conserved domain with a known gene. Phages potentially have two methods of donating phage genes back to their hosts; through generalised or specialised transduction. Generalised transduction results from non-productive infections where phages accidently package a head full of host DNA during the stage when their heads are being packaged and they inject this into a second host cell during a non-fatal infection. Specialised transduction in comparison results from the accidental acquisition of a host gene resulting from imprecise excision from a host which would occur during lysogenic induction. Although this area has been poorly studied there is some evidence for both generalised and specialised transduction in cyanophages [85].

Despite little direct evidence of lysogeny in marine cyanophages the relationship between host and phage genes can be established from phylogenetic analyses. When host genes are acquired by phages, they generally drift from having the GC composition of their hosts to that of the phage genome. This difference is much clearer in *Synechococcus*-phage relationships because *Synechococcus* genomes have a GC % of around 60% compared to the phages which have a GC% of around 40%. The GC of *psbA* in *Synechococcus* phages has drifted to a value between the average host and phage GC% so is around 50%. These differences are less clear in *Prochlorococcus* as it tends to have a similar CG% to the phages which infect it and thus phylogenetic analysis can be dominated by homoplasies (the same mutation happening independently).

All of the robust phylogenetic analyses that have been performed on metabolic phage genes that are shared between hosts and phages suggest that phages have generally picked up host genes on limited occasions and this has been followed by radiation has within the phage populations for example see Millard et al. 2005 [53].

There is nothing known about the biology and molecular basis of lysogeny or pseudolysogeny in T4 type cyanomyoviruses. Indirect evidence for the abundance of lysogens was obtained from studies on inducing wild populations of cyanobacteria and quantifying the number of potential phages using epifluorescence. This work demonstrated that more temperate phages could be induced in winter when the number of cyanobacterial hosts was low and so conditions were hostile for phages in the lytic part of their life cycle. Other studies have suggested that the apparent resistance *Synechococcus* shows to viral infection may be due to lysogenic infection [3]. It is also clear that the phosphate status of cyanobacteria influences the dynamics of integration [86]. During nutrient starvation cyanoviruses enter their hosts but do not lyse the cells, their genes are expressed during this period (Clokie et al., unpublished). The cells are

35

lysed when phosphate is added back into the media. It is not known exactly how cyanophage DNA is integrated into the cell during this psuedolysogenic period but this may be a time in which genes may be donated and integrated from the phage genome to that of the host.

Despite a lack direct evidence for phage-mediated gene transfer, it is likely that transduction is a major driver in cyanobacterial evolution as the other methods of evolution are not available to them. In the open oceans DNA is present at such low levels (0.6 - 88 μg liter^{-1}) that it is probably too dilute for frequent transformation [87]. Also both *Synechococcus* and *Prochlorococcus* appear to lack plasmids and transposons rendering conjugation an unlikely method for the acquisition of new genes. The large number of bacteriophages present in the oceans as well as the observation that phage-like particles appear to be induced from marine cyanobacteria, along with phage-like genes found in cyanobacterial genomes suggests that transduction is evident as a mechanism of evolution.

The genetic advantages that the T4-like cyanomyoviruses may confer to their hosts were listed in a recent review, but in brief they are: (1) prophages may function as transposons, essentially acting as foci for gene rearrangements, (2) they may interrupt genes through silencing non-essential gene functions, (3) they may confer resistance to infection from other phages, (4) they may excise and kill closely related strains, (5) they may cause increased fitness by the presence of physiologically important genes or (6) the phages may silence host genes.

In summary, it is difficult to pin down the exact contribution that T4-like cyanoviruses play in microbial evolution but their abundance, modes of infection and genetic content imply that they may be extremely important for cyanobacterial evolution. Their contribution will become clearer as more genomes are sequenced and as genetic systems are developed to experiment with model systems.

The impact of cyanomyoviruses on host populations

The two major biotic causes of bacterial mortality in the marine environment are phage-induced lysis and protistan grazing, currently efforts are being made to assess the relative impacts of these two processes on marine cyanobacterial communities. Accurate information is difficult to obtain for the oligotrophic oceans because of intrinsically slow rate processes [88]. It must also be borne in mind that there are likely to be extensive interactions between the two processes e.g. phage-infected cells might less or more attractive to grazers, phage-infected cells might be less or more resistant to digestion in the food vacuole and phages themselves might be subject to grazing. Estimates of the relative effects of phage-induced lysis and grazing on marine cyanobacterial

assemblages vary widely e.g. [89-91] and this probably reflects the fact the two processes do vary widely on both temporal and spatial scales.

A number of methods have been developed to assess viral activity in aquatic systems, but all suffer from a variety of limitations such as extensive sample manipulation or poorly constrained assumptions [92,93]. The application of these approaches to studying cyanomyovirus impact on *Synechococcus* populations has produced widely varying results. Waterbury and Valois [3] calculated that between 0.005% (at the end of the spring bloom) and 3.2% (during a *Synechococcus* peak in July) of the *Synechococcus* population was infected on a daily basis. Another study [94] indicated that as many as 33% of the *Synechococcus* population would have to have been lysed daily at one of the sampling stations. A subsequent study using the same approach [95] yielded figures for the proportion of the *Synechococcus* community infected ranging from 1 - 8% for offshore waters, but in nearshore waters only 0.01 - 0.02% were lysed on a daily basis. Proctor and Fuhrman [96] found that, depending on the sampling station, between 0.8% and 2.8% of cyanobacterial cells contained mature phage virions and making the questionable assumption that phage particles were only visible for 10% of the infection cycle, it was calculated that percentage of infected cells was actually ten-fold greater than the observed frequency.

An important consideration in attempting to establish the impact of cyanomyoviruses on their host populations is to ask at what point the infection rate becomes a significant selection pressure on a population, leading either to the succession of intrinsically resistant strains, or the appearance of resistant mutants. It has been calculated that the threshold would occur between 10^2 and 10^4 cells ml^{-1} [10] and this is in agreement with data from natural *Synechococcus* populations that suggest that a genetically homogeneous population would start to experience significant selection pressure when it reached a density of between 10^3 and 10^4 cells ml^{-1} [97].

The community ecology of cyanomyovirus-host interactions is complicated by a number of factors including the genetic diversity of phages and hosts, protistan grazing and variations in abiotic factors (e.g. light, nutrients, temperature). Thus simple modelling of predator-prey dynamics is not possible. However, a "kill the winner" model [92,98] in which the best competitor will become subject to infection has gained widespread acceptance. Recently, marine phage metagenomic data have been used to test theoretical models of phage communities [99] and the rank-abundance curve for marine phage communities is consistent with a power law distribution in which the dominant phage keeps changing and in which host ecotypes at very low numbers evade phage predation. A variety of studies have looked at spatio-temporal variations in cyanomyovirus populations. The earliest studies showed that cyanomyovirus abundance changed through an annual cycle [3] and with distance from shore, season and depth [94]. The ability to look at the diversity of cyanomyovirus population using *g20* primers revealed that maximum diversity in a stratified water column was correlated with maximum *Synechococcus* population density [30] and changes in phage clonal diversity were observed from the surface water down to the deep chlorophyll maximum in the open ocean [28]. Marston and Sallee [35] found temporal changes in both the abundance, overall composition of the cyanophage community and the relative abundance of specific *g20* genotypes In Rhode Island's coastal waters. Sandaa and Larsen [34] also observed seasonal variations in the abundance of cyanophages and in cyanomyovirus community composition in Norwegian coastal waters. Cyanomyovirus abundance and depth distribution was monitored over an annual cycle in the Gulf of Aqaba [40]. Cyanophages were found throughout the water column to a depth of 150 m, with a discrete maximum in the summer months and at a depth of 30 m. Whilst it is clear from all these studies that cyanomyovirus abundance and community composition changes on both a seasonal and spatial basis, little is know about short term variations. However, one study in the Indian Ocean showed that phage abundance peaked at around 0100 at a depth of 10 m, but the temporal variation was not as strong at greater depths [84]. It may well be the case that infection by cyanomyoviruses is a diel phenomenon as phage adsorption to host is light-dependent for several marine cyanomyoviruses studied [100]. A similar observation for the freshwater cyanomyovirus AS-1 [101]. There is currently only one published study that describes attempts to look at the co-variation in the composition of *Synechococcus* and cyanomyovirus communities to establish whether they were co-dependent [102]. In the Gulf of Aqaba, Red Sea, a succession of *Synechococcus* genotypes was observed over an annual cycle. There were large changes in the genetic diversity of *Synechococcus*, as determined by RFLP analysis of a 403 bp *rpoC1* gene fragment, which was reduced to one dominant genotype in July. The abundance of co-occurring cyanophages capable of infecting marine *Synechococcus* was determined by plaque assays and their genetic diversity was determined by denaturing gradient gel electrophoresis analysis of a 118 bp *g20* gene fragment. The results indicate that both abundance and genetic diversity of cyanophage covaried with that of *Synechococcus*. Multivariate statistical analyses show a significant relationship between cyanophage assemblage structure and that of *Synechococcus*. All these observations are consistent with cyanophage infection being a major controlling factor in cyanobacterial diversity and succession.

Analysis of the impact of cyanomyoviruses on host populations has been based on the assumption that they follow the conventional infection, replication and cell lysis life cycle, but there is some evidence to suggest that this may not always be the case. There is one particularly controversial area of phage biology and that is the topic of pseudolysogeny. There are in fact a variety of definitions of pseudolysogeny in the literature reflecting some quite different aspects of phage life history, but the one adopted here is "the presence of a temporarily non-replicating phage genome (a preprophage) within a poorly replicating bacterium" (S. Abedon - personal communication). The cyanobacterial hosts exist in an extremely oligotrophic environment posing constant nutritional stress and are exposed to additional environmental challenges such as light stress that may lead to rates of growth and replication that are far from maximal. There is evidence that obligately lytic *Synechococcus* phages can enter such a pseudolysogenic state. When phage S-PM2 (a myovirus) was used to infect *Synechococcus* sp. WH7803 cells grown in phosphate-replete or phosphate-deplete media there was no change in the adsorption rate constant, but there was an apparent 80% reduction in the burst size under phosphate-deplete conditions and similar observations were made with two other obligately lytic *Synechococcus* myoviruses, S-WHM1 and S-BM1 [86]. However, a more detailed analysis revealed this was due to a reduction in the proportion of cells lysing. 100% of the phosphate-replete cells lysed, compared to only 9% of the phosphate-deplete cells, suggesting that the majority of phosphate deplete cells were pseudolysogens.

From very early on in the study of marine cyanomyoviruses it was recognized that phage-resistance was likely to be an important feature of the dynamics of phage-host interactions. Waterbury and Valois [3] found that coastal *Synechococcus* strains were resistant to their co-occurring phages and suggested that the phage population was maintained by a small proportion of cells sensitive to infection. For well studied phage-host systems resistance is most commonly achieved by mutational loss of phage receptor on the surface of the cell, though there are other mechanisms of resistance to phage infection e.g. [103]. Stoddard et al. [104] used a combination of 32 genetically distinct cyanomyoviruses and four host strains to isolate phage-resistant mutants. Characterization of the mutants indicated that resistance was most likely due to loss or modification of receptor structures. Frequently, acquisition of resistance to one phage led to cross-resistance to one or more other phages. It is thought that mutation to phage resistance may frequently involve a fitness cost and this trade-off allows the coexistence of more competitive phage-sensitive and less competitive phage-resistant strains (for review see [105]). The cost of phage resistance in marine cyanobacteria has been investigated by Lennon et al. [106] using phylogenetically distinct *Synechococcus* strains and phage-resistant mutants derived from them. Two approaches were used to assess the cost of resistance (COR); measurement of alterations in maximum growth rate and competition experiments. A COR was found in roughly 50% of cases and when detected resulted in a ~20% reduction in relative fitness. Competition experiments suggested that fitness costs were associated with the acquisition of resistance to particular phages. A COR might be expected to be more clearly observed when strains are growing in their natural oligotrophic environment. The acquisition of resistance to one particular cyanophage, S-PM2, is associated with a change in the structure of the lipopolysaccharide (LPS) (E. Spence - personal communication).

A variety of observations arising from genomic sequencing have emphasized the role of alterations in the cell envelope in the speciation *Prochlorococcus* and *Synechococcus* strains, presumably as a result of selection pressures arising from phage infection or protistan grazing. An analysis of 12 *Prochlorococcus* genomes [107] revealed a number of highly variable genomic islands containing many of the strain-specific genes. Amongst these genes the greatest differentiator between the most closely related isolates were genes related to outer membrane synthesis such as acyltransferases. Similar genomic islands, containing the majority of strain-specific genes, were identified through an analysis of the genomes of 11 *Synechococcus* strains [108]. Among the island genes with known function the predominant group were those encoding glycosyl transferases and glycoside hydrolases potentially involved in outer membrane/cell wall biogenesis. The cyanomyovirus P-SSM2 was found to contain 24 LPS genes that form two major clusters [25]. It was suggested that these LPS genes might be involved in altering the cell surface composition of the infected host during pseudolysogeny to prevent infection by other phages. The same idea could apply to a normal lytic infection and could be extended to protection against protistan grazing. Similarly, cyanomyovirus S-PM2 encodes a protein with an S-layer homology domain. S-layers are quasi-crystalline layers on the bacterial cell surface and so this protein, known to be expressed in the infected cell as one of the earliest and most abundantly transcribed genes [56], may have a protective function against infection or grazing.

The potential value of continuing research on the "eco-genomics" of cyanophages

Eco-genomics is defined as the application of molecular techniques to ecology whereby biodiversity is considered at the DNA level and this knowledge is then used to understand the ecology and evolutionary processes of ecosystems. Cyanophage genomes encode a huge body of unexplored biodiversity which needs to be understood to further extend our knowledge of cyanophage-cyanobacteria interactions and thus to fully appreciate the multiple roles that cyanophages play in influencing bacterial evolution, physiology and biogeochemical cycling.

As cyanophage genomes are stripped down versions of essential gene combinations an understanding of their genomics will assist in defining key host genes that are essential for phage reproduction. As many of the host genes encoded in phage genomes have an unknown function in their hosts, the study of phage genomes will impinge positively on our understanding of cyanobacterial genomes. The other major spin-off from researching the products encoded by phage genomes is the discovery of novel enzymes or alternative versions existing enzymes with novel substrate specificities. This is likely to be of major importance to the biotechnology and pharmaceutical industries.

As more phage genomes and metagenomes are sequenced, the core set of phage genes will be refined and the extent of phage encoded host metabolic and other accessory genes will be revealed. We would expect to find specific environments selecting particular types of genes. This research area is often referred to as 'fishing expeditions' especially by grant panels. However it is analogous to the great collections of plants and animals that occurred during the 19th Century. These data were collected over a long period of time and it was only subsequently that scientists understood patterns of evolution, biogeography, variance and dispersal. This is an exciting time to be mining cyanophage genomes as metagenomic analysis of the viral fraction from marine ecosystems has suggested that there is little restriction to the types of genes that bacteriophages can carry [109]. These data will likely provide the bedrock on which generations of scientists can interpret and make sense of.

To drive our understanding of cyanophage genomes forward however there needs to a concerted effort to capitalise on the sequence libraries that are being collected from both phage metagenomes and phage genomes. Sequencing even large data-sets is now comparatively easy and sequence information should be seen as the exciting starting point rather than the endpoint. To determine the function of the reservoir of genes will require extensive biochemical, chemical and molecular biological investigations as well as physiological experimentation.

Currently when new T4-like cyanophage genes are identified using bioinformatic approaches, they are compared to T4 and their function is deduced on the basis of known genes in the T4 genome. In order to really progress with understanding the role of the genes which have no homology (and to confirm the homology in genes where an identity can be hypothesised) a genetic system needs to be developed where cyanophages can be mutated. This will take extensive research effort and hopefully international research groups will come together in the way that researchers on the T phages in the 1960 s did to gradually piece together to determine the function of the genes that constitute largest reservoir of genetic diversity on earth.

Conclusions

The study of the "photosynthetic" cyanomyoviruses has revealed novel and important facets of the phage-host relationship that were not apparent from previous studies with heterotrophic systems. However, in common with all the T4-like phages there is much work to do in ascribing functions to the many genes lacking known homologues. It is probable that many of these genes are involved in the subtle manipulation of the physiology of the infected cell and are likely to be of potential importance in biotechnology as well as being intrinsically interesting. However, there are three main features specific to marine cyanomyovirus biology that require further substantial attention. At present there has been little more than speculation and theoretical modelling on the contribution of host-derived genes to cyanomyovirus fitness and it is important to develop experimental approaches that will enable us to assess the contribution the genes make to the infection process. There is also the related topic of evaluating the role of these phages as agents of horizontal gene transfer and assessing their contribution to cyanobacterial adaptation and evolution. Furthermore, from the ecological perspective we are still a long way from being able to assess the true impact of these cyanomyoviruses on natural populations of their hosts. It is likely that these cyanomyoviruses will remain an important feature of research in both phage biology and marine ecology for a considerable while to come.

Abbreviations

PBPs: phycobilin-bearing phycobiliproteins; APC: allophycocyanin; PC: phycocyanin; PE: phycoerytherin; Chl a: chlorophyll a; nm: nanometer; GOS: global ocean sampling; Q-PCR: quantitative polymerase chain reaction; nr: non redundant; ORF(s): open reading frame(s); LPS: lipopolyscacchride; PSII: photosystem II.

Author details

[1]Department of Infection, Immunity and Inflammation, Maurice Shock Medical Sciences Building, University of Leicester, PO Box 138, Leicester, LE1 9HN, UK. [2]Department of Biological Sciences, University of Warwick, Gibbet Hill Road, Coventry, CV4 7AL, UK.

Authors' contributions

ADM, MRJC & NHM contributed to the original drafts of the manuscript and approved the final version.

Competing interests

The authors declare that they have no competing interests.

Received: 22 June 2010 Accepted: 28 October 2010
Published: 28 October 2010

References

1. Kutter E, Kellenberger E, Carlson K, Eddy S, Neitzel J, Messinger L, North J, Guttman B: **Effects of bacterial growth conditions and physiology on T4 infection.** In *Molecular biology of bacteriophage T4.* Edited by: Karam JD. Washington D.C.: American Society for Microbiology; 1994:406-418.
2. Wilson WH, Joint IR, Carr NG, Mann NH: **Isolation and molecular characterization of 5 marine cyanophages propagated on** *Synechococcus* **sp. strain WH7803.** *Appl Enviro Microbiol* 1993, 59:3736-3743.
3. Waterbury JB, Valois FW: **Resistance to co-occurring phages enables marine** *Synechococcus* **communities to coexist with cyanophages abundant in seawater.** *Appl Enviro Microbiol* 1993, 59:3393-3399.
4. Suttle CA, Chan AM: **Marine cyanophages infecting oceanic and coastal strains of** *Synechococcus* **- abundance, morphology, cross-infectivity and growth-characteristics.** *Mar Ecol Progr* 1993, 92:99-109.
5. Mann NH, Clokie MRJ, Millard A, Cook A, Wilson WH, Wheatley PJ, Letarov A, Krisch HM: **The genome of S-PM2, a "photosynthetic" T4-type bacteriophage that infects marine** *Synechococcus* **strains.** *J Bacteriol* 2005, 187:3188-3200.
6. Shan J, Jia Y, Clokie MRJ, Mann NH: **Infection by the 'photosynthetic' phage S-PM2 induces increased synthesis of phycoerythrin in** *Synechococcus* **sp WH7803.** *FEMS Microbiology Letters* 2008, 283:154-161.
7. Angly FE, Felts B, Breitbart M, Salamon P, Edwards RA, Carlson C, Chan AM, Haynes M, Kelley S, Liu H, *et al:* **The marine viromes of four oceanic regions.** *PLoS Biology* 2006, 4:2121-2131.
8. Bench SR, Hanson TE, Williamson KE, Ghosh D, Radosovich M, Wang K, Wommack KE: **Metagenomic characterization of Chesapeake bay virioplankton.** *Appl Enviro Microbiol* 2007, 73:7629-7641.
9. Edwards RA, Rohwer F: **Viral metagenomics.** *Nature Reviews Microbiology* 2005, 3:504-510.
10. Mann NH: **Phages of the marine cyanobacterial picophytoplankton.** *FEMS Microbiology Reviews* 2003, 27:17-34.
11. Clokie MRJ, Mann NH: **Marine cyanophages and light.** *Environ Microbiol* 2006, 8:2074-2082.
12. Mann NH: **Phages of cyanobacteria.** In *The bacteriophages.*. 2 edition. Edited by: Calendar R. Oxford: Oxford University Press; 2006:517-533.
13. Goericke R, Welschmeyer NA: **The marine prochlorophyte** *Prochlorococcus* **contributes significantly to phytoplankton biomass and primary production in the Sargasso Sea.** *Deep Sea Res Part I-Oceanographic Research Papers* 1993, 40:2283-2294.
14. Liu HB, Nolla HA, Campbell L: *Prochlorococcus* **growth rate and contribution to primary production in the equatorial and subtropical North Pacific Ocean.** *Aquat Microb Ecol* 1997, 12:39-47.
15. Veldhuis MJW, Kraay GW, VanBleijswijk JDL, Baars MA: **Seasonal and spatial variability in phytoplankton biomass, productivity and growth in the northwestern Indian Ocean: The southwest and northeast monsoon, 1992-1993.** *Deep Sea Res Part I-Oceanographic Research Papers* 1997, 44:425-449.
16. Herdman M, Castenholz RW, Iteman I, Waterbury JB, Rippka R: **Subsection I. (Formerly Chroococcales Wettstein 1924, emend. Rippka, Deruelles, Waterbury, Herdman and Stanier 1979).** In *Bergey's Manual of Systematic Bacteriology. Volume 1.*. 2 edition. Edited by: Boone DR, Castenmholz RW, Garrity GM. New York, Berlin, Heidelberg.: Springer Publishers; 2001:493-514.
17. Chisholm SW, Frankel SL, Goericke R, Olson RJ, Palenik B, Waterbury JB, Westjohnsrud L, Zettler ER: *Prochlorococcus marinus* **nov. gen. nov. sp.: an oxyphototrophic marine prokaryote containing divinyl chlorophyll** *a* **and chlorophyll** *b.* *Arch Microbiol* 1992, 157:297-300.
18. Zwirglmaier K, Heywood JL, Chamberlain K, Woodward EMS, Zubkov MV, Scanlan DJ: **Basin-scale distribution patterns lineages in the Atlantic Ocean.** *Environ Microbiol* 2007, 9:1278-1290.
19. Zwirglmaier K, Jardillier L, Ostrowski M, Mazard S, Garczarek L, Vaulot D, Not F, Massana R, Ulloa O, Scanlan DJ: **Global phylogeography of marine** *Synechococcus* **and** *Prochlorococcus* **reveals a distinct partitioning of lineages among oceanic biomes.** *Environ Microbiol* 2008, 10:147-161.
20. Garczarek L, Dufresne A, Rousvoal S, West NJ, Mazard S, Marie D, Claustre H, Raimbault P, Post AF, Scanlan DJ, Partensky F: **High vertical and low**

horizontal diversity of *Prochlorococcus* ecotypes in the Mediterranean Sea in summer. *FEMS Microbiol Ecol* 2007, **60**:189-206.

21. Six C, Finkel ZV, Irwin AJ, Campbell DA: Light variability illuminates niche-partitioning among marine picocyanobacteria. *PLoS ONE* 2007, **2**:e1341.

22. Hambly E, Tetart F, Desplats C, Wilson WH, Krisch HM, Mann NH: A conserved genetic module that encodes the major virion components in both the coliphage T4 and the marine cyanophage S-PM2. *Proc Natl Acad Sci* 2001, **98**:11411-11416.

23. Weigele PR, Pope WH, Pedulla ML, Houtz JM, Smith AL, Conway JF, King J, Hatfull GF, Lawrence JG, Hendrix RW: Genomic and structural analysis of Syn9, a cyanophage infecting marine *Prochlorococcus* and *Synechococcus*. *Environ Microbiol* 2007, **9**:1675-1695.

24. Clokie MRJ, Thalassinos K, Boulanger P, Slade SE, Stoilova-McPhie S, Cane M, Scrivens JH, Mann NH: A proteomic approach to the identification of the major virion structural proteins of the marine cyanomyovirus S-PM2. *Microbiology* 2008, **154**:1775-1782.

25. Sullivan MB, Coleman ML, Weigele P, Rohwer F, Chisholm SW: Three *Prochlorococcus* cyanophage genomes: Signature features and ecological interpretations. *PLoS Biology* 2005, **3**:790-806.

26. Sandaa RA, Clokie M, Mann NH: Photosynthetic genes in viral populations with a large genomic size range from Norwegian coastal waters. *FEMS Microbiology Ecology* 2008, **63**:2-11.

27. Fuller NJ, Wilson WH, Joint IR, Mann NH: Occurrence of a sequence in marine cyanophages similar to that of T4 g20 and its application to PCR-based detection and quantification techniques. *Appl Enviro Microbiol* 1998, **64**:2051-2060.

28. Zhong Y, Chen F, Wilhelm SW, Poorvin L, Hodson RE: Phylogenetic diversity of marine cyanophage isolates and natural virus communities as revealed by sequences of viral capsid assembly protein gene g20. *Appl Enviro Microbiol* 2002, **68**:1576-1584.

29. Short CM, Suttle CA: Nearly identical bacteriophage structural gene sequences are widely distributed in both marine and freshwater environments. *Appl Enviro Microbiol* 2005, **71**:480-486.

30. Wilson WH, Fuller NJ, Joint IR, Mann NH: Analysis of cyanophage diversity and population structure in a south-north transect of the Atlantic ocean. *Bulletin de l'Institut océanographique, Monaco* 1999, **19**:209-216.

31. Frederickson CM, Short SM, Suttle CA: The physical environment affects cyanophage communities in British Columbia inlets. *Microb Ecol* 2003, **46**:348-357.

32. Dorigo U, Jacquet S, Humbert JF: Cyanophage diversity, inferred from g20 gene analyses, in the largest natural lake in France, Lake Bourget. *Appl Enviro Microbiol* 2004, **70**:1017-1022.

33. Wang K, Chen F: Genetic diversity and population dynamics of cyanophage communities in the Chesapeake Bay. *Aquat Microb Ecol* 2004, **34**:105-116.

34. Sandaa RA, Larsen A: Seasonal variations in virus-host populations in Norwegian coastal waters: Focusing on the cyanophage community infecting marine *Synechococcus* spp. *Appl Enviro Microbiol* 2006, **72**:4610-4618.

35. Marston MF, Sallee JL: Genetic diversity and temporal variation in the cyanophage community infecting marine *Synechococcus* species in Rhode Island's coastal waters. *Appl Enviro Microbiol* 2003, **69**:4639-4647.

36. Wilhelm SW, Carberry MJ, Eldridge ML, Poorvin L, Saxton MA, Doblin MA: Marine and freshwater cyanophages in a Laurentian Great Lake: Evidence from infectivity assays and molecular analyses of g20 genes. *Appl Enviro Microbiol* 2006, **72**:4957-4963.

37. Sullivan MB, Coleman ML, Quinlivan V, Rosenkrantz JE, Defrancesco AS, Tan G, Fu R, Lee JA, Waterbury JB, Bielawski JP, Chisholm SW: Portal protein diversity and phage ecology. *Environ Microbiol* 2008, **10**:2810-23.

38. Filee J, Tetart F, Suttle CA, Krisch HM: Marine T4-type bacteriophages, a ubiquitous component of the dark matter of the biosphere. *Proc Natl Acad Sci* 2005, **102**:12471-12476.

39. Comeau AM, Krisch HM: The capsid of the T4 phage superfamily: The evolution, diversity, and structure of some of the most prevalent proteins in the biosphere. *Mol Biol Evo* 2008, **25**:1321-1332.

40. Millard AD, Mann NH: A temporal and spatial investigation of cyanophage abundance in the Gulf of Aqaba, Red Sea. *Journal of the Marine Biological Association of the United Kingdom* 2006, **86**:507-515.

41. McDaniel LD, delaRosa M, Paul JH: Temperate and lytic cyanophages from the Gulf of Mexico. *Journal of the Marine Biological Association of the United Kingdom* 2006, **86**:517-527.

42. Sullivan MB, Waterbury JB, Chisholm SW: Cyanophages infecting the oceanic cyanobacterium *Prochlorococcus*. *Nature* 2003, **424**:1047-1051.

43. Millard AD, Zwirglmaier K, Downey MJ, Mann NH, Scanlan DJ: Comparative genomics of marine cyanomyoviruses reveals the widespread occurrence of Synechococcus host genes localized to a hyperplastic region: implications for mechanisms of cyanophage evolution. *Environ Microbiol* 2009, **11**:2370-2387.

44. Miller ES, Heidelberg JF, Eisen JA, Nelson WC, Durkin AS, Ciecko A, Feldblyum TV, White O, Paulsen IT, Nierman WC, et al: Complete genome sequence of the broad-host-range vibriophage KVP40: comparative genomics of a T4-related bacteriophage. *J Bacteriol* 2003, **185**:5220-5233.

45. Miller ES, Kutter E, Mosig G, Arisaka F, Kunisawa T, Ruger W: Bacteriophage T4 genome. *Microbiol Mol Biol Rev* 2003, **67**:86-156.

46. Desplats C, Dez C, Tetart F, Eleaume H, Krisch HM: Snapshot of the genome of the pseudo-T-even bacteriophage RB49. *J Bacteriol* 2002, **184**:2789-2804.

47. Keller B, Dubochet J, Adrian M, Maeder M, Wurtz M, Kellenberger E: Length and shape variants of the bacteriophage T4 head: mutations in the scaffolding core genes 68 and 22. *J Virol* 1988, **62**:2960-2969.

48. Keller B, Maeder M, Becker-Laburte C, Kellenberger E, Bickle TA: Amber mutants in gene 67 of phage T4. Effects on formation and shape determination of the head. *J Mol Biol* 1986, **190**:83-95.

49. Sullivan MB, Lindell D, Lee JA, Thompson LR, Bielawski JP, Chisholm SW: Prevalence and evolution of core photosystem II genes in marine cyanobacterial viruses and their hosts. *PLoS Biology* 2006, **4**:1344-1357.

50. Comeau AM, Bertrand C, Letarov A, Tetart F, Krisch HM: Modular architecture of the T4 phage superfamily: A conserved core genome and a plastic periphery. *Virology* 2007, **362**:384-396.

51. Millard AD, Gierga G, Clokie MRJ, Evans DJ, Hess WR, Scanlan DJ: An antisense RNA in a lytic cyanophage links *psbA* to a gene encoding a homing endonuclease. *ISME J* 2010, **4**:1121-1135.

52. Mann NH, Cook A, Millard A, Bailey S, Clokie M: Marine ecosystems: Bacterial photosynthesis genes in a virus. *Nature* 2003, **424**:741-741.

53. Millard A, Clokie MRJ, Shub DA, Mann NH: Genetic organization of the *psbAD* region in phages infecting marine *Synechococcus* strains. *Proc Natl Acad Sci* 2004, **101**:11007-11012.

54. DeLong EF, Preston CM, Mincer T, Rich V, Hallam SJ, Frigaard NU, Martinez A, Sullivan MB, Edwards R, Brito BR, et al: Community genomics among stratified microbial assemblages in the ocean's interior. *Science* 2006, **311**:496-503.

55. Williamson SJ, Rusch DB, Yooseph S, Halpern AL, Heidelberg KB, Glass JI, Andrews-Pfannkoch C, Fadrosh D, Miller CS, Sutton G, et al: The Sorcerer II Global Ocean Sampling Expedition: metagenomic characterization of viruses within aquatic microbial samples. *PLoS ONE* 2008, **3**:e1456.

56. Clokie MR, Shan J, Bailey S, Jia Y, Krisch HM, West S, Mann NH: Transcription of a 'photosynthetic' T4-type phage during infection of a marine cyanobacterium. *Environ Microbiol* 2006, **8**:827-835.

57. Sharon I, Tzahor S, Williamson S, Shmoish M, Man-Aharonovich D, Rusch DB, Yooseph S, Zeidner G, Golden SS, Mackey SR, et al: Viral photosynthetic reaction center genes and transcripts in the marine environment. *ISME Journal* 2007, **1**:492-501.

58. Rusch DB, Halpern AL, Sutton G, Heidelberg KB, Williamson S, Yooseph S, Wu D, Eisen JA, Hoffman JM, Remington K, et al: The Sorcerer II Global Ocean Sampling expedition: northwest Atlantic through eastern tropical Pacific. *PLoS Biol* 2007, **5**:e77.

59. Melis A: Photosystem-II damage and repair cycle in chloroplasts: what modulates the rate of photodamage? *Trends Plant Sci* 1999, **4**:130-135.

60. Bailey S, Clokie MRJ, Millard A, Mann NH: Cyanophage infection and photoinhibition in marine cyanobacteria. *Research in Microbiology* 2004, **155**:720-725.

61. Lindell D, Jaffe JD, Johnson ZI, Church GM, Chisholm SW: Photosynthesis genes in marine viruses yield proteins during host infection. *Nature* 2005, **438**:86-89.

62. Hellweger FL: Carrying photosynthesis genes increases ecological fitness of cyanophage in silico. *Environ Microbiol* 2009, **11**(6):1386-1394.

63. Bragg JG, Chisholm SW: Modeling the fitness consequences of a cyanophage-encoded photosynthesis gene. *PLoS ONE* 2008, **3**:e3550.

64. Mulo P, Eloranta T, Aro EM, Maenpaa P: Disruption of a spe-like open reading frame alters polyamine content and psbA-2 mRNA stability in the cyanobacterium Synechocystis sp. PCC 6803. *Bot Acta* 1998, **111**:71-76.

65. Bograh A, Gingras Y, Tajmir-Riahi HA, Carpentier R: The effects of spermine and spermidine on the structure of photosystem II proteins in relation to inhibition of electron transport. *FEBS Lett* 1997, **402**:41-44.

66. Ioannidis NE, Kotzabasis K: Effects of polyamines on the functionality of photosynthetic membrane in vivo and in vitro. *Biochim Biophys Acta* 2007, **1767**:1372-1382.

67. He Q, Dolganov N, Bjorkman O, Grossman AR: The high light-inducible polypeptides in *Synechocystis* PCC6803. Expression and function in high light. *J Biol Chem* 2001, **276**:306-314.

68. Lindell D, Sullivan MB, Johnson ZI, Tolonen AC, Rohwer F, Chisholm SW: Transfer of photosynthesis genes to and from *Prochlorococcus* viruses. *Proc Natl Acad Sci* 2004, **101**:11013-11018.

69. Cobley JG, Clark AC, Weerasurya S, Queseda FA, Xiao JY, Bandrapali N, D'Silva I, Thounaojam M, Oda JF, Sumiyoshi T, Chu MH: CpeR is an activator required for expression of the phycoerythrin operon (cpeBA) in the cyanobacterium Fremyella diplosiphon and is encoded in the phycoerythrin linker-polypeptide operon (cpeCDESTR). *Molecular Microbiology* 2002, **44**:1517-1531.

70. Grossman AR, Schaefer MR, Chiang GG, Collier JL: The phycobilisome, a light-harvesting complex responsive to environmental conditions. *Microbiol Rev* 1993, **57**:725-749.

71. Shan J: An investigation into the effect of cyanophage infection on photosynthetic antenna. *PhD thesis* University of Warwick, Dept Biological Sciences; 2008.

72. Dammeyer T, Bagby SC, Sullivan MB, Chisholm SW, Frankenberg-Dinkel N: Efficient phage-mediated pigment biosynthesis in oceanic cyanabacteria. *Curr Biol* 2008, **18**:442-448.

73. Bailey S, Melis A, Mackey KR, Cardol P, Finazzi G, van Dijken G, Berg GM, Arrigo K, Shrager J, Grossman A: Alternative photosynthetic electron flow to oxygen in marine *Synechococcus*. *Biochim Biophys Acta* 2008, **1777**:269-276.

74. Kruger NJ, von Schaewen A: The oxidative pentose phosphate pathway: structure and organisation. *Curr Opin Plant Biol* 2003, **6**:236-246.

75. Udvardy J, Borbely G, Juhasz A, Farkas GL: Thioredoxins and the redox modulation of glucose-6-phosphate dehydrogenase in Anabaena sp. strain PCC 7120 vegetative cells and heterocysts. *J Bacteriol* 1984, **157**:681-683.

76. Dai S, Schwendtmayer C, Schurmann P, Ramaswamy S, Eklund H: Redox signaling in chloroplasts: cleavage of disulfides by an iron-sulfur cluster. *Science* 2000, **287**:655-658.

77. Clarke AK, Campbell D: Inactivation of the petE gene for plastocyanin lowers photosynthetic capacity and exacerbates chilling-induced photoinhibition in the cyanobacterium *Synechococcus*. *Plant Physiol* 1996, **112**:1551-1561.

78. Kim SK, Makino K, Amemura M, Shinagawa H, Nakata A: Molecular analysis of the phoH gene, belonging to the phosphate regulon in *Escherichia coli*. *J Bacteriol* 1993, **175**:1316-1324.

79. Scanlan DJ, Mann NH, Carr NG: The response of the picoplanktonic marine cyanobacterium *Synechococcus* species WH7803 to phosphate starvation involves a protein homologous to the periplasmic phosphate-binding protein of Escherichia coli. *Mol Microbiol* 1993, **10**:181-191.

80. Keller S, Wage T, Hohaus K, Holzer M, Eichhorn E, van Pee KH: Purification and Partial Characterization of Tryptophan 7-Halogenase (PrnA) from *Pseudomonas fluorescens*. *Angew Chem Int Ed Engl* 2000, **39**:2300-2302.

81. Bryan MJ, Burroughs NJ, Spence EM, Clokie MR, Mann NH, Bryan SJ: Evidence for the intense exchange of MazG in marine cyanophages by horizontal gene transfer. *PLoS ONE* 2008, **3**:e2048.

82. Gross M, Marianovsky I, Glaser G: MazG – a regulator of programmed cell death in *Escherichia coli*. *Mol Microbiol* 2006, **59**:590-601.

83. Mann N, Carr NG, Midgley JEM: RNA-Synthesis and Accumulation of Guanine Nucleotides During Growth Shift Down in Blue-Green-Alga Anacystis-Nidulans. *Biochimica Et Biophysica Acta* 1975, **402**:41-50.

84. Clokie MRJ, Millard AD, Mehta JY, Mann NH: Virus isolation studies suggest short-term variations in abundance in natural cyanophage populations of the Indian Ocean. *Journal of the Marine Biological Association of the United Kingdom* 2006, **86**:499-505.

85. Clokie MR, Millard AD, Wilson WH, Mann NH: Encapsidation of host DNA by bacteriophages infecting marine *Synechococcus* strains. *FEMS Microbiol Ecol* 2003, **46**:349-352.

86. Wilson WH, Carr NG, Mann NH: The effect of phosphate status on the kinetics of cyanophage infection in the oceanic cyanobacterium *Synechococcus* sp WH7803. *Journal of Phycology* 1996, **32**:506-516.

87. Karl DM, Bailiff MD: The measurement and distribution of dissolved nucleic acids in aquaitc environments. *Limnology and Oceanography* 1989, **34**:543-558.

88. Suttle CA: Viruses in the sea. *Nature* 2005, **437**:356-361.

89. Kimmance SA, Wilson WH, Archer SD: Modified dilution technique to estimate viral versus grazing mortality of phytoplankton: limitations associated with method sensitivity in natural waters. *Aquat Microb Ecol* 2007, **49**:207-222.

90. Baudoux AC, Veldhuis MJW, Noordeloos AAM, van Noort G, Brussaard CPD: Estimates of virus- vs. grazing induced mortality of picophytoplankton in the North Sea during summer. *Aquat Microb Ecol* 2008, **52**:69-82.

91. Baudoux AC, Veldhuis MJW, Witte HJ, Brussaard CPD: Viruses as mortality agents of picophytoplankton in the deep chlorophyll maximum layer during IRONAGES III. *Limnol Oceanogr* 2007, **52**:2519-2529.

92. Thingstad TF, Bratbak G, Heldal M: Aquatic phage ecology. In *Bacteriophage ecology*. Edited by: Abedon ST. Cambridge: Cambridge University Press; 2008:251-280.

93. Weinbauer MG: Ecology of prokaryotic viruses. *FEMS Microbiol Rev* 2004, **28**:127-181.

94. Suttle CA, Chan AM: Dynamics and distribution of cyanophages and their effect on marine *Synechococcus* spp. *Appl Enviro Microbiol* 1994, **60**:3167-3174.

95. Garza DR, Suttle CA: The effect of cyanophages on the mortality of *Synechococcus* spp. and selection for UV resistant viral communities. *Microbiol Ecol* 1998, **36**:281-292.

96. Proctor LM, Fuhrman JA: Viral mortality of marine bacteria and cyanobacteria. *Nature* 1990, **343**:60-62.

97. Suttle CA: Cyanophages and their role in the ecology of cyanobacteria. In *The Ecology of Cyanobacteria*. Edited by: Whitton BA, Potts M. Dordrecht: Kluwer Academic Publishers; 2000:563-589.

98. Thingstad TF, Lignell R: Theoretical models for the control of bacterial growth rate, abundance, diversity and carbon demand. *Aquat Microb Ecol* 1997, **13**:19-27.

99. Hoffmann KH, Rodriguez-Brito B, Breitbart M, Bangor D, Angly F, Felts B, Nulton J, Rohwer F, Salamon P: Power law rank-abundance models for marine phage communities. *FEMS Microbiology Letters* 2007, **273**:224-228.

100. Jia Y: An investigation into the adsorption of cyanophages to their cyanobacterial hosts. *PhD* University of Warwick, Biological Sciences; 2008.

101. Kao CC, Green S, Stein B, Golden SS: Diel infection of a cyanobacterium by a contractile bacteriophage. *Appl Environ Microbiol* 2005, **71**:4276-4279.

102. Mühling M, Fuller NJ, Millard A, Somerfield PJ, Marie D, Wilson WH, Scanlan DJ, Post AF, Joint I, Mann NH: Genetic diversity of marine *Synechococcus* and co-occurring cyanophage communities: evidence for viral control of phytoplankton. *Environ Microbiol* 2005, **7**:499-508.

103. Hoskisson PA, Smith MCM: Hypervariation and phase variation in the bacteriophage 'resistome'. *Curr Biol* 2007, **10**:396-400.

104. Stoddard LI, Martiny JBH, Marston MF: Selection and characterization of cyanophage resistance in marine *Synechococcus* strains. *Appl Enviro Microbiol* 2007, **73**:5516-5522.

105. Kerr B, West J, Bohannan BJ: Bacteriophages: models for exploring basic principles of ecology. In *Bacteriophage ecology*. Edited by: Abedon ST. Cambridge: Cambridge University Press; 2008:31-63.

106. Lennon JT, Khatana SAM, Marston MF, Martiny JBH: Is there a cost of virus resistance in marine cyanobacteria? *ISME Journal* 2007, **1**:300-312.

107. Kettler GC, Martiny AC, Huang K, Zucker J, Coleman ML, Rodrigue S, Chen F, Lapidus A, Ferriera S, Johnson J, et al: Patterns and implications of gene gain and loss in the evolution of *Prochlorococcus*. *PLoS Genetics* 2007, **3**:2515-2528.

108. Dufresne A, Ostrowski M, Scanlan D, Garczarek L, Mazard S, Palenik B, Paulsen I, de Marsac N, Wincker P, Dossat C, et al: Unraveling the genomic mosaic of a ubiquitous genus of marine cyanobacteria. *Genome Biol* 2008, **9**:R90.

109. Dinsdale EA, Edwards RA, Hall D, Angly F, Breitbart M, Brulc JM, Furlan M, Desnues C, Haynes M, Li L, et al: Functional metagenomic profiling of nine biomes. *Nature* 2008, **455**:830-830.

doi:10.1186/1743-422X-7-291
Cite this article as: Clokie *et al.*: T4 genes in the marine ecosystem:
studies of the T4-like cyanophages and their role in marine ecology.
Virology Journal 2010 7:291.

Edgell *et al. Virology Journal* 2010, **7**:290
http://www.virologyj.com/content/7/1/290

VIROLOGY JOURNAL

Mobile DNA elements in T4 and related phages

David R Edgell[1*], Ewan A Gibb[1], Marlene Belfort[2]

Abstract

Mobile genetic elements are common inhabitants of virtually every genome where they can exert profound influences on genome structure and function in addition to promoting their own spread within and between genomes. Phage T4 and related phage have long served as a model system for understanding the molecular mechanisms by which a certain class of mobile DNA, homing endonucleases, promote their spread. Homing endonucleases are site-specific DNA endonucleases that initiate mobility by introducing double-strand breaks at defined positions in genomes lacking the endonuclease gene, stimulating repair and recombination pathways that mobilize the endonuclease coding region. In phage T4, homing endonucleases were first discovered as encoded within the self-splicing *td*, *nrdB* and *nrdD* introns of T4. Genomic data has revealed that homing endonucleases are extremely widespread in T-even-like phage, as evidenced by the astounding fact that ~11% of the T4 genome encodes homing endonuclease genes, with most of them located outside of self-splicing introns. Detailed studies of the mobile *td* intron and its encoded endonuclease, I-TevI, have laid the foundation for genetic, biochemical and structural aspects that regulate the mobility process, and more recently have provided insights into regulation of homing endonuclease function. Here, we summarize the current state of knowledge regarding T4-encoded homing endonucleases, with particular emphasis on the *td*/I-TevI model system. We also discuss recent progress in the biology of free-standing endonucleases, and present areas of future research for this fascinating class of mobile genetic elements.

Introduction

In the 20 years since the first review on mobile genetic elements in the T4 genome, significant progress has been made with respect to understanding the biology of T4-encoded homing endonucleases [1]. In particular, we now have a firm grasp of the DNA repair and recombination pathways that promote mobility of intron-encoded endonucleases [2-5]. We also know more about the molecular details that regulate protein-DNA interactions of the long-serving model homing endonuclease, I-TevI, providing intriguing insights into how the enzyme has adapted well to life in a genome rich in glucosylated hydroxymethylcytosine-containing DNA [6-8]. Perhaps one of the most surprising discoveries was the finding that T4 encodes 12 homing endonucleases that are not intron encoded, but instead are located in intergenic regions (Figure 1, Table 1). The so-called free-standing endonucleases belong to the GIY-YIG and HNH homing endonuclease families, and are termed *seg*

(similar to endonucleases encoded within group I introns) and *mob* (mobility) genes, respectively [9,10]. In recent years, the explosion of phage genome sequences has revealed that free-standing endonucleases are more widespread than their intron-encoded cousins (at least in T-even phage genomes), while at the same time confirming a long-held suspicion that T4 is an oddity among T-even-like phages, for no other phage comes close to encoding the 15 homing endonucleases that T4 does - representing 11% of its coding potential!

Our purpose in this review is to summarize the past 20 years of research into T4 homing endonucleases, with emphasis on the mechanisms involved in mobility, protein-DNA recognition, and the regulation of endonuclease function within the context of a host genome. Because mechanistic insights into endonuclease function stemming from studies on T4-encoded endonucleases will be generally applicable to endonucleases encoded within other T-even phage genomes, we will focus mainly on T4 endonucleases, discussing examples in other phage only when obvious differences are found. We also point out areas for future research where we are still largely ignorant, namely the mobility pathways utilized by the *mob* endonucleases,

* Correspondence: dedgell@uwo.ca
[1]Department of Biochemistry, Schulich School of Medicine & Dentistry, The University of Western Ontario, London, ON, N6A 5C1, Canada
Full list of author information is available at the end of the article

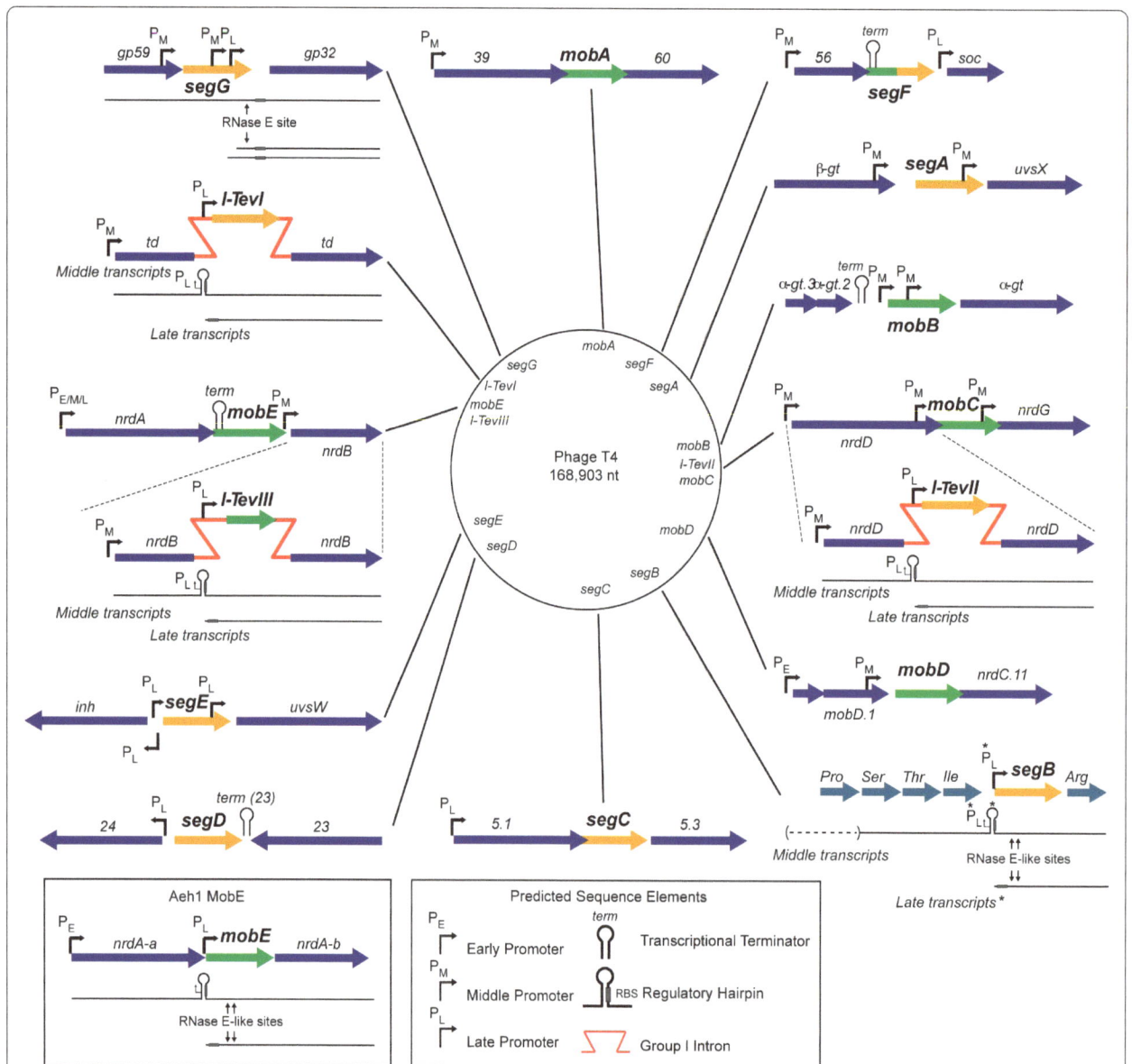

Figure 1 Schematic of the location of the fifteen homing endonuclease genes indicated on a genomic map of bacteriophage T4. For simplicity, each genomic segment is drawn with the endonuclease in the same orientation, with relevant regulatory elements indicated. The GIY-YIG endonucleases are shown in yellow, while the HNH-type endonucleases are green. The hybrid endonuclease *segF* is drawn with both colours. The bacteriophage Aeh1 *mobE* endonuclease, which is not part of the bacteriophage T4 genome, is set in a box. An asterix (*) marks a predicted late promoter upstream of SegB.

which nick rather than cleave their targets, transcriptional and translational regulation of endonucleases, and questions of an evolutionary nature dealing with the impact of endonuclease activity on phage genome structure and function.

Mechanisms of mobility
Pathways
The variable occurrence of the three T4 introns in other closely related T-even phage first suggested that these introns, in the *td*, *nrdB*, and *nrdD* (*sunY*) genes, are mobile genetic elements [11,12]. Shortly after these observations, mobility was demonstrated for the *td* and *nrdD* introns, and attributed to intron-encoded endonucleases that make a double-strand break (DSB) [13]. The first mechanistic insight came from the observation that intron insertion into the cleaved target, the so-called homing site, is accompanied by co-conversion of the flanking exon sequences [14]. Cleavage of target DNA by an intron endonuclease and co-conversion of flanking

Table 1 The homing endonucleases of phage T4

Endonuclease	Intron encoded or free-standing	Active	Family	Insertion Site	Target Gene	Reference
I-TevI	Intron	Yes	GIY-YIG	*td*	*td*	[14,28]
I-TevII	Intron	Yes	GIY-YIG	*nrdD*	*nrdD*	[7]
I-TevIII	Intron	Yes (RB3)	HNH	*nrdB*	*nrdB*	[19,20]
mobA	Free-standing	ND	HNH	*60.1/39*	*39*	D. Shub pers. comm..
mobB	Free-standing	ND	HNH	*α-gt/α-gt.2*	unknown	
mobC	Free-standing	ND	HNH	*nrdG/nrdD*	unknown	
mobD	Free-standing	ND	HNH	*nrdC.11/mobD.1*	unknown	
mobE	Free-standing	Yes	HNH	*nrdB/nrdA*	*nrdB*	[41,42]
segA	Free-standing	Yes	GIY-YIG	*uvsX/β-gt*	*uvsX*	[34]
segB	Free-standing	Yes	GIY-YIG	tRNA-Arg/tRNA-Ile	tRNA intergenic region	[38]
segC	Free-standing	Yes	GIY-YIG	*5.1/5.3*	of *5.1* and *5.3*	[37]
segD	Free-standing	ND	GIY-YIG	*23/24*	unknown	
segE	Free-standing	Yes	GIY-YIG	*inh/uvsW*	*uvsW*	[36]
segF	Free-standing	Yes	GIY-YIG	*soc/56*	*56*	[35]
segG	Free-standing	Yes	GIY-YIG	*32/59*	*32*	[39]

exon sequences are both features associated with mobile introns of eukaryotes [15], indicating a common mechanism for intron transfer. Indeed in both cases co-conversion of exon markers reflects the DSB being processed to a gap [16].

Because of the facile phage/bacterial genetic system, *td* intron homing has the best characterized group I intron inheritance pathway. Key studies involved defining both bacterial and phage functions that are required for the homing event as well as characterizing recombination intermediates [3,17]. Mobility depends on host or phage recombinase functions, RecA or UvsX, respectively. The process also uses phage-encoded exonuclease activities, single-stranded binding proteins (Gp32), DNA synthesis and repair functions, resolvase and ligase (Figure 2). In light of these dependencies, and exon co-conversion, it was concluded that introduction of the DSB is followed by exonucleolytic degradation [18], and that the processed 3' end invades the intron donor duplex and primes repair synthesis that results in copying of the intron into the recipient DNA. This process likely proceeds for at least some events via the DSB repair (DSBR) pathway, wherein a D-loop formed as the result of repair synthesis serves as a template for repair of the opposite strand [3,16] (Figure 2, left pathway). The two Holliday junctions formed during the repair process can be resolved to yield two intron-containing alleles: if the junction is cleaved in the crossover orientation flanking markers are exchanged, whereas if the junction is cleaved in the non-crossover orientation no exchange of flanking markers is observed. The T4 gene *49* product resolves these junctions [3].

Interestingly, homing is reduced but not abolished in gene *49* mutants. This ambiguous requirement for gp49 implies alternative resolution enzymes or additional homing pathways. The underrepresentation of crossover

events among the homing products favors alternative pathway(s), of which the synthesis-dependent strand annealing (SDSA) pathway is one (Figure 2, right pathway). The initial steps of the SDSA pathway are the same as those of DSBR, but unlike DSBR, Holliday junctions are not formed, circumventing the need for resolvase and resulting only in non-crossover products [3].

There is a close relationship between intron mobility and recombination-dependent replication in phage T4 [3]. Thus, intron homing occurs in mutants in which origin-dependent replication is disrupted [primase (gp61) and topoisomerase (gp39, gp52, gp60)], but is reduced in recombination-dependent replication mutants. The latter functions that play a role in homing include recombinase activities (UvsX, UvsY), single-stranded DNA-binding protein (gp32), exonucleolytic functions (RNaseH, DexA, 43Exo and possibly gp46, gp47), DNA polymerase (gp43) and its accessories (gp44, gp45, and gp62), helicase (gp41), the primase-helicase accessory (gp59), enzymes providing DNA precursors (eg. gp1) and DNA ligase (gp30) (Figure 2).

T4 RNase H, a 5'-3' exonuclease, T4 DNA exonuclease A (DexA) and the 3'-5' exonuclease activity of T4 DNA polymerase (*43*Exo) impact not only degradation, but also the homing efficiency and flanking marker coconversion [2]. The experiments implicating a role for these functions in intron homing provided the first direct evidence of a role for 3' ssDNA tails in T4 recombination. Together, the work that defined the involvement of phage accessories to the homing process demonstrates how a mobile intron harnesses phage replication, recombination and repair functions for its own propagation [2,3,17].

Although the above discussion is based on homology between donor and recipient, heterologous sequences

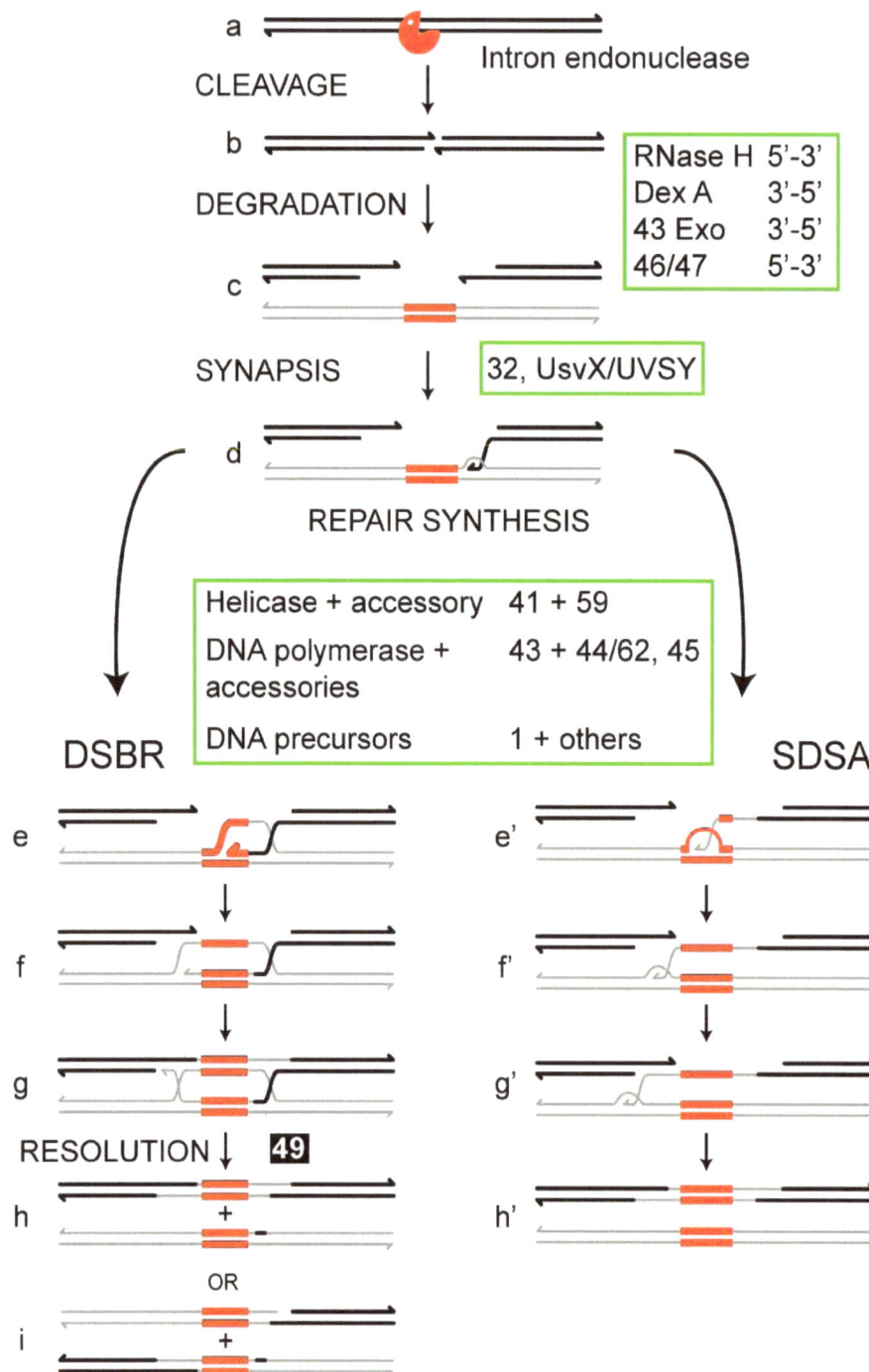

Figure 2 Alternative mechanisms for DSB-mediated intron homing. Subsequent to cleavage by the homing endonuclease (a), the recipient, intronless allele (thick lines) undergoes exonucleolytic degradation and homologous sequence alignment with an intron-containing donor (thin lines) (b, c). A 3' end of the recipient invades the donor, which serves as a template for repair synthesis (d). In the DSBR pathway (left), DNA synthesis through the intron (red) results in formation and expansion of a D-loop (e), which then serves as substrate for repair synthesis of the noninvading strand (f). Holliday junctions are resolved to produce either noncrossover (h) or crossover (i) products. During synthesis-dependent strand annealing (SDSA) (right), the displaced loop or bubble migrates with the replicative end as DNA synthesis proceeds through the intron (e' - g'). The newly synthesized strand is released from the donor and serves as template for repair synthesis of the noninvading strand (g' - h') to generate noncrossover products only (h'). Functions implicated in homing and their putative association with appropriate steps in the homing pathways are shown.

can also participate in DSB-mediated repair [4,5]. Extensive homology in one exon supports elevated homing levels when the other exon is absent, allowing analysis of "one-sided" events, which revealed illegitimate DSB repair. Recombination junctions at sites of microhomology and extensive nucleolytic degradation were evident. These observations suggest that illegitimate DSB repair may provide a means by which introns can invade ectopic sites, while lengthy resection may also be related to distal cleavage sites of the freestanding endonucleases, to be considered below.

Proteins: intron-encoded endonucleases

The intron-encoded endonucleases of the T-even phage genome are members of the GIY-YIG and HNH families. These families are characterized based on the catalytic cleavage domains, which are joined to DNA binding domains of varying specificities. The phage T4 *td*, *nrdB*, *nrdD* introns encode, respectively, the following endonucleases: I-TevI and I-TevII, both GIY-YIG endonucleases, and I-TevIII, a member of the HNH family. I-TevIII is, however, inactive on account of a large deletion, but a functional ortholog is found in phage RB3 [19,20]. The DNA-binding domains of both the phage-encoded endonuclease families appear to be architectually similar, in a beads-on-a-string arrangement, consisting of a variety of small protein modules that gives the proteins their specificity [21,22].

The best characterized of the T-even phage enzymes is the GIY-YIG *td* intron endonuclease I-TevI (Figure 3). The GIY-YIG family of endonucleases was first identified as representing sequence similarities in intron-encoded proteins of phage T4 and filmentous fungi [23]. Now, more than 20 years later, we know, mainly from multiple sequence alignments, of a large GIY-YIG superfamily of enzymes that nicks or cleaves DNA. This superfamily encompasses restriction enzymes, retrotransposons, and recombination and repair proteins, including UvrC, which performs nucleotide excision repair [24].

I-TevI and I-TevII, both GIY-YIG endonucleases, have several features in common. First, they have lengthy recognition sequences, spanning more than two helical turns of DNA; second, they induce conformational changes in the homing site during the substrate binding and cleavage process; third, they bind in the minor groove; and, finally they remain bound to the cleaved substrate [6,7,25]. Minor-groove binding is easily reconciled with T4 DNA being heavily modified in the major groove. Persistent binding to the cleavage product is also more than a curiosity, accounting for exon coconversion asymmetry. I-TevI, for example, remains bound to the exon II side of the homing site, resulting in coconversion biases in exon I, which is free for digestion by degradative nucleases [18].

The GIY-YIG module in I-TevI is 92 amino long, has five conserved motifs, of which GIY-$N_{10/11}$-YIG is the first (Motif A) and is thought to play a structural role. Motifs B, D and E contain conserved Arg, Glu and Asn residues, which function in catalysis [26]. The catalytic domain is joined to a lengthy, and distinct DNA-binding domain, which recognizes an expansive target sequence [6], [27]. The 28-kDa I-TevI recognizes a 38-bp target sequence, binding as a monomer. I-TevI cleaves intron-less DNA at sites 23 nt and 25 nt upstream of the intron insertion site (IS) to create a DSB, but how a monomeric enzyme cleaves two strands is not known [21]. In constrast, the HNH endonuclease I-TevIII of phage RB3, which is structurally and catalytically intact, acts as a dimer to make a DSB [20].

The monomeric I-TevI interacts with two regions of its 38-bp homing site [28]. The DNA-binding domain, which has an extended structure, winds around the primary binding region of 20 bp, centered on the intron IS [8] (Figure 3). This domain is joined via a long linker to the globular GIY-YIG-containing catalytic domain, which contacts the cleavage site (CS). The linker is 75-amino acids long, and has elements of structure, including a C-terminal zinc finger, which abuts the DNA-binding domain [29]. This linker is responsible for dynamic properties of I-TevI, and facilitates a dual role, namely to act as both an endonuclease or a transcriptional autorepressor [30,31].

I-TevI uses both sequence and distance determinants in selecting its CS [25]. Although the enzyme is generally tolerant of nucleotide changes in the homing site [6], it has a preference for both its natural cleavage sequence, and for the wild-type distance. If its CS is displaced from the optimal distance of 23 nt and 25 nt, I-TevI searches bidirectionally from its cleavage position to locate a preferred site, 5'-CX↑XX↓G-3', and cleaves at alternative distances, albeit with reduced efficiency [25,30,32]. The cleavage window extends from 5 bp upstream to 16 bp downstream of the normal cleavage site [25]. When a preferred site is not within the window, the enzyme defaults to the optimal distance and cleaves with reduced efficiency [25,32]. Most of the linker (except for ~20 N-terminal amino acids adjacent to the catalytic domain) and the zinc finger, serves as the distance determinant to constrain the catalytic domain, such that it is proximal to the cleavage site and promotes catalysis [29,31,32]. One of the functions of the linker is therefore to act as a "protein ruler", which we postulated to have evolved because I-TevI moonlights as an autorepressor, as described in section 3 [29-31]. Thus, the overall role of the I-TevI linker is to act as a communication device between the DNA-binding and catalytic GIY-YIG domains, such that they act in concert for DNA cleavage, but the DNA-binding domain

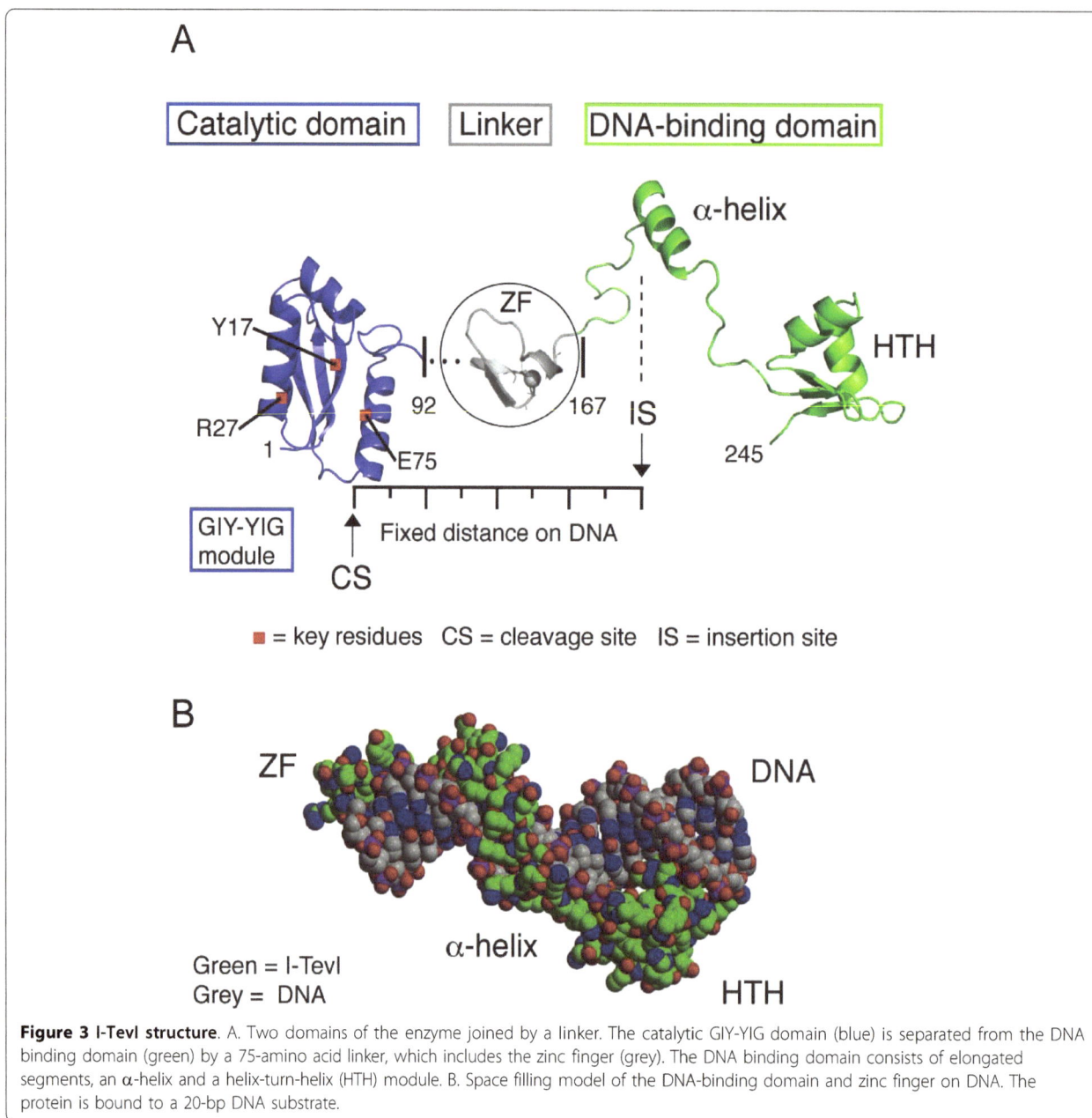

Figure 3 I-TevI structure. A. Two domains of the enzyme joined by a linker. The catalytic GIY-YIG domain (blue) is separated from the DNA binding domain (green) by a 75-amino acid linker, which includes the zinc finger (grey). The DNA binding domain consists of elongated segments, an α-helix and a helix-turn-helix (HTH) module. B. Space filling model of the DNA-binding domain and zinc finger on DNA. The protein is bound to a 20-bp DNA substrate.

acts independently when serving as a transcriptional repressor (Figure 4A) [31].

The zinc finger imparts another layer of regulation to I-TevI cleavage. In addition to the structural diversification that zinc fingers provide, an interesting aspect of these modules is their ability to be regulated by oxidation and reduction (redox) reactions [33]. Indeed, the zinc finger of I-TevI is redox responsive, and acts as a switch altering the ability of the enzyme to faithfully cleave its cognate substrate (Robbins, Smith and Belfort, in preparation). Under reducing conditions, the zinc finger is intact, active and accurate, whereas upon oxidation, the zinc is

lost, and I-TevI suffers comprised activity and fidelity. We speculate that oxidative stress may provide a signal to the enzyme, transduced via the zinc finger, to cleave at ectopic sites, and thereby to facilitate intron spread.

Proteins: Free-standing endonucleases

Of the 15 homing endonucleases encoded in the T4 genome, 12 are free standing and found in the inter-genic regions separating genes that are conserved amongst related phage genomes (Figure 1). Early work showed that SegA is a site-specific DNA endonuclease that generates a DSB with a 2-nucleotide 3' extension in

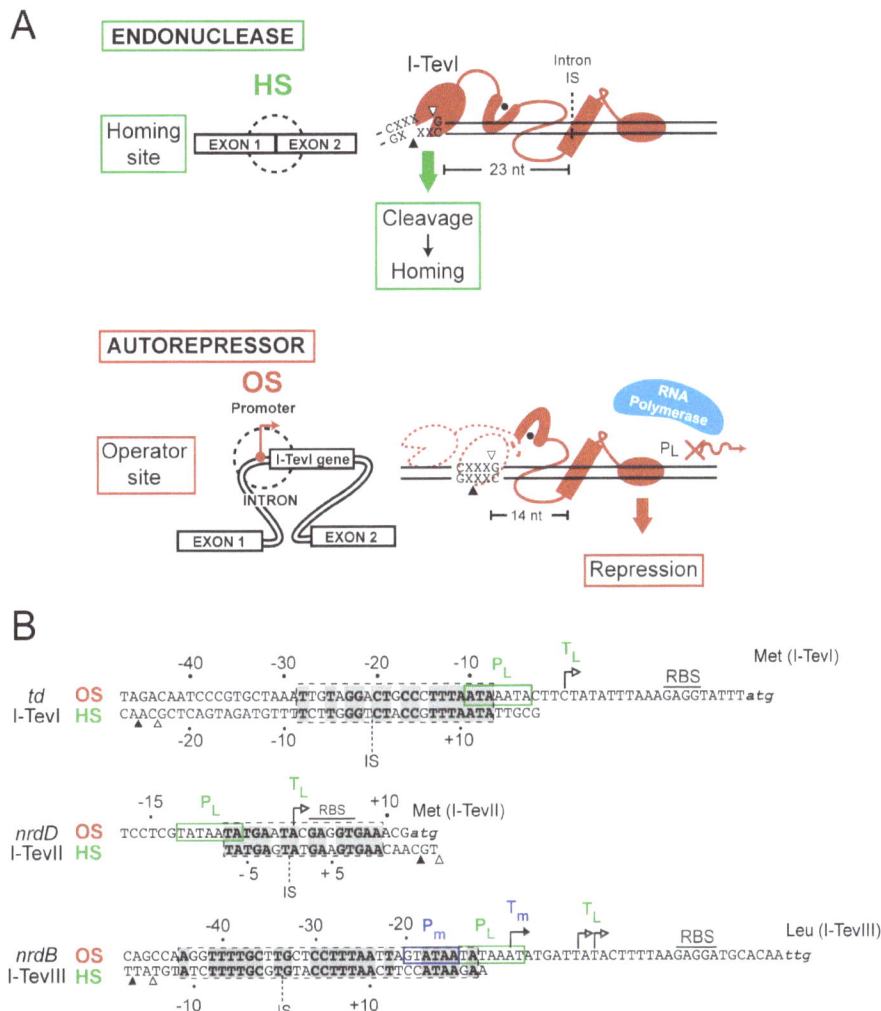

Figure 4 Dual function of I-TevI. A. I-TevI binds with equal affinity to the homing site (top) and operator site (bottom). The CS sequence at the natural distance in the homing site allows endonuclease cleavage, to initiate homing. In the operator site, there is no cleavage sequence at a suitable distance, resulting in autorepression, because I-TevI binding blocks the late promoter and transcription. B. Autorepression by T4 intron-encoded endonucleases. For each of the three endonucleases, I-TevI, I-TevII, and I-TevIII, the endonuclease's homing site (HS) is aligned with proven or putative the operator site (OS) upstream of the endonuclease ORF within the *td*, *nrdB*, and *nrdD* introns, respectively. The operator sites are indicated by dashed boxes, with bold-type nucleotides representing identity between the operator and homing sites. The position of the endonuclease's cleavage sites are indicated by open and black triangles. Green and blue boxes indicate late and middle T4 promoters, respectively, with corresponding transcription start sites indicated by right-facing arrows labeled with the same color.

the *uvsX* gene, consistent with its similarity to intron-encoded GIY-YIG endonucleases, which also generate 2-nt 3' extension [34]. Subsequent work demonstrated activity for a number of *seg* endonucleases (Table 1) [35-39]. Furthermore, *seg* endonucleases are inherited at high frequencies in the progeny of T4 and T2 mixed infections, showing that homing endonucleases could be mobile elements outside of a host group I intron or intein. The term intronless homing was coined to distinguish mobility of free-standing endonucleases from intron-encoded versions [35], and to emphasize one striking difference - the position of the enzyme's CS's

relative to the insertion site of the endonuclease gene. Unlike known intron-encoded endonucleases, which cleave within 25 bps of the intron insertion site, the CS of free-standing endonucleases are located hundreds or thousands of base pairs distant from the endonuclease gene. This separation of cleavage and insertion sites has important consequences for inheritance of a free-standing endonuclease, because exonucleolytic ressection of the DSB in the recipient genome must extend into regions of homology flanking the free-standing endonuclease in the donor genome in order to ensure that it is inherited in progeny phage [40]. Furthermore, sequencing of co-conversion tracts associated with mobility events of a number of free-standing endonucleases are consistent with DSB repair pathways [39,41].

Compared to the *seg* endonucleases, comparatively little is known about the five *mob* endonucleases of phage T4. Each of the *mob* genes consists of a well-defined HNH nuclease domain fused to a distinct C-terminal region, presumably the DNA-binding domain. Consistent with the presence of an HNH domain, both MobA and MobE nick one strand of their target substrates, and are inherited at high frequency in progeny of mixed infections [41,42]. Similar observations were made for I-HmuI and I-HmuII, HNH endonucleases encoded within group I introns of *Bacillus* phage [43-45]. One outstanding question regarding the HNH endonucleases is how does the introduction of a single-strand nick promote a recombination event. It is possible that nicks are converted to a recombinogenic DSB by collapse of a passing replication fork [46], or by subsequent processing of the nick by repair enzymes. However, persistent DSBs associated with endonuclease-generated nicks could not be detected by Southern blot analyses [43].

Regulation of homing endonuclease function
Transcriptional Regulation - Promoter choice
One potential impact on phage viability from an invading endonuclease stems from disruption of a coding sequence and another from perturbing the expression of host genes that neighbor the endonuclease insertion site by displacing existing promoters upon insertion. This is not a trivial concern, as many T4 promoters are located in intergenic regions, precisely the insertion sites of free-standing endonucleases [47]. Alternatively, an invading endonuclease can introduce additional promoters that enhance transcription of neighboring genes, or create antisense transcripts if the promoter is placed in the opposite transcriptional orientation to surrounding genes. Thus, in order to persist in the phage population, an invading homing endonuclease must successfully integrate into the host transcriptional program to minimize the impact on surrounding host genes.

The regulatory elements that govern expression of the three T4 intron-encoded endonucleases I-TevI, I-TevII, and I-TevIII were deduced soon after the discovery of the introns and endonucleases themselves [48]. Primer extension analyses located the middle and late promoters that drive expression of the three endonucleases, with the common theme that these transcripts are embedded within early or middle transcriptional units of the interrupted *td*, *nrdD*, and *nrdB* genes (Figure 1). In contrast, seven free-standing endonucleases appear to be promoter-less cassettes. In these cases, for *segC*, *segF*, *segG*, *mobA*, *mobC*, *mobD*, and *mobE*, the endonucleases harness upstream T4 promoters to become part of an existing polycistronic message. Because insertion of some of the *seg* and *mob* endonucleases displaced existing T4 transcription starts, promoters for downstream genes are embedded within the endonuclease's coding regions. For instance, the middle and late promoters that drive expression of the essential gene *32* are positioned within *segG* (formerly *32.1*) [49], with similar cases found for *segA*, *segE*, *mobB*, and *mobC*. This arrangement of embedded promoters favors retention of the endonuclease in the phage genome, because any deletion event that removed the essential host promoter would be detrimental to the phage.

Seven free-standing endonuclease genes (with one exception, *segD*) are all transcribed from promoters that lie in the non-coding regions upstream of the endonuclease ORF (*segB*, *segE*, and *mobB*), or in the 3' region of the gene immediately upstream of the endonuclease (*segA*, *segG*, *mobC* and *mobD*). Interestingly, these "endonuclease-specific" promoters are either middle (5 instances) or late (2 instances), with no occurrences of early promoters, suggesting that there is some advantage to expression of endonucleases >5 min post infection.

All of the homing endonuclease genes in T4 are present in the same transcriptional orientation as the surrounding genes, with one notable exception. The uncharacterized free-standing GIY-YIG endonuclease *segD* is oriented in the opposite transcriptional direction to the surrounding genes *23* and *24*, encoding the essential major capsid protein and vertex protein, respectively (Figure 1). This arrangement of *segD* with respect to the surrounding T4 genes is noteworthy in that bioinformatic searches failed to identify a *segD*-specific promoter. Thus, *segD* expression may depend on transcription events that initiate at either the late promoter upstream of the *inh* gene (~4.8 kb from *segD*) or one of two middle promoters upstream of or internal to *24.2* (~2.9 kb and ~2.4 kb from *segD*, respectively). Yet, transcripts initiated from these promoters would have to read through intrinsic transcriptional terminators up- and downstream of *segD*. Given the antisense orientation of *segD* and extensive transcriptional terminator in this region of the T4 genome, it is not unreasonable to assume that *segD* transcript levels are vanishingly low.

Intron-encoded endonucleases also function as transcriptional autorepressors
An added layer of transcriptional regulation was recently discovered for I-TevI [30]. In examining DNA sequence immediately upstream of the I-TevI ORF, strong similarity (15/20 nucleotide identity) was observed between a sequence that overlapped the late promoter that drives expression of I-TevI and the I-TevI homing site (Figure 4A &4B). A similar arrangement was also observed for I-TevII and I-TevIII, whereby potential binding sites

with similarity to the endonuclease's homing site over-lapped the middle or late promoters upstream of I-TevII and I-TevIII (Figure 4B). This arrangement of binding sites (operators) and promoters suggested that each of the T4 intron endonucleases also functioned as transcriptional autorepressors, regulating their expression by binding to operator sites to occlude the middle and late promoters from RNA polymerase. Indeed, I-TevI was shown to bind its operator site with the same affinity as its homing site, and functioned to downregulate expression of *lacZ* fused to the I-TevI late promoter during phage infection. Although not experimentally demonstrated for I-TevII or I-TevIII, it is likely that each endonuclease also functions like I-TevI to autoregulate its own expression.

One immediate question raised by the finding of an operator site was whether I-TevI cleaved the operator site with similar efficiency as its homing site, as I-TevI bound with similar affinity to the operator and homing sites. However, cleavage assays showed that I-TevI cleaved the operator site ~100-fold less efficiently than the homing site [30]. Reduction in I-TevI cleavage efficiency can be attributed to the lack of a critical 5'-CXXXG-3' sequence positioned appropriately to the I-TevI operator site (as described in section 2B). Interestingly, zinc finger mutants of I-TevI cleave the operator site more efficiently than the homing site substrate [31]. Zinc finger mutants, which have lost the ability to constrain cleavage to a fixed distance, can scan for a suboptimally placed 5'-CXXXG-3' sequence, which in the case of operator substrate lies at positions that would be equivalent to -15 through -19 of the homing site. Thus, the I-TevI zinc finger possesses two biological functions - to ensure that the enzyme cleaves at the optimal distance on homing site substrate to promote intron homing, and to prevent cleavage on the operator substrate to promote persistence of the *td* intron and I-TevI in the phage population (Figure 4).

Regulation by transcript processing

Transcriptional termination mediated by intrinsic or rho-independent terminators plays a key role in regulating the expression of T4 genes, and many intrinsic terminators have been computationally identified [47]. T4 terminators are very similar to *E. coli* intrinsic terminators, characterized by a GC-rich stem, a 4-nucleotide loop, and a poly(U) tract immediately downstream of the stem structure [50-52]. Interestingly, two free-standing endonucleases, *mobE* and *segF*, possess intrinsic transcriptional terminators in the 5' end of their coding regions [47,53]. The *mobE* endonuclease is inserted in the *nrdA/nrdB* intergenic region of a number of T-even phage, and its expression is dependent on promoters upstream of *nrdA*. The terminator internal to *mobE* was

predicted to be weak based on the length of the poly(U) tract [54], and RNase protection assays and mapping of 3' ends have shown that ~30% of transcripts terminate at the poly(U) tract that immediately follows the *mobE* terminator [42]. However, transcription of the essential *nrdB* gene downstream of *mobE* is not affected by the presence of the terminator in *mobE*, because a middle promoter is located in the intergenic space separating the 3' end of *mobE* and 5' end of *nrdB* [53]. One potential biological function of the *mobE* terminator is to limit read-through transcription from the *nrdA* promoter, modulating transcript levels of *nrdB* to coordinate synthesis of NrdB (the small subunit of aerobic ribonucleotide reductase) with that of NrdA (the large subunit) [53,55]. More speculatively, the terminator may also be a T4-specific adaptation to regulate *mobE* expression, reducing the amount of *mobE*-containing transcripts.

Similarly, post-transcriptional processing of T4 *segB* and *segG* by the host enzyme RNase E may be an adaptation to reduce endonuclease transcript levels [38,56]. An RNase E-like processing site was also described in the *mobE* transcript in the T-even-like phage Aeh1 that infects *Aeromonas hydrophila* [57]. For T4 *segB* and Aeh1 *mobE*, the extent and timing of RNase E processing is unknown, while for *segG*, RNase E processing has been shown to increase the stability of the downstream gene *32*, facilitating translation [57]. It should be noted, however, that RNase E processing does not appear to affect the ability of *segG* or *segB* to act as mobile elements, as both endonucleases are inherited at high frequency in the progeny of T4 × T2 co-infections [38,39].

Translational regulation - Involvement of RNA structures

The first hint that translational regulation was an important mechanism in the regulation of T4 homing endonucleases came from studies on the intron-encoded endonucleases I-TevI, I-TevII, and I-TevIII [48]. All three endonucleases possess a consensus Shine-Dalgarno sequence (or ribosome binding site, RBS) positioned approximately 8 nucleotides upstream of the AUG initation codon (Figure 5). However, a very stable RNA secondary structure sequesters the RBS such that translation would be very inefficient. For I-TevI, this RNA structure only forms on transcripts that initiate from early and middle promoters upstream of *td*, preventing translation of I-TevI at early and middle times during infection. A late promoter, immediately upstream of I-TevI, is positioned such that late transcripts do not include sufficient sequence to form the inhibitory RNA structure, freeing the RBS and facilitating translation of I-TevI at late times [30,48]. Similar arrangements of promoters and RNA secondary structures are found for I-TevII and I-TevIII, for T4 *segB*, and for *mobE* in phage Aeh1 (Figure 5) [30,38,48,57,58]. One departure from this mechanism of translational regulation

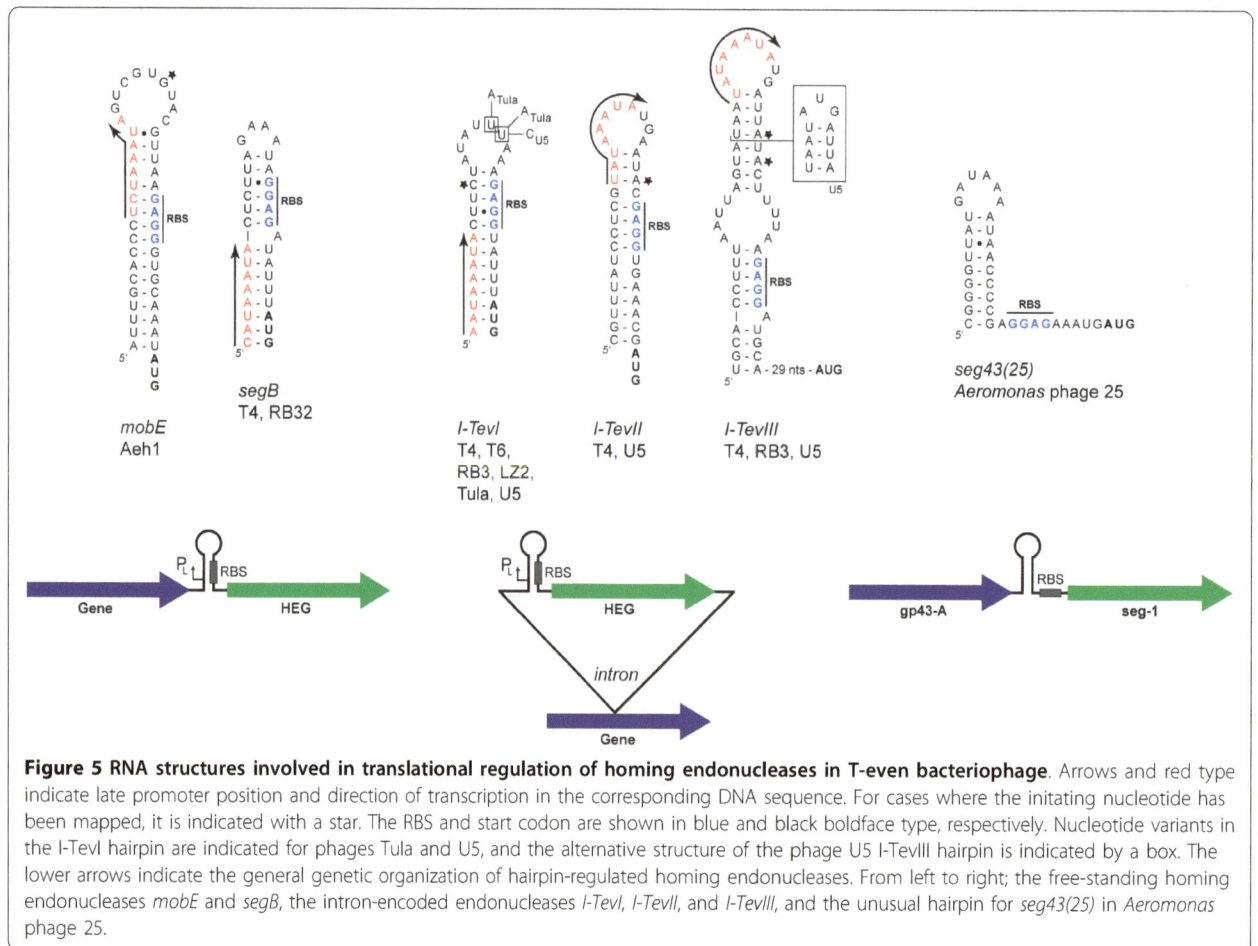

Figure 5 RNA structures involved in translational regulation of homing endonucleases in T-even bacteriophage. Arrows and red type indicate late promoter position and direction of transcription in the corresponding DNA sequence. For cases where the initating nucleotide has been mapped, it is indicated with a star. The RBS and start codon are shown in blue and black boldface type, respectively. Nucleotide variants in the I-TevI hairpin are indicated for phages Tula and U5, and the alternative structure of the phage U5 I-TevIII hairpin is indicated by a box. The lower arrows indicate the general genetic organization of hairpin-regulated homing endonucleases. From left to right; the free-standing homing endonucleases *mobE* and *segB*, the intron-encoded endonucleases *I-TevI*, *I-TevII*, and *I-TevIII*, and the unusual hairpin for *seg43(25)* in *Aeromonas* phage 25.

is found for a *segD*-like endonuclease, *seg43(25)*, in *Aeromonas* phage 25 [59]. Here, a predicted RNA hairpin with very high stability folds immediately upstream of the RBS for *seg43(25)*, but the hairpin does not sequester the RBS that is immediately downstream from the base of the stem (Figure 5). It remains to be determined if this arrangement results in translational regulation.

Translational repression by sequestration of the RBS by an RNA structure is not unique to homing endonucleases, and has been shown for T4 genes including *soc*, *e*, *39*, *25* and *26* [60-63]. Interestingly, one commonality shared by homing endonucleases and T4 genes regulated by this mechanism is the fact that they are all late genes, and present on long polycistronic transcripts that encode early and middle gene products. Such translational regulation of late gene products may represent a mechanism to temporally orchestrate translation of gene products on polycistronic messages.

Other potential translational regulation mechanisms
For the remaining free-standing homing endonucleases in phage T4, the translational control mechanisms are not obvious, and have yet to be addressed

experimentally. The free-standing endonuclease ORFs start with an AUG initiation codon, the lone exception being *mobA* that is predicted to start at a GUG codon. In addition, translation initiation regions (TIRs) that are a reasonable match to the T4 consensus can be identified upstream of only six of the twelve endonuclease genes (for *segB*, *segC*, *segD*, *segE*, *mobA*, and *mobC*). However, the RBSs are not positioned at the optimal distance of 6-9 nucleotides from the AUG codon. For instance, the predicted *segD* RBS is 2 nucleotides upstream of the AUG codon, whereas the *mobC* RBS is 27 nucleotodies upstream, questioning whether or not these sequences represent *bone-fide* translation start sites.

Six of the twelve free-standing endonucleases have no discernable TIRs, with five of the endonuclease ORFs overlapping the upstream ORFs. In the case of *mobE*, the AUG iniation codon of *mobE* overlaps by one nucleotide with the first of two termination codons of the upstream *nrdA* gene, creating the following arrangement UAAUGA. This arrangement of overlapping initiation and termination codons is suggestive of translational coupling, a mechanism where termination of the upstream protein is linked to initiation of translation the downstream protein [64]. Such arrangements are thought to provide a mechanism to control the relative amounts of protein products that function in the same biological processes. For instance, translational coupling regulates the production of the clamp loader proteins gp46 and gp62 [65-67]. In most cases, translational coupling results in a lower relative amount of the downstream gene product to the upstream gene product, likely due to the reduced frequency of translation re-initiation at an internal AUG codon. Thus, if translational coupling is the mechanism by which MobE is expressed, the *nrdA/mobE* overlap may represent a mechanism to limit translation of MobE endonuclease, in addition to the aforementioned transcriptional termination in *mobE*.

In other cases, the extent of overlap with the upstream ORF is more extreme, as evidenced for the homing endonucleases *segC* (20-bp overlap), *segF* (11-bp overlap), *mobA* (23-bp overlap), and *mobC* (24-bp overlap). Here, the mechanism of translational control is likely to be ribosome 'scanning', whereby the ribosome does not dissociate and diffuses along the mRNA within a seven-codon window up- or down-stream of the termination codon [68]. If an AUG codon is encountered within this window, translation is initiated but at a low frequency, resulting in lower protein levels of the downstream endonuclease relative to the upstream protein.

Why regulation?

Clearly, all of the mechanisms described above are negative regulatory mechanisms that function to downregulate the levels of homing endonucleases in T4-infected cells, suggesting that unregulated expression of endonucleases would be detrimental to T4. One obvious rationale for downregulating endonuclease function relates to the sequence-tolerant binding ability of homing endonucleases, and the potential for introducing (presumably) deleterious nicks and DSBs at ectopic sites throughout the T4 genome. Curiously, certain endonucleases, namely the intron-encoded endonucleases and *mobE* in phage Aeh1, are subject to multiple-layers of regulation, whereas other homing endonucleases are not as tightly regulated. Whether the intron-encoded endonucleases and *mobE* are more 'toxic' and require multi-layered regulation is an interesting possibility that requires experimental confirmation. Intriguingly, Kruezer and co-workers generated a T4 phage where the I-TevI operator site was deleted and replaced by a middle promoter, with no noticeable effect on phage viability [69]. Similarly, no endonuclease-mediated effect on phage viability was observed when the regulatory hairpin structure limiting translation of I-TevI to late in phage infection was deleted, facilitating translation of I-TevI at middle time points after phage infection [70].

However, recent work suggests that the stringent regulation of I-TevI is required to facilitate efficient splicing of the *td* intron and translation of full-length thymidylate synthase [70]. An interesting observation regarding I-TevI, and many other intron-encoded endonucleases, is that while the majority of the I-TevI ORF is located in a non-essential loop of the *td* intron, the 3' end of the ORF extends into the structured region of the intron and contributes key nucleotides that form critical secondary structures [71]. This observation led to the suggestion that translation of intron-encoded endonucleases from within the highly structured intron may interfere with intron folding and splicing [48]. Indeed, T4 phage mutant for the I-TevI translational regulatory hairpin exhibited a significant decrease in *td* intron splicing, an accumulation of unspliced *td* pre-mRNA, and a thymidine-dependent phenotype [70]. These observations suggest that one biological rationale for stringent, multi-layered regulation of intron-encoded endonucleases is a requirement to limit ribosome access to the intron core, ensuring proper splicing of the intron and function of the host gene interrupted by the intron [72,73].

One interesting commonality among all the regulatory mechanisms is the restriction of endonuclease function to middle or late times in the phage infective cycle [48,74]. We previously argued that such temporal regulation coordinates expression of the endonucleases with the DNA repair and replication machinery of phage T4, ensuring that the appropriate machinery and genome equivalents are present to repair endonuclease-mediated breaks and promote homing [30]. Delayed expression may also be a means to coincide homing endonuclease synthesis with recombination-dependent replication.

Evolution of Homing Endonucleases
Phage genomes as hosts for homing endonucleases

Phage T4 is an oddity among T-even phage, encoding 15 homing endonucleases. We pondered the significance of this observation in 2000, suggesting that the number of completely sequenced T-even phage genomes was too few to say anything definitive about endonuclease distribution [75]. In the intervening years, many more phage genomes have been sequenced [47,76-80], yet the trend holds - T4 remains the outlier, as most genomes have few endonuclease insertions. Based on these observations, it is tempting to conclude that there exists significant evolutionary pressure for phage to resist colonization by homing endonucleases. This conclusion, however, is at odds with the *in vitro* and *in vivo* characterization of endonuclease-mobility pathways that rely on extremely efficient DNA repair and recombination pathways to promote dissemination of homing endonucleases through populations of phage lacking them. Our understanding of factors that influence endonuclease mobility and retention in phage genomes is still in its infancy, but studies with eukaryotic homing endonuclease systems may provide some clues. In particular, studies on intron-encoded yeast homing endonucleases have elucidated a cyclical life cycle that allows homing endonuclease genes to escape degeneration and deletion due to lack of intron-less alleles for homing [81,82]. Related endonuclease life cycles have been proposed for transposition of homing endonucleases to new sites within a phage genome, allowing the endonuclease to escape deletion [41,83]. Moreover, recent modeling studies suggests that homing endonuclease genes could persist for significant time frames in the absence of homing sites and selection for a functional homing endonuclease [84].

One unifying characteristic of phage-encoded homing endonucleases is the observation that most endonuclease genes are inserted within or near phage genes that are functionally critical, such as DNA polymerases and ribonucleotide reductases. Targeting of functionally critical phage genes by homing endonucleases is an evolutionary strategy to maximize spread because very similar genes and target sites will be present in related genomes [75], and to minimize loss (see below). Moreover, the recognition sites of many homing endonucleases often encompass nucleotide sequence that corresponds to functionally critical amino acid (or RNA) residues of the host gene, often encoding an active site or essential region of the host gene [85,86]. It is also the case that homing endonucleases of different classes will target the same gene. For instance, the *nrdB* gene encoding the small subunit of aerobic ribonucleotide reductase is cleaved by both *mobE*, an HNH endonuclease of phage T4 [42], and by a unique endonuclease, *hef* (<u>h</u>oming <u>e</u>ndonuclease-like <u>f</u>unction), encoded in phage U5 [41]. Similarly, the anaerobic ribonucleotide reductase proteins, encoded by the *nrdD* and *nrdG* genes, are targeted by *seg* and *mob* homing endonucleases [41]. Given the relatively small size of phage genomes, and the fact that many phage genes are essential, it is not surprising that similar target sequences have independently been selected as recognition and cleavage sites by different classes of homing endonucleases.

The insertion sites of many self-splicing group I introns also correspond to functionally critical sequences in phage genomes. Insertion of the intron into a functionally critical region is thought to prevent deletion of the element from the phage genome, as only a precise deletion of the intron or intein will restore a functional host gene sequence, whereas an imprecise deletion would likely be lethal. The propensity for homing endonucleases and introns to target conserved sequences forms the core of a recently proposed evolutionary scenario termed collaborative homing, for the origin of mobile introns by recombination between an endonuclease-lacking intron

and a free-standing endonuclease that is "pre-adapted" to target the intron insertion site of the endonuclease-lacking intron, creating a highly efficient composite mobile genetic element [87,88]. The very similar *trans* homing pathway involves a free-standing homing endonuclase, *mobE*, mobilizing the defunct I-TevIII endonuclease and *nrdB* intron in phage T4 [42].

Impact of homing endonucleases on phage genome structure and function

Because homing endonucleases utilize DNA repair and recombination pathways to promote mobility, significant co-conversion of sequence flanking the endonuclease's cleavage site sequence is associated with endonuclease-mediated mobility [14,18,39]. This observation helps explain a long-known phenomenon of T-even phage biology, namely the exclusion of T2 markers from progeny of a T2 and T4 coinfection [89]. Marker exclusion was first described in 1974, but it was not until almost 30 years later that a link between homing endonucleases and marker exclusion was uncovered [35,36]. Strikingly, the recognition and cleavage sites of many characterized T4-encoded homing endonucleases correspond to sites in the T2 genome that are excluded from the progeny of a T2 and T4 coinfection. Cleavage of T2 by a T4-encoded endonuclease initiates a localized gene conversion event at the cleavage site that replaces T2 with T4 sequence, resulting in the exclusion of T2 markers from progeny. A similar marker exclusion phenomenon involving intron-encoded endonucleases was also observed in HMU phage of *Bacillus subtilis* [43,90]. Thus, homing endonucleases influence the distribution of sequences flanking their insertion site within populations of related phage, in essence promoting lateral gene transfer.

More dramatic effects on phage gene structure and function arise from what appear to be homing endonuclease transposition events, whereby an endonuclease gene has inserted into a site that is different from the insertion site of analogous homing endonucleases in related phage genomes. Such transposition-like insertions include the *mobE* insertion into *nrdA* large subunit gene of aerobic ribonucleotide reductase of *Aeromonas hydrophila* phage Aeh1 [91], the *mobA* insertion into the topoisomerase large subunit gene *60* of phage T4 [47], and a *seg43(25)* insertion associated with gene *43*, encoding a B-type DNA polymerase of *Aeromonas* phages 25 [88]. In the *mobE* and *mobA* cases, the homing endonuclease has inserted into a functionally critical region of the host gene, splitting the gene into separate coding regions as compared to related phage. For the *seg*-like insertion of phage 25, it is difficult to ascertain whether the split gene *43* structure arose by insertion of the *seg* homing endonuclease, because the related phage 44RR possess a different genetic arrangement, consisting of an intercistronic untranslated sequence (IC-UTS) that splits gene *43* into *43A* and *43B* [88]. Thus it is possible that the *seg* endonuclease invaded an already split gene created by the insertion of the IC-UTS.

Regardless of their origins, each insertion splits a contiguous coding region into two distinct polypeptides that must somehow reassemble to form a functional enzyme. Recent work has shown that the split *nrdA* gene of phage Aeh1 encodes a fully functional aerobic ribonucleotide reductase with activity similar to canonical enzymes that consist of a single NrdA polypeptide [91]. Similarly, the split *43A* and *43B* genes of phage 25 co-purify when overexpressed, and possess DNA polymerase activity [88]. Although the *mobA* insertion has not been studied in detail, phage T4 topoisomerase has long served as a model enzyme and possesses an unusual subunit structure with respect to other phage-encoded and bacterial topoisomerases [92], consistent with assembly of the split topoisomerase polypeptides to form a functional enzyme. How the split polypeptides assemble to form functional complexes in each of the enzyme systems is a fascinating structure and function question.

Conclusion

T-even phage have proven to be an attractive and tractable model system for studying the biology of homing endonucleases in the last 20 years, and we have learned much about the molecular details of mobility pathways and regulatory mechanisms. Many of these details are applicable to homing endonucleases in eukaryotic systems, and also have provided insight into the mobility pathways of other mobile elements such as inteins and group II introns. From a mechanistic perspective, how the *mob* endonucleases spread between genomes by nicking their target sites rather than introducing a double-strand break is an intriguing area of future research. From a genomic perspective, the relatively small size of phage genomes coupled with extraordinary advances in sequencing technology has revealed that homing endonuclease genes are widespread, but not as abundant as predicted based on laboratory experiments. It remains to be determined if more phage genome sequences can offer insight into evolutionary processes that regulate homing endonuclease distribution, as it already clear from existing sequences that complex regulatory mechanisms have evolved to control the expression of homing endonucleases. Clearly, there are interesting evolutionary forces at work, and experimentally manipulating regulatory controls will likely be required to understand the impact of homing endonuclease activity on phage genome structure and function.

Acknowledgements
The authors would like to acknowledge David Shub for thoughtful discussion. Funded by a CIHR Operating Grant MOP77779 and a NSERC Discovery Grant 311610-2005 to D.R.E, and by NIH Grants GM39422 and GM448844 to M.B.

Author details
[1]Department of Biochemistry, Schulich School of Medicine & Dentistry, The University of Western Ontario, London, ON, N6A 5C1, Canada. [2]Wadsworth Center, New York State Department of Health, Center for Medical Sciences, 150 New Scotland Ave., Albany, NY 12208, USA.

Authors' contributions
DRE, EAG and MB wrote the manuscript. All authors read and approved the final manuscript.

Competing interests
The authors declare that they have no competing interests.

Received: 26 May 2010 Accepted: 28 October 2010
Published: 28 October 2010

References
1. Belfort M: Phage T4 introns: self-splicing and mobility. *Annu Rev Genet* 1990, **24**:363-385.
2. Huang YJ, Parker MM, Belfort M: Role of exonucleolytic degradation in group I intron homing in phage T4. *Genetics* 1999, **153**:1501-1512.
3. Mueller JE, Clyman J, Huang YJ, Parker MM, Belfort M: Intron mobility in phage T4 occurs in the context of recombination-dependent DNA replication by way of multiple pathways. *Genes Dev* 1996, **10**:351-364.
4. Parker MM, Belisle M, Belfort M: Intron homing with limited exon homology. Illegitimate double-strand-break repair in intron acquisition by phage T4. *Genetics* 1999, **153**:1513-1523.
5. Parker MM, Court DA, Preiter K, Belfort M: Homology requirements for double-strand break-mediated recombination in a phage lambda-td intron model system. *Genetics* 1996, **143**:1057-1068.
6. Bryk M, Quirk SM, Mueller JE, Loizos N, Lawrence C, Belfort M: The *td* intron endonuclease I-*Tev*I makes extensive sequence-tolerant contacts across the minor groove of its DNA target. *EMBO J* 1993, **12**:2141-2149.
7. Loizos N, Silva GH, Belfort M: Intron-encoded endonuclease I-*Tev*II binds across the minor groove and induces two distinct conformational changes in its DNA substrate. *J Mol Biol* 1996, **255**:412-424.
8. Van Roey P, Waddling CA, Fox KM, Belfort M, Derbyshire V: Intertwined structure of the DNA-binding domain of intron endonuclease I-*Tev*I with its substrate. *EMBO J* 2001, **20**:3631-3637.
9. Sharma M, Ellis RL, Hinton DM: Identification of a family of bacteriophage T4 genes encoding proteins similar to those present in group I introns of fungi and phage. *Proc Natl Acad Sci USA* 1992, **89**:6658-6662.
10. Kutter E, Gachechiladze K, Poglazov A, Marusich E, Shneider M, Aronsson P, Napuli A, Porter D, Mesyanzhinov V: Evolution of T4-related phages. *Virus Genes* 1995, **11**:285-297.
11. Pedersen-Lane J, Belfort M: Variable occurrence of the nrdB intron in the T-even phages suggests intron mobility. *Science* 1987, **237**:182-184.
12. Quirk SM, Bell-Pedersen D, Tomaschewski J, Ruger W, Belfort M: The inconsistent distribution of introns in the T-even phages indicates recent genetic exchanges. *Nucleic Acids Res* 1989, **17**:301-315.
13. Quirk SM, Bell-Pedersen D, Belfort M: Intron mobility in the T-even phages: high frequency inheritance of group I introns promoted by intron open reading frames. *Cell* 1989, **56**:455-465.
14. Bell-Pedersen D, Quirk SM, Aubrey M, Belfort M: A site-specific endonuclease and co-conversion of flanking exons associated with the mobile *td* intron of phage T4. *Gene* 1989, **82**:119-126.
15. Zinn AR, Butow RA: Nonreciprocal exchange between alleles of the yeast mitochondrial 21S rRNA gene: Kinetics and involvement of a double-strand break. *Cell* 1985, **40**:887-895.
16. Szostak JW, Orr-Weaver TL, Rothstein RJ, Stahl FW: The double-strand-break repair model for recombination. *Cell* 1983, **33**:25-35.
17. Clyman J, Belfort M: *Trans* and *cis* requirements for intron mobility in a prokaryotic system. *Genes Dev* 1992, **6**:1269-1279.

18. Mueller JE, Smith D, Belfort M: Exon coconversion biases accompanying intron homing: battle of the nucleases. *Genes Dev* 1996, **10**:2158-2166.

19. Eddy SR, Gold L: The phage T4 *nrdB* intron: a deletion mutant of a version found in the wild. *Genes Dev* 1991, **5**:1032-1041.

20. Robbins JB, Stapleton M, Stanger MJ, Smith D, Dansereau JT, Derbyshire V, Belfort M: Homing endonuclease I-TevIII: dimerization as a means to a double-strand break. *Nucleic Acids Res* 2007, **35**:1589-1600.

21. Van Roey P, Derbyshire V: GIY-YIG homing endonucleases - Beads on a String. In *Homing Endonucleases and Inteins*. Edited by: Belfort MSB, Wood DW, Derbyshire V. Berlin: Springer-Verlag; 2005:67-83.

22. Shen BW, Landthaler M, Shub DA, Stoddard BL: DNA binding and cleavage by the HNH homing endonuclease I-Hmul. *J Mol Biol* 2004, **342**:43-56.

23. Michel F, Dujon B: Genetic exchanges between bacteriophage T4 and filamentous fungi? *Cell* 1986, **46**:323.

24. Dunin-Horkawicz S, Feder M, Bujnicki JM: Phylogenomic analysis of the GIY-YIG nuclease superfamily. *BMC Genomics* 2006, **7**:98.

25. Bryk M, Belisle M, Mueller JE, Belfort M: Selection of a remote cleavage site by I-TevI, the *td* intron-encoded endonuclease. *J Mol Biol* 1995, **247**:197-210.

26. Kowalski JC, Belfort M, Stapleton MA, Holpert M, Dansereau JT, Pietrokovski S, Baxter SM, Derbyshire V: Configuration of the catalytic GIY-YIG domain of intron endonuclease I-TevI: coincidence of computational and molecular findings. *Nucleic Acids Res* 1999, **27**:2115-2125.

27. Mueller JE, Smith D, Bryk M, Belfort M: Intron-encoded endonuclease I-TevI binds as a monomer to effect sequential cleavage via conformational changes in the *td* homing site. *EMBO J* 1995, **14**:5724-5735.

28. Bell-Pedersen D, Quirk SM, Bryk M, Belfort M: I-TevI, the endonuclease encoded by the mobile *td* intron, recognizes binding and cleavage domains on its DNA target. *Proc Natl Acad Sci USA* 1991, **88**:7719-7723.

29. Liu Q, Dansereau JT, Puttamadappa SS, Shekhtman A, Derbyshire V, Belfort M: Role of the interdomain linker in distance determination for remote cleavage by homing endonuclease I-TevI. *J Mol Biol* 2008, **379**:1094-1106.

30. Edgell DR, Derbyshire V, Van Roey P, LaBonne S, Stanger MJ, Li Z, Boyd TM, Shub DA, Belfort M: Intron-encoded homing endonuclease I-TevI also functions as a transcriptional autorepressor. *Nat Struct Mol Biol* 2004, **11**:936-944.

31. Liu Q, Derbyshire V, Belfort M, Edgell DR: Distance determination by GIY-YIG intron endonucleases: discrimination between repression and cleavage functions. *Nucleic Acids Res* 2006, **34**:1755-1764.

32. Dean AB, Stanger MJ, Dansereau JT, Van Roey P, Derbyshire V, Belfort M: Zinc finger as distance determinant in the flexible linker of intron endonuclease I-TevI. *Proc Natl Acad Sci USA* 2002, **99**:8554-8561.

33. Jakob U, Eser M, Bardwell JC: Redox switch of hsp33 has a novel zinc-binding motif. *J Biol Chem* 2000, **275**:38302-38310.

34. Sharma M, Hinton DM: Purification and characterization of the SegA protein of bacteriophage T4, an endonuclease related to proteins encoded by group I introns. *J Bacteriol* 1994, **176**:6439-6448.

35. Belle A, Landthaler M, Shub DA: Intronless homing: site-specific endonuclease SegF of bacteriophage T4 mediates localized marker exclusion analogous to homing endonucleases of group I introns. *Genes Dev* 2002, **16**:351-362.

36. Kadyrov FA, Shlyapnikov MG, Kryukov VM: A phage T4 site-specific endonuclease, SegE, is responsible for a non-reciprocal genetic exchange between T-even-related phages. *FEBS Lett* 1997, **415**:75-80.

37. Shcherbakov V, Granovsky I, Plugina L, Shcherbakova T, Sizova S, Pyatkov K, Shlyapnikov M, Shubina O: Focused genetic recombination of bacteriophage t4 initiated by double-strand breaks. *Genetics* 2002, **162**:543-556.

38. Brok-Volchanskaya VS, Kadyrov FA, Sivogrivov DE, Kolosov PM, Sokolov AS, Shlyapnikov MG, Kryukov VM, Granovsky IE: Phage T4 SegB protein is a homing endonuclease required for the preferred inheritance of T4 tRNA gene region occurring in co-infection with a related phage. *Nucleic Acids Res* 2008, **36**:2094-2105.

39. Liu Q, Belle A, Shub DA, Belfort M, Edgell DR: SegG endonuclease promotes marker exclusion and mediates co-conversion from a distant cleavage site. *J Mol Biol* 2003, **334**:13-23.

40. Edgell DR: Free-standing endonucleases of T-even phages: Free-loaders or functionaries? In *Homing endonucleases and Inteins*. Edited by Belfort M, Stoddard BL, Wood DW, Derbyshire V. Springer-Verlag; 2005:147-160.

41. Sandegren L, Nord D, Sjöberg BM: SegH and Hef: two novel homing endonucleases whose genes replace the *mobC* and *mobE* genes in several T4-related phages. *Nucleic Acids Res* 2005, **33**:6203-6213.

42. Wilson GW, Edgell DR: Phage T4 *mobE* promotes *trans* homing of the defunct homing endonuclease I-TevIII. *Nucleic Acids Res* 2009, **37**:7110-7123.

43. Goodrich-Blair H, Shub DA: Beyond homing: competition between intron endonucleases confers a selective advantage on flanking genetic markers. *Cell* 1996, **84**:211-221.

44. Landthaler M, Shub DA: The nicking homing endonuclease I-BasI is encoded by a group I intron in the DNA polymerase gene of the *Bacillus thuringiensis* phage Bastille. *Nucleic Acids Res* 2003, **31**:3071-3077.

45. Landthaler M, Shen BW, Stoddard BL, Shub DA: I-BasI and I-Hmul: two phage intron-encoded endonucleases with homologous DNA recognition sequences but distinct DNA specificities. *J Mol Biol* 2006, **358**:1137-1151.

46. Kuzminov A: Single-strand interruptions in replicating chromosomes cause double-strand breaks. *Proc Natl Acad Sci USA* 2001, **98**:8241-8246.

47. Miller ES, Kutter E, Mosig G, Arisaka F, Kunisawa T, Ruger W: Bacteriophage T4 genome. *Microbiol Mol Biol Rev* 2003, **67**:86-156.

48. Gott JM, Zeeh A, Bell-Pedersen D, Ehrenman K, Belfort M, Shub DA: Genes within genes: independent expression of phage T4 intron open reading frames and the genes in which they reside. *Genes Dev* 1988, **2**:1791-1799.

49. Loayza D, Carpousis AJ, Krisch HM: Gene *32* transcription and mRNA processing in T4-related bacteriophages. *Mol Microbiol* 1991, **5**:715-725.

50. Artsimovitch I, Landick R: Pausing by bacterial RNA polymerase is mediated by mechanistically distinct classes of signals. *Proc Natl Acad Sci USA* 2000, **97**:7090-7095.

51. Gusarov I, Nudler E: The mechanism of intrinsic transcription termination. *Mol Cell* 1999, **3**:495-504.

52. Yarnell WS, Roberts JW: Mechanism of intrinsic transcription termination and antitermination. *Science* 1999, **284**:611-615.

53. Tseng MJ, He P, Hilfinger JM, Greenberg GR: Bacteriophage T4 *nrdA* and *nrdB* genes, encoding ribonucleotide reductase, are expressed both separately and coordinately: characterization of the *nrdB* promoter. *J Bacteriol* 1990, **172**:6323-6332.

54. Christie GE, Farnham PJ, Platt T: Synthetic sites for transcription termination and a functional comparison with tryptophan operon termination sites in vitro. *Proc Natl Acad Sci USA* 1981, **78**:4180-4184.

55. Tseng MJ, Hilfinger JM, Walsh A, Greenberg GR: Total sequence, flanking regions, and transcripts of bacteriophage T4 *nrdA* gene, coding for alpha chain of ribonucleoside diphosphate reductase. *J Biol Chem* 1988, **263**:16242-16251.

56. Carpousis AJ, Mudd EA, Krisch HM: Transcription and messenger RNA processing upstream of bacteriophage T4 gene *32*. *Mol Gen Genet* 1989, **219**:39-48.

57. Gibb EA, Edgell DR: Multiple Controls Regulate the Expression of *mobE*, an HNH Homing Endonuclease Gene Embedded within a Ribonucleotide Reductase Gene of Phage Aeh1. *J Bacteriol* 2007, **189**:4648-4661.

58. Gibb EA, Edgell DR: An RNA Hairpin Sequesters the Ribosome Binding Site of Homing Endonuclease *mobE* Gene. *J Bacteriol* 2009, **191**:2409-2413.

59. Petrov VM, Nolan JM, Bertrand C, Levy D, Desplats C, Krisch HM, Karam JD: Plasticity of the gene functions for DNA replication in the T4-like phages. *J Mol Biol* 2006, **361**:46-68.

60. Barth KA, Powell D, Trupin M, Mosig G: Regulation of two nested proteins from gene 49 (recombination endonuclease VII) and of a lambda RexA-like protein of bacteriophage T4. *Genetics* 1988, **120**:329-343.

61. Macdonald PM, Kutter E, Mosig G: Regulation of a bacteriophage T4 late gene, soc, which maps in an early region. *Genetics* 1984, **106**:17-27.

62. McPheeters DS, Christensen A, Young ET, Stormo G, Gold L: Translational regulation of expression of the bacteriophage T4 lysozyme gene. *Nucleic Acids Res* 1986, **14**:5813-5826.

63. Gruidl ME, Chen TC, Gargano S, Storlazzi A, Cascino A, Mosig G: Two bacteriophage T4 base plate genes (*25* and *26*) and the DNA repair gene uvsY belong to spatially and temporally overlapping transcription units. *Virology* 1991, **184**:359-369.

64. Stahl FW, Crasemann JM, Yegian C, Stahl MM, Nakata A: Co-transcribed cistrons in bacteriophage T4. *Genetics* 1970, **64**:157-170.

65. Torgov MY, Janzen DM, Reddy MK: **Efficiency and frequency of translational coupling between the bacteriophage T4 clamp loader genes.** *J Bacteriol* 1998, **180**:4339-4343.

66. Trojanowska M, Miller ES, Karam J, Stormo G, Gold L: **The bacteriophage T4 *regA* gene: primary sequence of a translational repressor.** *Nucleic Acids Res* 1984, **12**:5979-5993.

67. Karam J, Bowles M, Leach M: **Expression of bacteriophage T4 genes *45*, *44*, and *62*. I. Discoordinate synthesis of the T4 45- and 44-proteins.** *Virology* 1979, **94**:192-203.

68. Adhin MR, van Duin J: **Scanning model for translational reinitiation in eubacteria.** *J Mol Biol* 1990, **213**:811-818.

69. Tomso DJ, Kreuzer KN: **Double-strand break repair in tandem repeats during bacteriophage T4 infection.** *Genetics* 2000, **155**:1493-1504.

70. Gibb EA, Edgell DR: **Better late than early: delayed translation of intron-encoded endonuclease I-TevI is required for efficient splicing of its host group I intron.** *Mol Microbiol* 2010, **78**:35-46.

71. Shub DA, Gott JM, Xu MQ, Lang BF, Michel F, Tomaschewski J, Pedersen-Lane J, Belfort M: **Structural conservation among three homologous introns of bacteriophage T4 and the group I introns of eukaryotes.** *Proc Natl Acad Sci USA* 1988, **85**:1151-1155.

72. Ohman-Heden M, Ahgren-Stalhandske A, Hahne S, Sjoberg BM: **Translation across the 5'-splice site interferes with autocatalytic splicing.** *Mol Microbiol* 1993, **7**:975-982.

73. Semrad K, Schroeder R: **A ribosomal function is necessary for efficient splicing of the T4 phage thymidylate synthase intron *in vivo*.** *Genes Dev* 1998, **12**:1327-1337.

74. Luke K, Radek A, Liu X, Campbell J, Uzan M, Haselkorn R, Kogan Y: **Microarray analysis of gene expression during bacteriophage T4 infection.** *Virology* 2002, **299**:182-191.

75. Edgell DR, Belfort M, Shub DA: **Barriers to intron promiscuity in bacteria.** *J Bacteriol* 2000, **182**:5281-5289.

76. Nolan JM, Petrov V, Bertrand C, Krisch HM, Karam JD: **Genetic diversity among five T4-like bacteriophages.** *Virol J* 2006, **3**:30.

77. Mann NH, Clokie MR, Millard A, Cook A, Wilson WH, Wheatley PJ, Letarov A, Krisch HM: **The genome of S-PM2, a "photosynthetic" T4-type bacteriophage that infects marine Synechococcus strains.** *J Bacteriol* 2005, **187**:3188-3200.

78. Miller ES, Heidelberg JF, Eisen JA, Nelson WC, Durkin AS, Ciecko A, Feldblyum TV, White O, Paulsen IT, Nierman WC, *et al*: **Complete genome sequence of the broad-host-range vibriophage KVP40: comparative genomics of a T4-related bacteriophage.** *J Bacteriol* 2003, **185**:5220-5233.

79. Desplats C, Dez C, Tetart F, Eleaume H, Krisch HM: **Snapshot of the genome of the pseudo-T-even bacteriophage RB49.** *J Bacteriol* 2002, **184**:2789-2804.

80. Crutz-Le Coq AM, Cesselin B, Commissaire J, Anba J: **Sequence analysis of the lactococcal bacteriophage bIL170: insights into structural proteins and HNH endonucleases in dairy phages.** *Microbiology* 2002, **148**:985-1001.

81. Goddard MR, Greig D, Burt A: **Outcrossed sex allows a selfish gene to invade yeast populations.** *Proc R Soc Lond B Biol Sci* 2001, **268**:2537-2542.

82. Goddard MR, Burt A: **Recurrent invasion and extinction of a selfish gene.** *Proc Natl Acad Sci USA* 1999, **96**:13880-13885.

83. Loizos N, Tillier ER, Belfort M: **Evolution of mobile group I introns: recognition of intron sequences by an intron-encoded endonuclease.** *Proc Natl Acad Sci USA* 1994, **91**:11983-11987.

84. Yahara K, Fukuyo M, Sasaki A, Kobayashi I: **Evolutionary maintenance of selfish homing endonuclease genes in the absence of horizontal transfer.** *Proc Natl Acad Sci USA* 2009, **106(1)**:8861-18866.

85. Edgell DR, Stanger MJ, Belfort M: **Coincidence of cleavage sites of intron endonuclease I-TevI and critical sequences of the host thymidylate synthase gene.** *J Mol Biol* 2004, **343**:1231-1241.

86. Scalley-Kim M, McConnell-Smith A, Stoddard BL: **Coevolution of a homing endonuclease and its host target sequence.** *J Mol Biol* 2007, **372**:1305-1319.

87. Bonocora RP, Shub DA: **A likely pathway for formation of mobile group I introns.** *Curr Biol* 2009, **19**:223-228.

88. Petrov VM, Ratnayaka S, Karam JD: **Genetic insertions and diversification of the PolB-type DNA polymerase (gp43) of T4-related phages.** *J Mol Biol* 2010, **395**:457-474.

89. Russell RL, Huskey RJ: **Partial exclusion between T-even bacteriophages: an incipient genetic isolation mechanism.** *Genetics* 1974, **78**:989-1014.

90. Landthaler M, Lau NC, Shub DA: **Group I intron homing in *Bacillus* phages SPO1 and SP82: a gene conversion event initiated by a nicking homing endonuclease.** *J Bacteriol* 2004, **186**:4307-4314.

91. Friedrich NC, Torrents E, Gibb EA, Sahlin M, Sjöberg BM, Edgell DR: **Insertion of a homing endonuclease creates a genes-in-pieces ribonucleotide reductase that retains function.** *Proc Natl Acad Sci USA* 2007, **104**:6176-6181.

92. Kreuzer KN, Neece SH: **Purification of the bacteriophage T4 type II DNA topoisomerase.** *Methods Mol Biol* 1999, **94**:171-177.

doi:10.1186/1743-422X-7-290
Cite this article as: Edgell *et al.*: Mobile DNA elements in T4 and related phages. *Virology Journal* 2010 **7**:290.

Hinton *Virology Journal* 2010, **7**:289
http://www.virologyj.com/content/7/1/289

VIROLOGY JOURNAL

REVIEW

Transcriptional control in the prereplicative phase of T4 development

Deborah M Hinton

Abstract

Control of transcription is crucial for correct gene expression and orderly development. For many years, bacteriophage T4 has provided a simple model system to investigate mechanisms that regulate this process. Development of T4 requires the transcription of early, middle and late RNAs. Because T4 does not encode its own RNA polymerase, it must redirect the polymerase of its host, *E. coli*, to the correct class of genes at the correct time. T4 accomplishes this through the action of phage-encoded factors. Here I review recent studies investigating the transcription of T4 prereplicative genes, which are expressed as early and middle transcripts. Early RNAs are generated immediately after infection from T4 promoters that contain excellent recognition sequences for host polymerase. Consequently, the early promoters compete extremely well with host promoters for the available polymerase. T4 early promoter activity is further enhanced by the action of the T4 Alt protein, a component of the phage head that is injected into *E. coli* along with the phage DNA. Alt modifies Arg265 on one of the two α subunits of RNA polymerase. Although work with host promoters predicts that this modification should decrease promoter activity, transcription from some T4 early promoters increases when RNA polymerase is modified by Alt. Transcription of T4 middle genes begins about 1 minute after infection and proceeds by two pathways: 1) extension of early transcripts into downstream middle genes and 2) activation of T4 middle promoters through a process called sigma appropriation. In this activation, the T4 co-activator AsiA binds to Region 4 of σ^{70}, the specificity subunit of RNA polymerase. This binding dramatically remodels this portion of σ^{70}, which then allows the T4 activator MotA to also interact with σ^{70}. In addition, AsiA restructuring of σ^{70} prevents Region 4 from forming its normal contacts with the -35 region of promoter DNA, which in turn allows MotA to interact with its DNA binding site, a MotA box, centered at the -30 region of middle promoter DNA. T4 sigma appropriation reveals how a specific domain within RNA polymerase can be remolded and then exploited to alter promoter specificity.

Background

Expression of the T4 genome is a highly regulated and elegant process that begins immediately after infection of the host. Major control of this expression occurs at the level of transcription. T4 does not encode its own RNA polymerase (RNAP), but instead encodes multiple factors, which serve to change the specificity of polymerase as infection proceeds. These changes correlate with the temporal regulation of three classes of transcription: early, middle, and late. Early and middle RNA is detected prereplicatively [previously reviewed in [1-6]], while late transcription is concurrent with T4 replication and discussed in another chapter. T4 early

transcripts are generated from early promoters (Pe), which are active immediately after infection. Early RNA is detected even in the presence of chloramphenicol, an antibiotic that prevents protein synthesis. In contrast, T4 middle transcripts are generated about 1 minute after infection at 37°C and require phage protein synthesis. Middle RNA is synthesized in two ways: 1) activation of middle promoters (Pm) and 2) extension of Pe transcripts from early genes into downstream middle genes.

This review focuses on investigations of T4 early and middle transcription since those detailed in the last T4 book [1,5]. At the time of that publication, early and middle transcripts had been extensively characterized, but the mechanisms underlying their synthesis were just emerging. In particular, *in vitro* experiments had just demonstrated that activation of middle promoters

Correspondence: dhinton@helix.nih.gov
Laboratory of Molecular and Cellular Biology, National Institute of Diabetes and Digestive and Kidney Diseases, National Institutes of Health, (Building 8, Room 2A-13) Bethesda, MD (20892-0830) USA

BioMed Central

requires a T4-modified RNAP and the T4 activator MotA [7,8]. Subsequent work has identified the needed RNAP modification as the tight binding of a 10 kDa protein, AsiA, to the σ^{70} subunit of RNAP [9-13]. In addition, a wealth of structural and biochemical information about *E. coli* RNAP [reviewed in [14-16]], MotA, and AsiA [reviewed in [2]] has now become available. As detailed below, we now have a much more mechanistic understanding of the process of prereplicative T4 transcription. To understand this process, we first start with a review of the host transcriptional machinery and RNAP.

The *E. coli* transcriptional machinery

E. coli RNAP holoenzyme, like all bacterial RNAPs, is composed of a core of subunits (β, β', α_1, α_2, and ω), which contains the active site for RNA synthesis, and a specificity factor, σ, which recognizes promoters within the DNA and sets the start site for transcription. The primary σ, σ^{70} in *E. coli*, is used during exponential growth; alternate σ factors direct transcription of genes needed during different growth conditions or times of stress [reviewed in [17-19]]. Sequence/function analyses of hundreds of σ factors have identified various regions and subregions of conservation. Most σ factors share similarity in Regions 2-4, the central through C-terminal portion of the protein, while primary σ factors also have a related N-terminal portion, Region 1.

Recent structural information, together with previous and ongoing biochemical and genetic work [reviewed in [14,15,20,21]], has resulted in a biomolecular understanding of RNAP function and the process of transcription. Structures of holoenzyme, core, and portions of the primary σ of thermophilic bacteria with and without DNA [15,16,22-28], and structures of regions of *E. coli* σ^{70} alone [29] and in a complex with other proteins [26,30] are now available. This work indicates that the interface between σ^{70} and core within the RNAP holoenzyme is extensive (Figure 1). It includes contact between a portion of σ Region 2 and a coiled/coil domain composed of β, β', an interaction of σ^{70} Region 1.1 within the "jaws" in the downstream DNA channel (where DNA downstream of the transcription start site will be located when RNAP binds the promoter), and an interaction between σ^{70} Region 4 and a portion of the β subunit called the β-flap.

For transcription to begin, portions of RNAP must first recognize and bind to double-stranded (ds) DNA recognition elements present within promoter DNA (Figure 1) [reviewed in [20]]. Each of the C-terminal domains of the α subunits (α-CTDs) can interact with an UP element, A/T rich sequences present between positions -40 and -60. Portions of σ^{70}, when present in RNAP, can interact with three different dsDNA elements. A helix-turn-helix, DNA binding motif in σ^{70} Region 4 can bind to the -35 element, σ^{70} Region 3 can bind to a -15TGn-13 sequence (TGn), and σ^{70} subregion 2.4 can bind to positions -12/-11 of a -10 element. Recognition of the -35 element also requires contact between residues in σ^{70} Region 4 and the β-flap in

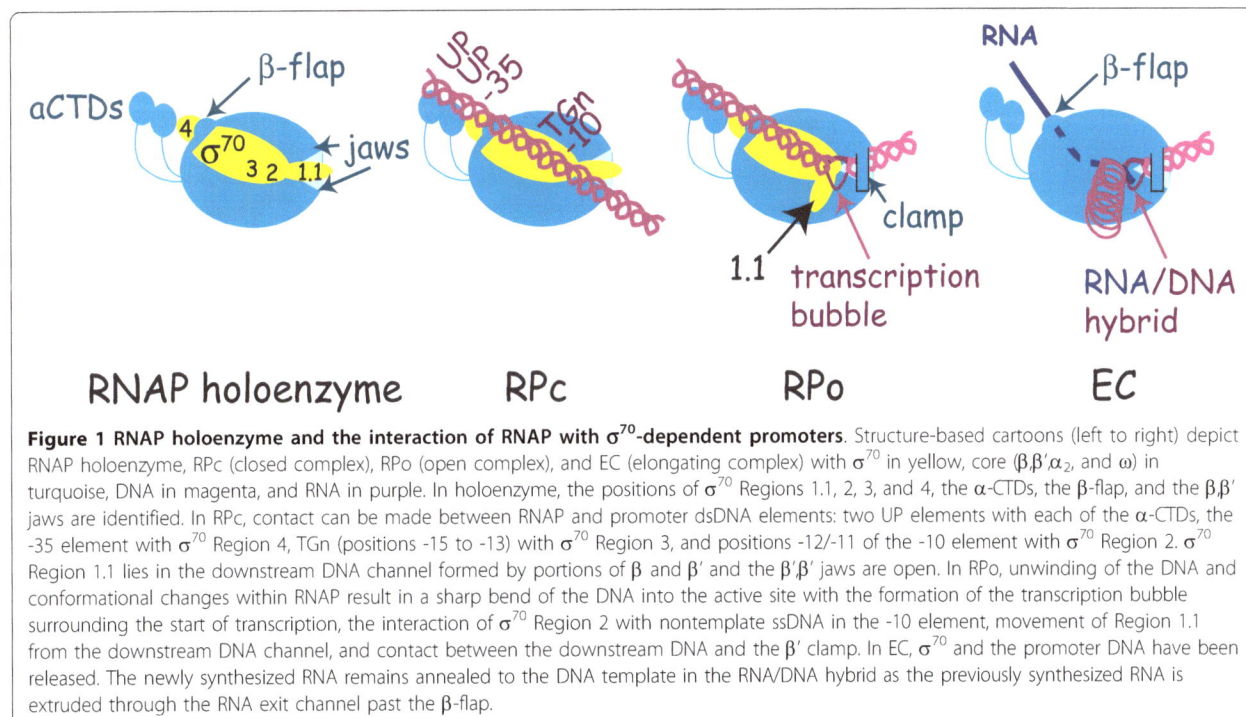

Figure 1 RNAP holoenzyme and the interaction of RNAP with σ70-dependent promoters. Structure-based cartoons (left to right) depict RNAP holoenzyme, RPc (closed complex), RPo (open complex), and EC (elongating complex) with σ70 in yellow, core (β,β'α2, and ω) in turquoise, DNA in magenta, and RNA in purple. In holoenzyme, the positions of σ70 Regions 1.1, 2, 3, and 4, the α-CTDs, the β-flap, and the β,β' jaws are identified. In RPc, contact can be made between RNAP and promoter dsDNA elements: two UP elements with each of the α-CTDs, the -35 element with σ70 Region 4, TGn (positions -15 to -13) with σ70 Region 3, and positions -12/-11 of the -10 element with σ70 Region 2. σ70 Region 1.1 lies in the downstream DNA channel formed by portions of β and β' and the β'β' jaws are open. In RPo, unwinding of the DNA and conformational changes within RNAP result in a sharp bend of the DNA into the active site with the formation of the transcription bubble surrounding the start of transcription, the interaction of σ70 Region 2 with nontemplate ssDNA in the -10 element, movement of Region 1.1 from the downstream DNA channel, and contact between the downstream DNA and the β' clamp. In EC, σ70 and the promoter DNA have been released. The newly synthesized RNA remains annealed to the DNA template in the RNA/DNA hybrid as the previously synthesized RNA is extruded through the RNA exit channel past the β-flap.

order to position σ^{70} correctly for simultaneous contact of the -35 and the downstream elements. Typically, a promoter only needs to contain two of the three σ^{70}-dependent elements for activity; thus, *E. coli* promoters can be loosely classified as -35/-10 (the major class), TGn/-10 (also called extended -10), or -35/TGn [reviewed in [20]].

The initial binding of RNAP to the dsDNA promoter elements usually results in an unstable, "closed" complex (RPc) (Figure 1). Creation of the stable, "open" complex (RPo) requires bending and unwinding of the DNA [31] and major conformational changes (isomerization) of the polymerase (Figure 1) [[32,33]; reviewed in [20]]. In RPo the unwinding of the DNA creates the transcription bubble from -11 to ~+3, exposing the single-stranded (ss) DNA template for transcription. Addition of ribonucleoside triphosphates (rNTPs) then results in the synthesis of RNA, which remains as a DNA/RNA hybrid for about 8-9 bp. Generation of longer RNA initiates extrusion of the RNA through the RNA exit channel formed by portions of β and β' within core. Since this channel includes the σ^{70}-bound β-flap, it is thought that the passage of the RNA through the channel helps to release σ from core, facilitating promoter clearance. The resulting elongation complex, EC, contains core polymerase, the DNA template, and the synthesized RNA (Figure 1) [reviewed in [34]]. The EC moves rapidly along the DNA at about 50 nt/sec, although the complex can pause, depending on the sequence [35]. Termination of transcription occurs either at an intrinsic termination signal, a stem-loop (hairpin) structure followed by a U-rich sequence, or a Rho-dependent termination signal [reviewed in [36,37]]. The formation of the RNA hairpin by an intrinsic terminator sequence may facilitate termination by destabilizing the RNA/DNA hybrid. Rho-dependent termination is mediated through the interaction of Rho protein with a rut site (Rho utilization sequence), an unstructured, sometimes C-rich sequence that lies upstream of the termination site. After binding

to the RNA, Rho uses ATP hydrolysis to translocate along the RNA, catching up with the EC at a pause site. Exactly how Rho disassociates a paused complex is not yet fully understood; the DNA:RNA helicase activity of Rho may provide a force to "push" RNAP off the DNA. Rho alone is sufficient for termination at some Rho-dependent termination sites. However, at other sites the termination process also needs the auxiliary *E. coli* proteins NusA and/or NusG [reviewed in [36]].

When present in intergenic regions, rut sites are readily available to interact with Rho. However, when present in protein-coding regions, these sites can be masked by translating ribosomes. In this case, Rho termination is not observed unless the upstream gene is not translated, for example, when a mutation has generated a nonsense codon. In such a case, Rho-dependent termination can prevent transcription from extending into the downstream gene. Thus, in this situation, which is called polarity [38], expression of both the upstream mutated gene and the downstream gene is prevented.

T4 early transcription
Early promoters
T4 only infects exponentially growing *E. coli*, and transcription of T4 early genes begins immediately after infection. Thus, for an efficient infection, the phage must rapidly redirect the σ^{70}-associated RNAP, which is actively engaged in transcription of the host genome, to the T4 early promoters. This immediate takeover is successful in part because most T4 early promoters contain excellent matches to the σ^{70}-RNAP recognition elements (-35, TGn, and -10 elements) and to the α-CTD UP elements (Figure 2; for lists of T4 early promoter sequences, see [4,5]). However, sequence alignments of T4 early promoters reveal additional regions of consensus, suggesting that they contain other bits of information that can optimize the interaction of host RNAP with the promoter elements. Consequently, unlike most host promoters that belong to the -35/-10, TGn/-10 or

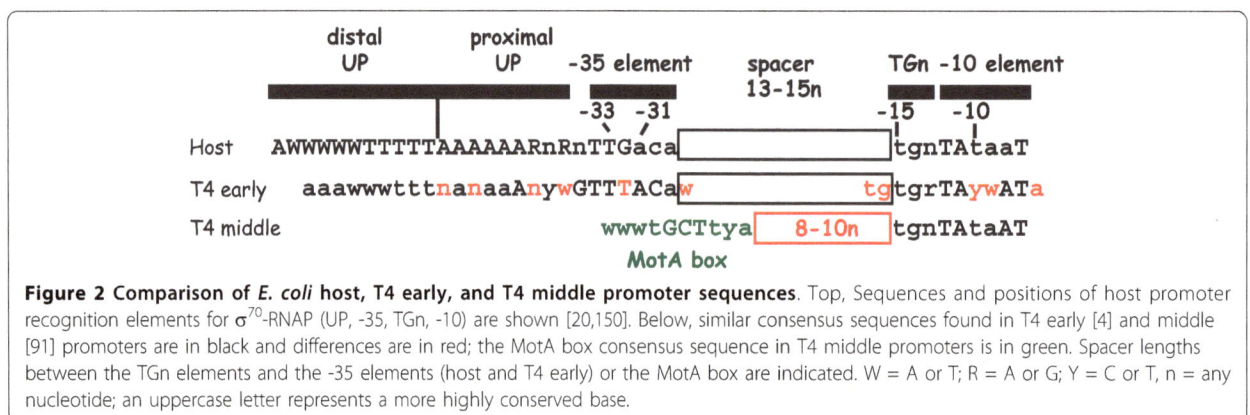

Figure 2 Comparison of *E. coli* host, T4 early, and T4 middle promoter sequences. Top, Sequences and positions of host promoter recognition elements for σ^{70}-RNAP (UP, -35, TGn, -10) are shown [20,150]. Below, similar consensus sequences found in T4 early [4] and middle [91] promoters are in black and differences are in red; the MotA box consensus sequence in T4 middle promoters is in green. Spacer lengths between the TGn elements and the -35 elements (host and T4 early) or the MotA box are indicated. W = A or T; R = A or G; Y = C or T, n = any nucleotide; an uppercase letter represents a more highly conserved base.

-35/TGn class, T4 early promoters can be described as "über" UP/-35/TGn/-10 promoters. Indeed, most T4 early promoters compete extremely well with the host promoters for the available RNAP [39] and are similar to other very strong phage promoters, such as T7 P_{A1} and λ P_L.

The T4 Alt protein

Besides the sheer strength of its early promoters, T4 has another strategy, the Alt protein, to establish transcriptional dominance [[40-43], reviewed in [1,4]]. Alt, a mono-ADP-ribosyltransferase, ADP-ribosylates a specific residue, Arg265, on one of the two α subunits of RNAP. In addition, Alt modifies a fraction of other host proteins, including the other RNAP subunits and host proteins involved in translation and cell metabolism. Alt is an internal phage head protein that is injected with the phage DNA. Consequently, Alt modification occurs immediately after infection and does not require phage protein synthesis. Each α subunit is distinct (one α interacts with β while the other interacts with β') and Alt modification is thought to specifically target a particular α, although which particular α is not known.

What is the purpose of Alt modification? The major Alt target, α Arg265, has been shown to be crucial for the interaction of an α-CTD with a promoter UP element [44-46] and with some host activators, including c-AMP receptor protein (CRP), a global regulator of *E. coli* [46,47]. Thus, an obvious hypothesis is that Alt simply impairs host promoters that either need these activators or are enhanced by α-CTD/UP element interaction. However, overexpression of Alt from a plasmid does not affect *E. coli* growth [40], and general transcription of *E. coli* DNA *in vitro* is not impaired when using Alt-modified RNAP [48]. Instead, it appears that Alt-modification is helpful because it increases the activity of certain T4 early promoters. This 2-fold enhancement of activity has been observed both *in vivo* [40,49] and *in vitro* [48]. How Alt-modification stimulates particular early promoters is not known, but it is clear that it is not simply due to their general strength. Other strong promoters, such as P_{tac}, T7 P_{A1} and P_{A2}, T5 P_{207}, and even some of the T4 early promoters, are unaffected when using Alt-modified RNAP [49]. Alt-mediated stimulation of a promoter is also not dependent on specific $σ^{70}$-dependent elements (-35, TGn, and -10 elements); some promoters with identical sequences in these regions are stimulated by Alt while others are not [49]. A comprehensive mutational analysis of the T4 early promoter $P_{8.1}$ and P_{tac} reveals that there is not a single, specific promoter position(s) responsible for the Alt effect. This result suggests that the mechanism of Alt stimulation may involve cross-talk between RNAP and more than one promoter region [50] or that ADP-ribosylation of α Arg265 is a secondary, less significant activity of Alt and additional

work on the importance of this injected enzyme is needed.

Continuing early strategies for T4 domination

Because T4 promoters are so efficient at out-competing those of the host, a burst of immediate early transcription occurs within the first minute of infection. From this transcription follows a wave of early products that continue the phage takeover of the host transcriptional machinery. One such product is the T4 Alc protein, a transcription terminator that is specific for dC-containing DNA, that is, DNA that contains unmodified cytosines. Consequently, Alc terminates transcription from host DNA without affecting transcription from T4 DNA, whose cytosines are hydroxymethylated and glucosylated [[51,52]; reviewed in [1,4]]. Alc directs RNAP to terminate at multiple, frequent, and discrete sites along dC-containing DNA. The mechanism of Alc is not known. Unlike other terminating factors, Alc does not appear to interact with either RNA or DNA, and decreasing the rate of RNA synthesis or RNAP pausing near an Alc termination site actually impairs Alc termination [51]. Mutations within an N-terminal region of the β subunit of RNAP, a region that is not essential for *E. coli* (dispensable region I), prevent Alc -mediated termination, suggesting that an interaction site for Alc may reside in this region [52].

T4 also encodes two other ADP-ribosylating enzymes, ModA and ModB, as early products. Like Alt, ModA modifies Arg265 of RNAP α [[53,48]; reviewed in [1,4]]. However, unlike Alt, ModA almost exclusively targets the RNAP α subunits. In addition, ModA modifies both α subunits so there is no asymmetry to ModA modification. Synthesis of ModA is highly toxic to *E. coli*. *In vitro*, ModA-modified RNAP is unable to interact with UP elements or to interact with CRP [cited in [40]] and is less active than unmodified RNAP when using either *E. coli* or T4 DNA [48]. Thus, it has been suggested that ModA helps to diminish both host and T4 early promoter activity, reprogramming the transcriptional machinery for the coming wave of middle transcription [48]. However, a deletion of the *modA* gene does not affect the rapid decrease in early transcription or the decrease in the synthesis of early gene products, which begins about 3 minutes post-infection [54]. This result suggests that the phage employs other as yet unknown strategies to stop transcription from early promoters. ModB, the other early ADP-ribosylating enzyme, targets host translation factors, the ribosomal protein S30 and trigger factor, which presumably helps to facilitate T4 translation [43].

Finally, many of the early transcripts include genes of unknown function and come from regions of the T4 genome that are not essential for infection of wild type (wt) *E. coli* under normal laboratory conditions.

Presumably, these genes encode phage factors that are useful under specific growth conditions or in certain strains. Whether any of these gene products aid T4 in its takeover of the host transcriptional machinery is not known.

The switch to middle transcription

Within a minute of infection at 37°C, some of the T4 early products mediate the transition from early to middle gene expression. As detailed below, the MotA activator and AsiA co-activator are important partners in this transition, since they direct RNAP to transcribe from middle promoters. In addition, the ComC-α protein, described later, may also have a role in the extension of early RNAs into downstream middle genes or the stability of such transcripts once they are formed.

As middle transcription begins, certain early RNAs decay rapidly after their initial burst of transcription. This arises from the activity of the early gene product RegB, an endoribonuclease, which specifically targets some T4 early mRNAs. For the mRNAs of MotA and RegB itself, a RegB cleavage site lies within the Shine-Dalgarno sequence; for ComC-α mRNA, the site is within AU-rich sequences upstream and downstream of this sequence [55]. The mechanism by which RegB recognizes and chooses the specific cleavage site is not yet known.

The onset of T4 middle transcription also finishes the process of eliminating host transcription by simply removing the host DNA template for RNAP. T4-encoded nucleases, primarily EndoII encoded by *denA* and EndoIV encoded by *denB*, selectively degrade the dC-containing host DNA ([56,57] and references therein). Thus, a few minutes after infection, there is essentially no host DNA to transcribe.

Transcription of middle genes from T4 middle promoters
Middle promoters

Middle genes primarily encode proteins needed for replication, recombination, and nucleotide metabolism; various T4-encoded tRNAs; and transcription factors that program the switch from middle to late promoter activation. Middle RNAs arise by 2 pathways: extension of early transcription into middle genes (discussed later) and the activation of T4 middle promoters by a process called σ appropriation [2]). To date, nearly 60 middle promoters have been identified (Table 1). Unlike early promoters, T4 middle promoters contain a host element, the σ^{70}-dependent -10 sequence, and a phage element, a MotA box, which is centered at -30 and replaces the σ^{70}-dependent -35 element present in T4 early promoters and most host promoters (Figure 2). In addition, about half of the middle promoters also contain TGn, the extended -10 sequence. Activation of the

Table 1 Positions of identified T4 middle promoters

Middle Promoter	Start site	Reference
PrIIB2	122	[99,141,142]
PrIIB1	377	[99,141,142]
PrIIA	2263	[141,142]
P39	5349	[99]
Pdex.2	10058	[91]
Pdda.1	11138	[91]
P56/69	16813	[99]
Pdam	17617	[91]
P61	19122	[100]
PuvsX	23752	[7]
PsegA	24460	[7]
P42	26320	[100]
P43	29933	[99,143]
P45	32626	[99,143]
P45.2	33257	[143]
P46i(2)	33803	[101]
P46i(1)	34394	[101]
P46	35014	[99,143]
P47	36576	[99,143]
Pαgt	38430	[91]
PmobB (Pαgt.1)	38682	[99]
Pαgt.4	39447	[91]
P55	40180	[100]
P55.8(2)	42542	[101]
P55.8	42805	[100]
PnrdG (P55.9)	43023	[100]
PmobC	43744	[101]
PnrdD+ (P49.1)	6440	[144]
PnrdC(2)	48465	[101]
PnrdC(1)	48492	[101]
PnrdC.7	53325	[101]
PmobD	57389	[101]
PmobD.3	58381	[101]
Ptk.3	61076	[101]
Pvs.7	64382	[101]
PipIII	66724	[101]
PtRNAE (PtRNAsc1)	72593	[99]
P57A	74877	[99]
P1	75393	[99]
PrnIB	109763	[91]
P24.3	110108	[91]
Phoc	111757	[91]
PuvsY	115371	[8,99,126,145]
P30	127234	[102]
P30.2	128355	[102]
P31	131540	[146]
Pcd	132839	[91]
PI-TevIII (PnrdBin)	138939	[147]
PnrdB+	139878	[148]
PnrdA	142726	[100]
Ptd	145142	[100]

Table 1 Positions of identified T4 middle promoters
(Continued)

P32	148057*	[149]
PsegG	148678	[91]
PdsbA	149873	[129]
PdsbA(2)	149951	[91]
P34i	153011	[99]
P52	65227	[91]
Pndd.3	166702	[91]

Position of transcription start refers to T4 sequence [[4]; http://phage.bioc.tulane.edu/] Promoter names in parentheses refer to previous designations.

phage middle promoters requires the concerted effort of two T4 early products, AsiA and MotA.

AsiA, the co-activator of T4 middle transcription

AsiA (*A*udrey *S*tevens *i*nhibitor or *a*nti-*s*igma *i*nhibitor) is a small protein of 90 residues. It was originally identified as a 10 kDa protein that binds very tightly to the σ^{70} subunit of RNAP [11,58,59] with a ratio of 1:1 [60]. Later work indicated that a monomer of AsiA binds to C-terminal portions of σ^{70}, Regions 4.1 and 4.2 [26,60-70]. In solution, AsiA is a homodimer whose self-interaction face is composed of mostly hydrophobic residues within the N-terminal half of the protein [65,71]. A similar face of AsiA interacts with σ^{70} [26], suggesting that upon binding to σ^{70}, a monomer of AsiA in the homodimer simply replaces its partner for σ^{70}. Curiously, the AsiA structure also contains a helix-turn-helix motif (residues 30 to 59), suggesting the possibility of an interaction between AsiA and DNA [71]. However, as yet, no such interaction has been detected.

Multiple contacts make up the interaction between AsiA and σ^{70} Region 4 (Figure 3A). The NMR structure (Figure 3B, right) reveals that 18 residues present in three α helices within the N-terminal half of AsiA (residues 10 to 42) contact 17 residues of σ^{70} [26]. Biochemical analyses have confirmed that AsiA residues E10, V14, I17, L18, K20, F21, F36, and I40, which contact σ^{70} Region 4 in the structure, are indeed important for the AsiA/σ^{70} interaction and/or for AsiA transcriptional function *in vitro* [72-74]. Of all of these residues, I17 appears to be the most important, and thus, has been termed "the linchpin" of the AsiA/σ^{70} Region 4 interaction [74]. A mutant AsiA missing the C-terminal 17 residues is as toxic as the full length protein when expressed *in vivo* [72,75], and even a mutant missing the C-terminal 44 residues is still able to interact with σ^{70} Region 4 and to co-activate transcription weakly [72]. These results are consistent with the idea that only the N-terminal half of AsiA is absolutely required to form a functional AsiA/σ^{70} complex. Together, the structural and biochemical work indicate that there is an extensive interface between the N-terminal half of

AsiA and σ^{70} Region 4, consistent with the early finding that AsiA copurifies with σ^{70} until urea is added to dissociate the complex [76].

The σ^{70} face of the AsiA/σ^{70} complex includes residues in Regions 4.1 and 4.2 that normally contact the -35 DNA element or the β-flap of core [26] (Figure 3). Mutations within Region 4.1 or Region 4.2, which are at or near the AsiA contact sites in σ^{70}, impair or eliminate AsiA function [77-79], providing biochemical evidence for these interactions. The structure of the AsiA/σ^{70} Region 4 complex also reveals that AsiA binding dramatically changes the conformation of σ^{70} Region 4, converting the DNA binding helix-turn-helix (Figure 3B, left) into one continuous helix (Figure 3B, right). Such a conformation would be unable to retain the typical σ^{70} contacts with either the -35 DNA or with the β-flap. Thus, the association of AsiA with σ^{70} should inhibit the binding of RNAP with promoters that depend on recognition of a -35 element. Indeed, early observations showed that AsiA functions as a transcriptional inhibitor at most promoters *in vitro* [9,10], blocking RPc formation [60], but TGn/-10 promoters, which are independent of a RNAP/-35 element contact, are immune to AsiA [62,66,80]. However, this result is dependent on the buffer conditions. In the presence of glutamate, a physiologically relevant anion that is known to facilitate protein-protein and protein-DNA interactions [81,82], extended incubations of AsiA-associated RNAP with -10/-35 and -35/TGn promoters eventually result in the formation of transcriptionally competent, open complexes that contain AsiA [72,83]. Under these conditions, AsiA inhibition works by significantly slowing the rate of RPo formation [83]. However, the formation of these complexes still relies on DNA recognition elements other than the -35 element (UP, TGn, and -10 elements), again demonstrating that AsiA specifically targets the interaction of RNAP with the -35 DNA.

Because AsiA strongly inhibits transcription from -35/-10 and -35/TGn promoters, expression of plasmid-encoded AsiA is highly toxic in *E. coli*. Thus, during infection, AsiA may serve to significantly inhibit host transcription. Although it might be reasonable to suppose that AsiA performs the same role at T4 early promoters, this is not the case. The shut-off of early transcription, which occurs a few minutes after infection, is still observed in a T4 *asiA*- infection [54], and early promoters are only modestly affected by AsiA *in vitro* [84]. This immunity to AsiA is probably due to the multiple RNAP recognition elements present in T4 early promoters (Figure 2). Thus, AsiA inhibition does not significantly contribute to the early to middle promoter transition. AsiA also does not help to facilitate the replacement of σ^{70} by the T4-encoded late σ factor, which is needed for T4 late promoter activity [85],

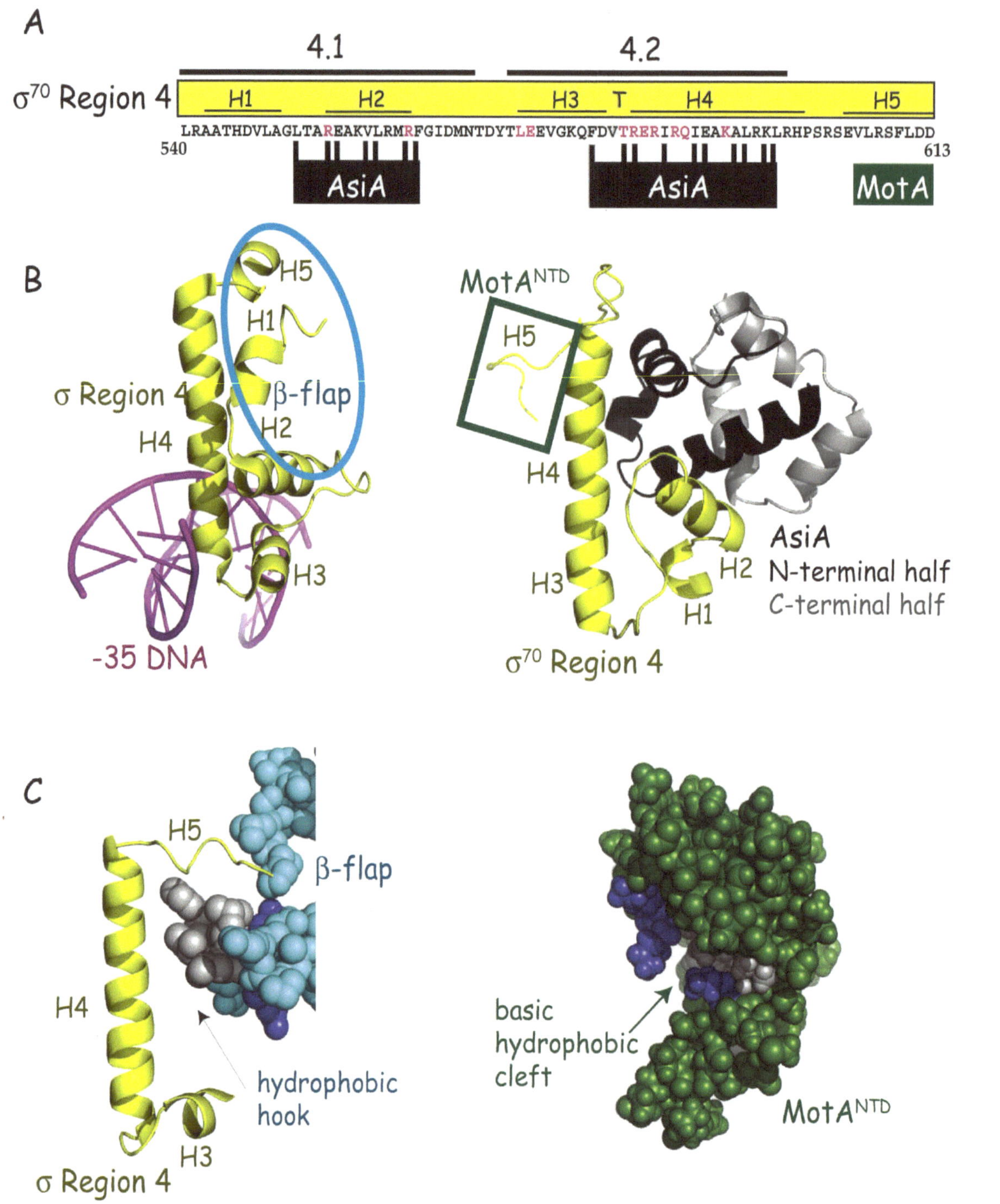

Figure 3 Interaction of σ⁷⁰ region 4 with -35 element DNA, the β-flap, AsiA and MotA. A) Sequence of σ⁷⁰ Region 4 (residues 540-613) with subregions 4.1 and 4.2; the α helices H1 through H5 with a turn (T) between H3 and H4 are shown. Residues of σ⁷⁰ that interact with the -35 element [25] are colored in magenta. Residues that interact with AsiA [26] or the region that interacts with MotA [97,104] is indicated. B) Structures showing the interaction of *T. aquaticus* σ Region 4 with -35 element DNA [25] (left, accession # 1KU7) and interaction of σ⁷⁰ Region 4 with AsiA [26] (right, accession # 1TLH). σ, yellow; DNA, magenta; AsiA, N-terminal half in black, C-terminal half in gray. On the left, the portions of σ that interact with the β-flap (σ residues in and near H1, H2, and H5) are circled in turquoise; on the right, H5, the far C-terminal region of σ⁷⁰ that interacts with MotA, is in the green square. C) Structures showing the interaction of *T. thermophilus* σ H5 with the β-flap tip [22] (left, accession # 1IW7) and the structure of MotA^NTD [94] (right, accession # 1I1S) are shown. On the β-flap (left) and MotA^NTD (right) structures, hydrophobic residues (L, I, V, or F) and basic residues (K or R) are colored in gray or blue, respectively. The interaction site at the β-flap tip is a hydrophobic hook, while the structure in MotA^NTD is a hydrophobic cleft.

indicating that AsiA is not involved in the middle to late promoter transition.

Although AsiA was originally designated as an "anti-sigma" factor and is still frequently referred to as such, it is important to note that it behaves quite differently from classic anti-sigma factors. Unlike these factors, its binding to σ^{70} does not prevent the σ^{70}/core interaction; it does not sequester σ^{70}. Instead it functions as a member of the RNAP holoenzyme. Consequently, AsiA is more correctly designated as a co-activator rather than an anti-sigma factor, and its primary role appears to be in activation rather than inhibition.

MotA, the transcriptional activator for middle promoters

The T4 *motA* (*m*odifier *o*f *t*ranscription) gene was first identified from a genetic selection developed to isolate mutations in T4 that increase the synthesis of the early gene product rIIA [86]. In fact, expression of several early genes increase in the T4 *motA*- infection, presumably because of a delay in the shift from early to middle transcription [87]. MotA is a basic protein of 211 amino acids, which is expressed as an early product [88]. The MotA mRNA is cleaved within its Shine-Dalgarno sequence by the T4 nuclease, RegB. Consequently, the burst of MotA protein synthesis, which occurs within the first couple minutes of infection [55], must be sufficient for all the subsequent MotA-dependent transcription.

MotA binds to a DNA recognition element, the MotA box, to activate transcription in the presence of AsiA-associated RNAP [7,8,11-13,89,90]. A MotA box consensus sequence of 5'(a/t)(a/t)(a/t)TGCTTtA3' [91] has been derived from 58 T4 middle promoters (Pm) (Table 1). This sequence is positioned 12 bp +/- 1 from the σ^{70}-dependent -10 element,-12TAtaaT-7 (Figure 2). MotA functions as a monomer [92-94] with two distinct domains [95]. The N-terminal half of the protein, MotA^{NTD} contains the trans-activation function [96-98]. The structure of this region shows five α-helices, with helices 1, 3, 4, and 5 packing around the central helix 2 [93]. The C-terminal half, MotA^{CTD}, binds MotA box DNA [97] and consists of a saddle-shaped, 'double wing' motif, three α-helices interspersed with six β-strands [94]. As information about MotA-dependent activation has emerged, it has become apparent that MotA differs from other activators of bacterial RNAP in several important aspects. The unique aspects of MotA are discussed below.

1) MotA tolerates deviations within the MotA box consensus sequence

Early work [[3,99]; reviewed in [1]] identified a highly conserved MotA box sequence of (a/t)(a/t)TGCTT(t/c)a with an invariant center CTT based on more than twenty T4 middle promoters. However, subsequent mutational analyses revealed that most single bp changes within the consensus sequence, even

within the center CTT, are well-tolerated for MotA binding and activation *in vitro* [100]. Furthermore, several active middle promoters have been identified whose MotA boxes deviate significantly from the consensus, confirming that MotA is indeed tolerant of bp changes *in vivo* [91,100-102].

An examination of the recognized base determinants within the MotA box has revealed that MotA senses minor groove moieties at positions -32 and -33 and major groove determinants at positions -28 and -29 [103]. (For this work, the MotA box was located at positions -35 to -26, its position when it is present 13 bp upstream of the -10 element.) In particular, the 5-Me on -29 T contributes to MotA binding. However, despite its high conservation, there seems to be little base recognition of -31 G:C, -30 C:G at the center of the MotA box. In wt T4 DNA, each cytosine in this sequence is modified by the presence of a hydroxy-methylated, glucosylated moiety at cytosine position 5. This modification places a large, bulky group within the major groove, making it highly unlikely that MotA could contact a major groove base determinant at these positions. In addition, MotA binds and activates transcription using unmodified DNA; thus, the modification itself cannot be required for function. However, for two specific sequences, DNA modification does seem to affect MotA activity. One case is the middle promoter upstream of gene 46, P46. The MotA box within P46 contains the unusual center sequence ACTT rather than the consensus GCTT. MotA binds a MotA box with the ACTT sequence poorly, and MotA activation of P46 *in vitro* using wt T4 DNA is significantly better than that observed with unmodified DNA [100]. These results suggest that DNA modification may be needed for full activity of the ACTT MotA box motif. On the other hand, when using unmodified DNA *in vitro*, MotA binds a MotA box with a center sequence of GATT nearly as well as one with the consensus GCTT sequence, and a promoter with the GATT motif is fully activated by MotA *in vitro*. However, several potential T4 middle promoter sequences with a GATT MotA box and an excellent σ^{70}-dependent -10 element are present within the T4 genome, but these promoters are not active [100]. This result suggests that the cytosine modification opposite the G somehow "silences" GATT middle promoter sequences.

2) MotA is not a strong DNA-binding protein

In contrast to many other well-characterized activators of *E. coli* RNAP, MotA has a high apparent dissociation constant for its binding site (100 - 600 nM [92,103,104]), and a large excess of MotA relative to DNA is needed to detect a MotA/DNA complex in a gel retardation assay or to detect protein protection of the DNA in footprinting assays [90]. In contrast, stoichiometric levels of MotA are sufficient for transcription *in vitro* [90]. These results are inconsistent with the idea that the tight binding of MotA to a middle promoter recruits AsiA-associated RNAP for transcription. In fact, in nuclease protection assays, MotA binding to the MotA box of a middle promoter is much stronger in the presence of AsiA and RNAP than with MotA alone [89,90]. Furthermore, in contrast to the sequence deviations permitted within the MotA box, nearly all middle promoters have a stringent requirement for an excellent match to the σ^{70}-dependent -10 element [91,100,101]. This observation suggests that the interaction of σ^{70} Region 2.4 with its cognate -10 sequence contributes at least as much as MotA binding to the MotA box in the establishment of a stable RNAP/MotA/AsiA/Pm complex.

3) The MotA binding site on σ^{70} is unique among previously characterized activators of RNAP

Like many other characterized activators, MotA interacts with σ^{70} residues within Region 4 to activate transcription. However, other activators target basic σ^{70} residues from 593 to 603 within Region 4.2 that are immediately C-terminal to residues that interact specifically with the -35 element DNA [27,105-112] [Figure 3A; reviewed in [113]]. In contrast, the interaction site for MotA is a hydrophobic/acidic helix (H5) located at the far C-terminus of σ^{70} (Figure 3A). MotA^{NTD} interacts with this region *in vitro* and mutations within σ^{70} H5 impair both MotA binding to σ^{70} and MotA-dependent transcription [77,97,104]. In addition, a mutation within H5 restores infectivity of a T4 *motA*- phage in a particular strain of *E. coli*, TabG [114], which does not support T4 *motA*-growth [115].

Recent structural and biochemical work has indicated that a basic/hydrophobic cleft within MotA^{NTD} contains the molecular face that interacts with σ^{70} H5 (Figure 3C, right). Mutation of MotA residues K3, K28, or Q76, which lie in this cleft, impair the ability of MotA to interact with σ^{70} H5 and to activate transcription, and render the protein incapable of complementing a T4 *motA*- phage for growth [104]. Interestingly, substitutions of MotA residues D30, F31, and D67, which lie on another exposed surface outside of this cleft, also have deleterious effects on the interaction with σ^{70}, transcription, and/or phage viability [98,104]. These residues are contained within a hydrophobic, acidic patch, which may also be involved in MotA activation or another unidentified function of MotA.

The process of sigma appropriation

The mechanism of MotA-dependent activation occurs through a novel process, called sigma appropriation [reviewed in [2]]. Insight into this process began with the finding that some middle promoters function *in vitro* with RNAP alone. The middle promoter P$_{uvsX}$, which is

positioned upstream of the T4 recombination gene *uvsX*, is such a promoter [13]. This promoter is active because it has UP elements and a perfect -10 element to compensate for its weak homology to a σ^{70} -35 sequence. (It should be noted that significant activity of P$_{uvsX}$ and other middle promoters in the absence of MotA/AsiA is only seen when using unmodified DNA, because the modification present in T4 DNA obscures needed major grove contacts for RNAP.) Using unmodified P$_{uvsX}$ DNA, it has been possible to investigate how the presence of MotA and AsiA alone and together affect the interactions between RNAP and a middle promoter [72,89,90,103]. The RPo formed by RNAP and P$_{uvsX}$ exhibits protein/DNA contacts that are similar to those seen using a typical -35/-10 promoter; addition of MotA in the absence of AsiA does not significantly alter these contacts. As expected, addition of AsiA without MotA inhibits the formation of a stable complex. However, in the presence of both MotA and AsiA, a unique RPo is observed. This MotA/AsiA activated complex has the expected interactions between RNAP and the -10 element, but it has unique protein-DNA interactions upstream of the -10 element. In particular, σ^{70} Region 4 does not make its usual contacts with the -35 element DNA; rather MotA binds to the MotA box that overlaps the -35 sequence. As expected, when using fully ADP-ribosylated RNAP there is an abrupt loss of footprint protection just upstream of the MotA box in P$_{uvsX}$, consistent with the loss of UP element interactions when both α-CTD's are modified; when using RNAP that has not been ADP-ribosylated, the UP elements in P$_{uvsX}$ are protected.

Taken together, these biochemical studies argued that within the activated complex, σ^{70} Region 2.4 binds tightly to the σ^{70}-dependent -10 element, but the MotA/MotA box interaction is somehow able to replace the contact that is normally made between σ^{70} Region 4 and the -35 DNA (Figure 4) [89,103]. The subsequent AsiA/σ^{70} Region 4 structure [26] (Figure 3B, right) shows just how this can be done. Through its multiple contacts with σ^{70} residues in Regions 4.1 and 4.2, AsiA remodels Region 4 of σ^{70}. When the AsiA/σ^{70} complex then binds to core, σ^{70} Region 4 is incapable of forming its normal contacts with -35 element DNA (Figure 3B, left). In addition, the restructuring of σ^{70} Region 4 prevents its interaction with the β-flap, allowing the far C-terminal region H5 of σ^{70} to remain available for its interaction with MotA. Consequently, in the presence of AsiA-associated RNAP, MotA can interact both with the MotA box and with σ^{70} H5 [77,97,104].

Recent work has suggested that additional portions of AsiA, MotA and RNAP may be important for σ appropriation. First, the C-terminal region of AsiA (residues 74-90) may contribute to activation at P$_{uvsX}$ by directly

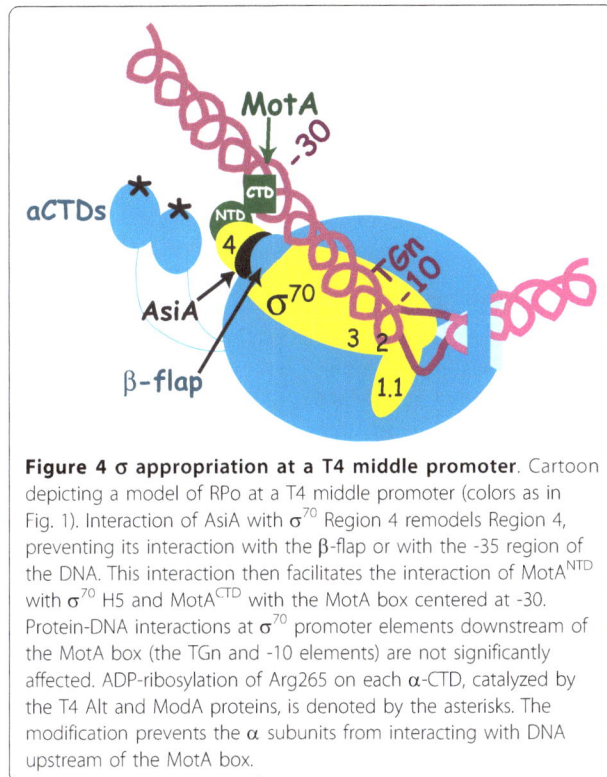

Figure 4 σ appropriation at a T4 middle promoter. Cartoon depicting a model of RPo at a T4 middle promoter (colors as in Fig. 1). Interaction of AsiA with σ[70] Region 4 remodels Region 4, preventing its interaction with the β-flap or with the -35 region of the DNA. This interaction then facilitates the interaction of MotA[NTD] with σ[70] H5 and MotA[CTD] with the MotA box centered at -30. Protein-DNA interactions at σ[70] promoter elements downstream of the MotA box (the TGn and -10 elements) are not significantly affected. ADP-ribosylation of Arg265 on each α-CTD, catalyzed by the T4 Alt and ModA proteins, is denoted by the asterisks. The modification prevents the α subunits from interacting with DNA upstream of the MotA box.

interacting both with the β-flap and with MotA[NTD]. In particular, the AsiA N74D substitution reduces an AsiA/β-flap interaction observed in a 2-hybrid assay and impairs the ability of AsiA to inhibit transcription from a -35/-10 promoter *in vitro* [116]. This mutation also renders AsiA defective in co-activating transcription from P$_{uvsX}$ *in vitro* if it is coupled with a σ[70] F563Y substitution that weakens the interaction of AsiA with σ[70] Region 4 [117]. On the other hand, an AsiA protein with either a M86T or R82E substitution has a reduced capacity to interact with MotA[NTD] in a 2-hybrid assay and yields reduced levels of MotA/AsiA activated transcription from P$_{uvsX}$ *in vitro* [118]. The M86 and R82 mutations do not affect the interaction of AsiA with σ[70] or with the β-flap, and they do not compromise the ability of AsiA to inhibit transcription [118], suggesting that they specifically affect the interaction with MotA. These results argue that AsiA serves as a bridge, which connects σ[70], the β-flap, and MotA. However, in other experiments, MotA/AsiA activation of P$_{uvsX}$ is not affected when using AsiA proteins with deletions of this C-terminal region (Δ79-90 and Δ74-90), and even AsiA Δ47-90 still retains some ability to co-activate transcription [72]. Furthermore, the C-terminal half of the AsiA ortholog of the *vibrio* phage KVP40 (discussed below) has little or no sequence homology with its T4 counterpart yet in the presence of T4 MotA and *E. coli* RNAP, it effectively co-activates transcription from P$_{uvsX}$ *in*

vitro [119], and NMR analyses indicate that the addition of MotA to the AsiA/σ[70] Region 4 complex does not significantly perturb chemical shifts of AsiA residues [104]. Thus, further work is needed to clarify the role of the of AsiA C-terminal region. Finally, very recent work has shown that the inability of T4 *motA* mutants to plate on the TabG strain arises from a G1249D substitution within β, thereby implicating a region of β that is distinct from the β-flap in MotA/AsiA activation [120]. This mutation is located immediately adjacent to a hydrophobic pocket, called the Switch 3 loop, which is thought to aid in the separation of the RNA from the DNA-RNA hybrid as RNA enters the RNA exit channel [28]. The presence of the β G1249D mutation specifically impairs transcription from T4 middle promoters *in vivo*, but whether the substitution directly or indirectly affects protein-protein interactions is not yet known [120]. Taken together, these results suggest that MotA/AsiA activation employs multiple contacts, some of which are essential under all circumstances (AsiA with σ[70] Regions 4.1 and 4.2, MotA with σ[70] H5) and some of which may provide additional contacts perhaps under certain circumstances to strengthen the complex.

Concurrent work with the T4 middle promoter P$_{rIIB2}$ has yielded somewhat different findings than those observed with P$_{uvsX}$ [121]. P$_{rIIB2}$ is a TGn/-10 promoter that does not require an interaction between σ[70] Region 4 and the -35 element for activity. Thus, the presence of AsiA does not inhibit RPo formation at this promoter. An investigation of the complexes formed at P$_{rIIB2}$ using surface plasmon resonance revealed that MotA and AsiA together stimulate the initial recognition of the promoter by RNAP. In addition, *in vitro* transcription experiments indicated that MotA and AsiA together aid in promoter clearance, promoting the formation of the elongating complex. Thus, MotA may activate different steps in initiation, depending on the type of promoter. However, there is no evidence to suggest that the protein/protein and protein/DNA contacts are significantly different with different middle promoters.

Interestingly, AsiA binds rapidly to σ[70] when σ[70] is free, but binds poorly, if at all, to σ[70] that is present in RNAP [122]. The inability of AsiA to bind to σ[70] within holoenzyme may be useful for the phage because it ties the activation of middle promoters to the efficiency of early transcription. This stems from the fact that σ[70] is usually released from holoenzyme once RNAP has cleared a promoter [[123] and references therein]. Since there is an excess of core relative to σ factors, there is only a brief moment for AsiA to capture σ[70]. Consequently, the more efficiently the T4 early promoters fire, the more opportunities are created for AsiA to bind to σ[70], which then leads to increased MotA/AsiA-dependent middle promoter transcription.

Sigma appropriation in other T4-type phages

Although hundreds of activators of bacterial RNAP are known, the T4 MotA/AsiA system represents the first identified case of sigma appropriation. A search for MotA and AsiA orthologs has revealed several other T4-type phage genomes that contain both *motA* and *asiA* genes [[124] and http://phage.bioc.tulane.edu/]. These range from other coliphages (RB51, RB32, and RB69) to more distantly related phages that infect *aeromonas* (PHG25, PHG31, and 44RR) and *acinetobacter* (PHG133). In addition, orthologs for *asiA* have also been found in the genomes of the *vibrio* phages KVP40 and NT1 and the *aeromonas* phages PHG65 and Aeh1, even though these genomes do not have a recognizable *motA*. The KVP40 AsiA protein shares only 27% identity with its T4 counterpart. However, it inhibits transcription by *E. coli* RNAP alone and co-activates transcription with T4 MotA as effectively as T4 AsiA [119]. Thus, it may be that KVP40 and other phages that lack a MotA sequence homolog, do in fact have a functional analog of the MotA protein. Alternatively, the KVP40 AsiA may serve only as an inhibitor of transcription.

No examples of sigma appropriation outside of T4-type phage have been discovered. Although sequence alignments suggested that the *E. coli* anti-sigma protein Rsd, which also interacts with σ^{70}, may be a distant member of the AsiA family [119], a structure of the Rsd/sigma Region 4 complex is not consistent with this idea [30]. Recent work has identified a protein (CT663) involved in the developmental pathway of the human pathogen *Chlamydia trachomatis* that shares functional features with AsiA [125]. It binds both to Region 4 of the primary σ (σ^{66}) of *C. trachomatis* and to the β-flap of core, and it inhibits σ^{66}-dependent transcription. More importantly, like AsiA, it works by remaining bound to the RNAP holoenzyme rather than by sequestering σ^{66}.

Transcription of middle genes by the extension of early transcripts

Even though the expression of middle genes is highly dependent on the activation of middle promoters, isolated mutations within *motA* and *asiA* are surprisingly not lethal. Such mutant phage show a DNA delay phenotype, producing tiny plaques on wt *E. coli* [11,87]. The replication defect reflects the reduced level of T4 replication proteins, whose genes have MotA-dependent middle promoters. In addition, two T4 replication origins are driven by MotA-dependent transcription from the middle promoters, P_{uvsY} and P_{34i} [126]. However, deletion of either *motA* [127] or *asiA* [54] is lethal. Recent work suggests that leakiness of the other nonsense and temperature sensitive mutations provide enough protein for minimal growth [120].

Besides MotA-dependent promoters, middle RNA is also generated by the extension of early transcripts into middle genes. This is because most, if not all, middle genes are positioned downstream of early gene(s) and early promoters. Production of this extended RNA is time-delayed relative to the RNA from the upstream "immediate early (IE)" gene. Thus, middle RNA generated from this extension was originally designated "delayed early" (DE), since it cannot be synthesized until the elongating RNAP reaches the downstream gene(s). Early work (reviewed in [1]) classified genes as IE, DE, or middle based on when and under what conditions the RNA or the encoded protein was observed. IE RNA represents transcripts that are detected immediately after infection and do not require phage protein synthesis. DE RNA requires phage protein synthesis, but this RNA and DE gene products are still detected in a T4 *motA*- infection. In contrast, the expression of genes that were classified as "middle" is significantly reduced in a T4 motA- infection. In addition, while both DE and "middle" RNA arise after IE transcription, the peak of the RNA that is substantially dependent on MotA is slightly later and lasts somewhat longer than the DE peak. However, it should be noted that these original designations of genes as DE or middle are now known to be somewhat arbitrary. Many, if not all, of these genes are transcribed from both early and middle promoters. In fact, while a microarray analysis investigating the timing of various prereplicative RNAs [128] was generally consistent with known Pe and Pm promoters [4], there were a number of discrepancies, especially between genes that were originally classified as either "DE" or "middle". Thus, it is now clear that both the extension of early transcripts and the activation of middle promoters is important for the correct level of middle transcription.

Early experiments [summarized in [1]] offered evidence that DE RNA synthesis might require a T4 system to overcome Rho-dependent termination sites located between IE and DE genes. First, the addition of chloramphenicol at the start of a T4 infection prevents the generation of DE RNAs, indicating a requirement for protein synthesis and suggesting that phage-encoded factor(s) might be needed for the extension of IE RNAs. Second, in a purified *in vitro* system using RNAP and T4 DNA, both IE and DE RNA are synthesized unless the termination factor Rho is added. Addition of Rho restricts transcription to IE RNA, indicating that Rho-dependent termination sites are located upstream of DE genes. Third, DE RNA from a specific promoter upstream of gene 32 is not observed in a T4 *motA*-infection, suggesting that MotA itself may be needed to form or stabilize this DE RNA [129]. It is unlikely that a MotA-dependent gene product, rather than MotA, is responsible for this effect, since the DE transcripts are synthesized before or simultaneously with the activation of middle promoters. Finally, wt T4 does not grow in particular *rho* mutant alleles, called *nusD*, that produce Rho proteins with altered activity, and the level of certain DE RNAs and DE gene products in T4/*nusD* infections is depressed. An initial interpretation of this result was that there is more Rho-dependent termination in a *nusD* allele, which then depresses the level of DE RNA. T4 suppressors that grow in *nusD* were subsequently isolated and found to contain mutations within the T4 *comC-α* (also called *goF*) gene [130,131], which expresses an early product.

Given all of these findings, it was postulated that T4 uses an anti-termination system, perhaps like the N or Q systems of phage λ [reviewed in [132]], to actively prevent Rho-dependent termination and that MotA, ComC-α, or another protein is involved in this process. However, *comC-α* is not essential, and the addition of amino acid analogs, which would generate nonfunctional proteins, has been shown to be sufficient for the synthesis of at least certain DE RNAs [reviewed in [1]]. These results suggest that at least in some cases, translation is simply needed to prevent polarity; consequently, the process of translation itself, rather than a specific factor (s), is sufficient to inhibit Rho termination. If so, the loss of DE RNA observed in the presence of Rho *in vitro* would be due to the lack of coupled transcription/translation. Thus, when the upstream gene is being translated in an infection *in vivo*, Rho RNA binding sites would be occluded by ribosomes and consequently unavailable.

More recent work has suggested that Rho may affect DE RNA *in vivo* because of its ability to bind RNA rather than its termination activity [133,134]. Sequencing of the *rho* gene in six *nusD* alleles has revealed that in five cases, the *rho* mutation lies within the RNA-binding site of Rho. Furthermore, the addition of such a mutant Rho protein to an *in vitro* transcription system does not produce more termination but rather results in an altered and complicated pattern of termination. There is actually less termination at legitimate Rho-dependent termination sites, but in some cases, more termination at other sites. Unexpectedly, increasing the amount of the mutant Rho proteins rescues T4 growth in a *nusD* allele, a result that is not compatible with the mutant Rho promoting more termination. In addition, expression of the Rop protein, an RNA-binding protein encoded by the pBR322 plasmid, also rescues T4 growth in *nusD*.

Taken together, these results have led to another hypothesis to explain DE RNA. In this model, T4 DE transcripts *in vivo* are susceptible to nuclease digestion and require a process to limit this degradation. Active translation can prevent this nuclease attack, thus explaining the loss of DE RNA in the presence of chloramphenicol. In addition, a protein that can bind RNA, such as wt Rho, Rop, or perhaps the mutated T4 ComC-α, may also be useful. Thus, the *nusD* Rho proteins are defective not because they terminate IE transcripts more effectively, but because they have lost the ability of wt Rho to bind and somehow protect the RNA. However, it should be noted that as of yet, there is no evidence identifying a particular nuclease(s) involved in this model. Furthermore, the function of wt *comC-α* or exactly how Rho or Rop "protect" DE RNA is not known. Recent work has shown that both transcription termination and increased mRNA stability by RNA-binding proteins are involved in the regulation of gene expression in eukaryotes and their viruses [135,136]. A thorough investigation of these processes in the simple T4 system could provide a powerful tool to understanding this mode of gene regulation.

Conclusion

T4 regulates its development and the timed expression of prereplicative genes by a sophisticated process. In the past few years, we have learned how T4 employs several elegant strategies, from encoding factors to alter the host RNAP specificity to simply degrading the host DNA, in order to overtake the host transcriptional machinery. Some of these strategies have revealed unexpected and fundamentally significant findings about RNAP. For example, studies with T4 early promoters have challenged previous ideas about how the α-CTDs of RNAP affect transcription. Work with host promoters argued that contact between the α-CTDs of RNAP and promoter UP elements or certain activators increases transcription; in particular, α residue Arg265 was crucial for this interaction. Thus, one would expect that modification of Arg265 would depress transcription. However, the activity of certain T4 early promoters actually increases when Arg265 of one of the two RNAP α subunits is ADP-ribosylated. This finding underscores our limited understanding of α-CTD function and highlights how T4 can provide a tool for investigating this subunit of RNAP.

The T4 system has also revealed a previously unknown method of transcription activation called sigma appropriation. This process is characterized by the binding of a small protein, T4 AsiA, to Region 4 of the σ^{70} subunit of RNAP, which then remodels this portion of polymerase. The conformation of Region 4 in the AsiA/σ^{70} Region 4 structure differs dramatically from that seen in other structures of primary σ factors and demonstrates that Region 4 has a previously unknown flexibility. Furthermore, studies with the T4

MotA activator have identified the far C-terminal region of σ^{70} as a target for activation. Prior to the T4 work, it was thought that this portion of σ^{70}, which is normally embedded within the β-flap "hook" of core, is unavailable. Based on the novel strategy T4 employs to activate its middle promoters, we now know how a domain within RNAP can be remodeled and then exploited to alter promoter specificity. It may be that other examples of this type of RNAP restructuring will be uncovered.

The core subunits of bacterial RNAP are generally conserved throughout biology both in structure and in function [reviewed in [137,138]]. In addition, it is now apparent that eukaryotic RNAP II employs protein complexes that function much like σ factors to recognize different core promoter sequences [[139,140] and references therein]. Thus, the T4 system, which is simple in components yet complex in details, provides an amenable resource for answering basic questions about the complicated process of transcriptional regulation. Using this system, we have been able to uncover at a molecular level many of the protein/protein and protein/DNA interactions that are needed to convert the host RNAP into a RNAP that is dedicated to the phage. This work has given us "snapshots" of the transcriptionally competent protein/DNA complexes generated by the actions of the T4 proteins. The challenge in the future will be to understand at a detailed mechanistic level how these interactions modulate the various "nuts and bolts" of the RNAP machine.

List of abbreviations
bp: base pair(s); ds: double-stranded; ss: single-stranded; RPo: open complex; RPc: closed complex; R or RNAP: RNA polymerase; P: promoter; TGn: -15TGn-13 (extended -10 motif); Pe: T4 early promoter; Pm: T4 middle promoter; rNTPs: ribonucleoside triphosphates; wt: wild type.

Acknowledgements
I thank T. James, K. Decker, L. Knipling, R. Bonocora, M. Hsieh, and C. Philpott for helpful discussions and R. Bonocora for help with the design of Figure 3. This research was supported by the Intramural Research Program of the NIH, National Institute of Diabetes and Digestive and Kidney Diseases.

Authors' contributions
DH is solely responsible for this manuscript.

Competing interests
The author declares that they have no competing interests.

Received: 4 June 2010 Accepted: 28 October 2010
Published: 28 October 2010

References
1. Stitt B, Hinton DM: **Regulation of middle-mode transcription.** In *Molecular biology of bacteriophage T4.* Edited by: Karam JD, Drake J, Kreuzer KN, Mosig G, Hall D, Eiserling F, Black L, Spicer E, Kutter E, Carlson K, Miller ES. Washington, D.C.: American Society for Microbiology; 1994:142-160.
2. Hinton DM, Pande S, Wais N, Johnson XB, Vuthoori M, Makela A, Hook-Barnard I: **Transcriptional takeover by sigma appropriation: remodelling of the sigma70 subunit of Escherichia coli RNA polymerase by the** bacteriophage T4 activator MotA and co-activator AsiA. *Microbiology* 2005, **151**:1729-1740.
3. Brody E, Rabussay D, Hall D: **Regulation of transcription of prereplicative genes.** In *Bacteriophage T4.* Edited by: Mathews CK, Kutter EM, Mosig G, Berget PB. Washington, D. C.: American Society for Microbiology; 1983:174-183.
4. Miller ES, Kutter E, Mosig G, Arisaka F, Kunisawa T, Ruger W: **Bacteriophage T4 genome.** *Microbiol Mol Biol Rev* 2003, **67**:86-156.
5. Wilkens K, Ruger W: **Transcription from early promoters.** In *Molecular Biology of Bacteriophage T4.* Edited by: Karam JD, Drake JW, Kreuzer KN, Mosig G, Hall DH, Eiserling FA, Black LW, Spicer EK, Kutter E, Carlson K, Miller ES. Washington, D. C.: American Society for Microbiology; 1994:132-141.
6. Weisberg R, Hinton DM, Adhya S: **Transcriptional Regulation in Bacteriophage.** In *Encyclopedia of Virology.* 3 edition. Edited by: Mahy BWJ, van Regenmortel MHV. Oxford: Elsevier; 2008:174-186, 174-186.
7. Hinton DM: **Transcription from a bacteriophage T4 middle promoter using T4 motA protein and phage-modified RNA polymerase.** *J Biol Chem* 1991, **266**:18034-18044.
8. Schmidt RP, Kreuzer KN: **Purified MotA protein binds the -30 region of a bacteriophage T4 middle-mode promoter and activates transcription in vitro.** *J Biol Chem* 1992, **267**:11399-11407.
9. Stevens A: **New small polypeptides associated with DNA-dependent RNA polymerase of Escherichia coli after infection with bacteriophage T4.** *Proc Natl Acad Sci USA* 1972, **69**:603-607.
10. Stevens A: **An inhibitor of host sigma-stimulated core enzyme activity that purifies with DNA-dependent RNA polymerase of E. coli following T4 phage infection.** *Biochem Biophys Res Commun* 1973, **54**:488-493.
11. Ouhammouch M, Orsini G, Brody EN: **The asiA gene product of bacteriophage T4 is required for middle mode RNA synthesis.** *J Bacteriol* 1994, **176**:3956-3965.
12. Ouhammouch M, Adelman K, Harvey SR, Orsini G, Brody EN: **Bacteriophage T4 MotA and AsiA proteins suffice to direct Escherichia coli RNA polymerase to initiate transcription at T4 middle promoters.** *Proc Natl Acad Sci USA* 1995, **92**:1451-1455.
13. Hinton DM, March-Amegadzie R, Gerber JS, Sharma M: **Bacteriophage T4 middle transcription system: T4-modified RNA polymerase; AsiA, a sigma 70 binding protein; and transcriptional activator MotA.** *Methods Enzymol* 1996, **274**:43-57.
14. Browning DF, Busby SJ: **The regulation of bacterial transcription initiation.** *Nat Rev Microbiol* 2004, **2**:57-65.
15. Murakami KS, Darst SA: **Bacterial RNA polymerases: the wholo story.** *Curr Opin Struct Biol* 2003, **13**:31-39.
16. Young BA, Gruber TM, Gross CA: **Views of transcription initiation.** *Cell* 2002, **109**:417-420.
17. Paget MS, Helmann JD: **The sigma70 family of sigma factors.** *Genome Biol* 2003, **4**:203.
18. Gruber TM, Gross CA: **Multiple sigma subunits and the partitioning of bacterial transcription space.** *Annu Rev Microbiol* 2003, **57**:441-466.
19. Campbell EA, Westblade LF, Darst SA: **Regulation of bacterial RNA polymerase sigma factor activity: a structural perspective.** *Curr Opin Microbiol* 2008, **11**:121-127.
20. Hook-Barnard IG, Hinton DM: **Transcription Initiation by Mix and Match Elements: Flexibility for Polymerase Binding to Bacterial Promoters.** *Gene Regulation and Systems Biology* 2007 [http://la-press.com/article.php?article_id=481:275-293].
21. Helmann JD: **RNA polymerase: a nexus of gene regulation.** *Methods* 2009, **47**:1-5.
22. Vassylyev DG, Sekine S, Laptenko O, Lee J, Vassylyeva MN, Borukhov S, Yokoyama S: **Crystal structure of a bacterial RNA polymerase holoenzyme at 2.6 A resolution.** *Nature* 2002, **417**:712-719.
23. Murakami KS, Masuda S, Campbell EA, Muzzin O, Darst SA: **Structural basis of transcription initiation: an RNA polymerase holoenzyme-DNA complex.** *Science* 2002, **296**:1285-1290.
24. Murakami KS, Masuda S, Darst SA: **Structural basis of transcription initiation: RNA polymerase holoenzyme at 4 A resolution.** *Science* 2002, **296**:1280-1284.
25. Campbell EA, Muzzin O, Chlenov M, Sun JL, Olson CA, Weinman O, Trester-Zedlitz ML, Darst SA: **Structure of the bacterial RNA polymerase promoter specificity sigma subunit.** *Mol Cell* 2002, **9**:527-539.

26. Lambert LJ, Wei Y, Schirf V, Demeler B, Werner MH: **T4 AsiA blocks DNA recognition by remodeling sigma(70) region 4.** *Embo J* 2004, **23**:2952-2962.

27. Jain D, Nickels BE, Sun L, Hochschild A, Darst SA: **Structure of a ternary transcription activation complex.** *Mol Cell* 2004, **13**:45-53.

28. Vassylyev DG, Vassylyeva MN, Perederina A, Tahirov TH, Artsimovitch I: **Structural basis for transcription elongation by bacterial RNA polymerase.** *Nature* 2007, **448**:157-162.

29. Malhotra A, Severinova E, Darst SA: **Crystal structure of a sigma 70 subunit fragment from E. coli RNA polymerase.** *Cell* 1996, **87**:127-136.

30. Patikoglou GA, Westblade LF, Campbell EA, Lamour V, Lane WJ, Darst SA: **Crystal Structure of the Escherichia coli Regulator of sigma(70), Rsd, in Complex with sigma(70) Domain 4.** *J Mol Biol* 2007, **372**:649-659.

31. Sasse-Dwight S, Gralla JD: **KMnO4 as a probe for lac promoter DNA melting and mechanism in vivo.** *J Biol Chem* 1989, **264**:8074-8081.

32. Kontur WS, Saecker RM, Davis CA, Capp MW, Record MT Jr: **Solute probes of conformational changes in open complex (RPo) formation by Escherichia coli RNA polymerase at the lambdaPR promoter: evidence for unmasking of the active site in the isomerization step and for large-scale coupled folding in the subsequent conversion to RPo.** *Biochemistry* 2006, **45**:2161-2177.

33. Saecker RM, Tsodikov OV, McQuade KL, Schlax PE Jr, Capp MW, Record MT Jr: **Kinetic studies and structural models of the association of E. coli sigma(70) RNA polymerase with the lambdaP(R) promoter: large scale conformational changes in forming the kinetically significant intermediates.** *J Mol Biol* 2002, **319**:649-671.

34. Erie DA: **The many conformational states of RNA polymerase elongation complexes and their roles in the regulation of transcription.** *Biochim Biophys Acta* 2002, **1577**:224-239.

35. Landick R: **The regulatory roles and mechanism of transcriptional pausing.** *Biochem Soc Trans* 2006, **34**:1062-1066.

36. Ciampi MS: **Rho-dependent terminators and transcription termination.** *Microbiology* 2006, **152**:2515-2528.

37. Gilmour DS, Fan R: **Derailing the locomotive: transcription termination.** *J Biol Chem* 2008, **283**:661-664.

38. Adhya S, Gottesman M: **Control of transcription termination.** *Annu Rev Biochem* 1978, **47**:967-996.

39. Wilkens K, Ruger W: **Characterization of bacteriophage T4 early promoters in vivo with a new promoter probe vector.** *Plasmid* 1996, **35**:108-120.

40. Koch T, Raudonikiene A, Wilkens K, Ruger W: **Overexpression, purification, and characterization of the ADP-ribosyltransferase (gpAlt) of bacteriophage T4: ADP-ribosylation of E. coli RNA polymerase modulates T4 "early" transcription.** *Gene Expr* 1995, **4**:253-264.

41. Horvitz HR: **Bacteriophage T4 mutants deficient in alteration and modification of the Escherichia coli RNA polymerase.** *J Mol Biol* 1974, **90**:739-750.

42. Horvitz HR: **Control by bacteriophage T4 of two sequential phosphorylations of the alpha subunit of Escherichia coli RNA polymerase.** *J Mol Biol* 1974, **90**:727-738.

43. Depping R, Lohaus C, Meyer HE, Ruger W: **The mono-ADP-ribosyltransferases Alt and ModB of bacteriophage T4: target proteins identified.** *Biochem Biophys Res Commun* 2005, **335**:1217-1223.

44. Ross W, Gosink KK, Salomon J, Igarashi K, Zou C, Ishihama A, Severinov K, Gourse RL: **A third recognition element in bacterial promoters: DNA binding by the alpha subunit of RNA polymerase.** *Science* 1993, **262**:1407-1413.

45. Gaal T, Ross W, Blatter EE, Tang H, Jia X, Krishnan VV, Assa-Munt N, Ebright RH, Gourse RL: **DNA-binding determinants of the alpha subunit of RNA polymerase: novel DNA-binding domain architecture.** *Genes Dev* 1996, **10**:16-26.

46. Murakami K, Fujita N, Ishihama A: **Transcription factor recognition surface on the RNA polymerase alpha subunit is involved in contact with the DNA enhancer element.** *EMBO J* 1996, **15**:4358-4367.

47. Zou C, Fujita N, Igarashi K, Ishihama A: **Mapping the cAMP receptor protein contact site on the alpha subunit of Escherichia coli RNA polymerase.** *Mol Microbiol* 1992, **6**:2599-2605.

48. Tiemann B, Depping R, Gineikiene E, Kaliniene L, Nivinskas R, Ruger W: **ModA and ModB, two ADP-ribosyltransferases encoded by bacteriophage T4: catalytic properties and mutation analysis.** *J Bacteriol* 2004, **186**:7262-7272.

49. Wilkens K, Tiemann B, Bazan F, Ruger W: **ADP-ribosylation and early transcription regulation by bacteriophage T4.** *Adv Exp Med Biol* 1997, **419**:71-82.

50. Sommer N, Salniene V, Gineikiene E, Nivinskas R, Ruger W: **T4 early promoter strength probed in vivo with unribosylated and ADP-ribosylated Escherichia coli RNA polymerase: a mutation analysis.** *Microbiology* 2000, **146**:2643-2653.

51. Kashlev M, Nudler E, Goldfarb A, White T, Kutter E: **Bacteriophage T4 Alc protein: a transcription termination factor sensing local modification of DNA.** *Cell* 1993, **75**:147-154.

52. Severinov K, Kashlev M, Severinova E, Bass I, McWilliams K, Kutter E, Nikiforov V, Snyder L, Goldfarb A: **A non-essential domain of Escherichia coli RNA polymerase required for the action of the termination factor Alc.** *J Biol Chem* 1994, **269**:14254-14259.

53. Tiemann B, Depping R, Ruger W: **Overexpression, purification, and partial characterization of ADP-ribosyltransferases modA and modB of bacteriophage T4.** *Gene Expr* 1999, **8**:187-196.

54. Pene C, Uzan M: **The bacteriophage T4 anti-sigma factor AsiA is not necessary for the inhibition of early promoters in vivo.** *Mol Microbiol* 2000, **35**:1180-1191.

55. Sanson B, Uzan M: **Dual role of the sequence-specific bacteriophage T4 endoribonuclease RegB. mRNA inactivation and mRNA destabilization.** *J Mol Biol* 1993, **233**:429-446.

56. Hirano N, Ohshima H, Takahashi H: **Biochemical analysis of the substrate specificity and sequence preference of endonuclease IV from bacteriophage T4, a dC-specific endonuclease implicated in restriction of dC-substituted T4 DNA synthesis.** *Nucleic Acids Res* 2006, **34**:4743-4751.

57. Carlson K, Lagerback P, Nystrom AC: **Bacteriophage T4 endonuclease II: concerted single-strand nicks yield double-strand cleavage.** *Mol Microbiol* 2004, **52**:1403-1411.

58. Stevens A, Rhoton JC: **Characterization of an inhibitor causing potassium chloride sensitivity of an RNA polymerase from T4 phage-infected Escherichia coli.** *Biochemistry* 1975, **14**:5074-5079.

59. Orsini G, Ouhammouch M, Le Caer JP, Brody EN: **The asiA gene of bacteriophage T4 codes for the anti-sigma 70 protein.** *J Bacteriol* 1993, **175**:85-93.

60. Adelman K, Orsini G, Kolb A, Graziani L, Brody EN: **The interaction between the AsiA protein of bacteriophage T4 and the sigma70 subunit of Escherichia coli RNA polymerase.** *J Biol Chem* 1997, **272**:27435-27443.

61. Severinova E, Severinov K, Fenyo D, Marr M, Brody EN, Roberts JW, Chait BT, Darst SA: **Domain organization of the Escherichia coli RNA polymerase sigma 70 subunit.** *J Mol Biol* 1996, **263**:637-647.

62. Colland F, Orsini G, Brody EN, Buc H, Kolb A: **The bacteriophage T4 AsiA protein: a molecular switch for sigma 70-dependent promoters.** *Mol Microbiol* 1998, **27**:819-829.

63. Severinov K, Muir TW: **Expressed protein ligation, a novel method for studying protein-protein interactions in transcription.** *J Biol Chem* 1998, **273**:16205-16209.

64. Sharma UK, Ravishankar S, Shandil RK, Praveen PV, Balganesh TS: **Study of the interaction between bacteriophage T4 asiA and Escherichia coli sigma(70), using the yeast two-hybrid system: neutralization of asiA toxicity to E. coli cells by coexpression of a truncated sigma(70) fragment.** *J Bacteriol* 1999, **181**:5855-5859.

65. Urbauer JL, Adelman K, Urbauer RJ, Simeonov MF, Gilmore JM, Zolkiewski M, Brody EN: **Conserved regions 4.1 and 4.2 of sigma(70) constitute the recognition sites for the anti-sigma factor AsiA, and AsiA is a dimer free in solution.** *J Biol Chem* 2001, **276**:41128-41132.

66. Pahari S, Chatterji D: **Interaction of bacteriophage T4 AsiA protein with Escherichia coli sigma70 and its variant.** *FEBS Lett* 1997, **411**:60-62.

67. Simeonov MF, Bieber Urbauer RJ, Gilmore JM, Adelman K, Brody EN, Niedziela-Majka A, Minakhin L, Heyduk T, Urbauer JL: **Characterization of the interactions between the bacteriophage T4 AsiA protein and RNA polymerase.** *Biochemistry* 2003, **42**:7717-7726.

68. Sharma UK, Chatterji D: **Both regions 4.1 and 4.2 of E. coli sigma(70) are together required for binding to bacteriophage T4 AsiA in vivo.** *Gene* 2006, **376**:133-143.

69. Minakhin L, Niedziela-Majka A, Kuznedelov K, Adelman K, Urbauer JL, Heyduk T, Severinov K: **Interaction of T4 AsiA with its target sites in the RNA polymerase sigma70 subunit leads to distinct and opposite effects on transcription.** *J Mol Biol* 2003, **326**:679-690.

70. Dove SL, Hochschild A: **Bacterial two-hybrid analysis of interactions between region 4 of the sigma(70) subunit of RNA polymerase and the transcriptional regulators Rsd from Escherichia coli and AlgQ from Pseudomonas aeruginosa.** *J Bacteriol* 2001, **183**:6413-6421.

71. Urbauer JL, Simeonov MF, Urbauer RJ, Adelman K, Gilmore JM, Brody EN: **Solution structure and stability of the anti-sigma factor AsiA: implications for novel functions.** *Proc Natl Acad Sci USA* 2002, **99**:1831-1835.

72. Pal D, Vuthoori M, Pande S, Wheeler D, Hinton DM: **Analysis of regions within the bacteriophage T4 AsiA protein involved in its binding to the sigma70 subunit of E. coli RNA polymerase and its role as a transcriptional inhibitor and co-activator.** *J Mol Biol* 2003, **325**:827-841.

73. Lambert LJ, Schirf V, Demeler B, Cadene M, Werner MH: **Flipping a genetic switch by subunit exchange.** *Embo J* 2001, **20**:7149-7159.

74. Gilmore JM, Bieber Urbauer RJ, Minakhin L, Akoyev V, Zolkiewski M, Severinov K, Urbauer JL: **Determinants of Affinity and Activity of the Anti-Sigma Factor AsiA.** *Biochemistry* 2010.

75. Sharma UK, Praveen PV, Balganesh TS: **Mutational analysis of bacteriophage T4 AsiA: involvement of N- and C-terminal regions in binding to sigma(70) of Escherichia coli in vivo.** *Gene* 2002, **295**:125-134.

76. Stevens A: **A salt-promoted inhibitor of RNA polymerase isolated from T4 phage-infected E. coli.** In *RNA Polymerase*. Edited by: Losick R, Chamberlin M. Cold Spring Harbor, N. Y.: Cold Spring Harbor Laboratory; 1976:617-627.

77. Baxter K, Lee J, Minakhin L, Severinov K, Hinton DM: **Mutational Analysis of sigma(70) Region 4 Needed for Appropriation by the Bacteriophage T4 Transcription Factors AsiA and MotA.** *J Mol Biol* 2006, **363**:931-944.

78. Minakhin L, Camarero JA, Holford M, Parker C, Muir TW, Severinov K: **Mapping the molecular interface between the sigma(70) subunit of E. coli RNA polymerase and T4 AsiA.** *J Mol Biol* 2001, **306**:631-642.

79. Gregory BD, Nickels BE, Garrity SJ, Severinova E, Minakhin L, Urbauer RJ, Urbauer JL, Heyduk T, Severinov K, Hochschild A: **A regulator that inhibits transcription by targeting an intersubunit interaction of the RNA polymerase holoenzyme.** *Proc Natl Acad Sci USA* 2004, **101**:4554-4559.

80. Severinova E, Severinov K, Darst SA: **Inhibition of Escherichia coli RNA polymerase by bacteriophage T4 AsiA.** *J Mol Biol* 1998, **279**:9-18.

81. Leirmo S, Harrison C, Cayley DS, Burgess RR, Record MT Jr: **Replacement of potassium chloride by potassium glutamate dramatically enhances protein-DNA interactions in vitro.** *Biochemistry* 1987, **26**:2095-2101.

82. Zou LL, Richardson JP: **Enhancement of transcription termination factor rho activity with potassium glutamate.** *J Biol Chem* 1991, **266**:10201-10209.

83. Orsini G, Kolb A, Buc H: **The Escherichia coli RNA polymerase.anti-sigma 70 AsiA complex utilizes alpha-carboxyl-terminal domain upstream promoter contacts to transcribe from a -10/-35 promoter.** *J Biol Chem* 2001, **276**:19812-19819.

84. Orsini G, Igonet S, Pene C, Sclavi B, Buckle M, Uzan M, Kolb A: **Phage T4 early promoters are resistant to inhibition by the anti-sigma factor AsiA.** *Mol Microbiol* 2004, **52**:1013-1028.

85. Kolesky S, Ouhammouch M, Brody EN, Geiduschek EP: **Sigma competition: the contest between bacteriophage T4 middle and late transcription.** *J Mol Biol* 1999, **291**:267-281.

86. Mattson T, Richardson J, Goodin D: **Mutant of bacteriophage T4D affecting expression of many early genes.** *Nature* 1974, **250**:48-50.

87. Mattson T, Van Houwe G, Epstein RH: **Isolation and characterization of conditional lethal mutations in the mot gene of bacteriophage T4.** *J Mol Biol* 1978, **126**:551-570.

88. Uzan M, Brody E, Favre R: **Nucleotide sequence and control of transcription of the bacteriophage T4 motA regulatory gene.** *Mol Microbiol* 1990, **4**:1487-1496.

89. Hinton DM, March-Amegadzie R, Gerber JS, Sharma M: **Characterization of pre-transcription complexes made at a bacteriophage T4 middle promoter: involvement of the T4 MotA activator and the T4 AsiA protein, a sigma 70 binding protein, in the formation of the open complex.** *J Mol Biol* 1996, **256**:235-248.

90. March-Amegadzie R, Hinton DM: **The bacteriophage T4 middle promoter PuvsX: analysis of regions important for binding of the T4 transcriptional activator MotA and for activation of transcription.** *Mol Microbiol* 1995, **15**:649-660.

91. Stoskiene G, Truncaite L, Zajanckauskaite A, Nivinskas R: **Middle promoters constitute the most abundant and diverse class of promoters in bacteriophage T4.** *Mol Microbiol* 2007, **64**:421-434.

92. Cicero MP, Alexander KA, Kreuzer KN: **The MotA transcriptional activator of bacteriophage T4 binds to its specific DNA site as a monomer.** *Biochemistry* 1998, **37**:4977-4984.

93. Li N, Sickmier EA, Zhang R, Joachimiak A, White SW: **The MotA transcription factor from bacteriophage T4 contains a novel DNA-binding domain: the 'double wing' motif.** *Mol Microbiol* 2002, **43**:1079-1088.

94. Li N, Zhang W, White SW, Kriwacki RW: **Solution structure of the transcriptional activation domain of the bacteriophage T4 protein, MotA.** *Biochemistry* 2001, **40**:4293-4302.

95. Finnin MS, Hoffman DW, Kreuzer KN, Porter SJ, Schmidt RP, White SW: **The MotA protein from bacteriophage T4 contains two domains. Preliminary structural analysis by X-ray diffraction and nuclear magnetic resonance.** *J Mol Biol* 1993, **232**:301-304.

96. Gerber JS, Hinton DM: **An N-terminal mutation in the bacteriophage T4 motA gene yields a protein that binds DNA but is defective for activation of transcription.** *J Bacteriol* 1996, **178**:6133-6139.

97. Pande S, Makela A, Dove SL, Nickels BE, Hochschild A, Hinton DM: **The bacteriophage T4 transcription activator MotA interacts with the far-C-terminal region of the sigma70 subunit of Escherichia coli RNA polymerase.** *J Bacteriol* 2002, **184**:3957-3964.

98. Finnin MS, Cicero MP, Davies C, Porter SJ, White SW, Kreuzer KN: **The activation domain of the MotA transcription factor from bacteriophage T4.** *Embo J* 1997, **16**:1992-2003.

99. Guild N, Gayle M, Sweeney R, Hollingsworth T, Modeer T, Gold L: **Transcriptional activation of bacteriophage T4 middle promoters by the motA protein.** *J Mol Biol* 1988, **199**:241-258.

100. Marshall P, Sharma M, Hinton DM: **The bacteriophage T4 transcriptional activator MotA accepts various base-pair changes within its binding sequence.** *J Mol Biol* 1999, **285**:931-944.

101. Truncaite L, Piesiniene L, Kolesinskiene G, Zajanckauskaite A, Driukas A, Klausa V, Nivinskas R: **Twelve new MotA-dependent middle promoters of bacteriophage T4: consensus sequence revised.** *J Mol Biol* 2003, **327**:335-346.

102. Truncaite L, Zajanckauskaite A, Nivinskas R: **Identification of two middle promoters upstream DNA ligase gene 30 of bacteriophage T4.** *J Mol Biol* 2002, **317**:179-190.

103. Sharma M, Marshall P, Hinton DM: **Binding of the bacteriophage T4 transcriptional activator, MotA, to T4 middle promoter DNA: evidence for both major and minor groove contacts.** *J Mol Biol* 1999, **290**:905-915.

104. Bonocora RP, Caignan G, Woodrell C, Werner MH, Hinton DM: **A basic/hydrophobic cleft of the T4 activator MotA interacts with the C-terminus of E. coli sigma70 to activate middle gene transcription.** *Mol Microbiol* 2008, **69**:331-343.

105. Kuldell N, Hochschild A: **Amino acid substitutions in the -35 recognition motif of sigma 70 that result in defects in phage lambda repressor-stimulated transcription.** *J Bacteriol* 1994, **176**:2991-2998.

106. Li M, Moyle H, Susskind MM: **Target of the transcriptional activation function of phage lambda cI protein.** *Science* 1994, **263**:75-77.

107. Nickels BE, Dove SL, Murakami KS, Darst SA, Hochschild A: **Protein-protein and protein-DNA interactions of sigma70 region 4 involved in transcription activation by lambda cI.** *J Mol Biol* 2002, **324**:17-34.

108. Rhodius VA, Busby SJ: **Interactions between activating region 3 of the Escherichia coli cyclic AMP receptor protein and region 4 of the RNA polymerase sigma(70) subunit: application of suppression genetics.** *J Mol Biol* 2000, **299**:311-324.

109. Lonetto MA, Rhodius V, Lamberg K, Kiley P, Busby S, Gross C: **Identification of a contact site for different transcription activators in region 4 of the Escherichia coli RNA polymerase sigma70 subunit.** *J Mol Biol* 1998, **284**:1353-1365.

110. Landini P, Busby SJ: **The Escherichia coli Ada protein can interact with two distinct determinants in the sigma70 subunit of RNA polymerase according to promoter architecture: identification of the target of Ada activation at the alkA promoter.** *J Bacteriol* 1999, **181**:1524-1529.

111. Bhende PM, Egan SM: **Genetic evidence that transcription activation by RhaS involves specific amino acid contacts with sigma 70.** *J Bacteriol* 2000, **182**:4959-4969.

112. Wickstrum JR, Egan SM: **Amino acid contacts between sigma 70 domain 4 and the transcription activators RhaS and RhaR.** *J Bacteriol* 2004, **186**:6277-6285.

113. Decker KB, Hinton DM: **The secret to 6S: regulating RNA polymerase by ribo-sequestration.** *Mol Microbiol* 2009, **73**:137-140.

114. Pulitzer JF, Coppo A, Caruso M: **Host–virus interactions in the control of T4 prereplicative transcription. II. Interaction between tabC (rho) mutants and T4 mot mutants.** *J Mol Biol* 1979, **135**:979-997.

115. Cicero MP, Sharp MM, Gross CA, Kreuzer KN: **Substitutions in bacteriophage T4 AsiA and Escherichia coli sigma(70) that suppress T4 motA activation mutations.** *J Bacteriol* 2001, **183**:2289-2297.

116. Yuan AH, Nickels BE, Hochschild A: **The bacteriophage T4 AsiA protein contacts the beta-flap domain of RNA polymerase.** *Proc Natl Acad Sci USA* 2009, **106**:6597-6602.

117. Yuan AH, Hochschild A: **Direct activator/co-activator interaction is essential for bacteriophage T4 middle gene expression.** *Mol Microbiol* 2009, **74**:1018-1030.

118. Yuan AH, Hochschild A: **Direct activator/co-activator interaction is essential for bacteriophage T4 middle gene expression.** *Mol Microbiol* 2009.

119. Pineda M, Gregory BD, Szczypinski B, Baxter KR, Hochschild A, Miller ES, Hinton DM: **A family of anti-sigma70 proteins in T4-type phages and bacteria that are similar to AsiA, a Transcription inhibitor and co-activator of bacteriophage T4.** *J Mol Biol* 2004, **344**:1183-1197.

120. James TD, Cashel M, Hinton DM: **A mutation within the β subunit of Escherichia coli RNA polymerase impairs transcription from bacteriophage T4 middle promoters.** *J Bacteriol* 2010.

121. Adelman K, Brody EN, Buckle M: **Stimulation of bacteriophage T4 middle transcription by the T4 proteins MotA and AsiA occurs at two distinct steps in the transcription cycle.** *Proc Natl Acad Sci USA* 1998, **95**:15247-15252.

122. Hinton DM, Vuthoori S: **Efficient inhibition of Escherichia coli RNA polymerase by the bacteriophage T4 AsiA protein requires that AsiA binds first to free sigma70.** *J Mol Biol* 2000, **304**:731-739.

123. Wade JT, Struhl K: **Association of RNA polymerase with transcribed regions in Escherichia coli.** *Proc Natl Acad Sci USA* 2004, **101**:17777-17782.

124. Miller ES, Heidelberg JF, Eisen JA, Nelson WC, Durkin AS, Ciecko A, Feldblyum TV, White O, Paulsen IT, Nierman WC, *et al*: **Complete genome sequence of the broad-host-range vibriophage KVP40: comparative genomics of a T4-related bacteriophage.** *J Bacteriol* 2003, **185**:5220-5233.

125. Rao X, Deighan P, Hua Z, Hu X, Wang J, Luo M, Liang Y, Zhong G, Hochschild A, Shen L: **A regulator from Chlamydia trachomatis modulates the activity of RNA polymerase through direct interaction with the beta subunit and the primary sigma subunit.** *Genes Dev* 2009, **23**:1818-1829.

126. Menkens AE, Kreuzer KN: **Deletion analysis of bacteriophage T4 tertiary origins. A promoter sequence is required for a rifampicin-resistant replication origin.** *J Biol Chem* 1988, **263**:11358-11365.

127. Benson KH, Kreuzer KN: **Role of MotA transcription factor in bacteriophage T4 DNA replication.** *J Mol Biol* 1992, **228**:88-100.

128. Luke K, Radek A, Liu X, Campbell J, Uzan M, Haselkorn R, Kogan Y: **Microarray analysis of gene expression during bacteriophage T4 infection.** *Virology* 2002, **299**:182-191.

129. Carpousis AJ, Mudd EA, Krisch HM: **Transcription and messenger RNA processing upstream of bacteriophage T4 gene 32.** *Mol Gen Genet* 1989, **219**:39-48.

130. Sanson B, Uzan M: **Sequence and characterization of the bacteriophage T4 comC alpha gene product, a possible transcription antitermination factor.** *J Bacteriol* 1992, **174**:6539-6547.

131. Chiurazzi M, Pulitzer JF: **Characterisation of the bacteriophage T4 comC alpha 55.6 and comCJ mutants. A possible role in an antitermination process.** *FEMS Microbiol Lett* 1998, **166**:187-195.

132. Dodd IB, Shearwin KE, Egan JB: **Revisited gene regulation in bacteriophage lambda.** *Curr Opin Genet Dev* 2005, **15**:145-152.

133. Washburn RS, Stitt BL: **In vitro characterization of transcription termination factor Rho from Escherichia coli rho(nusD) mutants.** *J Mol Biol* 1996, **260**:332-346.

134. Sozhamannan S, Stitt BL: **Effects on mRNA degradation by Escherichia coli transcription termination factor Rho and pBR322 copy number control protein Rop.** *J Mol Biol* 1997, **268**:689-703.

135. Glisovic T, Bachorik JL, Yong J, Dreyfuss G: **RNA-binding proteins and post-transcriptional gene regulation.** *FEBS Lett* 2008, **582**:1977-1986.

136. Zhang Z, Klatt A, Henderson AJ, Gilmour DS: **Transcription termination factor Pcf11 limits the processivity of Pol II on an HIV provirus to repress gene expression.** *Genes Dev* 2007, **21**:1609-1614.

137. Ebright RH: **RNA polymerase: structural similarities between bacterial RNA polymerase and eukaryotic RNA polymerase II.** *J Mol Biol* 2000, **304**:687-698.

138. Werner F: **Structural evolution of multisubunit RNA polymerases.** *Trends Microbiol* 2008, **16**:247-250.

139. Freiman RN, Albright SR, Zheng S, Sha WC, Hammer RE, Tjian R: **Requirement of tissue-selective TBP-associated factor TAFII105 in ovarian development.** *Science* 2001, **293**:2084-2087.

140. Isogai Y, Keles S, Prestel M, Hochheimer A, Tjian R: **Transcription of histone gene cluster by differential core-promoter factors.** *Genes Dev* 2007, **21**:2936-2949.

141. Daegelen P, Brody E: **The rIIA gene of bacteriophage T4. II. Regulation of its messenger RNA synthesis.** *Genetics* 1990, **125**:249-260.

142. Daegelen P, Brody E: **The rIIA gene of bacteriophage T4. I. Its DNA sequence and discovery of a new open reading frame between genes 60 and rIIA.** *Genetics* 1990, **125**:237-248.

143. Hsu T, Karam JD: **Transcriptional mapping of a DNA replication gene cluster in bacteriophage T4. Sites for initiation, termination, and mRNA processing.** *J Biol Chem* 1990, **265**:5303-5316.

144. Barth KA, Powell D, Trupin M, Mosig G: **Regulation of two nested proteins from gene 49 (recombination endonuclease VII) and of a lambda RexA-like protein of bacteriophage T4.** *Genetics* 1988, **120**:329-343.

145. Gruidl ME, Mosig G: **Sequence and transcripts of the bacteriophage T4 DNA repair gene uvsY.** *Genetics* 1986, **114**:1061-1079.

146. Nivinskas RG, Raudonikene AA, Guild N: **A new early gene in front of the middle gene 31 of bacteriophage T4: cloning and expression.** *Mol Biol (Mosk)* 1989, **23**:739-749.

147. Gott JM, Zeeh A, Bell-Pedersen D, Ehrenman K, Belfort M, Shub DA: **Genes within genes: independent expression of phage T4 intron open reading frames and the genes in which they reside.** *Genes Dev* 1988, **2**:1791-1799.

148. Tseng MJ, He P, Hilfinger JM, Greenberg GR: **Bacteriophage T4 nrdA and nrdB genes, encoding ribonucleotide reductase, are expressed both separately and coordinately: characterization of the nrdB promoter.** *J Bacteriol* 1990, **172**:6323-6332.

149. Belin D, Mudd EA, Prentki P, Yi-Yi Y, Krisch HM: **Sense and antisense transcription of bacteriophage T4 gene 32. Processing and stability of the mRNAs.** *J Mol Biol* 1987, **194**:231-243.

150. Estrem ST, Ross W, Gaal T, Chen ZW, Niu W, Ebright RH, Gourse RL: **Bacterial promoter architecture: subsite structure of UP elements and interactions with the carboxy-terminal domain of the RNA polymerase alpha subunit.** *Genes Dev* 1999, **13**:2134-2147.

doi:10.1186/1743-422X-7-289
Cite this article as: Hinton: Transcriptional control in the prereplicative phase of T4 development. *Virology Journal* 2010 **7**:289.

Geiduschek and Kassavetis *Virology Journal* 2010, **7**:288
http://www.virologyj.com/content/7/1/288

VIROLOGY JOURNAL

Transcription of the T4 late genes

E Peter Geiduschek[*], George A Kassavetis[*]

Abstract

This article reviews the current state of understanding of the regulated transcription of the bacteriophage T4 late genes, with a focus on the underlying biochemical mechanisms, which turn out to be unique to the T4-related family of phages or significantly different from other bacterial systems. The activator of T4 late transcription is the *gene 45 protein* (gp45), the sliding clamp of the T4 replisome. Gp45 becomes topologically linked to DNA through the action of its clamp-loader, but it is not site-specifically DNA-bound, as other transcriptional activators are. Gp45 facilitates RNA polymerase recruitment to late promoters by interacting with two phage-encoded polymerase subunits: gp33, the co-activator of T4 late transcription; and gp55, the T4 late promoter recognition protein. The emphasis of this account is on the sites and mechanisms of actions of these three proteins, and on their roles in the formation of transcription-ready open T4 late promoter complexes.

Introduction

T4 late genes are transcribed from simple promoters consisting of an 8-base pair TATA box placed ~1 helical DNA turn upstream of the transcriptional start site (the location of the bacterial σ^{70}-family RNA polymerase (RNAP) promoter -10 site). A significant AT base pair preponderance characterizes the segment immediately downstream of the TATA box that strand-separates when the late promoter opens for initiation of transcription; there is no sequence conservation at the position corresponding to the bacterial promoter -35 site.

Fifty of these sites are listed for the T4 genome [1,2]. The consensus first proposed by Christensen and Young [3] is tightly adhered to overall (Figure 1), perfectly so at 32 sites, with A(-13) in place of T at nine sites and other single deviations from consensus at the remaining sites, with two exceptions, (one a TA→AT change). Variant T4 late promoters are used for (basal) transcription *in vitro* [4] and a number of variant promoters have also been associated with RNA 5" ends *in vivo* [5,6] (Three cautionary notes: 1) these 50 sites have not all been identified as promoters that are active *in vivo*; 2) some of the RNA 5" ends that have been mapped to putative promoters were specified by primer extension analysis, which does not distinguish between 5" ends generated by bona fide initiation and endonucleolytic processing; 3) the relative rates of initiation at consensus and

variant T4 late promoters *in vivo* have not been determined.) While all early and middle transcripts have the same polarity, that is, counterclockwise in the standard representation of the T4 genetic map, and complementary to the DNA *l* strand [7], late transcripts have either polarity. At several sites, both T4 DNA strands are transcribed at different times of the multiplication cycle [8,9].

Transcription initiating at these simple promoters requires the function of T4 genes 33 and 55. These two genes hold a special place in the history of molecular biology, because they are the first master regulators of a developmental program of gene expression to have been discovered [10]. Both genes encode RNAP-binding proteins [11,12]: the gene 55 protein (gp55) is the smallest and one of the most highly divergent members of the σ^{70} family [13-15], while gp33 has no recognizable homology with σ proteins. The phenotypes of cells infected with 33⁻ and 55⁻ phage are, however, not the same. In the absence of gene 55 function, late genes are not transcribed. In contrast, some late transcription eventually materializes, and late proteins are also made at reduced levels, in cells infected with gene 33-defective phage. These differences of phenotype of gene 33 and gene 55 mutants reflect the different mechanisms of action of gp33 and gp55 in transcription, as discussed below. Late transcription normally also requires DNA replication [10,16] and is, in fact, coupled to concurrent DNA synthesis [17].

* Correspondence: epg@ucsd.edu; gak@ucsd.edu
Division of Biological Sciences, Section of Molecular Biology, University of California, San Diego, La Jolla, CA 92093-0634, USA

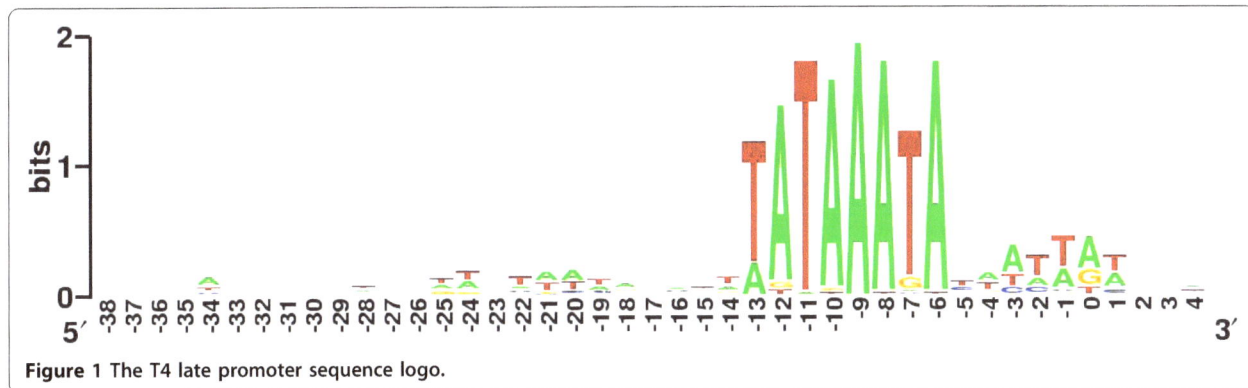

Figure 1 The T4 late promoter sequence logo.

The coupling of late transcription to DNA replication is enforced by the action of gp30, the T4 DNA ligase [18]. Single-strand breaks make T4 DNA subject to nucleolytic attack, but protecting against that degradation by knocking out the exonuclease function encoded by gene 46 generates a situation in which late transcription occurs in the absence of DNA replication (e.g., in the absence of T4 DNA polymerase (gp43) function) [19,20]. Thus, the just-specified gene $30^-/43^-/46^-$ triple mutant serves as a platform for finding proteins that are not only required for T4 DNA replication but have an additional direct role in late transcription. Those experiments clearly identify the involvement of gp45, the sliding clamp processivity factor of the T4 DNA polymerase holoenzyme, in T4 late transcription [21]. (That this approach does not equally clearly identify the involvement of the gp44/62 clamp loader complex in T4 late transcription is puzzling, as discussed further on.)

In summary, the primary direct roles in T4 late transcription are played by three proteins–gp55, gp33 and gp45–and by a transient form of the T4 DNA template that is generated in the process of replication. The focus of the rest of this account is on explaining the mechanisms of action of these components.

Gp55

Gp55 is a very small, highly diverged σ^{70}-family protein (Figure 2). The σ^{70}/σ^A subunits of the bacterial RNAPs comprise 4 globular domains (σ_1, σ_2, σ_3 and σ_4; Figure 3) that are widely separated on the surface of the RNAP holoenzymes. When σ detaches from the RNAP core, these domains swap their sites of interaction with the β and β" RNAP subunits for internal contacts and assume a compact structure [22,23]. The σ structural domains also correspond with segments of sequence conservation (segments 1.1, 1.2; 2.1-2.4; 2.5 and 3.1; 4.1 and 4.2;[15]). Discernible similarity of gp55 with σ^{70} is confined to domain 2 [13-15], which provides the principal RNAP core-binding and -10 DNA site-recognition functions of σ proteins (involving conserved sequence segments 2.2

and 2.4, respectively) [24-26]. Since a direct determination of gp55 structure is not yet at hand, what follows pieces together the information that can be derived from site-directed mutagenesis, analysis of function and interactions *in vitro*, and consideration of amino acid sequence conservation.

Gp55 is the promoter recognition subunit of the T4 late gene-transcribing RNAP holoenzyme [27] and confers the ability to execute basal level accurately initiating transcription on unmodified and exhaustively σ-stripped *E. coli* RNAP core. This basal transcription by gp55•RNAP is sensitive to ionic strength, and greatly reduced at lower temperature or when relaxed DNA is used as template in place of supercoiled plasmid DNA [27-31].

Initial binding of gp55•RNAP to DNA is not highly specific, in the sense that it does not greatly favor promoters relative to non-promoter sequence. (What this means operationally is that, for example, DNase I footprints of initially forming closed T4 late promoter complexes are not discernible above the background of non-specific DNA binding under conditions that are satisfactory for analysis of closed σ^{70}•RNAP promoter complexes) [32,33]. In contrast, open T4 late promoter complexes are site-specific, stable and readily detected by footprinting [32,34]. The acquisition of additional sequence discrimination on promoter opening implies sequence-specific recognition of some feature of the open promoter (perhaps its separated non-transcribed DNA strand) by gp55, but this has not been demonstrated directly.

The σ segment 2.2-equivalent RNAP core-binding motif of gp55 has been inferred on the basis of alanine-scan mutants analyzed for RNAP core-binding, basal and activated transcription [35]. This segment of gp55 is highly conserved (Figure 2). Extension of the alignment and secondary structure prediction suggests that residues ~42-122 constitute the σ_2-equivalent domain of gp55. Conservation of sequence among gp55 homologues extends outside this segment (Figure 2). In

```
                    1                          2.1              2.2                 2.3
           |<---------------->|       |<-------------->|  |<--------------->|  |<---->
          10        20        30        40        50        60        70        80        90
       ...|....|....|....|....|....|....|....|....|....|....|....|....|....|....|....|....|....|....|.
T4     MSETKP-KYNYVNNKELLQAIIDWKTELANNKDP-NKVVRQNDTIGLAIMLIAEGLSKRFNFSGYTQSWKQEMIADGIEASIKGLHNFDETKYKNPHA
RB69   *T*Q**-*************AG**E**K**L*****-**II******************************************************N****
JS98   *T*I*LT***********K**TQ****RE*T**-**I*****V*****************************************************D****
RB49   **------D*******YEE*CR**QQI*AQ----GRQ*PMS*Q******Q*SKN**R*********S*T*RE***D*****AV***I*************
RB43   *T-**I----*AD*DK*YPVMCK**QQIRETG*-----RKMP*EL*I***N**H**AR*Y*NR*SED**MD*D***S*T*S*********T*VYG
RB16   *T-**I----*AD*DK*YPVMCK**QQIRETG*-----RKMP*DL*I***N**H**AR*Y*NR*SED**MD*D***S*T*S*********T*VYG
PHG133 *AN------**D**K*YED*CT**QACYEAG----QK*QMPNS**D**ID**R*F*SFHK**R**ED**E**VG*A**IVV*Y*DR**********

44RR   *N--------**D**A*Y*N*CK**QDMREAG----HHIKMP*S**ID*LK**K*FTGYWK*T****N**DG*V**AV*SV****V****HN******
PHG25  *K--------**D*EV*Y*N*CK**QDMRDAG----HHIKMP*A**ID**K**K*FTGYYK*A****N**DG**S*AV**VV***I****DN******
AEH1   *A-------***D*E**Y**FVA**EK*KE*P-----TE*MPEF**G***K****FTHHWR*NR**DT**EA*VGAA*DVCLRYCK***TQR*S****
PHG65  *T-------V**D*DK*IAD*VQ**KDR*EDP-----PKKMS*SL****N*VN**TEYYR*RR**DV**EN*VSEAQ*ILVRKI*K***N*FD*A**
NT1    *KN------H**D**A*YASLCE**E*MKTAPE--DVH*PVPNFVAES*LK*S*N**R*Y*****AT**E***G*A**HCLRYIK***TE*******
KVP40  *KN------H**D**A*YASLCE**D*MNK*P*---ATIPVP*VVAIS**K*S*N**R*Y*****AT**E***G*A**HCLRYIK***TD**N****
S-PM2  *TR**N-*EY****DF*T**VEYRKRVQLAEKEGRPRP*VTNYL*ECFLK**TH**YKP**VN*--MFRED**C****NCLQYID***PE*SS**F*


        2.3        2.4        2.5                    3                        4
       |<---|<--------->|<------>|          |<--------------->|        |<----->|
          100       110       120       130       140       150       160       170       180
       ...|....|....|....|....|....|....|....|....|....|....|....|....|....|....|....|....|....|....|.
T4     YITQACFNAFVQRIKKERKEVAKKYSYFVHNVYDSRDDDMVALVDETFIQDIYDKMTHYEESTYRTPGAEKKSVVDDS-PSLDFLYEAND------
RB69   ***R*******************************H*****************************A*KA*****E*S**-*******DDN------
JS98   ***M****************M***********AN****GIA************Q**T*LVKA**SD*-*AESKV-GE****GTQ*------
RB49   ******************TAT*A**K****H****E**A*CQIA**A*******INN***VNKPKAPKVED*LTEEL***QF**IEN-------
RB43   **NK**WQ***T**LY*K**N****K**LEH***D*T**T*IA*********H**LNQ***AKKPK--D*IE***E*-*T*EMFL----------
RB16   **NK**WQ***T**LY*K**N****K**LEH***D*A**TSIA*********H**LNQ***AKKPK--D*IE***E*-*T*EMFL----------
PHG133 ***MI*N*C*I***TQ*K*QA*T***********PE*ASMA**D********QKF*T*LNKKAN---NQIENLPDT**S**F*EENESTD---

44RR   *********IGY**S*KR*M*T*R**LT****EH*S**SRIA**G****H**LVV**A*IKSKK------KDKPEL*N**NF**DEPET-----
PHG25  ******YR**IG***Y*KH*M*I**R**LT****EH****SR****N****HG*LVA**S*LKSKK------KDKEDV*N**SF**DDPED-----
AEH1   **SMI*AR**FN***FMKQ*E*A**K**LEH*H*GE*E**AK***VA*YN*MMT*ANDF*AKRKIK*K------KPKQLGG*S**EDDE*DSE----
PHG65  *V*MIAARC*FDELR**K*KE*T*NR**LEC***GD*E**GEM**PD*YL*LV**VSS****VKKKQKD-----K*EDIGE**W**DLEESEDETES
NT1    ***RI***********QDT*I**KV*LCDSSNYSM*--GEEA*YE*L*QMLERVNS**T*KKDKAQ----PDEKNEPLN*QKF*N---------
KVP40  ***RI***********DT*I**KV*LCDSSNYTVE--GEET*FE*L*QMLERVNS**SGKTAKAQ----PDEQKEPLN*QKF*D----------
S-PM2  *F**IIYY**LR**Q**K*QLEI*GKILERSGHQEVMYTEKFEG*MAGMNMS*SD*GSIK*NIETRMNR--------------------
```

Figure 2 Amino acid sequence conservation of gp55. All T4-related phage genomes sequenced to date (see [59], which is a review by Petrov, et al., in this series) contain readily identifiable gp55 homologues [81]. Four segments of sequence conservation can be noted. The central and largest segment 2 allows the distant relationship to domain 2 of σ^{70} to be discerned, primarily through correspondence with σ^{70} conserved segments 2.1 and 2.2 and secondary structure. The presumption that segment 2.4 harbors the late promoter recognition element of gp55 is speculative. Conserved segment 4 is the sliding clamp-binding epitope. Conserved segments 1 and 3 share no recognizable sequence similarity with σ^{70}. Whether they correspond functionally with σ segment 1.1/1.2 and 3.1, respectively, is not known. The numbering of residues is continuous for the T4 protein. Amino acid sequences of the T4, RB14 and RB32 proteins are identical; only T4 is listed. RB49 and phi-1 gp55 are also identical except for Q30 (RB49)→E30 (phi-1); only RB49 is listed. A secondary structure prediction from HHpred, with α-helices as cylinders, is shown below the alignment. Vertical lines at the side cluster phages infecting (top to bottom): *E. coli* (133 was isolated as an *Acinetobacter* phage); *Aeromonas* species; and *Vibrio* species. The more divergent S-PM2 protein is the only representative of the completely sequenced cyanobacterial phages that has been included for this presentation. (The cyanobacterial RNAPs constitute a separate clade in the phylogeny of the multisubunit enzymes, as do the archaeal RNAPs and the individual eukaryotic nuclear RNAPs I-V.)

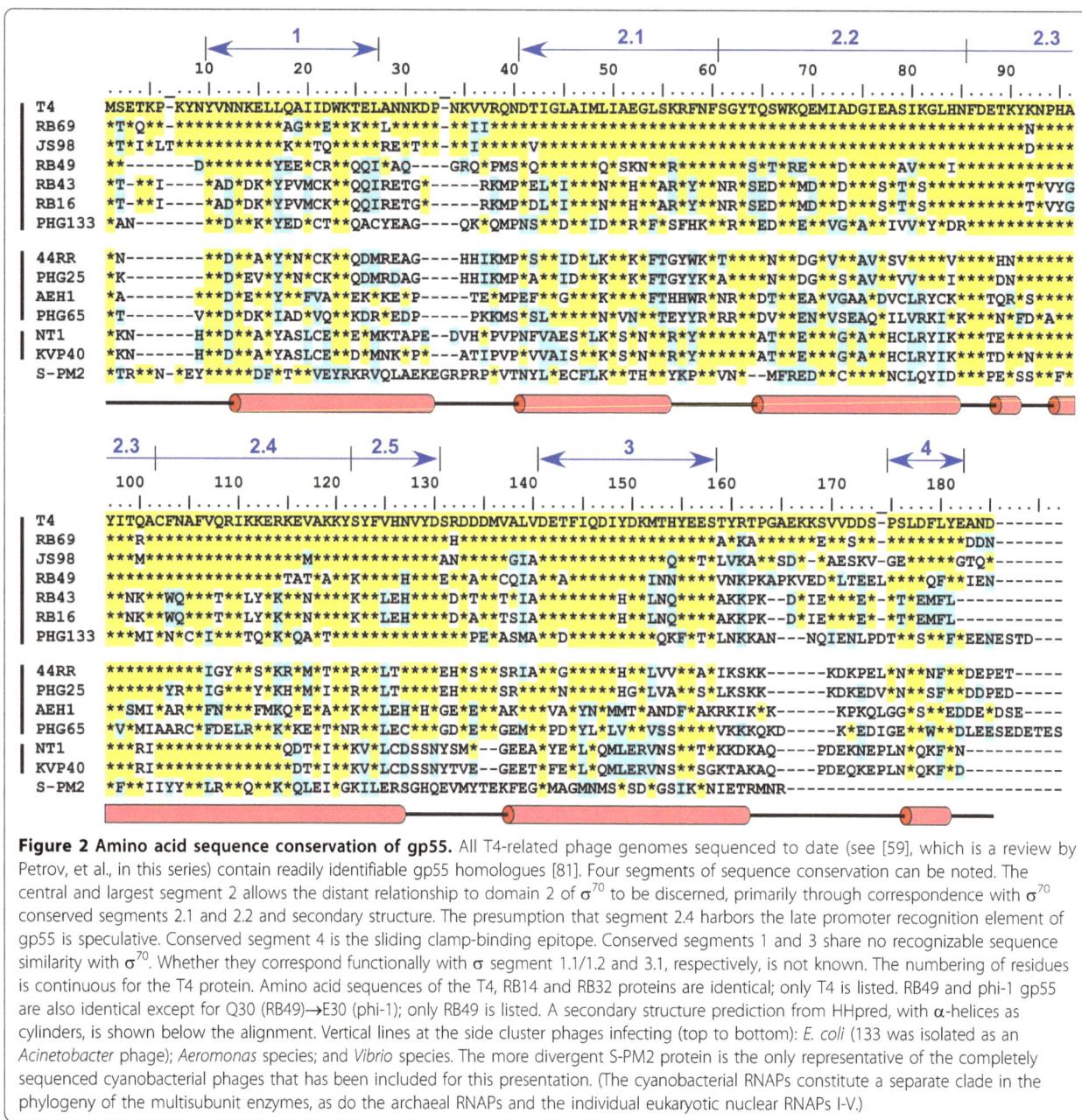

particular, absolute conservation of aromatic residues at N-proximal positions 10 and 23 of segment 1 is notable, as is conservation of sequence for residues ~141 - 156 (segment 3; numbering refers to the T4 protein) implying essential gp55 functions that might be related to σ^{70} segments 1.1 and 3.1, respectively. Sequence of a short hydrophobic and acidic C-terminal segment of gp55 is also conserved. This is the sliding clamp-binding epitope of T4 gp55 [36,37] and its conservation suggests that ability of the late gene-transcribing RNAP holoenzyme to bind the sliding clamp is a widely shared function of T4-related family phages. A 17-residue segment connecting the C-terminal epitope of gp55 to the rest of the protein is highly divergent in sequence and of varying length even among phages infecting *E. coli*. In the case of the T4 protein, extensive amino acid substitutions as well as insertions of a flexible (Ser-Gly) linker and small deletions do not eliminate the ability to support sliding clamp-activated late transcription [33]. This gp55 segment may be an unstructured linker connecting the sliding clamp-interacting C-terminus with the RNAP core-bound rest of the protein, somewhat comparable with the flexible linker that connects the N- and C-terminal domains of the RNAP α subunits [38].

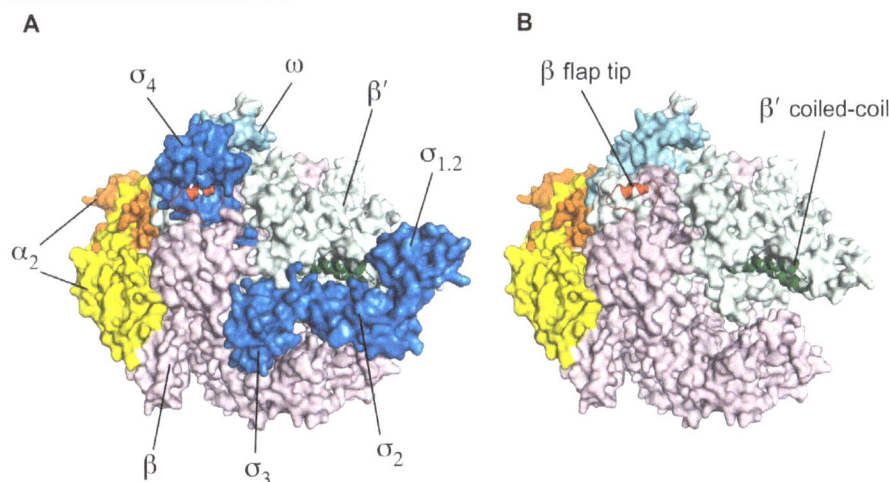

Figure 3 Bacterial RNAP holoenzyme. A. The *Thermus aquaticus* RNAP holoenzyme. The β (pink), β" (pale green), α_2 (yellow, orange; without their C-terminal domains) and ω (cyan) subunits are identified, and the β subunit flap (red), which is the attachment site of σ domain 4 and gp33, as well as the β" coiled-coil (green), which is the docking site of σ domain 2 and gp55, are emphasized. σ domains 1.2, 2, 3 and 4 (dark blue) are identified. **B**. The same, with σ removed (i.e., RNAP core, but with the coordinates of the holoenzyme) (Adapted from [26]).

Gp33

The 112-residue gp33 binds to the flap tip of the RNAP β subunit [39]. This is also the RNAP core attachment site of σ domain 4, which recognizes the -35 promoter element. Thus, gp33 can be thought of as a σ_4 mimic, and gp55 together with gp33 as a split σ. On the other hand, the β flap, which juts out over the RNA exit pore of the elongating transcription complex, is also the attachment site of other effectors of transcription, notably the phage λQ protein and other regulators of transcriptional elongation and termination. Moreover, gp33 does not recognize DNA sequence (and no sequence recognition is required since T4 late promoters do not have an upstream/-35 element). Instead, gp33 represses basal transcription [40,41] by diminishing promoter as well as general non-specific DNA binding. Binding of RNAP to DNA ends and DNA end-initiating transcription is exempt from this inhibition [42].

Conservation of amino acid sequence among gp33 homologues is primarily confined to individual residues in the C-terminal two-thirds of these proteins, which include the RNAP core binding site and the C-terminal sliding clamp-binding epitope (Figure 4). A recently completed determination of the structure of a gp33 complex with the *E. coli* RNAP β flap [43] and modeling into the *Thermus* RNAP structures [24,25] accounts for this conservation in terms of protein-protein contacts in this complex, suggests additional gp33:RNAP core interactions [43] and rationalizes extensive mutational analysis of gp33:RNAP binding and function [33,39]. The N-proximal one-third of gp33 is highly variable, entirely missing in homologues from other *E. coli*-infecting T4-related phages. There is no discernible similarity

of amino acid sequence between gp33 and σ proteins, but the new structure allows functional correspondences between individual gp33 and σ^{70} domain 4 residues to be seen.

It has been proposed that when it binds to the β flap, gp33 occludes a non-specific DNA-binding site on RNAP core, that this RNAP core site also interacts nonspecifically with DNA upstream of the T4 late promoter's -10 element and, in so doing, contributes to the promoter affinity of gp55•RNAP without contributing to selectivity [42]. The exemption of DNA-end-initiating transcription from inhibition by gp33 is presumed to be a direct consequence of its mechanism: binding to, and initiating transcription at, linear DNA template ends involves threading those ends through the downstream DNA channel for access to the catalytic center of RNAP, out of contact with β flap-bound gp33 and the upstream-facing part of RNAP.

Gp45

Gp45 is the T4 representative of the sliding clamp proteins. Sliding clamps are six-domain rings with a central hole large enough to accommodate a DNA helix: head-to-tail homodimers of 3-domain subunits in the case of the *E. coli* replisome's β protein; homotrimers of 2-domain subunits in the case of gp45 and the eukaryotic PCNA (*p*roliferating *c*ell *n*uclear *a*ntigen); homo- or heterotrimers of 2-domain subunits in the case of archaeal PCNA (for a review, see [44] and [45], which is an article by Mueser, et al., in this series). α-helices with a net positive charge line the central cavity and antiparallel β sheets with a net negative charge form the periphery of sliding clamps. Pseudo-6-fold symmetry axes run

```
                 10        20        30        40        50        60        70
         ...|....|....|....|....|....|. ___...|....|....|._...|....|....|....|....|.
T4       MTQFSLNDIRPVDETGLSEKELSIKKEKDEIAKLLD---RQENGFIIEKMVEEF-GMSYLEATTAFLEENSIPETQFAKF
RB69     *********K********Q**AV*HD**D*****---********S***Q*-************************
JS98     ***-****F----------DGAVNTASEPVHV*VN---K*Q**LD**AF**AE-*C****A**W*******GN**RY
RB49     *IGC*MLN------------------EHAV*AAN---KTA**LA***L*V*E-*LT*****Q*V****EYGHYQ*Y
Phi1     *LN-----------------EHAV*AAN---KTA**LA***LAV*E-*LT*****Q*V****EYGHYQ*Y
RB43     *SE-------------------IGN---KT*V*LY**NL*ASE-*AT*M***LQWMD***DYSMLN*T
PHG133   *QC-*T*EV---------------HKVLHDATS---K*ASEK**RE*A*T-*CL****A*KW***SGFT*A*YQRY

44RR     *SN-----------------PNN*S*---KTS**LE**AL*VKE-*LT*M**CIQW****LEISNCQ*Y
PHG25    *SN-----------------PNN*N*---KTA**IE**NL*KT--*LT***CVQW****LEINSCH*Y
Aeh1     *AE-------------------IS---KIDFSMKV*ERAR*K-EL*LI*SCLEIA**MD*DPNDIP**
PHG65    *AT-------------------DI*---KV*FSKRV*EIAI*T-*A*LV*SCFVVAD*MQVD*EHIPAH
NT1      *MK-*-**Y-----------A****LGDLVPPIKGSL**QISLE**EIYNSCPE*T****CLF*I*QYDYDISR*PYL
KVP40    *MK-*-E*Y-----------VL***LGDLVPPIKGSL**QISLE**EIYNSCPA*T****CLF*I*QYDYDISR*PYL
S-PM2    **E-**E*----------------*FMT---AAKFSQEV*RL*LNNSD*N*ID*VVHYC*V*E*EIDSVS*L

                 80        90        100       110
         ...|....|....|....|....|....|._...|....
T4       IPSGIIEKIQSEAIDENLLRPSVVRCEKTN-TLDFLL--
RB69     *****V***T*******M****A*G***-*****--
JS98     L*A***D**MN***D*****MA*TQ***-*****--
RB49     **VS**D**TQ*C*VNRT**KG*I-D*P**-***I----
Phi1     **VS**D**TQ*C*VNRT**KG*I-D*P**-***I----
RB43     V*KA**D**SD***KN******AKDHT*TQ***DFM--
PHG133   L*TS***A*KL***E*R*VA**M*NYHQSA-****I*NA

44RR     **RAL*D*LSQ*C*ESDM****I**SMTR*-*****M--
PHG25    **RAV*D*LSK*CV*SDM****IAKSMTR*-S****M--
Aeh1     *YPALRD**EE*G*ECRTVK*T------H*A**A**E--
PHG65    *HTVLK***RI**V------------------*----
NT1      *NQTLKD**ET***N*KTI*SRFGSS--SD-LGQWI---
KVP40    *NQTLRD**ET***N*KTI*SRFGSA--SD-LGQWV---
S-PM2    *SKPLK**LKFD*QKL*FMKKT------SRAK*MLV---
```

Figure 4 The limited sequence conservation of gp33. The presentation of the sequence alignment follows Figure 2. Amino acid sequences of the T4 and RB14 proteins are identical; RB32 gp33 differs only by E50→K; only the T4 protein is listed. RB43 and RB16 gp33 are identical and only RB43 is listed. A secondary structure prediction from HHpred is shown below the alignment.

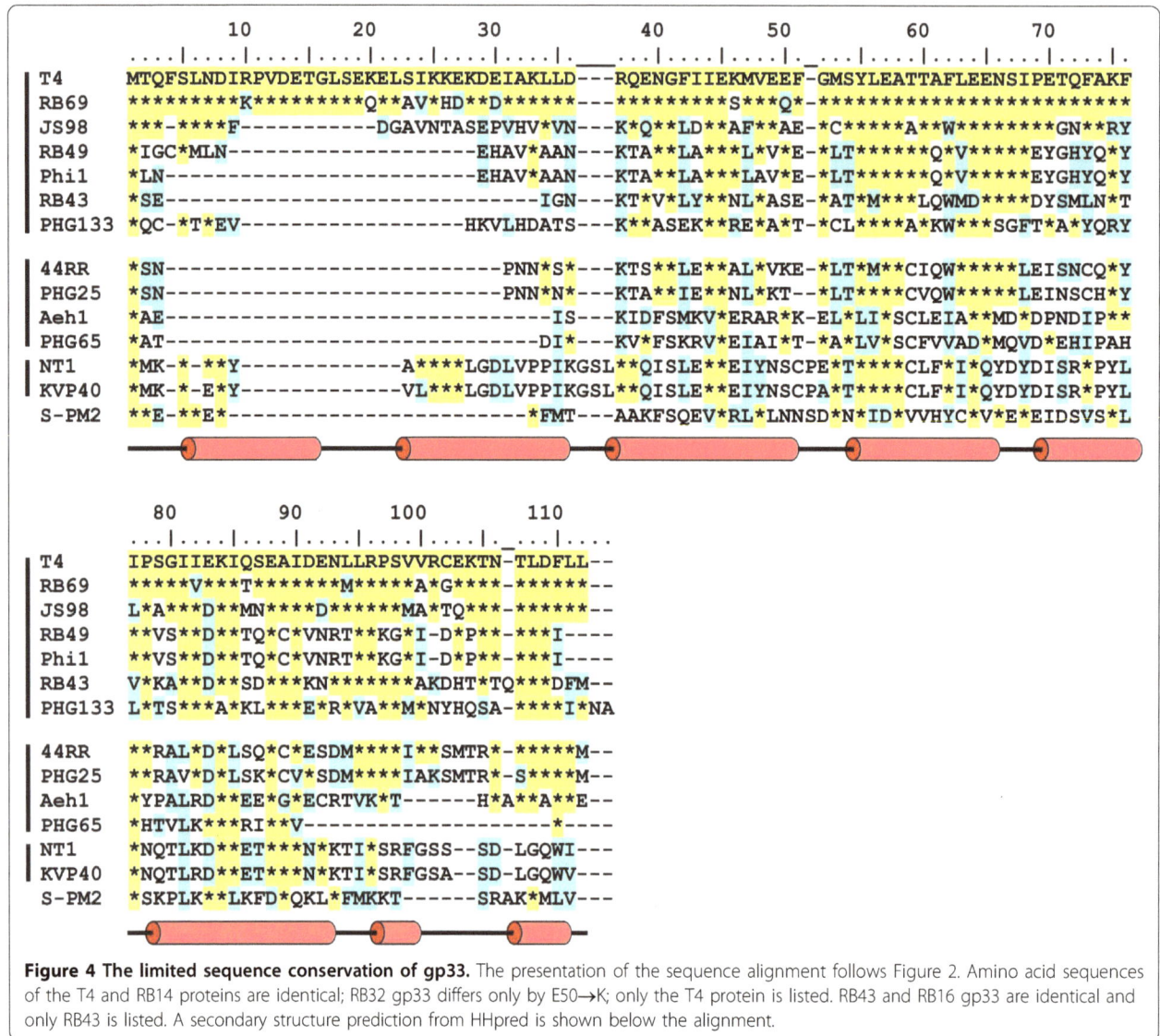

through the centers of the sliding clamps, except for the case of gp45, whose C-proximal domain of each protomer is somewhat shorter than the N-proximal domain, generating a form that is closer to triangular than hexagonal (i.e., with 3-fold symmetry instead of 6-fold pseudo-symmetry) [46,47].

The lateral faces of the sliding clamp are chemically distinctive; notably, the lateral face with the protruding C-terminus presents a hydrophobic patch on each protomer that serves as a binding site for the numerous and functionally diverse ligands that sliding clamps tether to DNA. (The sliding clamps are, for that reason, also aptly referred to as sliding toolbelts.) The ligands of the T4 sliding clamp include its clamp loader, the gp44/62 complex, and the highly similar hydrophobic and acidic C-termini of gp43, gp55 and gp33. For gp43, this interaction establishes processive DNA chain elongation

(by confining DNA polymerase to the one-dimensional space of the DNA thread (see [45], by Mueser, et al., this series).

Crystal structures of sliding clamps show them all as closed rings. In contrast, detailed analysis shows that the gp45 trimer in solution is open at one monomer interface and out of plane, somewhat like a split-ring lock washer [48]. All sliding clamps require loading factors that mount them on to DNA at double-strand-single-strand/primer-template junctions in an ATP hydrolysis-requiring process. The gp44/62 complex is the T4 clamp loader and it also loads gp45 on to DNA at nicks. Since their lateral faces are not identical, there are two distinguishable orientations of sliding clamps on DNA. The DNA strand with the 3"OH end determines the orientation of the clamp loader and, in turn, of the loaded sliding clamp. Thus, in the case of clamp loading at a DNA nick, for example, switching the strand that is interrupted reverses the orientation of the sliding clamp on DNA and therefore the polarity of its protein interactions. The same face of gp45 that attaches to the clamp loader also binds gp43 and, as argued below, the gp55- and gp33-containing T4 late RNAP holoenzyme.

The RB69 sliding clamp (81% identity of amino acid sequence with the T4 protein) has been co-crystallized with its ligand, the 11 C-terminal residues of the DNA polymerase [47]. The structure of the complex shows attachment of the hydrophobic 11-mer to the already referred to hydrophobic patch on the gp45 face with the protruding C-end of the protein (the C face), with only one of the three available sites occupied in each gp45 trimer. This is also the ligand-interaction mode of other sliding clamps [44,47,49]. In contrast, the preferred binding site of the C-terminal epitope of T4 gp43 in solution is the open gp45 inter-monomer interface [50]. (The gp45 ring being closed in crystals, that site would not be available for complex formation.) Thus, at least two different attachment sites on gp45 are apparently available for its gp43, gp55, gp33 and clamp loader partners. These sites do not offer the same affinity to their ligands, but they may both play roles in clamp loading, replication and/or transcription.

Gp45 sliding along DNA can be detected by DNA-protein photochemical cross-linking as occupancy of interior DNA sites that is dependent on a DNA-loading site, a clamp loader and ATP [51]. Experiments of that type show that gp55 tracks along DNA as a gp45 ligand [52]. This implies an ability of the sliding clamp to confer a mode of promoter searching that is dominated by processive one-dimensional scanning along the DNA thread. A snakes-and-ladders game model has dominated thinking about how proteins find their sites on genomes [53]. Sliding clamp-facilitated promoter searching is more-snakes-less-ladders. Whether facilitating promoter searching increases transcriptional activity depends on whether it is rate-limiting. This is unlikely to be the case for basal (gp33-independent) transcription, for which promoter opening is slow, as described below, but is not excluded for activated transcription, which is marked by very rapid promoter opening [32].

T4 sliding clamps must be loaded onto DNA by their clamp loaders in order to execute their functions in DNA replication and transcription. It is puzzling, therefore, that gene 44 and 62 *amber* mutations are clearly and nearly absolutely replication-defective (D0 phenotype) [10], but that the requirement for gp44/62 complex function in T4 late transcription was not clearly identified by the analysis that established the essential role of gp45 [21]. As referred to below, macromolecular crowding agents, such as poly(ethyleneglycol), allow gp45 to escape total reliance on the clamp loader for

activating DNA replication by gp43 and T4 late transcription [54,55]. The bacterial cytoplasm is a macromolecularly crowded medium, suggesting that these observations may have some physiological relevance, but they do not account for differences of effect of clamp-loader mutations on replication and late transcription [21]. The explanation of these differences may instead reside in the existence of additional interactions of the T4 clamp loader with the T4 replisome.

Other genes and functions

The T4 genome encodes more than 300 proteins, many with unknown or barely explored function. Several of these genes and functions relate to viral transcription and they have been most recently referred to in the detailed 2003 overview of the T4 genome [2]. As pointed out there, the functions of most of these proteins probably relate to early and middle viral transcription (see [56], which is a review by Hinton in this series) and to shutting off host transcription under conditions (such as nutrient limitation and stress) that are very different from those that were used for the classical analysis of the T4 multiplication cycle in early log phase cells. There is nothing new regarding them to report in the context of this chapter, with the possible exception of DsbA. *dsbA*, which first came to attention as the immediately upstream-lying and translationally coupled ORF to gene 33 [40], encodes an ~10 kDa DNA-binding protein, for which specific A/T rich DNA-binding sites overlapping two late promoters were identified but with surprisingly low affinity (in the μM range for K_d at moderate ionic strength) [57,58]. Finding *dsbA* to be a non-essential gene [2] has not encouraged further analysis in the T4 late transcription *in vitro* system, but genome sequencing in the T4-related phage family (see [59], which is a review by Petrov, et al., in this series) brings an interesting feature of *dsbA* to light. As already mentioned, the N-terminal 1/3 of gp33 is highly divergent among T4-related phages; even homologues from phage that are all capable of infecting *E. coli* lack the N-terminal 20-30 codons of the T4 protein. Nevertheless, *dsbA* genes are widely distributed and the *dsbA*-gene 33 ORF overlap, indicating translational coupling, is conserved, suggesting a significant role for *dsbA*, possibly relating to gene 33 and late transcription, that remains to be discovered. Our tentative examination of this issue has not been encouraging: under the standard conditions of the *in vitro* transcription system [32,33] no effect of DsbA on gp33-repressed or gp33/sliding clamp-activated transcription was discerned (V. Jain, unpublished observation).

The mechanism of activation

The 8-bp T4 late promoter resembles σ[70] extended -10 promoters in that DNA sequence recognition is confined to the downstream site at which promoter opening is initiated and proceeds in the absence of a σ_4-equivalent domain (in the case of the T4 late RNAP) and without requiring σ_4 participation (in the case of σ^{70}•RNAP). Gp55 dictates specifically initiating transcription at late promoters by unmodified *E. coli* RNAP core (RNAP^U) and by the T4-modified core enzyme (RNAP^{T4}), whose α subunits are ADP-ribosylated in both C-terminal domains (CTD) at Arg265. As already mentioned, transcription is more active on supercoiled than on relaxed (nicked circular or linear) DNA, at higher temperature and at lower ionic strength [27-31], generally in keeping with the activities of most weak bacterial promoters. Kinetic analysis of transcriptional initiation by gp55•RNAP^{T4} at the consensus gene 23 promoter in linear DNA (limited to a single temperature and in a single reaction medium) indicates weak promoter binding and relatively slow promoter opening [32].

Promoter opening by σ^{70} family RNAPs is temperature-dependent, to a significant degree adjusted to bacterial lifestyle in the sense that it operates at higher temperature in thermophiles than in mesophiles [60], and it is a reversible process [61,62]: when the λP_R and *gal* P1 promoters (to take one example each of a strong -35/-10 promoter and an extended -10 promoter) are opened at 37°C and brought to 0°C they close (although that process can be relatively slow, implying the existence of a significant kinetic barrier). In contrast, the T4 late promoter opens thermo-irreversibly: while it does not open at 0°C (even on a multi-hour time scale) it does not close at 0°C once it has been opened at higher temperature. The kinetic block has been suggested to lie on the promoter-closing pathway [63].

Activated transcription requires the participation of DNA-mounted gp45 and RNAP-bound gp33. The critical observations leading to the current understanding of activated transcription were made with an *in vitro* system that was designed to allow concurrent leading-strand DNA synthesis and late transcription, using a plasmid DNA template with a uniquely placed single-strand break serving as the initiation site for DNA synthesis. It was relatively promptly found that transcriptional activation in this *in vitro* system does not require DNA replication but does require the participation of three T4 replication proteins, the gp44/62 complex and gp45, ATP or dATP hydrolysis (ATP-γ-S, the very slowly hydrolyzing ATP analog blocking activation), and RNAP from T4-infected cells. Activation is not supported by gp55•RNAP^U, and absolutely requires gp33 [41].

The DNA template's single-strand break, which is essential for transcriptional activation, has the properties of an enhancer in that it can be placed close to, or at kbp separation from the promoter, but with the special constraint that the DNA break has to be in the non-transcribed strand of the activated promoter, so that switching the nicked strand switches the polarity of transcriptional activation [30]. The general mode of action of the enhancer was established by showing that it acts strictly *in cis* and that it requires a continuous, unobstructed path to the promoter [64]. The gp44/62 complex having been established as the non-processive DNA-loading factor for gp45 at about the same time [65-68], and DNA nicks being candidate loading sites for gp45, it was probable at this point [64] that the required continuous DNA path allows gp45 to slide from its DNA-loading site to the promoter. That this is the case was established by showing that gp45 becomes a stably bound part of the activated promoter complex, and is located at its upstream end [34], tethered there by the C-termini of gp55 and gp33 [36], as already mentioned.

Loading gp45 onto DNA at nicks does not require the gp32 single-stranded DNA-binding protein. However, primer-template junctions are more efficient gp45-loading sites in the presence of gp32 than are DNA nicks. The transcription-activating primer-template junction also has a polarity constraint: it must be located downstream of its target promoter [69]. The existence of this constraint establishes that the same lateral face of the sliding clamp interacts with T4 DNAP and with the late gene-transcribing gp55•gp33•RNAP holoenzyme. In contrast, the DNA-nick gp45-loading site can be located upstream or downstream of its target promoter [64]. This is a reflection of the ability of the gp45 clamp to slide across a DNA break, whereas it does not slide efficiently across single-stranded DNA [69]. In the presence of macromolecular crowding agents such as high molecular weight poly(ethyleneglycol) (PEG), gp45 can activate transcription and replication in the absence of the clamp loader [54,55]. Activation under these conditions also dispenses with the need for a nick or primer-template loading site as well as ATP hydrolysis, and functions with relaxed closed circular as well as blunt-end linear DNA. The requirement for gp33 and gp55 is retained. Needless to say, this finding also establishes gp45 as the activator of late transcription [55].

These facts about the sliding clamp-activated T4 late promoter complex suffice for the construction of a composite partial molecular model (Figure 5) based on the structure of the *Thermus aquaticus* (*Taq*) RNAP-fork junction complex [26], the just-recently determined structure of gp33 in complex with the β subunit flap domain and ~100-residue dispensable region (DR)II of *E. coli* RNAP [43], and gp45 [46]. The DNase I footprint of the activated and basal open promoter complexes differ by a 13 bp extension at the upstream end, almost exactly the DNA span of the sliding clamp (see also [70]). Thus, the sliding clamp must be pressed close to

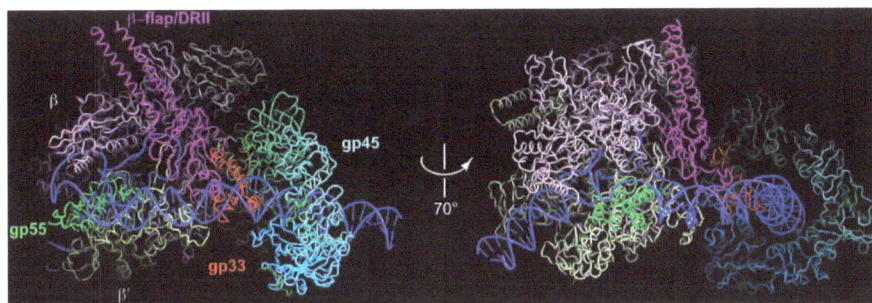

Figure 5 A composite partial molecular model of the sliding clamp docking on an RNAP:promoter complex. The structure of the RB69 sliding clamp [47] has been docked against a *Taq* RNAP holoenzyme fork junction promoter DNA complex [25]. Evidence from site-specific DNA-protein photochemical cross-linking and DNA footprinting [34] specifies that the sliding clamp abuts RNAP. Gp33 is placed in the model in accordance with the recent determination of its structure in complex with the *E. coli* β flap and DRII (amino acids 831-1057) by K-A.F. Twist and S.A. Darst [43][K-A.F. Twist, P. Deighan, S. Nechaev, A. Hochschild, E.P. Geiduschek & S.A. Darst, in preparation] and a complete structural model of *E. coli* RNAP based on a combination of approaches [82]. Placement of the C-end of gp33 in proximity to DNA is consistent with evidence from site-specific DNA-protein cross-linking [34]. The rotational orientation of gp45 is arbitrary, but is likely to be constrained by the interacting RNAP surface and also by the short tether to gp33. The location of the C-end of gp33 on the sliding clamp in the T4 late promoter complex is not known; a C-terminal 11-mer of phage RB69 DNA polymerase from the structure in [47] has not been removed and is barely visible, but its relevance to the late promoter complex is unclear, as discussed in the text. Residues 44-123 of gp55, comprising its RNAP core- and DNA-biding sites, have been modeled based on homology with σ^{70} domain 2 [26] and docked onto the β" subunit coiled-coil. Colors of components are indicated in the Figure. (Images provided by K-A. Twist and S.A. Darst and reproduced with their permission.)

RNAP core on DNA, with the α subunit C-terminal domains pushed out of the way. The only segment of gp55 that is represented in Figure 5 is region *2.1-2.4* (amino acids *44-123*, Figure 2, modeled by homology with *Taq* σ^{70} domain 2 [26]) attached to the β" subunit coiled-coil. The model is consistent with gp33 (presumably in a C-proximal segment) lying within cross-linking proximity of DNA (~1 nm) at bp -39 and -36/-35 of the activated T4 late promoter complex [71], although it does not bind sequence-specifically to it.

The functional consequences of attachment of the sliding clamp to the upstream end of RNAP in the activated late promoter complex through its interactions with hydrophobic and acidic motifs at the C-termini of gp55 and gp33 are a greatly increased overall rate of promoter opening. Kinetic analysis within a simplified 2-step framework for bacterial promoters [61,62,72] (Figure 6) indicates that the sliding clamp increases the effective affinity of the initially forming closed promoter complex (K_B) and the phenomenological first order rate constant for the subsequent step(s) of promoter opening (k_2) for a combined ~300-fold activation (measured at 30°C, with RNAP^{T4}) [32]. Basal transcription is repressed about one order of magnitude by gp33 (e.g., [42]); relative to this lowest activity of the

gp33•gp55•RNAP^{T4} holoenzyme, the sliding clamp mediates a >1,000-fold activation [32] (*Footnote* 1, which is embedded in the text below). The notion that tethering the promoter complex to DNA would increase its effective affinity is intuitively uncomplicated; that gp45 also lowers the activation energy barrier for promoter opening by holding on to gp55 and gp33 is less so; what follows suggests that this effect is probably mediated by gp33. Changes of promoter activity of this magnitude generate the emergence of qualitatively new properties. For example, avid association of the gp45-activated RNAP complex with DNA allows open promoter complexes to form in competition with high concentrations of the polyanionic competitor heparin [33].

(*Footnote* 1, A technical note: the above kinetic scheme adequately describes basal transcription with its characteristically slow promoter opening, and serves to parametrize a simple kinetic analysis of the just-cited work [32]. The principal result of that analysis–that the activator increases the second order rate constant for forming the open promoter complex by several hundred-fold relative to basal transcription and even more relative to gp33-repressed transcription, and that this increase results from a combination of tighter promoter binding and faster promoter opening–is not in question. However, the kinetic scheme is probably an inadequate representation of gp45-activated transcription, which is characterized by very rapid promoter opening and low selectivity, so that formation of the closed but precisely positioned promoter complex may not come to equilibrium.)

The highly similar C-terminal sliding clamp-binding motifs of gp55, gp33 and DNA polymerase (gp43) can

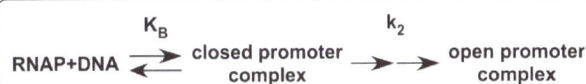

Figure 6 A simplified 2-step model for kinetic analysis of the formation of initiation-ready open promoter complexes.

be freely interchanged; replacing both C-terminal motifs of gp55 and gp33 with the C-terminal motif of gp43 leaves transcriptional activation *in vitro* quantitatively unchanged [33]. While this eliminates the possibility that their C-ends direct gp55 and gp33 to different binding sites on gp45, it does not settle the question of where, on the sliding clamp, these sites are located. The open interface of the gp45 trimer is the preferred binding site of gp43; while a sliding clamp cannot be simultaneously open at two sites, binding by both the gp55 and gp33 termini to separate clamp subunit interfaces is conceivable if at least one ligand seals its opening. Alternatively, even identical C-terminal motifs might occupy non-identical binding sites on gp45 (e.g., one ligand inserted into a monomer interface and the other attached to a lateral face hydrophobic patch) under the steric constraint that is imposed by gp33 and gp55 attachment to RNAP core.

The sliding clamp activator is held by two "arms" that extend from the gp33•gp55•RNAP. Separately detaching each of these arms has drastically different consequences for transcriptional activation: gp33:clamp binding is absolutely essential, while eliminating gp55-binding reduces but does not eliminate activation [33,36]. Conversely, gp45 exerts little or no activating effect on basal transcription by gp55•RNAP (*Footnote 2*, which is embedded in the text below). "One-armed" partial activation of transcription by gp45 (i.e., in the absence of the gp45:gp55 interaction) is also sensitive to inhibition by heparin [33]. This probably reflects a loss of late promoter binding affinity (K_B) due to the lost gp45:gp55 interaction.

(*Footnote 2.* Another technical note: these effects are more readily noted with $RNAP^{T4}$ than with the unmodified *E. coli* RNAP, most probably because of the effect of modifying the αCTD after T4 infection: ADPribosylation at Arg265 in the DNA-binding helix of the αCTD eliminates or at least reduces DNA binding; DNA binding by the αCTD may interfere with gp45 access to gp33 more effectively in the case of "one-armed activation" (that is, activation by the sliding clamp connected to the RNAP holoenzyme only through the C-end of gp33) than in the case of bivalent attachment to the C-ends of both gp55 and gp33; ADPribosylation may eliminate or diminish the competition.)

Gp45 is the least stable of the sliding clamps [73,74] perhaps reflecting the fact that it is partly open in solution, and its DNA-tracking state is accordingly relatively transient [51,73]. This is proposed to be the mechanistic basis of the coupling of T4 late transcription to concurrent DNA replication *in vivo* [75]. The DNA-loading sites of sliding clamps are transient intermediates of replication: they are continuously created, predominantly by lagging strand DNA synthesis, and consumed as DNA discontinuities are sealed by ligation.

Interrupting ongoing DNA replication quickly leads to a loss of clamp-loading sites, followed soon thereafter by a loss of DNA-loaded sliding clamps as they fall off DNA. This can be prevented if DNA ligation is also blocked and the resulting DNA breaks are stabilized against degradation–precisely the conditions under which T4 late gene expression becomes independent of DNA replication *in vivo*, as already described.

It is a common cellular strategy to make the expression of certain genes contingent on genome replication. Linking these separate processes involves symbolic communication provided by signaling pathways. Employing the DNA-loaded sliding clamp as the activator of T4 late transcription instead allows the state of DNA replication to be communicated directly through the availability of sliding clamp-loading sites, and dispenses with symbolically mediated signaling. One can think of the strategy as an instance of elegant streamlining or as a primitive relic.

Phages of the T4 family

Sequenced genomes of T4-related phages (see [59], which is a review by Petrov, et al., in this series) infecting a wide range of bacterial hosts (*E. coli, Acinetobacter, Vibrio, Aeromonas*, marine cyanobacteria) permit a glance at the prevalence of the transcription system of which T4 is the prototype. Gene 45 and 55 homologues are members of the core gene set of this family of phages [76,77]. Strong conservation of amino acid sequence for extended segments of gp55, including its putative σ domain 2, have been commented on above; the hydrophobic C-terminal motif is also retained in gp55 homologues (Figure 2). Thus, it appears probable that a late transcription system based on gene 55 and the sliding clamp is a general feature of the multiplication cycles of the T4-related family of phages. Indeed highly similar consensus sequences have been identified (*in silico*) for *Vibrio* phage KVP40, *Aeromonas* phage 44RR, and the marine cyanophage S-PM2, and a closely related consensus (a/gC at positions -13/-12 in place of TA) has been found for the *Aeromonas* phage Aeh1 [77-79]. The role of gp33 homologues (Figure 4) is less obvious. Bivalent tethering of the late RNAP holoenzymes of the T4-related phages to their sliding clamps should suffice to generate activation by increasing the effective avidity of promoter binding. The coliphage gp33 homologues are identifiable as RNAP core- and sliding clamp-binding proteins and so are the *Aeromonas* phage homologues, with the exception of phage 65. Whether the two vibriophages, phage 65, and cyanobacterial phage SPM-2 homologues bind their conjugate sliding clamps is not made obvious by their sequences and consequently the mechanism of their participation in late transcription cannot be guessed by inspection.

Speculation about coupling of late transcription to concurrent DNA replication as a general feature of the multiplication cycles of these phages is on even shakier ground. Coupling is proposed to arise as a consequence of the instability of the DNA-mounted state of T4 gp45. The T4-related sliding clamps are all 3-domain PCNA-like rather than 2-domain bacterial type proteins, but whether they generally fall off DNA equally readily remains to be determined. Another feature of the T4 late transcription system is the high sequence similarity of the C-termini of gp43, gp55 and gp33 [69]. This is not a conserved feature of all the phages of this family. Thus, the sites of attachment of gp55, gp33 and DNA polymerase homologues to their conjugate sliding clamps may vary.

If gp55 and gp33 are primarily "merely" deviant σ domains 2 and 4, why are they invariably encoded by widely separated and separately regulated genes? Why is there no fused late-transcription σ? Some suggestions for why this hypothetical fusion protein does not exist in nature or, at any rate, has not been found, can be offered: 1) Physically separating these two domains weakens their competitive advantage for binding to RNAP core, and modulates the competition between middle and late transcription. If a hypothetical gp55-gp33 fusion protein has a great RNAP core-binding advantage over σ^{70} and AsiA (the co-activator of T4 middle gene expression), then the dosage and timing of its production relative to the initiation of DNA replication become critical design elements of the viral multiplication cycle. In the extreme case, sufficiently premature and abundant production of the fusion protein might prevent DNA replication and shut down transcription. 2) The "split-σ" gp55/gp33 combination is a device for bivalent tethering of the sliding clamp to the late promoter, which optimizes late transcription. One way of approaching these questions is to design appropriate composite proteins and examine their modes of action and interaction *in vitro*. Experiments along those lines favor the first of these explanations and tend to discount the second: 1) Fused gp55-gp33 proteins with the gp55 sliding clamp-binding domain consequently internal instead of C-terminal are functional for sliding clamp-activated T4 late transcription so long as the length of the connector joining gp55 to the RNAP β flap-binding domain of gp33 is optimized. 2) The corresponding RNAP holoenzyme with its fused pseudo-σ subunit is almost completely inactive for basal transcription as a consequence of repression by its C-terminal gp33 domain. In that sense (essentially complete activator-dependence), the gp55-gp33 fusion version of the T4 late RNAP holoenzyme resembles $\sigma^{54} \cdot$RNAP. 3) When gp33 is covalently linked to gp55,

suppression of basal transcription still depends on ability to bind to the β flap. 4) Fusing gp33 to gp55 generates an effective competitor against $\sigma^{70} \cdot$RNAP transcription at a strong -35/-10 type promoter [V. Jain & EPG, unpublished observations].

Coupling transcription of selected genes to specific states of the cell-division cycle, including S phase, is a ubiquitous strategy of cells and it ubiquitously engages signaling pathways, that is, molecular systems for generating messengers and interpreting messages. The mechanism that couples transcription of the viral late genes to replication in the T4 multiplication cycle elegantly dispenses with (or, depending on perspective, is too primitive for) symbolic communication, instead directly using universal components of cellular DNA replication, the primer-template junction and the clamp-loading factors, as generators of activation and the ubiquitous sliding clamp as the activator. It is puzzling that this efficient and direct regulatory device should be restricted to T4 and perhaps other members of the T4-related phage family. In fact, it has been possible to design a sliding clamp-activation domain fusion protein that generates clamp loader-dependent transcriptional activation of eukaryotic RNAP II *in vitro* [80]. Nevertheless, other instances of the use of this direct and simple mechanism for coupling transcriptional regulation to DNA replication in nature have not been found.

Acknowledgements
Research in our laboratory on the T4 late genes has been supported by a long-running grant from the National Institute of General Medical Sciences.

Authors' contributions
EPG and GAK composed this review. Both authors have read and approved the final manuscript.

Competing interests
The authors declare that they have no competing interests.

Received: 29 July 2010 Accepted: 28 October 2010
Published: 28 October 2010

References
1. Karam JD, Editor-in-Chief: *Molecular biology of bacteriophage T4* Washington, DC: American Society for Microbiology; 1994.
2. Miller ES, Kutter E, Mosig G, Arisaka F, Kunisawa T, Rüger W: **Bacteriophage T4 genome.** *Microbiol Mol Biol Rev* 2003, **67**:86-156.
3. Christensen AC, Young ET: **T4 late transcripts are initiated near a conserved DNA sequence.** *Nature* 1982, **299**:369-371.
4. Kassavetis GA, Zentner PG, Geiduschek EP: **Transcription at bacteriophage T4 variant late promoters. An application of a newly devised promoter-mapping method involving RNA chain retraction.** *J Biol Chem* 1986, **261**:14256-14265.
5. Williams KP, Kassavetis GA, Herendeen DR, Geiduschek EP: **Regulation of late-gene expression.** In *Molecular Biology of Bacteriophage T4.* Edited by: Karam JD. Washington, D.C.: American Society for Microbiology; 1994:161-175.
6. Vaiskunaite R, Miller A, Davenport L, Mosig G: **Two new early bacteriophage T4 genes, repEA and repEB, that are important for DNA replication initiated from origin E.** *J Bacteriol* 1999, **181**:7115-7125.

7. Guha A, Szybalski W: Fractionation of the complementary strands of coliphage T4 DNA based on the asymmetric distribution of the poly U and poly U,G binding sites. *Virology* 1968, **34**:608-616.
8. Geiduschek EP, Grau O: RNA-Polymerase and Transcription. In *First International Lepetit Colloquium. Volume RNA-Polymerase and Transcription.* Edited by: Silverstri L. Florence: North-Holland Publishing Co; 1969:190-203.
9. Jurale C, Kates JR, Colby C: Isolation of double-stranded RNA from T4 phage infected cells. *Nature* 1970, **226**:1027-1029.
10. Epstein RH, Bolle A, Steinberg CM, Kellenberger E, Boy De la Tour E: Physiological studies of conditional lethal mutants of bacteriophage T4D. *Cold Spring Harbor Symposium* 1963, **XXVIII**:375-394.
11. Horvitz HR: Polypeptide bound to the host RNA polymerase is specified by T4 control gene 33. *Nat New Biol* 1973, **244**:137-140.
12. Ratner D: Bacteriophage T4 transcriptional control gene 55 codes for a protein bound to Escherichia coli RNA polymerase. *J Mol Biol* 1974, **89**:803-807.
13. Gribskov M, Burgess RR: Sigma factors from E. coli, B. subtilis, phage SP01, and phage T4 are homologous proteins. *Nucleic Acids Res* 1986, **14**:6745-6763.
14. Helmann JD, Chamberlin MJ: Structure and function of bacterial sigma factors. *Annu Rev Biochem* 1988, **57**:839-872.
15. Lonetto M, Gribskov M, Gross CA: The sigma 70 family: sequence conservation and evolutionary relationships. *J Bacteriol* 1992, **174**:3843-3849.
16. Wiberg JS, Dirksen ML, Epstein RH, Luria SE, Buchanan JM: Early enzyme synthesis and its control in E. coli infected with some amber mutants of bacteriophage T4. *Proc Natl Acad Sci USA* 1962, **48**:293-302.
17. Riva S, Cascino A, Geiduschek EP: Coupling of late transcription to viral replication in bacteriophage T4 development. *J Mol Biol* 1970, **54**:85-102.
18. Bolle A, Epstein RH, Salser W, Geiduschek EP: Transcription during bacteriophage T4 development: requirements for late messenger synthesis. *J Mol Biol* 1968, **33**:339-362.
19. Riva S, Cascino A, Geiduschek EP: Uncoupling of late transcription from DNA replication in bacteriophage T4 development. *J Mol Biol* 1970, **54**:103-119.
20. Cascino A, Riva S, Geiduschek EP: DNA Ligation and the Coupling of T4 Late Transcription to Replication. *Cold Spring Harb Symp Quant Biol* 1970, **35**:213-220.
21. Wu R, Geiduschek EP: The role of replication proteins in the regulation of bacteriophage T4 transcription. II. Gene 45 and late transcription uncoupled from replication. *J Mol Biol* 1975, **96**:539-562.
22. Callaci S, Heyduk E, Heyduk T: Core RNA polymerase from E. coli induces a major change in the domain arrangement of the sigma 70 subunit. *Mol Cell* 1999, **3**:229-238.
23. Schwartz EC, Shekhtman A, Dutta K, Pratt MR, Cowburn D, Darst S, Muir TW: A full-length group 1 bacterial sigma factor adopts a compact structure incompatible with DNA binding. *Chem Biol* 2008, **15**:1091-1103.
24. Vassylyev DG, Sekine S, Laptenko O, Lee J, Vassylyeva MN, Borukhov S, Yokoyama S: Crystal structure of a bacterial RNA polymerase holoenzyme at 2.6 A resolution. *Nature* 2002, **417**:712-719.
25. Murakami KS, Masuda S, Darst SA: Structural basis of transcription initiation: RNA polymerase holoenzyme at 4 A resolution. *Science* 2002, **296**:1280-1284.
26. Murakami KS, Masuda S, Campbell EA, Muzzin O, Darst SA: Structural basis of transcription initiation: an RNA polymerase holoenzyme-DNA complex. *Science* 2002, **296**:1285-1290.
27. Kassavetis GA, Elliott T, Rabussay DP, Geiduschek EP: Initiation of transcription at phage T4 late promoters with purified RNA polymerase. *Cell* 1983, **33**:887-897.
28. Kassavetis GA, Geiduschek EP: Defining a bacteriophage T4 late promoter: bacteriophage T4 gene 55 protein suffices for directing late promoter recognition. *Proc Natl Acad Sci USA* 1984, **81**:5101-5105.
29. Elliott T, Geiduschek EP: Defining a bacteriophage T4 late promoter: absence of a "-35" region. *Cell* 1984, **36**:211-219.
30. Herendeen DR, Kassavetis GA, Barry J, Alberts BM, Geiduschek EP: Enhancement of bacteriophage T4 late transcription by components of the T4 DNA replication apparatus. *Science* 1989, **245**:952-958.
31. Williams KP: Transcriptional effects of viral proteins that bind host RNA polymerase, PhD Thesis. *PhD Thesis* University of California, San Diego; 1991.

32. Kolesky SE, Ouhammouch M, Geiduschek EP: The mechanism of transcriptional activation by the topologically DNA-linked sliding clamp of bacteriophage T4. *J Mol Biol* 2002, **321**:767-784.
33. Nechaev S, Geiduschek EP: Dissection of the bacteriophage T4 late promoter complex. *J Mol Biol* 2008, **379**:402-413.
34. Tinker RL, Williams KP, Kassavetis GA, Geiduschek EP: Transcriptional activation by a DNA-tracking protein: structural consequences of enhancement at the T4 late promoter. *Cell* 1994, **77**:225-237.
35. Wong K, Kassavetis GA, Léonetti JP, Geiduschek EP: Mutational and functional analysis of a segment of the sigma family bacteriophage T4 late promoter recognition protein gp55. *J Biol Chem* 2003, **278**:7073-7080.
36. Sanders GM, Kassavetis GA, Geiduschek EP: Dual targets of a transcriptional activator that tracks on DNA. *EMBO J* 1997, **16**:3124-3132.
37. Wong K, Geiduschek EP: Activator-sigma interaction: A hydrophobic segment mediates the interaction of a sigma family promoter recognition protein with a sliding clamp transcription activator. *J Mol Biol* 1998, **284**:195-203.
38. Blatter EE, Ross W, Tang H, Gourse RL, Ebright RH: Domain organization of RNA polymerase alpha subunit: C-terminal 85 amino acids constitute a domain capable of dimerization and DNA binding. *Cell* 1994, **78**:889-896.
39. Nechaev S, Kamali-Moghaddam M, André E, Léonetti JP, Geiduschek EP: The bacteriophage T4 late-transcription coactivator gp33 binds the flap domain of Escherichia coli RNA polymerase. *Proc Natl Acad Sci USA* 2004, **101**:17365-17370.
40. Williams KP, Müller R, Rüger W, Geiduschek EP: Overproduced bacteriophage T4 gene 33 protein binds RNA polymerase. *J Bacteriol* 1989, **171**:3579-3582.
41. Herendeen DR, Williams KP, Kassavetis GA, Geiduschek EP: An RNA polymerase-binding protein that is required for communication between an enhancer and a promoter. *Science* 1990, **248**:573-578.
42. Nechaev S, Geiduschek EP: The role of an upstream promoter interaction in initiation of bacterial transcription. *EMBO J* 2006, **25**:1700-1709.
43. Twist K-AF: Structural studies of factors that affect the transcription cycle; microcin J25, Lambda Q and T4 GP33. *PhD Thesis* Rockefeller University, New York; 2009.
44. Indiani C, O'Donnell M: The replication clamp-loading machine at work in the three domains of life. *Nat Rev Mol Cell Biol* 2006, **7**:751-761.
45. Mueser TC, Hinerman JM, Devos JM, Boyer RA, Williams KJ: Structural analysis of Bacterophage T4 DNA replication. *Virology J* 2010.
46. Moarefi I, Jeruzalmi D, Turner J, O'Donnell M, Kuriyan J: Crystal structure of the DNA polymerase processivity factor of T4 bacteriophage. *J Mol Biol* 2000, **296**:1215-1223.
47. Shamoo Y, Steitz TA: Building a replisome from interacting pieces: sliding clamp complexed to a peptide from DNA polymerase and a polymerase editing complex. *Cell* 1999, **99**:155-166.
48. Millar D, Trakselis MA, Benkovic SJ: On the solution structure of the T4 sliding clamp (gp45). *Biochemistry* 2004, **43**:12723-12727.
49. Georgescu RE, Yurieva O, Kim SS, Kuriyan J, Kong XP, O'Donnell M: Structure of a small-molecule inhibitor of a DNA polymerase sliding clamp. *Proc Natl Acad Sci USA* 2008, **105**:11116-11121.
50. Trakselis MA, Alley SC, Abel-Santos E, Benkovic SJ: Creating a dynamic picture of the sliding clamp during T4 DNA polymerase holoenzyme assembly by using fluorescence resonance energy transfer. *Proc Natl Acad Sci USA* 2001, **98**:8368-8375.
51. Tinker RL, Kassavetis GA, Geiduschek EP: Detecting the ability of viral, bacterial and eukaryotic replication proteins to track along DNA. *EMBO J* 1994, **13**:5330-5337.
52. Tinker-Kulberg RL, Fu TJ, Geiduschek EP, Kassavetis GA: A direct interaction between a DNA-tracking protein and a promoter recognition protein: implications for searching DNA sequence. *EMBO J* 1996, **15**:5032-5039.
53. von Hippel PH, Berg OG: Facilitated target location in biological systems. *J Biol Chem* 1989, **264**:675-678.
54. Reddy MK, Weitzel SE, von Hippel PH: Assembly of a functional replication complex without ATP hydrolysis: a direct interaction of bacteriophage T4 gp45 with T4 DNA polymerase. *Proc Natl Acad Sci USA* 1993, **90**:3211-3215.
55. Sanders GM, Kassavetis GA, Geiduschek EP: Use of a macromolecular crowding agent to dissect interactions and define functions in transcriptional activation by a DNA-tracking protein: bacteriophage T4 gene 45 protein and late transcription. *Proc Natl Acad Sci USA* 1994, **91**:7703-7707.

56. Hinton DM: Transcriptional control in the prereplicative phase of T4 development. *Virol J* 2010, **7**:289.
57. Gansz A, Kruse U, Rüger W: Gene product dsbA of bacteriophage T4 binds to late promoters and enhances late transcription. *Mol Gen Genet* 1991, **225**:427-434.
58. Sieber P, Lindemann A, Boehm M, Seidel G, Herzing U, van der Heusen P, Muller R, Rüger W, Jaenicke R, Rösch P: Overexpression and structural characterization of the phage T4 protein DsbA. *Biol Chem* 1998, **379**:51-58.
59. Petrov VM, Ratnayaka S, Nolan JM, Miller ES, Karam JD: Genomes of the T4-related bacteriophages as windows on microbial genome evolution. *Virol J* 2010, **7**:292.
60. Schroeder LA, deHaseth PL: Mechanistic differences in promoter DNA melting by Thermus aquaticus and Escherichia coli RNA polymerases. *J Biol Chem* 2005, **280**:17422-17429.
61. Record MT Jr, Reznikoff WS, Craig ML, McQuade KL, Schlax PJ, et al: In *Escherichia coli and Salmonella*. 2 edition. Edited by: Neidhardt FC. Washington, D.C.: ASM Press; 1996:792-821.
62. deHaseth PL, Zupancic ML, Record MT Jr: RNA polymerase-promoter interactions: the comings and goings of RNA polymerase. *J Bacteriol* 1998, **180**:3019-3025.
63. Kamali-Moghaddam M, Geiduschek EP: Thermoirreversible and thermoreversible promoter opening by two Escherichia coli RNA polymerase holoenzymes. *J Biol Chem* 2003, **278**:29701-29709.
64. Herendeen DR, Kassavetis GA, Geiduschek EP: A transcriptional enhancer whose function imposes a requirement that proteins track along DNA. *Science* 1992, **256**:1298-1303.
65. Stukenberg PT, Studwell-Vaughan PS, O'Donnell M: Mechanism of the sliding beta-clamp of DNA polymerase III holoenzyme. *J Biol Chem* 1991, **266**:11328-11334.
66. Capson TL, Benkovic SJ, Nossal NG: Protein-DNA cross-linking demonstrates stepwise ATP-dependent assembly of T4 DNA polymerase and its accessory proteins on the primer-template. *Cell* 1991, **65**:249-258.
67. Gogol EP, Young MC, Kubasek WL, Jarvis TC, von Hippel PH: Cryoelectron microscopic visualization of functional subassemblies of the bacteriophage T4 DNA replication complex. *J Mol Biol* 1992, **224**:395-412.
68. Munn MM, Alberts BM: The T4 DNA polymerase accessory proteins form an ATP-dependent complex on a primer-template junction. *J Biol Chem* 1991, **266**:20024-20033.
69. Sanders GM, Kassavetis GA, Geiduschek EP: Rules governing the efficiency and polarity of loading a tracking clamp protein onto DNA: determinants of enhancement in bacteriophage T4 late transcription. *EMBO J* 1995, **14**:3966-3976.
70. Georgescu RE, Kim SS, Yurieva O, Kuriyan J, Kong XP, O'Donnell M: Structure of a sliding clamp on DNA. *Cell* 2008, **132**:43-54.
71. Tinker RL, Sanders GM, Severinov K, Kassavetis GA, Geiduschek EP: The COOH-terminal domain of the RNA polymerase alpha subunit in transcriptional enhancement and deactivation at the bacteriophage T4 late promoter. *J Biol Chem* 1995, **270**:15899-15907.
72. McClure WR: Mechanism and control of transcription initiation in prokaryotes. *Annu Rev Biochem* 1985, **54**:171-204.
73. Fu TJ, Sanders GM, O'Donnell M, Geiduschek EP: Dynamics of DNA-tracking by two sliding-clamp proteins. *EMBO J* 1996, **15**:4414-4422.
74. Yao N, Turner J, Kelman Z, Stukenberg PT, Dean F, Shechter D, Pan ZQ, Hurwitz J, O'Donnell M: Clamp loading, unloading and intrinsic stability of the PCNA, beta and gp45 sliding clamps of human, E. coli and T4 replicases. *Genes Cells* 1996, **1**:101-113.
75. Geiduschek EP, Fu TJ, Kassavetis GA, Sanders GM, Tinker-Kulberg RL: Transcriptional Activation by a Topologically Linkable Protein: Forging a Connection Between Replication and Gene Activity. *Nucleic Acids and Molecular Biology* 1997, **11**:135-150.
76. Comeau AM, Bertrand C, Letarov A, Tetart F, Krisch HM: Modular architecture of the T4 phage superfamily: a conserved core genome and a plastic periphery. *Virology* 2007, **362**:384-396.
77. Nolan JM, Petrov V, Bertrand C, Krisch HM, Karam JD: Genetic diversity among five T4-like bacteriophages. *Virol J* 2006, **3**:30.
78. Miller ES, Heidelberg JF, Eisen JA, Nelson WC, Durkin AS, Ciecko A, Feldblyum TV, White O, Paulsen IT, Nierman WC, Lee J, Szczypinski B, Fraser CM: Complete genome sequence of the broad-host-range vibriophage KVP40: comparative genomics of a T4-related bacteriophage. *J Bacteriol* 2003, **185**:5220-5233.
79. Mann NH, Clokie MR, Millard A, Cook A, Wilson WH, Wheatley PJ, Letarov A, Krisch HM: The genome of S-PM2, a "photosynthetic" T4-type bacteriophage that infects marine Synechococcus strains. *J Bacteriol* 2005, **187**:3188-3200.
80. Ouhammouch M, Sayre MH, Kadonaga JT, Geiduschek EP: Activation of RNA polymerase II by topologically linked DNA-tracking proteins. *Proc Natl Acad Sci USA* 1997, **94**:6718-6723.
81. Genomes of the T4-like phages. [http://phage.bioc.tulane.edu].
82. Opalka N, Brown J, Lane WJ, Twist KA, Landick R, Asturias FJ, Darst SA: Complete Structural Model of Escherichia coli RNA Polymerase from a Hybrid Approach. *PLOS Biology* 2010.

doi:10.1186/1743-422X-7-288
Cite this article as: Geiduschek and Kassavetis: **Transcription of the T4 late genes**. *Virology Journal* 2010 **7**:288.

Uzan and Miller *Virology Journal* 2010, **7**:360
http://www.virologyj.com/content/7/1/360

VIROLOGY JOURNAL

REVIEW   Open Access

Post-transcriptional control by bacteriophage T4: mRNA decay and inhibition of translation initiation

Marc Uzan[1], Eric S Miller[2]*

Abstract

Over 50 years of biological research with bacteriophage T4 includes notable discoveries in post-transcriptional control, including the genetic code, mRNA, and tRNA; the very foundations of molecular biology. In this review we compile the past 10 - 15 year literature on RNA-protein interactions with T4 and some of its related phages, with particular focus on advances in mRNA decay and processing, and on translational repression. Binding of T4 proteins RegB, RegA, gp32 and gp43 to their cognate target RNAs has been characterized. For several of these, further study is needed for an atomic-level perspective, where resolved structures of RNA-protein complexes are awaiting investigation. Other features of post-transcriptional control are also summarized. These include: RNA structure at translation initiation regions that either inhibit or promote translation initiation; programmed translational bypassing, where T4 orchestrates ribosome bypass of a 50 nucleotide mRNA sequence; phage exclusion systems that involve T4-mediated activation of a latent endoribonuclease (PrrC) and cofactor-assisted activation of EF-Tu proteolysis (Gol-Lit); and potentially important findings on ADP-ribosylation (by Alt and Mod enzymes) of ribosome-associated proteins that might broadly impact protein synthesis in the infected cell. Many of these problems can continue to be addressed with T4, whereas the growing database of T4-related phage genome sequences provides new resources and potentially new phage-host systems to extend the work into a broader biological, evolutionary context.

Introduction

The temporal ordering of bacteriophage T4 development is assured, in great part, by the cascade activation of three different classes of promoters (see [1,2] in this series). However, control of phage development is also exercised at the post-transcriptional level, in particular by mechanisms of mRNA destabilization and translation inhibition [see earlier reviews [3-6]]. In this review we detail advances in understanding these processes, and summarize some of the other posttranscriptional processes that occur in T4-infected cells.

Posttranscriptional control by mRNA decay

Endoribonuclease RegB and its role in inactivating phage early mRNAs

The end of the early period, 5 minutes after infection at 30°C, is marked by a strong decline in the synthesis of many early proteins. This inhibition is due to the abrupt shut-down of the early promoters by a mechanism that is not completely understood [7,8]. In addition, the phage-encoded RegB endoribonuclease (T4 *regB* gene) functionally inactivates many early transcripts and expedites their degradation. As described below, this role of RegB is accomplished in part, with the cooperation of the host endoribonucleases RNase E and RNase G and the T4 polynucleotide kinase, PNK.

The T4 RegB RNase exhibits unique properties. It generates cuts in the middle of GGAG/U sequences located in the intergenic regions of early genes, mostly in translation initiation regions. In fact, the GGAG motif is one of the most frequent Shine-Dalgarno sequences encountered in T4. Some efficient RegB cuts have also been detected at GGAG/U within coding sequences. RegB cleavages can be detected very soon after infection, earlier than 45 seconds at 30°C [5,9-14].

The RegB endonuclease requires a co-factor to act efficiently. When assayed *in vitro*, RegB activity is extremely low but can be stimulated up to 100-fold by the ribosomal protein S1, depending on the RNA substrate [9,15,16].

* Correspondence: eric_miller@ncsu.edu
[2]Department of Microbiology, North Carolina State University, Raleigh, 27695-7615, NC, USA
Full list of author information is available at the end of the article

Functional inactivation of mRNA by RegB

The consequence of RegB cleavage within translation initiation regions is the functional inactivation of the transcripts. The synthesis of a number of early proteins starts immediately after infection and reaches a maximum in four minutes before declining abruptly thereafter. In *regB* mutant infections, several of these early proteins continue to be synthesized for a longer time, resulting in twice the accumulation as compared to when RegB is functional. The abrupt arrest of synthesis of these proteins at ~4 min postinfection with wild-type phage results both from the sudden inhibition of early transcription and the functional inactivation of mRNA targets by RegB. However, in addition to down-regulating the translation of many early T4 genes RegB-mediated mRNA processing stimulates the synthesis of a few middle proteins, such as the phage-induced DNA polymerase, encoded by T4 gene *43* [11,12].

RegB accelerates early mRNA breakdown

RegB accelerates the degradation of most early, but not middle or late mRNAs. Indeed, bulk early mRNA is stabilized about 3-fold in a *regB* mutant compared to wild-type infection. After ~3 min post-infection, mRNAs decay with a constant half-life of about 8 minutes for the remainder of the growth period at 30°C, irrespective of the presence or the absence of a functional RegB nuclease [11]. The host RNase E plays an important role in T4 mRNA degradation throughout phage development [17]. Total T4 RNA synthesized during the first two minutes of infection of the temperature-sensitive *rne* host mutant is stabilized 3-fold at non-permissive temperatures. When both genes, *regB* and *rne*, are mutationally inactivated, bulk early T4 mRNA is stabilized 8 to 10-fold (half-life of 50 min at 43°C), showing that both T4 RegB and host RNase E endonucleases are major actors in T4 early mRNA turnover (B. Sanson & M. Uzan, unpublished results).

RegB could accelerate mRNA decay by increasing the number of entry sites for one or the other of the two host 3' exoribonucleases, RNase II and RNase R, which can attack the mRNA from the 3'-phosphate terminus left after RegB cleavage. An alternative pathway was suggested by the finding that some endonucleolytic cleavages within A-rich sequences depend upon RegB primary cuts a short distance upstream. This was interpreted as meaning that RegB triggers a degradation pathway that involves a cascade of endonucleolytic cuts in the 5' to 3' orientation [12]. The host endoribonucleases, RNase G and RNase E, are responsible for cutting at secondary sites, with RNase G playing a major role [14]. This finding appeared paradoxical since these two endonucleases have a marked preference for RNA substrates bearing a monophosphate at their 5' extremities [18-20], while RegB produces 5'-hydroxyl RNA

termini. Therefore, we suspected that T4 infection induced an activity able to phosphorylate the 5'-OH left by RegB, and the best candidate for filling this function is the phage-encoded 5' polynucleotide kinase/3' phosphatase (PNK). This enzyme catalyzes both the phosphorylation of 5'-hydroxyl polynucleotide termini and the hydrolysis of 3'-phosphomonoesters and 2':3'-cyclic phosphodiesters. Indeed, Durand *et al.* (2008; unpublished data) showed that the secondary cleavages are abolished in an infection with a phage that carries a deletion of the *pseT* gene, encoding PNK. In addition, many cleavages detected over a distance of 200 nucleotides downstream of the initial RegB cut (mostly generated by RNase E and a few by RNase G), disappear or are strongly weakened in the PNK mutant infection. The availability of a mutant affected only in the phosphatase activity (*pseT1*) made it possible to show that the phosphatase activity of PNK also contributes to mRNA destabilization from the 3' terminus. This presumably occurs through the conversion of 3'-phosphate into 3'-hydroxyl termini, making RNAs better substrates for polynucleotide phosphorylase, the only host 3' exoribonuclease that requires a 3'-hydroxyl terminus to act efficiently. The total inactivation of PNK increases the stability of some RegB-processed transcripts (Durand *et al.* 2008, unpublished data). Thus, both the kinase and phosphatase activities of PNK control the degradation of some RegB-processed transcripts from the 5' and the 3' extremities, respectively. This shows that the status of the 5' and 3' RNA extremities plays a major role in mRNA degradation (see also [21]). This was the first time a direct role was ascribed to T4 PNK in the utilization of phage mRNAs. In bacteriophage T4, as in other phages and bacteria where this enzyme is found, PNK is involved in tRNA repair, together with the RNA ligase, in response to cleavage catalyzed by host enzymes [22,23] (and see below). Durand's finding should prompt one to consider that, in addition to a role in RNA repair, prokaryotic PNKs might participate in the regulation of mRNA degradation.

The data presented above show that RNase G, a paralogue of RNase E in *E. coli*, participates in the processing and decay of several phage transcripts [14] (Durand *et al.* 2008, unpublished data). Nevertheless, it seems clear that it does not have the same general effect on phage mRNA as RNase E. The plating efficiency of T4 is reduced only by 30% on a strain deficient in RNase G (*rng*::Tn5) relative to a wild-type strain (Durand *et al.* 2008, unpublished data).

The RegB/S1 target site

It has been obvious since the initial discovery of RegB activity that not all intergenic GGAG sequences are cleaved by this RNase [13,24], suggesting that the motif is necessary but not sufficient for cleavage. RNA secondary structure protects against cleavage and several phage mRNAs that carry an intergenic GGAG/U motif are resistant to the nuclease, including a few early, most middle and all late transcripts [11]. These GGAG-containing mRNAs are not substrates of the enzyme either *in vitro* or *in vivo* [11].

A SELEX (systematic evolution of ligands by exponential enrichment; [25]) experiment, based on the selection of RNA molecules cleaved by RegB in the presence of the ribosomal protein S1, led to the selection of RNA molecules that all contained the GGAG tetranucleotide [26] and no other conserved sequence or structural motif. However, in most cases, the GGAG sequence was found in the 5' portion of the randomized region, suggesting that the nucleotide composition 3' to this conserved motif plays a role. More recently, by using classical molecular genetic techniques, Durand *et al.* [9] showed that this was indeed the case. The strong intergenic RegB cleavage sites share the following consensus: GG*AGRAYARAA, where R is a purine (often an A, leading to an A-rich sequence 3' to the very conserved GGAG motif) and Y a pyrimidine (the star indicates the site of cleavage) [9]. This unusually long nuclease recognition motif is reminiscent of cleavage sites for some mammalian endoribonucleases that function with auxiliary factors. One possible model assumes that the auxiliary factors bind the long nucleotide sequence and recruit the endonuclease [27]. Durand et al. [9] provided evidence that RegB alone recognizes the trinucleotide GGA, which it cleaves very inefficiently, irrespective of its nucleotide sequence context, and that stimulation of the cleavage activity by S1 depends on the base composition immediately 3' to -GGA-.

RegB catalysis and structure

The bacteriophage T4 RegB endoribonuclease is a basic, 153-residue protein. Although its amino acid sequence is unrelated to any other known RNase, it was shown to be a cyclizing ribonuclease of the Barnase family, producing 5'-hydroxyl and cyclic 2',3'-phosphodiester termini, with two histidines (in positions 48 and 68) as potent catalytic residues [28].

NMR was used to solve the structure of RegB and to map its interactions with two RNA substrates. Despite the absence of any sequence homology and a different organization of the active site residues, RegB shares structural similarities with two *E. coli* ribonucleases of the toxin/antitoxin family: YoeB and RelE [29]. YoeB and RelE are involved in the inactivation of mRNA translated under nutritional stress conditions [30,31]. Interestingly, like RegB, RelE, and in some cases YoeB recognize triplets on mRNAs, which they cleave between the second and third nucleotides. It has been proposed that RegB, RelE and YoeB are members of a newly recognized structural and functional family of ribonucleases specialized in mRNA inactivation within the ribosome [29] (Figure 1).

How does S1 activate the RegB cleavage reaction?

The *E. coli* S1 ribosomal protein is an RNA-binding protein required for the translation of virtually all the cellular mRNAs [32]. It contains six homologous regions, each of about 70 amino acids, called S1 modules (or

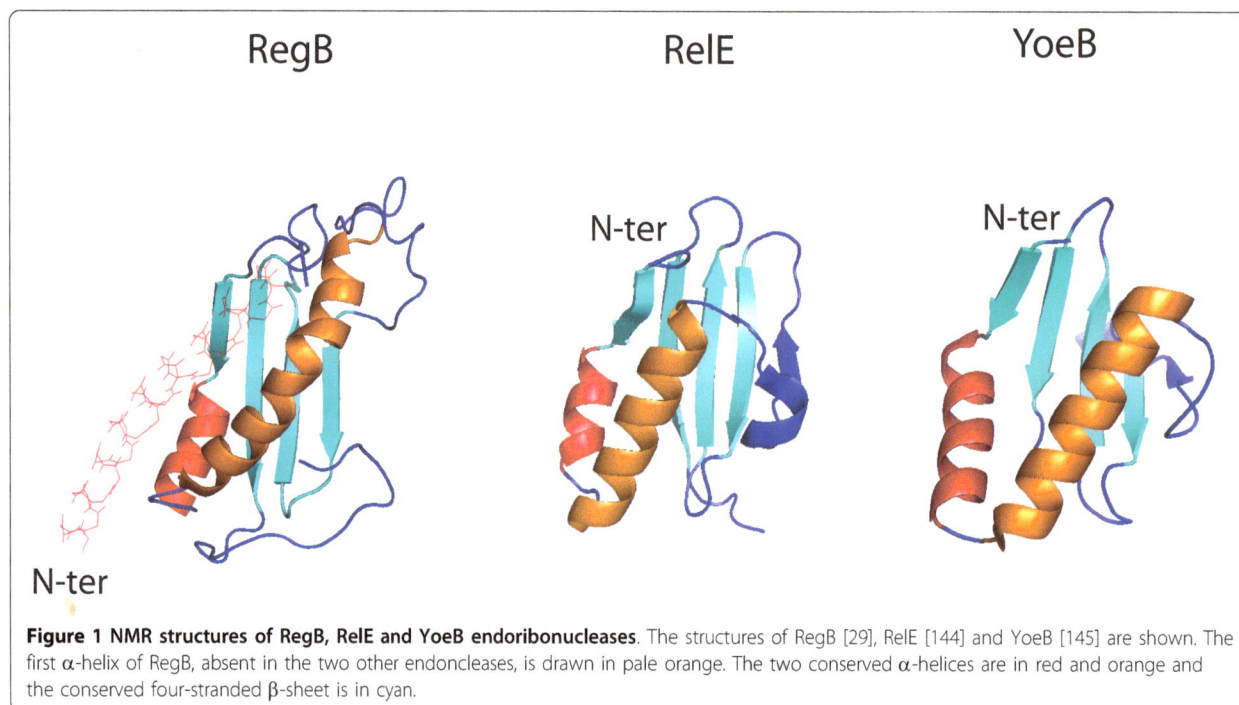

Figure 1 NMR structures of RegB, RelE and YoeB endoribonucleases. The structures of RegB [29], RelE [144] and YoeB [145] are shown. The first α-helix of RegB, absent in the two other endonucleases, is drawn in pale orange. The two conserved α-helices are in red and orange and the conserved four-stranded β-sheet is in cyan.

domains) connected by short linkers. S1 binds to ribosomes through its two N-terminal domains (modules 1-2) while mRNAs interact with the C-terminal domain made of the four other modules (3-4-5-6) [33]. S1-like modules are found in many proteins involved in the metabolism of RNA throughout evolution. The structure of these modules, (based on studies of the *E. coli* S1 protein itself as well as RNase E and PNPase), are predicted to belong to the OB-fold family [34-38].

The modules required in RegB activation have been identified. The C-terminal domain of S1 (including modules 3-4-5-6) stimulates the RegB reaction to the same extent as the full-length protein. Depending on the substrate, domain 6 can be removed without affecting the efficacy of the reaction. The smallest domain combination able to stimulate the cleavage reaction significantly is the bi-module 4-5 [9,39]. Interestingly, small angle X-ray scattering studies performed on the tri-module 3-4-5 showed that the two adjacent domains 4 and 5 are tightly associated, forming a rigid rod, while domain 3 has no or only a weak interaction with the others. This suggests that the S1 domains 4 and 5 cooperate to form an RNA binding surface able to interact with the nucleotides of RegB target sites. Module 3 could help stabilize the interaction with the RNA [34].

The 3' A-rich sequence that characterizes strong RegB sites (see above) plays a role in the mechanism of stimulation by S1. Indeed, directed mutagenesis experiments showed that the stimulation of RegB cleavage by S1 depends on nucleotides immediately 3' to the totally conserved GGA triplet. The closer the sequence is to the consensus shown above, the greater the stimulation by S1 [9]. The affinity of S1 for the A-rich sequence is not better than for any other RNA sequence (S. Durand and M. Uzan, unpublished data); suggesting that the function of this sequence is not simply to recruit S1 locally. Rather, specific interactions of S1 with the conserved sequence might make the G-A covalent bond more accessible to RegB. In support of this view, RegB alone (without S1) is able to perform efficient and specific cleavage in a small RNA carrying the GGAG sequence, provided the GGA triplet is unpaired and the fourth G nucleotide of the motif is partly constrained [15]. The RegB protein shows very weak affinity for its substrates [26,28] and in fact, no RegB-RNA complex can be visualized by gel shift experiments. However, in the presence of S1, RegB-RNA-S1 ternary complexes can form, suggesting that the first step in the S1 activation pathway involves S1 interaction with the RNA (S. Durand and M. Uzan, unpublished observations). Taken together, these observations suggest that through its interaction with the A-rich sequence 3' to the cleavage site, the S1 protein promotes a local constraint on the RNA, facilitating the association or reactivity of RegB.

As RegB is easily inhibited by RNA secondary structures, one possibility was that S1 stimulates RegB through its RNA unwinding ability [40,41]. However, Lebars *et al.* [15] provided evidence that does not support this hypothesis.

Whether S1 participates in the RegB reaction as a free protein or in association with the ribosome or other partners *in vivo* remains to be determined. However, the structural and mechanistic analogy of RegB to the two *E. coli* RNase toxins, YoeB and RelE [29], which depend on translating ribosomes for activity [30], and the efficiency of RegB cleavage *in vivo* very shortly after infection [13], favor the likelihood of ribosomes participating in RegB processing of mRNAs *in vivo*.

Regulation and distribution of the regB gene

The *regB* gene is transcribed from a typical early promoter that is turned off two to three minutes after infection. The *regB* gene is also regulated at the post-transcriptional level, suggesting that the production of this nuclease must be tightly regulated. Indeed, RegB efficiently cleaves its own transcript in the SD sequence, indicating that RegB controls its own synthesis. Three other cleavages of weaker efficiency occur in the *regB* coding sequence, which probably contribute to *regB* mRNA breakdown [10].

Despite the fact that the RegB nuclease seems dispensable for T4 growth, the *regB* gene is widely distributed among T4-related phages. The *regB* sequence was determined from 35 different T4-related phages. Thirty-two of these showed striking sequence conservation, while three other sequences (from RB69, TuIa and RB49) diverged significantly. As in T4, the SD sequence of these *regB* genes is GGAG, with only one case (RB49) of GGAU. When experimentally tested, this sequence was always found to be cleaved by RegB *in vivo*, suggesting that translational auto-control of *regB* is conserved in T4-related phages [42].

Mutants of *regB* are viable on laboratory *E. coli* strains, although their plaques are slightly smaller in minimal medium than those of the wild-type phage. Also, T4 *regB* mutants form minute plaques on the hospital *E. coli* strain CTr5x, with a plating efficiency of one third that on classical laboratory strains (M. Uzan, unpublished data).

What is the role of RegB in T4 development?

Early transcripts are synthesized in abundance immediately after infection, reflecting the exceptional strength of most T4 early promoters. In fact, effective promoter competition for RNA polymerase can be considered one of the first mechanisms leading to shut-off of host gene transcription. Abundant and stable phage early transcripts would compete for translation with the subsequently made middle and late transcripts. Therefore, a specific mechanism leading to early mRNA inactivation and increased rate of degradation should free the translation apparatus more rapidly and facilitate the transition between early and later phases of T4 gene expression [5]. Functional mRNA endonucleolytic inactivation is certainly a faster means to arrest ongoing translation and rapidly re-orient gene expression in response to changes in growth conditions or the stage of development. In this regard, it is striking that the two toxin endoribonucleases, RelE and YoeB, to which RegB shows strong structural similarities (Figure 1) [29], also allow swift inactivation of translated mRNAs in response to nutritional stress.

The finding that RegB shares structural and functional similarities with other toxin RNases that have antitoxin partners raises the possibility that an anti-RegB partner might be encoded by T4. On the other hand, RegB might not require an antitoxin to block its activity since its *in vivo* targets disappear through mRNA decay shortly after it acts in the infected cell.

T4 Dmd and E. coli RNase LS antagonism
T4 Dmd controls the stability of middle and late mRNAs
The T4 early *dmd* gene (*d*iscrimination of *m*essages for *d*egradation) encodes a protein that controls middle and late mRNA stability. Indeed, an amber mutation in *dmd* leads to strong inhibition of phage development. Protein synthesis is normal until the beginning of the middle period and collapses thereafter. A number of endonucleolytic cleavages can be detected in middle and late transcripts, which are not present in wild-type phage infection. Consistent with this observation, the accumulation of these RNA species drops dramatically and the chemical and functional half-lives of several middle and late transcripts were shown to be shortened [43-46]. The host RNA chaperone, Hfq, seems to enhance the deleterious effect of the *dmd* mutation [47]. These data strongly suggest that the arrest of protein synthesis in T4 *dmd* mutants is the consequence of mRNA destabilization and that the function of the Dmd protein is to inhibit an endoribonuclease that targets middle and late transcripts.

The endoribonuclease responsible for middle and late mRNA destabilization in the *dmd* mutant is of host origin as shown by the fact that a late mRNA (*soc*) produced from a plasmid in uninfected bacteria undergoes the same cleavages as those observed after infection by a *dmd* mutant phage [43,48]. Yonesaki's group further showed that this RNase activity depends on a new endonuclease, RNase LS, for *l*ate gene *s*ilencing in T4. Several *E. coli* mutants able to support the growth of a *dmd* mutant phage were isolated, among which, two very efficiently reversed the *dmd* phenotype. Both mutations were mapped within the ORF *yfjN*, which was renamed *rnlA* [44,45,48].

Biochemical characterization of RNase LS
Purified his-tagged RnlA protein cleaves the late *soc* transcript *in vitro* at only one site among the three usually observed *in vivo* after infection with *dmd* mutant phage. This cleavage is inhibited by purified Dmd protein [49]. Thus, RnlA has an RNase activity that responds directly to Dmd. Whether RnlA has targets in other T4 mRNAs remains to be determined.

Biochemical experiments showed that RNase LS activity is associated with a large complex whose MW was estimated to be more than 1,000 kDa. More than 10 proteins participate in the complex. Two of them were identified: RnlA and triose phosphate isomerase. The latter is present in stoichiometric amounts relative to RnlA and binds very tightly to it [45,49]. Interestingly, a mutation in the gene for triose phosphate isomerase is able to partially allow the growth of a T4 *dmd* mutant, suggesting that RnlA and triose phosphate isomerase functionally interact. It is unclear whether RNase LS carries only one RNase activity (presumably that of the RnlA protein) or more, and if the activity of RnlA is modulated by other components of the complex.

The multi-protein complex that constitutes RNase LS is not simply a modification of the host degradosome to contain the RnlA protein during T4 infection, since the *dmd* phenotype is not reversed in infection of an RNase E host mutant (*rne*Δ131) unable to assemble the degradosome [48].

The specificity of RNase LS and coupling with translation
The specificity and mode of action of RNase LS are not yet understood. Most of the ~30 cleavages analyzed in various middle and late transcripts occur 3' to a pyrimidine in single-stranded RNA. Also, nucleotides 3' to the cleavage site might play a role. Apart from these observations, no sequence or structural motif seems to be shared by the RNase LS target sites [43,44,50,51].

The presence of ribosomes loaded on the mRNA seems to be required for some RNase LS sites to be efficiently cut. The ribosomes may be either translating or pausing at a nonsense codon. In the later case, new cleavage sites by RNase LS appear at some distance (20-25 nucleotides) downstream of the stop codon [44,48,51]. It has been suggested that ribosomes act through their RNA unwinding property, maintaining the RNA in a locally single-stranded conformation. In the absence of translation, a number of potential RNase LS sites would be masked by secondary structure [51]. Whether this is the only role of the ribosome in RNase LS activation is an open question.

The role of RNase LS in E. coli
A mutation in the *E. coli rnlA* gene, whether a point mutation or an insertion, leads to reduction in the size of colonies on minimum medium, but has no effect on growth in rich medium. Growth of *rnlA* mutants is

however dramatically affected in rich medium supplemented with high sodium chloride concentrations, thus providing a phenotype for *rnlA* mutants. RNA is stabilized by 30% on average in an *rnlA* mutant. RNase LS was shown to participate in the degradation of specific mRNAs as reflected by the prolonged functional lifetime of several mRNAs in the *rnlA* mutant. The *rpsO*, *bla* and *cya* mRNAs are stabilized 2 to 3-fold, in the *rnlA* mutant, while other transcripts are unaffected. The greater stability of *cya* mRNA (adenylate cyclase) in an *rnlA* mutant might indirectly account for the sensitivity of *rnlA* cells to NaCl [45,52]. In addition to moderately controlling the decay of some bacterial transcripts, it is possible that the first function of RNase LS is host defense against phage propagation and Dmd is a phage response to overcome the host defense.

Other activities implicated in RNA decay during T4 infection

The *E. coli* poly(A) polymerase (PAP), encoded by the *pcnB* gene, adds poly(A) tails to the 3' ends of *E. coli* mRNAs and contributes to the destabilization of transcripts [53]. T4 mRNAs are probably not polyadenylated. Indeed, it has been found that after infection with the closely related bacteriophage T2, host poly(A) polymerase activity is inhibited [54]. Also, no poly(A) extension could be detected at the 3' end of the *soc* and *uvsY* transcripts after infection with T4 [55], suggesting that bacteriophage T4 infection also leads to PAP inhibition. This could, for example, occur through ADP-ribosylation of the protein.

Growth of bacteriophage T4 on an *E. coli* strain carrying the *rne*Δ131 mutation, which is unable to assemble the RNA degradosome, is unchanged relative to infection of a wild-type strain [48] (also, S. Durand and M. Uzan, unpublished data). However, the *rne*Δ131 mutation has no effect on the growth of *E. coli* either, despite affecting the stability of several individual transcripts [56-59]. Therefore, the question of whether the degradosome plays a role in the turnover of some T4 mRNAs or is modified after infection remains open. Similarly, whether the host RNA pyrophosphohydrolase, RppH [21,60] is implicated in T4 mRNA turnover has not yet been determined.

Infection with bacteriophage T4 expedites host mRNA degradation. The two long-lived *E. coli* mRNAs, *lpp* and *ompA*, are dramatically destabilized after infection with T4. The host endonucleases, RNases E and G, are responsible for this increased rate of degradation [61]. Phage-induced host mRNA destabilization requires the degradosome. Indeed, the *lpp* mRNA is not destabilized after infection of a strain that carries a nonsense mutation in the middle of the *E. coli rne* gene (encoding RNase E), leading to a protein unable to assemble the

degradosome. A viral factor is also involved, since a phage carrying the Δ*tk2* deletion that removes an 11.3 kbp region of the T4 genome, from the *tk* gene to ORF *nrdC.2*, loses the ability to destabilize host transcripts. The gene implicated has not yet been identified [61]. There is certainly an advantage for a virulent phage to accelerate host mRNA degradation immediately after infection, as this provides ribonucleotides for nucleic acid synthesis, frees the translation apparatus for viral mRNAs, and facilitates the transition from host to phage gene expression.

A list of the several endoribonucleases and other enzymes involved in mRNA degradation and modification during T4 infection is presented in Table 1.

Inhibition of translation initiation
RegA translational repression
Inhibition of middle transcription, some 12-15 minutes post-infection at 30°C, is concomitant with the strong activation of late transcription [62]. This is the consequence of competition among sigma factors and changing the promoter specificity of the modified host RNA polymerase. Indeed, transcription initiation at T4 late promoters requires the phage-encoded late σ-factor, gp55, which replaces the major host σ70, and the T4-encoded gp33, which ensures coupling of late transcription with ongoing viral DNA replication [1,62-64]. Superimposed on this transcriptional regulation, the translation of a number of transcripts is inhibited by the RegA translational repressor. This small, 122 amino acid protein competes with the ribosome for binding to the translation initiation regions of approximately 30 mRNAs [65]

RegA protein
The crystal structure of T4 RegA is a homodimer, with symmetrical pairs of salt bridges between Arg-91 and Glu-68 and pairs of hydrogen bonds between Thr-92 of both subunits [66] (Figure 2). The monomer subunit has an alpha-helical core and two anti-parallel beta sheet regions. Two of the beta strands in the four-stranded beta sheet region B were identified by Kang *et al.* [65] as having amino acid sequences similar to RNP-1 and RNP-2 that are well characterized RNA-binding motifs. In addition, two pairs of lysines, K7-K8 and K41-K42 are in the same position in the proposed RegA RNP-1 domain [66] as they occur in the U1A RNA-binding protein, where they comprise basic "jaws" that straddle the RNA. However, none of the *regA* mutations identified in either T4 or phage RB69 prior to the availability of the RegA structure affected these lysine residues [65]. Structure-guided mutagenesis summarized below also did not implicate the lysines or the RNP-like domains in direct RNA binding by RegA.

Concurrent with the T4 RegA structure determination, E. Spicer's group reported a terminal deletion

Table 1 Enzymes involved in mRNA degradation and modification during T4 infection

Enzyme	Origin	Reaction catalyzed. Main properties	Role in T4 development
RNase E	*E. coli*	Endonuclease. Produces 5′-P termini. Activated by 5′-monophosphorylated RNA. Scaffold of the degradosome	Major role in mRNA degradation throughout the phage developmental cycle.
RNase G	*E. coli*	Endonuclease. Produces 5′-P termini. Activated by 5′-monophosphorylated RNA.	Cuts in the 5′ regions of some early RegB processed transcripts.
RegB	T4	Sequence-specific endonuclease. Produces 5′-OH termini. Requires S1 r-protein as co-factor	Inactivates early transcripts by cleaving in Shine-Dalgarno sequences. Expedites early mRNA degradation.
RNase LS	*E. coli*	Endonuclease. Its activity depends on *rnlA* and *rnlB* loci. Associated in a multiprotein compex.	Cleaves within T4 middle and late transcripts and expedites their degradation.
RNase II RNase R Polynucleotide phosphorylase	*E. coli*	3′-5′ exonucleases. PNPase requires 3′-OH termini; the other two are indifferent to the nature of the 3′ terminus.	Degrade mRNAs. The relative contribution of each RNase has not been determined.
PrrC	*E. coli*	tRNA^{lys} anticodon nuclease. Normally silent in *E. coli* but activated by the T4-encoded Stp polypeptide.	Deleterious to T4 propagation if Pnk or Rli1 enzymes are inactivated.
Polynucleotide kinase (PNK)	T4	Phosphorylation of 5′-OH polynucleotide termini. Hydrolysis of 3′-terminal phosphomonoesters and of 2′,3′-cyclic phosphodiesters	Counteracts, together with T4 RNA ligase 1, host tRNA anticodon nuclease PrrC. Makes RegB-processed RNA substrates for RNases E and G.
Dmd	T4	An early product that binds the RnlA protein, a member of RNase LS	Antagonist of RNase LS
Poly(A) polymerase	*E. coli*	Addition of poly(A) tails to the 3′ end of RNAs	Probably inactivated after T4 infection
RNA pyrophospho-hydrolase (RppH)	*E. coli*	Hydrolysis of a pyrophosphate moiety from the 5′-triphosphorylated primary transcripts.	Not yet investigated

mutant having residues 1 - 109 that bound RNA with reduced affinity, with 28% of the free energy of binding attributed to the terminal 10% of the protein [67]. It was also shown by proteolytic cleavage of free RegA, and RegA bound to an RNA oligonucleotide (the gene

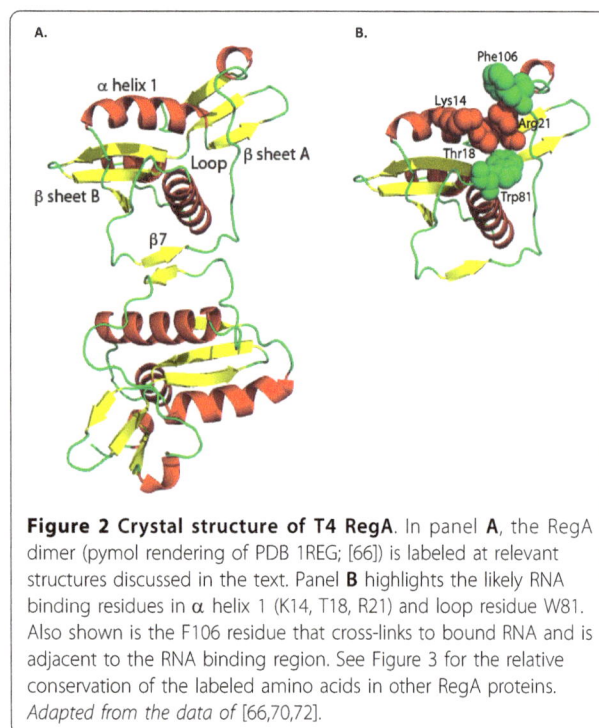

Figure 2 Crystal structure of T4 RegA. In panel **A**, the RegA dimer (pymol rendering of PDB 1REG; [66]) is labeled at relevant structures discussed in the text. Panel **B** highlights the likely RNA binding residues in α helix 1 (K14, T18, R21) and loop residue W81. Also shown is the F106 residue that cross-links to bound RNA and is adjacent to the RNA binding region. See Figure 3 for the relative conservation of the labeled amino acids in other RegA proteins. *Adapted from the data of [66,70,72].*

44 operator), that conformational change in RegA upon RNA binding affected access to the C-terminal region. The C-terminal region is part of beta sheet region A of RegA [66], appears to be solvent-exposed, and thus potentially could interact with RNA in some manner. However, with the RegA structure available, targeted substitutions in the protein would reveal that specific RNA recognition likely occurs in an entirely different region of the protein.

Structure-guided mutagenesis of RegA was undertaken to evaluate some of these findings and for understanding the specific interactions for RNA binding. Binding stoichiometry of RegA:gene *44* RNA complexes, gluteraldehyde cross-linking of RegA, and mutagenesis of amino acids in the inter-subunit interface showed that T4 RegA is a dimer in solution (as also revealed in the crystal structure), but binds RNA as monomer [68]. A 1:1 RNA:RegA monomer stoichiometry was independently shown using electrospray ionization mass spectrometry [69]. Mutagenesis of Arg91 again suggested that at least some residues in the C-terminal region are involved in subunit interactions and in RNA recognition [66-68]; Arg91 appears more relevant for RNA binding, whereas Thr92 is more relevant for dimerization. Spicer and colleagues further demonstrated that 19 mutations substituting amino acids in T4 RegA surface residues of both beta structures, including residues similar to the RNP-1 and RNP-2 motifs proposed by Kang *et al.* [66], as well as the two paired lysines, had essentially no affect on RNA binding affinity or on RegA structure [70]. Together with mutations in helix-A, and interpretation of mutations in T4 and RB69 *regA* that were isolated prior to the structure determination [71], a somewhat unique RNA-binding helix-loop groove (or "pocket") of RegA was proposed to provide the primary RNA recognition element for the protein. Modeling of the 78% conserved phage RB69 RegA protein showed that it also likely contains this unique RNA binding structure [72]. Exposed residues on helix-A (i.e., Lys14, Thr18, Arg21) are conserved and substitutions reduce RNA binding substantially. Additionally, a conserved loop Trp81 to Ala81 substitution in both proteins abolishes RNA binding [72]. Phe106, earlier shown to crosslink with bound RNA, is positioned in a loop bordering the other end of the helix and further defines the apparent binding pocket [67,70,72]. Figure 2 summarizes these findings.

In summary, biochemical and structural studies of T4 and RB69 RegA have led from inferences of possible motifs in RNA binding to structure guided mutagenesis revealing a unique protein pocket or groove that, in the monomer form, accommodates the many different mRNAs that RegA proteins bind to cause translational repression. The apparent binding domain and exposed amino acids are largely conserved in RegA proteins from diverse phages sequenced to date (Figure 3). As for gp32 and gp43, a RegA-RNA complex has not been structurally resolved and additional analysis of RegA-RNA interactions in the helix-loop groove would be of interest.

RegA RNA operators

Early genetic and translational repression assays confirmed that RegA binding sites on mRNA overlap the AUG translation initiation codon, or are located immediately 5' to the AUG, and occluding the site reduces formation of the ternary translation initiation complex; decay of the repressed messages is then enhanced [65]. The lack of clear sequence conservation or secondary structure to define RegA binding sites in the ~30 mRNAs repressed, prompted use of RNA SELEX with T4 RegA to capture high-affinity RNA ligands. This RNA binding site selection was thus performed in the absence of constraints imposed on the sequence by 30S ribosome subunits that bind the same region of mRNA for translation initiation [73]. Emerging from multiple rounds of SELEX was an RNA consensus sequence of 5'-aaAAUUGUUAUGUAA-3' that bound RegA with an apparent Kd of 5 nM (the lower case 5' bases were already present in the starting, non-variable regions of the RNA). The sequence showed no apparent structure using nuclease or base-modifying chemical probes and is consistent with earlier observations that biologically relevant RegA binding sites lack clear RNA secondary

structure. Although the T4 RegA SELEX sequence is similar to mRNA sequences repressed by RegA (i.e., T4 gene *rIIB*, AAAAUUAUGUAC; gene *44*, AAAUUAUGAUU; *dexA*, AAAAUUUAAUGUUU), there was no exact match between it and the repressed T4 messages [73]. These findings emphasize that T4 RegA binding sites are A+U rich; include an AUG and a 5' poly(A) tract; lack apparent structure; and in general, illustrate how an RNA binding determinant has evolved for occurring on many different mRNAs where fMet-tRNA and the 30S ribosome subunit also bind.

RNA sequences bound by phage RB69 RegA have also been examined [65,72,74,75]. Translational repression occurs at RNAs from both phages, although binding affinities displayed by the two proteins are different *in vivo* and *in vitro*; a hierarchy of early and middle genes repressed by T4 RegA is also seen with RB69 RegA. For RB69 RegA, the protein protected a region between the gene *44* and gene *45* Shine-Dalgarno and AUG, but not the initiator AUG itself [72]. The protein would still compete for the same binding site as the ribosome. Using a stringent but reduced number of selection cycles, RNA SELEX was performed using immobilized RB69 RegA and a variable sequence of 14 bases [75]. The selected RB69 RegA RNAs were predominately 5"AAUAAUAAUAAnA-3', which also did not contain a conserved AUG but were clearly A+U rich. As discussed by Dean *et al.* [75], a stop codon (i.e, UAA) for an upstream gene within the ribosome binding site region of the adjacent downstream gene, may contribute a relevant sequence for RNA recognition by RegA proteins. All of these findings emphasize the range of RegA repression efficiencies at different sites, lack of RNA structure in binding sites, and the variable mRNA sequences to which the protein binds.

Specific autocontrol of translation: gp32 and gp43

Besides the two general post-transcriptional regulators, RegA and RegB, the T4 DNA unwinding protein, gp32, and the DNA polymerase, gp43, both involved in DNA replication, recombination and repair, autogenously regulate their translation.

Control of gene 32 translation and mRNA degradation

Gene 32 encodes a single-stranded DNA binding protein (gp32) essential for replication, recombination and repair of T4 DNA. It appears after a few minutes of infection, reaches a maximum around the 12-14th minute and declines thereafter. In addition to being temporally regulated at the transcriptional level, gp32 inhibits its own translation when the protein accumulates in excess over its primary ligand, single-stranded DNA. This regulation is achieved through binding of gp32 to a pseudoknot RNA structure located 5' in region 67 nucleotides upstream of the gene 32 translation

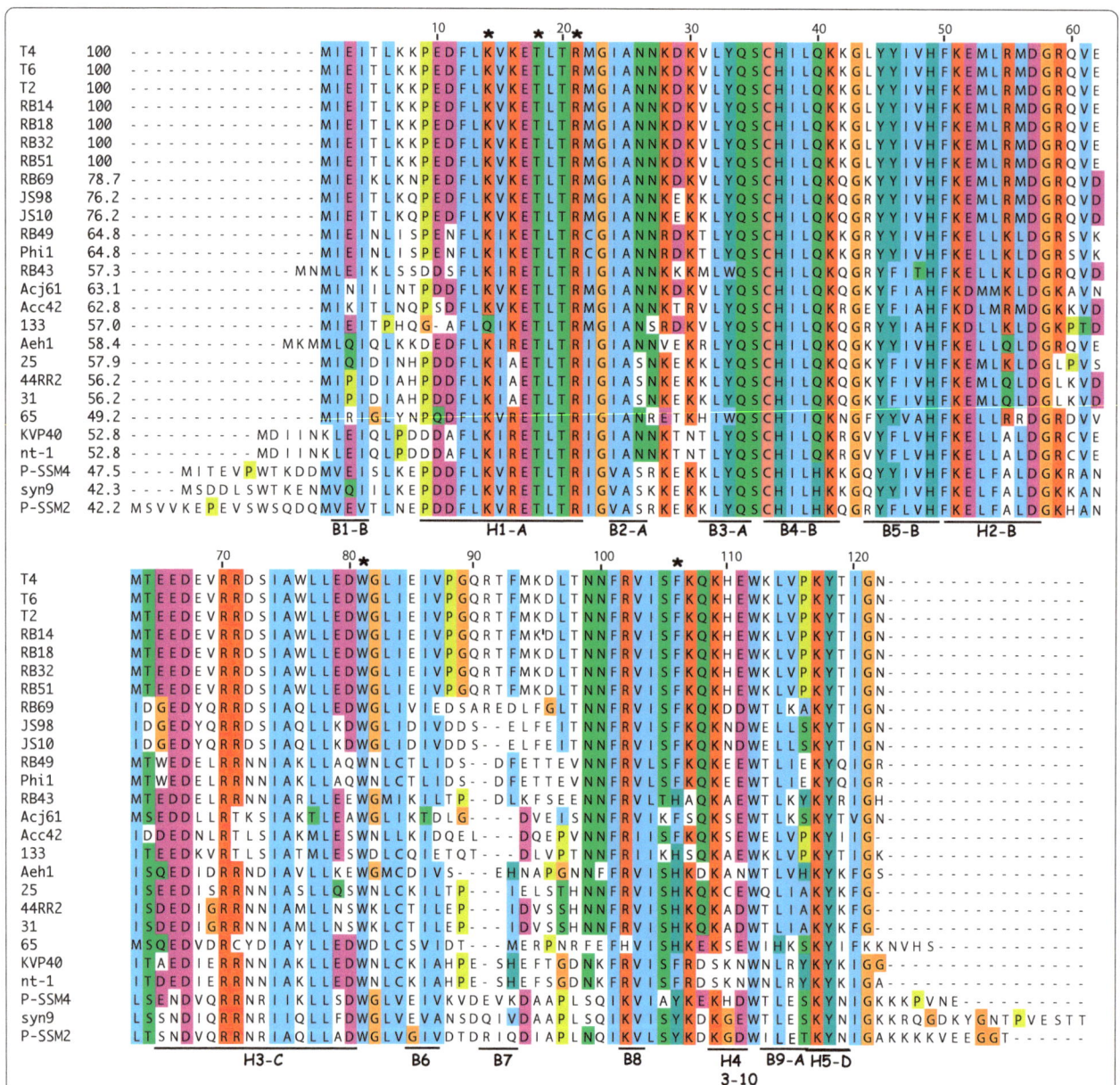

Figure 3 Aligned RegA proteins of 26 T4-related phages. *regA* is immediately distal to gene *62* in the core DNA replication gene cluster of all T4-related genomes sequenced to date. Identity relative to T4 RegA is in column 2, aligned amino acids are shown using ClustalW colors, and dashes are gaps in the alignment. Residues numbered above the sequences reference the T4 protein. Asterisks mark the amino acids cited in the text as involved in RNA binding. At the bottom of the alignment are underlined structural elements of the protein from PDB 1REG [66]. Sequences were obtained from GenBank or the T4-like phage genome browser (http://phage.ggc.edu/).

initiation codon. This binding is thought to nucleate cooperative binding through an unstructured A+U-rich sequence (including several UUAA(A) repeats 3' to the pseudoknot) that overlaps the ribosome binding site [3,6,65].

Gp32 is a Zn(II) metalloprotein with three distinct binding domains [76]. To date, the structure of full-length gp32 has not been determined, nor has the protein in complex with RNA been structurally examined.

It has been presumed that DNA and RNA are alternative ligands that bind in the same cleft. Although there is substantial study of gp32 interactions with ssDNA, and with proteins of the DNA replication apparatus, few studies have investigated either the RNA pseudoknot in the mRNA autoregulatory site or the molecular details of gp32-RNA interactions. NMR analysis of the phage T2 gene 32 pseudoknot revealed two A-form helices coaxially stacked, with two loops separating the two helical structures [77] (Figure 4). A related translational regulatory structure is present in gene 32 leader mRNA of the phylogenetically related T4-type phage RB69 [78]. In this case, sequence alignment, chemical- and RNase-sensitivity, and gp32-RNA footprinting revealed mRNA operator similarities and differences that explain overlapping yet distinct RNA-binding properties by the two gene 32 proteins [78]. However, the T4-type coliphage RB49 genome sequence revealed no conserved pseudoknot or an A+U-rich sequence near the predicted ribosome binding site of its gene 32 mRNA [79]. More thorough study of translational autocontrol by gp32 in diverse T4-related phages is needed. To date, the T4-type phage gene 32 RNA pseudoknot may still be the only viral example of this structure used in autoregulation of translation. The various biological roles of viral RNA pseudoknots was well reviewed by Brierley *et al.* [80].

The gene 32 transcripts are more stable than any other T4 mRNAs. A half-life of 15 minutes was measured at 30°C and, under derepression conditions (in a T4 gene 32 mutant infection unable to achieve translation repression), the half-life can reach 30 minutes [81,82], indicating that translation of the gene 32 mRNA positively affects its stability. All the gene 32 mRNA species are processed by RNase E, 71 nucleotides upstream of the translation initiation codon of the gene [83,84]. In addition to the cleavage at -71, two other major cleavages were identified, one far upstream in the polycistronic transcripts (-1340) and the other at the end of the coding sequence of gene 32 (+831) [85,86]. The conservation of all three RNase E processing sites in 5 different T4-related phages, in spite of significant changes in the organization of the upstream regions, suggests that these cleavages play an important role in controlling expression of gene 32 and/or its upstream genes [86]. The new 3' ends created by RNase E processing are potential entry sites for the host 3'-5' exoribonucleases. In fact, portions of the transcript upstream of the -71 and -1340 cleavage sites were shown to be rapidly degraded [84,85].

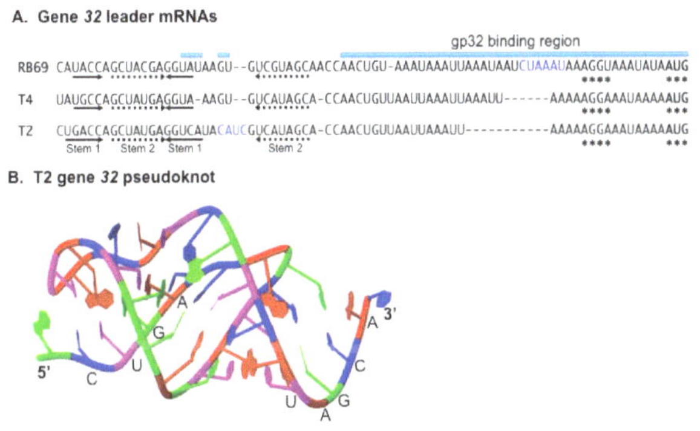

The RNase E cleavage at +831 has no consequences on the functional decay of the gene 32 mRNA, while it affects the chemical decay [17]. It is noteworthy that this RNase E site is very close to the translation termination codon of gene 32. The *E. coli* ribosomal protein S15, encoded by the *rpsO* gene, autogenously regulates its own translation. The *rpsO* transcript carries a pseudoknot in its translational operator [87], like the T4 32 mRNA. Also, a strong RNase E cleavage site, involved in *rpsO* mRNA decay, lies at the end of the structural gene, in close proximity of the translation termination codon. Interestingly, ribosomes were shown to inhibit this distal RNase E cleavage [88]. On this basis, it is tempting to suggest that a ribosome that reaches the end of gene 32 transcript would hinder the accessibility of the distal RNase E site to RNase E. Thus, gene 32 transcripts that undergo RNase E processing at this site might be only those that have been already translationally inactivated, e.g., under repression conditions (excess of gp32 over single stranded DNA). This situation would promote rapid elimination of the untranslated gene 32 transcripts.

Autocontrol of gene 43 translation

Like gp32, T4 DNA polymerase (gp43) is an autoregulatory translational repressor protein; it binds an RNA operator sequence that includes a hairpin about 40 bases upstream of its translation initiation codon and sequence that overlaps the ribosome binding site [89]. Most T4 gene 43 transcripts are synthesized early during infection and have a half-life of approximately 3 minutes, yet it is these transcripts on which the polymerase exerts translational repression when not engaged in DNA replication [65].

gp43 RNA-binding determinants

The structure of the closely related gp43 DNA polymerase of phage RB69 serves as an excellent model for α DNA polymerases that are conserved across phylogenetic domains [90,91]. Due to the availability of the RB69 gp43 structure, more recent RNA binding studies have been conducted using this protein and its RNA operator.

RB69 operator RNA chemically crosslinks with gp43 in the DNA binding "palm" domain, but other sites and

Figure 4 Gene 32 translational repression site. In Panel **A** the leader mRNA for autogenous gp32 binding is shown for RB69, T4 and T2. The important TIR nucleotides are underscored with asterisks, the base-paired regions of the 5' pseudoknot are marked with arrows, and the T4 and RB69 regions bound by gp32 in protection assays are overlined [78]. Short nucleotide insertions in RB69 or T2 relative to T4 are in blue. Dashes (gaps) are inserted for alignment. Panel **B** is a cartoon-ribbon diagram of the T2 gene 32 pseudoknot diagramed in panel A that was obtained by multidimensional NMR methods [77]. Two A-form coaxially stacked stems are apparent. 5' and 3' terminal nucleotides are labeled. Jmol rendering used database entry 2 tpk. *Figure was derived and adapted primarily from data in* [77,78].

residues protected from protease when the protein is bound to specific RNA were distributed across domains of the polymerase. These numerous affects were attributable to either direct interactions, or conformational changes induced by RNA binding [92]. As for the gp32-RNA interactions, full appreciation of the contacts and conformational changes during binding of gp43 to its specific RNA target will require solution or crystal structure of gp43-RNA complexes.

Gene 43 mRNA autoregulatory site

The gene *43* RNA operator includes an upstream hairpin, but there is no evidence that it forms a pseudoknot structure like that of the gene *32* binding site. While the T4 hairpin-loop operator is 18 bases and that of RB69 is 16 bases, the top 10 bases are identical, including nucleotides in the loop [93]. The -UAAC- loop sequence of the T4 & RB69 operators were also the predominant bases selected in the first RNA SELEX experiment that used gp43 for RNA binding site characterization [25]; it will be interesting to see whether any phage gp43 proteins closely related to the T4 protein have the SELEX major variant loop sequence (-CAAC-) in their native, autoregulatory RNA hairpins. Phage RB49 contains -UAAA- in its RNA loop, and various repression and RNA-protein interaction assays point to the 3' AC and AA loop bases as especially relevant for binding by these three phage proteins; however, some T4-related phages encode gp43 DNA polymerases that do not autoregulate translation [92-94].

Other T4 post-transcriptional control systems

RNA structure at translation initiation regions

RNA structure influences translation initiation of T4 mRNAs, especially as they target protein binding in translational repression (i.e., gp32 and gp43 above; [65,95]). In addition, some T4 mRNAs form intramolecular RNA structures that directly contribute to translation initiation efficiency of the respective mRNAs. Only a few advances have been made in the last decade on these *cis*-acting RNAs, which are briefly summarized here. We should note that no riboswitch system [96,97] or small, *trans*-acting regulatory RNA has been functionally characterized from T4; maybe some of the genome sequences of T4-related phages will suggest good candidates for these types of RNAs. Two small RNAs, RNAC and RNAD, are transcribed from the T4 tRNA region, but their biological roles are unknown [95].

Examples of inhibitory RNA structures at translation initiation regions include mRNAs encoded by T4 genes *e*, *soc*, *49*, and I-TevI [65]. In each case, the Shine-Dalgarno and/or the AUG start codon are sequestered in an RNA helix that reduces 30S subunit binding in forming the ternary translation initiation complex [98]. The well-documented case for gene *e* (T4 lysozyme) is that

early during infection longer transcripts are made that extend into *e* and if translated could potentially lead to premature cell lysis. However, the longer transcripts clearly form the inhibitory RNA structure [98], reducing synthesis of lysozyme 100-fold [99] relative to transcripts lacking RNA structure. Transcripts initiated from either of two T4 late promoters immediately upstream of the ribosome binding site lack the 5' portion of the gene *e* mRNA inhibitory structure and are well translated. Although there is no additional analysis of the *e* mRNA structures, similar leader RNA sequences are predicted from genome sequences of closely related T4-type phages, and each has a T4-type late promoters in upstream region the encodes the 5' strand of the RNA structure. Therefore, early translation of these lysis genes may also be inhibited by intramolecular RNA structures (Figure 5).

The T4 thymidylate synthase gene (*td*) contains an intron, wherein the intron encodes a homing endonuclease, I-TevI [100]. Similar to gene *e*, early and middle period transcripts that extend through the *td* 5' exon

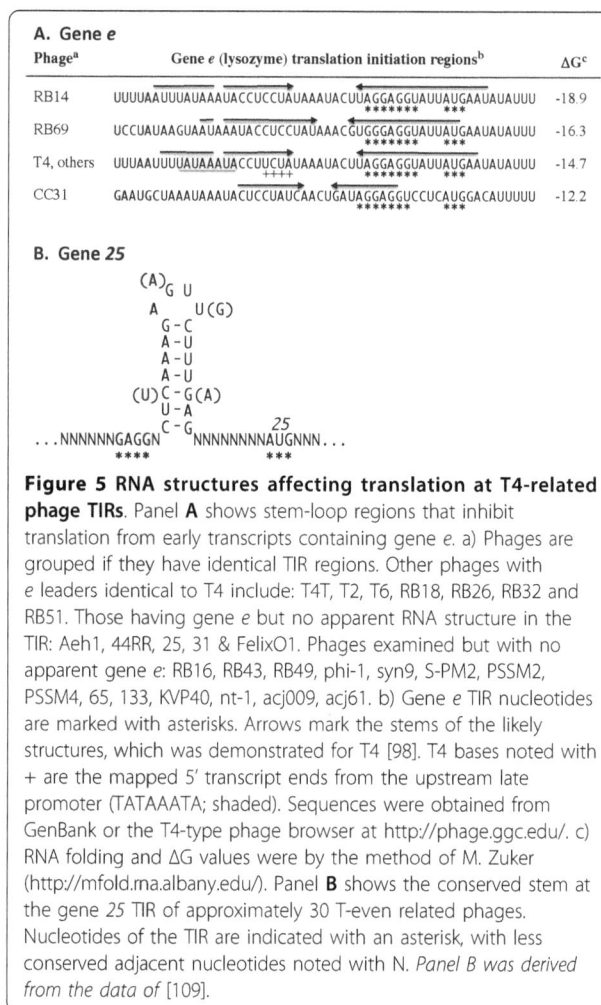

Figure 5 RNA structures affecting translation at T4-related phage TIRs. Panel **A** shows stem-loop regions that inhibit translation from early transcripts containing gene *e*. a) Phages are grouped if they have identical TIR regions. Other phages with *e* leaders identical to T4 include: T4T, T2, T6, RB18, RB26, RB32 and RB51. Those having gene *e* but no apparent RNA structure in the TIR: Aeh1, 44RR, 25, 31 & FelixO1. Phages examined but with no apparent gene *e*: RB16, RB43, RB49, phi-1, syn9, S-PM2, PSSM2, PSSM4, 65, 133, KVP40, nt-1, acj009, acj61. b) Gene *e* TIR nucleotides are marked with asterisks. Arrows mark the stems of the likely structures, which was demonstrated for T4 [98]. T4 bases noted with + are the mapped 5' transcript ends from the upstream late promoter (TATAAATA; shaded). Sequences were obtained from GenBank or the T4-type phage browser at http://phage.ggc.edu/. c) RNA folding and ΔG values were by the method of M. Zuker (http://mfold.ma.albany.edu/). Panel **B** shows the conserved stem at the gene *25* TIR of approximately 30 T-even related phages. Nucleotides of the TIR are indicated with an asterisk, with less conserved adjacent nucleotides noted with N. *Panel B was derived from the data of* [109].

and into the intron do not yield translated I-TevI because of the tightly regulated, sequestered ribosome binding site [101-103]. Recently, Edgell and colleagues [104] showed that deletion of nucleotides that comprise the late promoter proximal 5' portion of the RNA structure leads to increased levels of I-Tev1 throughout infection that is translated from upstream-initiated early and middle promoters. In the presence of added thymidine, the mutant phage (ΔHP) showed no reduction in T4 viability or burst size attributable to increased translation initiation of I-TevI mRNA. However, a series of phage growth, RT-PCR, and tRNA suppressor assays, led Gibb and Edgell [104] to conclude that tight regulation of I-TevI translation initiation by the RNA secondary structure increases intron splicing. That is, the structure reduces ribosome loading and movement through the intron RNA, thereby promoting structure formation (P6, P6a and P7) in the intron. Loss of translation inhibition disrupts intron RNA folding and splicing, and prevents proper accumulation of thymidylate synthase [104]. Similar structures predicted to cause negative translational regulation in I-TevI RNAs have been identified in T4-related phages, and also in the translation initiation regions of phage I-TevII and I-TevIII homing endonuclease genes [105,106]. Stand-alone homing endonucleases (not located within an intron) also have RNA structures that have been shown (*Aeromonas* phage Aeh1 *mobE*; [106]) or implicated (T4 *segB*; [107]) to reduce translation initiation.

Intramolecular RNA structures in T4 mRNAs also have been shown to improve translation initiation. Of particular note are T4 genes *38* and *25* [108]. In these cases, suboptimal, extended spacing between the Shine-Dalgarno and AUG start codon is brought to a functional distance by an RNA secondary structure between the SD and AUG. For gene *38* mRNA, the spacing of 22 nucleotides is reduced to 5 nucleotides with the structure; for gene *25* the structure reduces the spacing from 27 to 11 nucleotides [108]. Mutations in the intervening sequence that destabilize the structure reduce translation initiation efficiency. More recently, Malys and Nivinskas [109] used reporter assays of the gene *25* TIR region fused to *lacZ*, in conjunction with DMS probing of the intervening RNA structure, to confirm the "split" RBS-SD arrangement and its use for effective expression of gene *25*. Phylogenetic evaluations of 38 T4-related phages revealed that the close T-even phages all have the intervening RNA structure in the split TIR configuration, but more distant, non-coliform T4-related phages lack this arrangement (Figure 5). This suggested an evolutionary history for the gene *25* split TIR, along with the enhancing, intervening RNA structure, where the arrangement arose after the close T-even phages diverged from other members of the phage group [109].

T4 exclusion and the mechanism of bacterial PrrC anticodon nuclease

T4 mutants defective in polynucleotide kinase (Pnk) or RNA ligase 1 (Rli1) grow normally on *E. coli* laboratory strains, but are restricted on some *E. coli* hospital strains. The restrictive hosts are referred to as *prr*⁺ for T4 *pnk⁻* or *rli⁻* mutants. T4 intergenic suppressors of the restriction of *pnk* or *rli* mutants on *prr*⁺ hosts define the T4 *stp* locus (see early reports cited [23,110]). The system of growth restriction results from activation of a host anticodon RNase (ACNase), PrrC, by the phage-encoded Stp protein. The bacterial PrrC RNase cuts within the tRNA^{Lys} anticodon loop, upstream of the wobble nucleotide and causes the arrest of phage protein synthesis and phage growth. T4 has evolved a tRNA repair mechanism to escape this restriction by way of the phage-induced polynucleotide kinase - 3' phosphatase (*pnk* gene), which converts the tRNA 5'-hydroxyl and 3'-phosphate termini left by PrrC, into 5'-phosphate and 3'-hydroxyl ends. Subsequently, the T4 RNA ligase 1 rejoins the tRNA ends. Stp, Pnk and Rnl1 are all under the delayed early mode of expression [23], meaning that restoration of the cleaved tRNA^{Lys} takes place early during infection. The *E. coli prrC* gene is located within a group of genes that encode type Ic restriction-modification (R-M) proteins, *Eco*prrI, in the order *hsdM-hsdS-prrC-hsdR* (or *prrABCD*). The Hsd enzymes are assembled in a multimeric complex, HsdR$_2$M$_2$S [23,110-112].

Stp alleviates type Ic restriction and activates the tRNA^{Lys} ACNase

Although only 26 residues long, the Stp polypeptide is necessary and sufficient to elicit the tRNA^{Lys} ACNase activity and mutations in the *stp* gene abolish activation of the ACNase. Expression of Stp protein from a plasmid also elicits ACNase activity in an uninfected *prr*⁺ strain [113].

Stp alleviates *Eco*prrI-mediated DNA restriction, indicating that this protein targets the *Eco*prrI complex rather than just PrrC directly. Several observations support this explanation: a) Growth of the lambdoid phage, HK022, propagated on *prr*⁰ cells, is heavily restricted upon plating on a *prr*⁺ strain; b) Expression of Stp from a plasmid in *prr*⁺ cells alleviates this restriction; c) *prr*⁺ cells do not restrict growth of phage HK022 prepared on a *prr*⁺ host expressing Stp. This strongly suggests that Stp inhibits *Eco*prrI restriction enzyme but does not affect the modification activity. Also, the fact that *Eco*R124I, another type Ic R-M system that does not include an ACNase, is inhibited by Stp, strongly supports the above conclusion. Stp is specific for type Ic R-M systems; it has no effect on the type Ia R-M systems, *Eco*KI and *Eco*BI [113].

The N-proximal 18 amino acids of Stp protein are probably involved in the interaction with *Eco*prrI since a number of missense *stp* mutants deficient in ACNase activation have been detected among revertants of T4 *pnk⁻* or *rli⁻* mutants that are able to grow on the *E. coli prr⁺* host. The majority of these suppressors cluster between residues 4 and 14 in the N-terminal part of the Stp polypeptide. In contrast, a deletion of 8 codons from the C-terminus only moderately decrease the two activities of Stp. Alignment of Stp sequences from eight T4-related phages with that of T4 reveals an almost absolute conservation of the 18 N-terminal residues, whereas polymorphism is evident in the remainder of the polypeptide. In most cases, the amino acids important for ACNase activation are also implicated in *Eco*prrI inhibition, suggesting some shared features [113]. We direct the reader to the primary literature by Kaufmann and colleagues [113] that hypothesizes on the evolutionary history of Stp in counteracting host DNA restriction enzymes, while also activating host ACNase.

Interaction of PrrC with EcoprrI and mechanism of ACNase activation

It appears that PrrC is maintained in a latent, inactive form, due to its association with the *Eco*prrI proteins. Antibodies against the closely related *Eco*R124I R-M system co-immunoprecipitate the PrrC protein. Conversely, antiserum against PrrC precipitates the HsdR (PrrD) protein [114,115].

Activation of latent ACNase in *prr⁺* cell extracts requires both Stp and GTP and is likely accompanied by GTP hydrolysis since addition of the non-hydrolysable analogue, GTPγS, is inhibitory. DNA is another positive effector of ACNase. Indeed, if the cell extract is treated with DNase I, activation by Stp is abolished. The activating DNA must carry cleavable (unmodified) *Eco*prrI restriction sites to be effective in ACNase activation. This led to the proposal that Stp activates the latent ACNase when its *Eco*prrI partner is tethered to *Eco*prrI DNA substrates [114-116].

Induction of *prrC* from a multicopy plasmid elicits ACNase activity in uninfected *E. coli* cells or in cultured mammalian cells [116,117]. This occurs in the absence of any other *prr* genes. In *E. coli*, this core ACNase is highly labile ($t_{1/2}$ < 1 minute at 30°C) while the ACNase found in extracts of *prr⁺* cells is rather stable, indicating that the association with the Hsd proteins stabilizes PrrC [115]. In crude cell extracts as well as with a partially purified leaky mutant form of PrrC (more stable than the wild-type enzyme; see [22]) core ACNase is not affected by Stp and is indifferent to the presence of DNA. This suggests that the role of these two effectors is to alleviate the Hsd masking effect on PrrC [116]. dTTP and other pyrimidine nucleotides, but not GTP or ATP, stimulate core ACNase activity at physiological concentrations, most probably by stabilizing the protein. ATP, GTP and dTTP bind to the NTP-binding domain

of PrrC (see below) [22,116]. Unexpectedly, GTP is inhibitory. The reason why core ACNase does not respond to GTP like the holoenzyme is unclear. Although this nucleotide binds PrrC (see below), it is possible that the GTPase catalytic site becomes active only when PrrC is associated with the Hsd component. However, it must be noted that the purified PrrC used in this study bears a leaky mutation, D222E, which confers a higher stability to the protein and permits its purification. Unfortunately, this mutation lies in the Walker B motif, which might affect the GTPase activity [22,116].

Several pyrimidine nucleotides are able to activate the latent ACNase *in the absence* of Stp, but at concentrations far above those required to protect the core ACNase or to UV-crosslink with PrrC (see below). dTTP is the most potent of them [116]. Like for Stp, the activation by dTTP requires GTP hydrolysis and *Eco*prrI DNA substrate. However, unlike Stp, which targets the Hsd complex, dTTP targets PrrC directly. The physiological meaning of this alternative mode of activation is not clear. It has been interpreted to mean that ACNase may be mobilized under cellular stress conditions not related to T4 infection. An alternative, but not exclusive, model assumes that dTTP is an obligatory co-activator working in concert with Stp. Because dTTP binds PrrC with high affinity, trace amounts of this nucleotide in crude extracts would be sufficient to allow latent ACNase activation upon Stp addition. Excess dTTP would by-pass the requirement for Stp [22].

PrrC structure, domain organization and distribution

The N-proximal two-thirds of the PrrC protein (ca. 265 residues out of 396) harbors a nucleotide-binding site and is thought to mediate activation of the latent ACNase (Figure 6). It features motifs that resemble those found in typical ABC-transporter ATPases: a somewhat degenerated ABC signature motif, Walker A (phosphate-loop) and Walker B motifs and an H-motif that contains a highly conserved His (the linchpin His) [22]. Mutations in the universally conserved residues of the Walker A motif of PrrC abolish ACNase activity [116]. ATP, GTP and dTTP bind PrrC to this region, as a mutation lying immediately upstream of the ABC signature severely decreases the ability of the protein to UV-crosslink with all three nucleotides. Because dTTP activates ACNase by targeting PrrC directly and requires GTP hydrolysis, the binding sites for the two nucleotides are likely different in PrrC oligomer. GTP and ATP likely share the same site. The interaction of dTTP with PrrC departs from that of the two other nucleotides in several respects. Mutations in the N-proximal Walker A motif, in the ABC signature sequence, or in the linchpin His do not affect the binding of ATP or GTP while they abolish dTTP binding to PrrC. Also, dTTP affinity to PrrC is three orders of magnitude

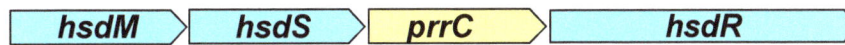

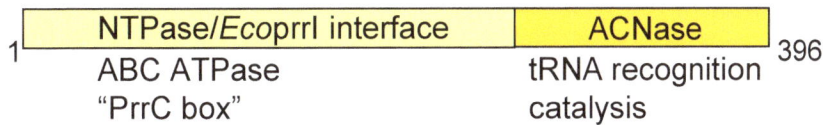

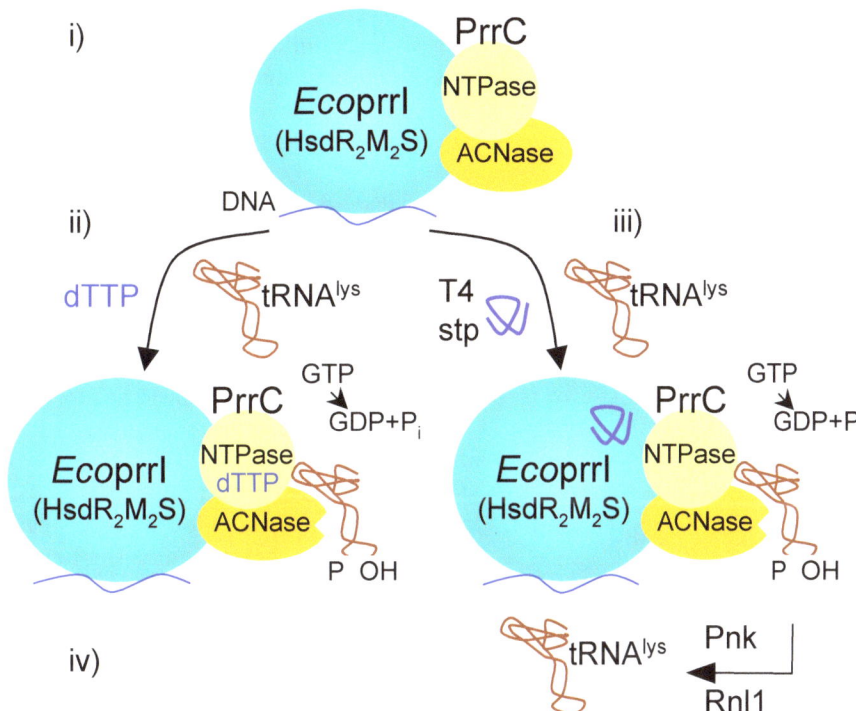

Figure 6 PrrC tRNA anticodon nuclease and T4 exclusion system. A) The *E. coli hsd* gene cluster includes *prrC*. **B)** PrrC has N-terminal two thirds NTPase and *Eco*prrI interaction domains and, starting at residue 265, C-terminal tRNA recognition and ribonuclease (ACNase) catalytic domains. **C)** Current model for tRNA cleavage and T4 exclusion. i) PrrC, minimally as a head-to-tail dimer (tetramer and hexamer oligomers are possible) associates with *Eco*prrI on DNA, as an inactive, latent endoribonuclease. ii) In one of two allosteric activation mechanisms, *Eco*prrI-PrrC-DNA complex binds increased levels of dTTP and with GTP hydrolysis activated ACNase cleaves the anticodon of tRNA^{Lys}. iii) During infection, the small, T4-encoded polypeptide stp binds *Eco*prrI activating tRNA^{Lys} ACNase. iv) T4 repairs the cleaved tRNA at the 2′,3′cyclic phosphate and 5′ OH using polynucleotide kinase (Pnk) and RNA ligase (Rnl1). *Figure adapted from the publications of Kaufmann and colleagues* [111,114,116,119].

higher than that of the two other nucleotides. Furthermore, a mild heat inactivation of ACNase has little consequence on ATP or GTP binding but abolishes dTTP binding. This suggests that the dTTP binding site is distinct and is sensitive to small changes in PrrC structure [22,116].

The C-terminal third of PrrC is implicated in tRNA recognition and catalysis. Several missense mutations affecting ACNase activity are located in close proximity

here (Figure 6). These were selected as mutations conferring the ability to survive the lethal overproduction of PrrC [118]. Because most of these substitutions are clustered in a short sequence highly conserved in a subset of the known PrrC homologues (residues 287 to 303) [22,118,119], the behavior of an 11-residue peptide (residues 284 to 294) was examined for RNA substrate interactions. This peptide forms UV-induced crosslinks with tRNA^{Lys} anticodon stem-loop analogs and inhibits the

ACNase activity of PrrC. Introducing certain substitutions in the peptide that are known to inactivate full-length PrrC, or shortening it by one amino acid from either end, leads to strong decrease in its ability to inhibit ACNase and to UV-crosslink with the anticodon stem-loop substrates [119]. Thus, this sequence is likely a part of the PrrC protein that interacts with the tRNA. In addition, substitutions in Arg320, Glu324 and His356 that are 100% conserved in all known PrrC homologs and suspected to participate in the acid-base catalytic mechanism, completely abolish ACNase activity. Null mutations in Arg320 and Glu324 can be rescued chemically by small molecules, indicating that the ACNase deficiency does not arise from a change in the structure of the protein, but rather from the lack of the correct amino acid side chain. This is compatible with the notion that at least two of the three conserved residues are implicated in catalysis [22].

Orthologues of *prrC* were found in 19 distantly related bacteria, all linked to genes for type Ic R-M enzymes. All of the orthologue proteins share in their N-terminal domains the NTP-binding site and a sequence of 15 residues called the "PrrC box". Also, their C-terminal domains contain the catalytic amino acid triad mentioned above. Thus, the PrrC proteins form a family whose members are strongly suspected not only to possess anticodon nuclease activity (as shown to be the case for those encoded by *Haemophilus influenzae* and *Streptococcus mutans* [120]), but also to be regulated like *E. coli* PrrC. However, their substrates may vary since the sequence involved in tRNA recognition varies among the PrrC proteins [22,119].

ACNase activity co-elutes from a gel filtration column with a homo-oligomer of ca. 200 kDa, suggesting that active PrrC could be a tetramer. Glutaraldehyde protein-protein crosslinking experiments confirm this, as mostly dimers and tetramers are produced [22]. Klaiman *et al.* [119] showed evidence suggesting that the C-terminal region of PrrC, involved in tRNA recognition, interacts with the substrate as a parallel dimer. Thus, while the N-terminal domain of PrrC is expected to associate in a head-to-tail dimer, by analogy with known structures of ABC transporter ATPases, the C-terminal region seems to dimerize in opposite orientation. To account for this situation, Klaiman et al. [119] proposed a model in which the PrrC subunits are associated in a unique tetramer conformation. Clearly, additional structural studies are necessary to elucidate the oligomeric structure of PrrC.

ACNase specificity

Kaufmann and colleagues have shown that the tRNA^{Lys} anticodon stem-loop region plays a prominent role in PrrC recognition: (a) PrrC, when overproduced in cells, cleaves other tRNAs in addition to tRNA^{Lys}. The anticodon sequences of all these secondary tRNA substrates share sequence similarities with that of tRNA^{Lys} [118]. (b) Expression of PrrC in human HeLa cells elicits cleavage of intracellular tRNA^{Lys3} that shares with the *E. coli* tRNA^{Lys} the same anticodon loop sequence [117]. (c) Most mutations in the tRNA^{Lys} anticodon sequence make the resulting tRNAs very poor substrates for ACNase. One of them, however, (U35 -> C leading to UCU anticodon) leads to relaxed site specificity as new cleavages occur upstream and downstream of the usual cleavage site [121]. (d) A chimeric, unmodified, tRNA^{Arg1} carrying the UUU lysine anticodon instead of its own anticodon, is as efficiently cleaved as the unmodified tRNA^{Lys} [121]. (e) PrrC quite efficiently cleaves a fragment of the tRNA^{Lys} encompassing only the anticodon loop and the first 5 base pairs of the associated stem (17 nucleotides altogether) [122]. (f) Cleavage of tRNA^{Lys} that lacks either of the two modifications of the uridine wobble base (2-thio- and 5-methylaminomethyl) is severely affected. Interestingly, three substitutions of PrrC Asp287 (D287Q, D287H and D287N), known to reduce the efficiency of cleavage of normally modified *E. coli* tRNA^{Lys}, reverse the negative effect of the hypo-modifications of the wobble base. This strongly supports the notion that Asp287 directly contacts the modified wobble base. Experiments carried out with the anticodon stem-loop (17-mer) as substrates reinforce this conclusion. Indeed, Jiang *et al.* [122] showed that the wobble base modification present in the anticodon stem-loop derived from mammalian tRNA^{Lys3} (5-methoxycarbonyl-2-thiouridine instead of 5-methylaminomethyl-2-thiouridine) is inhibitory to ACNase activity. However, D287H PrrC, poorly active on the fully modified *E. coli* anticodon stem-loop counterpart, overcomes this inhibitory effect [121,122]. (g) The influence of the stem stability and of the three different modifications in these anticodon stem-loop structures was examined in great detail. The picture that emerges is the following. A stable stem is inhibitory to ACNase activity. Some breathing of the duplex seems necessary, possibly to facilitate conformational changes of the tRNA upon interaction with PrrC. Also, PrrC seems to favor base modifications that help stack the anticodon nucleotides into an A-RNA conformation [122]. Thus, three elements are recognized by PrrC: the anticodon sequence, the base modifications and base-pairing of the stem.

Although the anticodon stem-loop region of tRNA^{Lys} is the predominant element of PrrC specificity, other sequence and/or structural elements of tRNA^{Lys} seem to be involved. This is indicated by the fact that chimeric tRNAs, other than the tRNA^{Arg1}, carrying the lysine anticodon, are not substrates for PrrC. Also, any substitution of the discriminator nucleotide (A73) of the tRNA^{Lys}, a major identity element of LysRS that lies in the acceptor arm, reduces, though moderately, the ACNase cleavage efficiency. Furthermore, trimming the 3'-terminal ACCA overhang nucleotides has little effect on ACNase activity but relaxes the cleavage site specificity in a manner similar to the U35 -> C mutation [121]. These data suggest additional interactions between PrrC and the acceptor region of tRNA^{Lys}.

Gathering the data into a model

Taken together, the above data suggest the following cascade of events. A few minutes after infection of *prr*+ *E. coli* cells, the T4-encoded Stp polypeptide binds the bacterial *Eco*prrI component and inhibits its DNA restriction activity (Figure 6). This modifies the *Eco*prrI/ PrrC interaction, inducing a change in PrrC conformation that unmasks ACNase activity. This process requires GTP hydrolysis. dTTP, bound to PrrC, is a co-activator with Stp. Its role could be to stabilize PrrC that would otherwise be labile in its activated conformation. The tRNA^{Lys} anticodon is then bound and cleaved by the respective ACNase regions. But the phage provides the healing (Pnk) and sealing (Rnl1) enzymes required to restore the affected tRNA, allowing the phage to escape the cellular defense. The phage exclusion mechanism depicted here and the way the phage wards off this cellular defense revealed an intimate physiological link between restriction-modification regulation and translational activity. The distribution of PrrC homologs in unrelated bacteria and their systematic link with type Ic R-M systems, suggest that the PrrC proteins have a cellular function not related to phage infection, possibly to disable protein synthesis under conditions of stress that affect activity of type I DNA restriction endonucleases [22,111,120].

Cellular RloC proteins

Using a bioinformatical approach, Davidov *et al.* [120] recently found a new class of PrrC homologs called RloC (restriction linked orf). RloC proteins are widespread in bacteria, although they are not present in *E. coli* and only one was found in Archaea and none in Eukarya. Genes for some of these proteins were first characterized as linked to genes for type I or III R-M enzymes in *Campylobacter jejuni* [123] however, now only a minority of the *rloC* genes map to R-M loci. RloC orthologues share with *E. coli* PrrC the presence of ATPase motifs in the N-termini and the amino acid triad thought to constitute the catalytic site in the C-termini. This structural homology is accompanied by a functional homology: when expressed in *E. coli*, RloC from the thermophilic *Geobacillus kaustophilus* exhibits "ACNase" activity. Also, alanine substitutions of the three amino acids of the triad abolish RloC ACNase. However, RloC differs from PrrC in several respects: (1) RloC substrate is still uncertain but it is not tRNA^{Lys}; (2) RloC ACNase actually *excises* the wobble nucleotide

rather than just cleaves upstream; and (3) Like the other RloC orthologues, the *G. kaustophilus* protein is larger than PrrC because the N-terminal NTPase domain is interrupted by a large coiled-coil fragment that's similar to sequences found in proteins implicated in DNA repair. This fragment contains a typical "zinc hook" motif able to co-ordinate Zn^{+2} ions. Mutations in the zinc-hook motif lead to increased ACNase activity and conversely, Zn^{+2} ions are inhibitory [120].

The RloC proteins show quite interesting and new properties that lead to several questions.

a) Is the RloC-dependent ACNase normally maintained in a latent, inactive form that is activated upon phage infection?
b) Since RloC excises the wobble nucleotide, is there a phage that repairs this lesion? If not, this would be an efficient mechanism of cellular defense against phages.
c) Are there stress conditions, unrelated to phage infection, that elicit RloC ACNase activation?
d) Are the RloC proteins associated with restriction-modification proteins? If so, do they respond to the presence of DNA?

By analogy with the PrrC ACNase, Kaufmann and colleagues [120] speculate that, in addition to conferring a mechanism of phage exclusion, the RloC proteins couple DNA damage that occurs under stress conditions, to translation inactivation *via* tRNA cleavage. Their model is based on two main observations: a) some proteins containing zinc-hook/coiled-coil domains are implicated in DNA repair; and b) DNA damage leads to alleviation of type Ia and Ic restriction enzymes, a process aimed at protecting unmodified, newly synthesized DNA during the process of repair and recovery from damaged DNA. The model assumes that the RloC protein would sense DNA damage signals via its zinc-hook and would convey activation to the ACNase domain, possibly via conformation changes driven by NTP hydrolysis. Such a model requires demonstrating a link between RloC proteins and DNA.

T4 exclusion by Gol-activated proteolysis of EF-Tu

Translation elongation is targeted in the T4 *gol-lit* phage exclusion system [23,110]. Inhibition of translation occurs when T4 gene *23* (the major head protein) is translated during infection of *E. coli* cells that harbor the defective prophage e14. The e14 element carries the *lit* gene, which encodes a latent protease that, somewhat similar to allosteric activation of latent PrrC ACNase activity, is active on EF-Tu when the so-called *gol* region of gene *23* is translated. Biochemical analyses of Lit/Gol/ EF-Tu interactions have revealed the process by which phage exclusion occurs through proteolysis of EF-Tu.

A short 29-residue region of the gp23 polypeptide defines Gol function, but a more stable interaction with Ef-Tu appears to occur with 100 amino acids from the first ¼ of gp23. Scanning mutagenesis showed 13 residues in a 20 amino acid core region of Gol to be most important for its activity [124]. Binding of Gol to EF-Tu is required to promote Lit reactivity. By binding to domains II and III of EF-Tu, the Gol peptide promotes Lit-mediated hydrolysis of EF-Tu between Gly59 - Ile60. Binding of Gol peptide is preferential for the open EF-Tu:GDP complex, and binding itself inhibits the EF-Tu GTPase of domain I. When Gol is bound to EF-Tu, it appears that EF-Tu domain I is more accessible to Lit, leading to "substrate-assisted" or "cofactor-induced" activation of cleavage by the protease [124,125]. Lit is a zinc metallo-protease with the active site motif HEXXH of this protease class, but Gol does not contribute directly to active site residues [124-126]. Kleanthous and colleagues [124] have noted that the gp23 Gol region is the most conserved region, in the overall conserved gp23 major head protein of sequenced T4-related phages. They suggest that gp23 of these phages interacts with EF-Tu of all the respective hosts. While Gol interactions may be broadly relevant for translation and folding of this extremely abundant capsid protein, other extant prophage-encoded, Lit-type proteases may also elicit "cellular suicide" via Lit/Gol/EF-Tu proteolytic assemblies.

Programmed translational bypassing

Topoisomerase of phage T4 is encoded by three genes: *39*, *60* and *52*. Most type II topoisomerases are comprised of two distinct subunits (i.e., *gyrA* and *gyrB* of DNA gyrase) that are assembled as tetrameric A_2B_2 enzymes. The adjacent T4 genes *39* and *60* are separated by 1010 nucleotides that include an apparently defective HNH homing endonuclease gene (*mobA*) and ORF *60.1* [95] (see [127] for a recent summary). Following their respective translation, gp39 and gp60 assemble to comprise the "gyrB-like" large, ATP-hydrolyzing subunit of the T4 topoisomerase. In all other T4-related phages sequenced to date, this subunit is encoded by a single open reading frame that is typically annotated as gene *39* (Figure 7).

An interesting, post-transcriptional feature in this region of the T4 genome that has received considerable attention is the presence of a 50 nucleotide "intervening sequence" in the 5' coding region of gene *60* that is transcribed into mRNA but is not translated into gp60. The ribosome "hops" or "bypasses" the extra 50 bases in the mRNA to produce the gp60 polypeptide in a process termed programmed translational bypassing [128,129]. In all other T4-related phages, not only are genes *39* and *60* joined as a single gene (they lack the *mobA - 60.1*

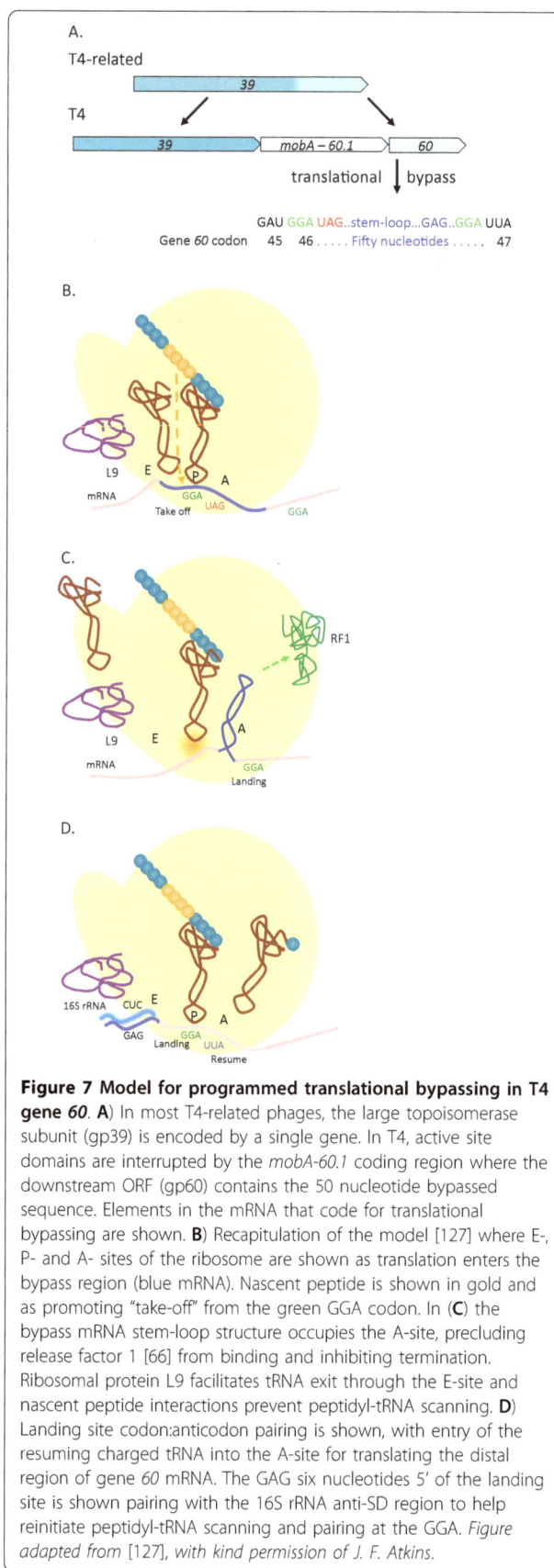

Figure 7 Model for programmed translational bypassing in T4 gene 60. A) In most T4-related phages, the large topoisomerase subunit (gp39) is encoded by a single gene. In T4, active site domains are interrupted by the *mobA-60.1* coding region where the downstream ORF (gp60) contains the 50 nucleotide bypassed sequence. Elements in the mRNA that code for translational bypassing are shown. **B)** Recapitulation of the model [127] where E-, P- and A- sites of the ribosome are shown as translation enters the bypass region (blue mRNA). Nascent peptide is shown in gold and as promoting "take-off" from the green GGA codon. In (**C**) the bypass mRNA stem-loop structure occupies the A-site, precluding release factor 1 [66] from binding and inhibiting termination. Ribosomal protein L9 facilitates tRNA exit through the E-site and nascent peptide interactions prevent peptidyl-tRNA scanning. **D)** Landing site codon:anticodon pairing is shown, with entry of the resuming charged tRNA into the A-site for translating the distal region of gene *60* mRNA. The GAG six nucleotides 5' of the landing site is shown pairing with the 16S rRNA anti-SD region to help reinitiate peptidyl-tRNA scanning and pairing at the GGA. *Figure adapted from* [127], *with kind permission of J. F. Atkins.*

insertion), but none appears to have the intervening gap nucleotides that would suggest programmed translational bypassing. The process appears to be unique to T4 gene *60*, but its study has shed new light on the mechanisms of translation. Atkins, Gesteland and colleagues at the University of Utah have studied many of the features that promote programmed translational bypassing by *E. coli* ribosomes on this unique T4 mRNA; the reader is urged to read further details through their primary research articles and reviews on the topic [129-131]. It is important to emphasize that experiments elucidating processes affecting gene *60* translational bypassing have provided new insights to the general mechanisms of mRNA decoding, including the roles of mRNA sequence and structure, peptidyl-tRNA interactions within the ribosome, occupancy of ribosome decoding sites, and many other features of translating ribosomes.

Figure 7 summarizes the major components of the T4 programmed translational bypass in gene *60* transcripts, as elaborated by the Utah group. Some of principal features include: a domain of the nascent gp60 polypeptide preceding the hop, the glycine 46 codon GGA at the end of the initial ORF (the "take-off site"), a UGA stop codon in the gap right after the GGA take-off codon, a stem-loop mRNA structure with a stabilizing tetraloop that is formed by gap RNA nucleotides, the peptidyl-tRNA$_2^{Gly}$ occupying the P site of the ribosome, rRNA:mRNA interactions during scanning of the gap by the ribosome, the L9 subunit of the ribosome, and the 50 nucleotide distal GGA codon after the gap nucleotides (the "landing site") [127,132-138]. Protein fusions, mass spectrometry, targeted mutations and a number of analyses have combined to address the roles of each component, leading to an approximately 50% efficiency of ribosomes bypassing the intervening 50 nucleotides to land correctly at the downstream GGA codon. Translation then resumes as the next UUA codon and cognate tRNA enter the A site. Again, each of these features shed new light not only on mechanisms of translational bypassing and aspects of "re-programming" the basic genetic code, but also on the dynamics and numerous interactions occurring in all translating ribosomes. It will be interesting to see whether instances of programmed translational bypassing occur in other genes of the many T4-related bacteriophages.

ADP-ribosyltransferases in post-transcriptional control
T4 encodes three enzymes that covalently modify proteins via ADP-ribosylation during the infection cycle: Alt, ModA and ModB. Alt is injected with phage DNA to immediately initiate ADP-ribosylation of one of the α-subunits (at arginine 265) of RNA polymerase, and by about 4 minutes post-infection newly synthesized ModA completes modification of both α-subunits at the same

arginine. The biochemistry of T4-directed ADP-ribosylation of RNA polymerase and its impact on phage promoter selection have been reviewed [2,139].

Other *E. coli* proteins have been recognized as undergoing ADP-ribosylation during T4 infection, which primarily appears to be due to the activity of Alt and ModB. Alt ADP-ribosylates β, β' and σ subunits of RNA polymerase and also other host proteins. The modifications also include proteins of the translation apparatus, as shown by Ruger and colleagues using the purified proteins [140]. An *in vitro* system incubated with total *E. coli* proteins (cell extract), purified Alt or ModB, and ^{32}P [NAD$^+$] showed with mass spectrometry that ADP-ribosylation occurred on as many as 27 proteins by Alt and on approximately 8 proteins by ModB [141]. For Alt, these included EF-Tu, trigger factor, prolyl-tRNA synthetase and GroEL that are known to have important roles in translation or protein folding. ModB also ADP-ribosylates EF-Tu and trigger factor, as well as ribosomal protein S1 [142]. For trigger factor (a chaperone of newly translated proteins), arginine 45 is ADP-ribosylated by ModB, but not by Alt, which must target a different, as yet unidentified amino acid [141]. Arg45 lies in the Phe44-Arg45-Lys46 domain that interacts with ribosomal protein L23, and thus might affect ribosome conformation and translation, as certainly would modifications to EF-Tu and S1. Early studies noted rapid & immediate shut-off of host mRNA translation during T4 infection [143], but no clear mechanism has been elucidated. The identification of pivotal proteins in the translation apparatus as targets of T4 ADP-ribosyltransferases, together with the observed delivery of Alt into the cell with the injected DNA and the lethality to the cell of over-expressed ModB [142], suggest that mechanistic studies into the impact of these T4 enzymes on translation of host and phage mRNAs is warranted.

Conclusions
Although post-transcriptional control in T4 development and gene expression has been appreciated and studied for decades, many of the molecular details, especially for specific RNA-protein interactions, have yet to be resolved. For most, crystal or solution structures of bound mRNA-repressor or RNA-nuclease complexes would significantly advance our understanding of complex formation and substrate interactions in catalysis. While clearly germane to T4 and the large diversity of T4-related bacteriophages in the biosphere, continued study of post-transcriptional processes directed by these phages will provide new advances in the biochemistry pertinent to all cellular systems. Undoubtedly new antivirals and anti-microbials targeting these and related systems in pathogens can be anticipated.

Acknowledgements
We thank Virology Journal for their support of this series on phage T4 and its relatives.

Author details
[1]Acides Nucléiques et Biophotonique, FRE 3207 CNRS-Université Pierre & Marie Curie, 4 Place Jussieu, 75252 PARIS cedex 05, France. [2]Department of Microbiology, North Carolina State University, Raleigh, 27695-7615, NC, USA.

Authors' contributions
MU and ESM wrote the manuscript and approved the final version.

Competing interests
The authors declare that they have no competing interests.

Received: 4 November 2010 Accepted: 3 December 2010
Published: 3 December 2010

References
1. Geiduschek EP, Kassavetis GA: **Transcription of the T4 late genes.** *Virol J* 2010, **7**:288.
2. Hinton DM: **Transcriptional control in the prereplicative phase of T4 development.** *Virol J* 2010, **7**:289.
3. Miller ES, Kutter E, Mosig G, Arisaka F, Kunisawa T, Ruger W: **Bacteriophage T4 genome.** *Microbiol Mol Biol Rev* 2003, **67**:86-156.
4. Sanson B, Uzan M: **Post-transcriptional controls in bacteriophage T4: roles of the sequence-specific endoribonuclease RegB.** *FEMS Microbiol Rev* 1995, **17**:141-150.
5. Uzan M: **Bacteriophage T4 RegB endoribonuclease.** *Methods Enzymol* 2001, **342**:467-480.
6. Uzan M: **RNA processing and decay in bacteriophage T4.** *Prog Mol Biol Transl Sci* 2009, **85**:43-89.
7. Orsini G, Igonet S, Pene C, Sclavi B, Buckle M, Uzan M, Kolb A: **Phage T4 early promoters are resistant to inhibition by the anti-sigma factor AsiA.** *Mol Microbiol* 2004, **52**:1013-1028.
8. Pene C, Uzan M: **The bacteriophage T4 anti-sigma factor AsiA is not necessary for the inhibition of early promoters in vivo.** *Mol Microbiol* 2000, **35**:1180-1191.
9. Durand S, Richard G, Bisaglia M, Laalami S, Bontems F, Uzan M: **Activation of RegB endoribonuclease by S1 ribosomal protein requires an 11 nt conserved sequence.** *Nucleic Acids Res* 2006, **34**:6549-6560.
10. Ruckman J, Parma D, Tuerk C, Hall DH, Gold L: **Identification of a T4 gene required for bacteriophage mRNA processing.** *New Biol* 1989, **1**:54-65.
11. Sanson B, Hu RM, Troitskaya E, Mathy N, Uzan M: **Endoribonuclease RegB from bacteriophage T4 is necessary for the degradation of early but not middle or late mRNAs.** *J Mol Biol* 2000, **297**:1063-1074.
12. Sanson B, Uzan M: **Dual role of the sequence-specific bacteriophage T4 endoribonuclease RegB. mRNA inactivation and mRNA destabilization.** *J Mol Biol* 1993, **233**:429-446.
13. Uzan M, Favre R, Brody E: **A nuclease that cuts specifically in the ribosome binding site of some T4 mRNAs.** *Proc Natl Acad Sci USA* 1988, **85**:8895-8899.
14. Zajanckauskaite A, Truncaite L, Strazdaite-Zieliene Z, Nivinskas R: **Involvement of the Escherichia coli endoribonucleases G and E in the secondary processing of RegB-cleaved transcripts of bacteriophage T4.** *Virology* 2008, **375**:342-353.
15. Lebars I, Hu RM, Lallemand JY, Uzan M, Bontems F: **Role of the substrate conformation and of the S1 protein in the cleavage efficiency of the T4 endoribonuclease RegB.** *J Biol Chem* 2001, **276**:13264-13272.
16. Ruckman J, Ringquist S, Brody E, Gold L: **The bacteriophage T4 regB ribonuclease. Stimulation of the purified enzyme by ribosomal protein S1.** *J Biol Chem* 1994, **269**:26655-26662.
17. Carpousis AJ, Krisch HM: **mRNA processing and degradation.** In *Molecular Biology of Bacteriophage T4.* Edited by: Karam JD, Drake JD, Kreuzer KN, Mosig G, Hall DW, Eiserling FA, Black LW, Spicer EK, Kutter E, Carlson K, Miller ES. Washington, D.C.: American Society for Microbiology; 1994:193-205.
18. Mackie GA: **Ribonuclease E is a 5'-end-dependent endonuclease.** *Nature* 1998, **395**:720-723.
19. Mackie GA: **Stabilization of circular rpsT mRNA demonstrates the 5'-end dependence of RNase E action in vivo.** *J Biol Chem* 2000, **275**:25069-25072.
20. Tock MR, Walsh AP, Carroll G, McDowall KJ: **The CafA protein required for the 5'-maturation of 16 S rRNA is a 5'-end-dependent ribonuclease that has context-dependent broad sequence specificity.** *J Biol Chem* 2000, **275**:8726-8732.
21. Deana A, Celesnik H, Belasco JG: **The bacterial enzyme RppH triggers messenger RNA degradation by 5' pyrophosphate removal.** *Nature* 2008, **451**:355-358.
22. Blanga-Kanfi S, Amitsur M, Azem A, Kaufmann G: **PrrC-anticodon nuclease: functional organization of a prototypical bacterial restriction RNase.** *Nucleic Acids Res* 2006, **34**:3209-3219.
23. Snyder L, Kaufmann G: **T4 phage exclusion mechanims.** In *Molecular Biology of Bacteriophage T4.* Edited by: Karam JD, Drake JD, Kreuzer KN, Mosig G, Hall DW, Eiserling FA, Black LW, Spicer EK, Kutter E, Carlson K, Miller ES. Washington, DC: American Society for Microbiology; 1994:391-396.
24. Uzan M, Brody E, Favre R: **Nucleotide sequence and control of transcription of the bacteriophage T4 motA regulatory gene.** *Mol Microbiol* 1990, **4**:1487-1496.
25. Tuerk C, Gold L: **Systematic evolution of ligands by exponential enrichment: RNA ligands to bacteriophage T4 DNA polymerase.** *Science* 1990, **249**:505-510.
26. Jayasena VK, Brown D, Shtatland T, Gold L: **In vitro selection of RNA specifically cleaved by bacteriophage T4 RegB endonuclease.** *Biochemistry* 1996, **35**:2349-2356.
27. Dodson RE, Shapiro DJ: **Regulation of pathways of mRNA destabilization and stabilization.** *Prog Nucleic Acid Res Mol Biol* 2002, **72**:129-164.
28. Saida F, Uzan M, Bontems F: **The phage T4 restriction endoribonuclease RegB: a cyclizing enzyme that requires two histidines to be fully active.** *Nucleic Acids Res* 2003, **31**:2751-2758.
29. Odaert B, Saida F, Aliprandi P, Durand S, Crechet JB, Guerois R, Laalami S, Uzan M, Bontems F: **Structural and functional studies of RegB, a new member of a family of sequence-specific ribonucleases involved in mRNA inactivation on the ribosome.** *J Biol Chem* 2007, **282**:2019-2028.
30. Gerdes K, Christensen SK, Lobner-Olesen A: **Prokaryotic toxin-antitoxin stress response loci.** *Nat Rev Microbiol* 2005, **3**:371-382.
31. Yamaguchi Y, Inouye M: **mRNA interferases, sequence-specific endoribonucleases from the toxin-antitoxin systems.** *Prog Mol Biol Transl Sci* 2009, **85**:467-500.
32. Sorensen MA, Fricke J, Pedersen S: **Ribosomal protein S1 is required for translation of most, if not all, natural mRNAs in Escherichia coli in vivo.** *J Mol Biol* 1998, **280**:561-569.
33. Subramanian AR: **Structure and functions of ribosomal protein S1.** *Prog Nucleic Acid Res Mol Biol* 1983, **28**:101-142.
34. Aliprandi P, Sizun C, Perez J, Mareuil F, Caputo S, Leroy JL, Odaert B, Laalami S, Uzan M, Bontems F: **S1 ribosomal protein functions in translation initiation and ribonuclease RegB activation are mediated by similar RNA-protein interactions: an NMR and SAXS analysis.** *J Biol Chem* 2008, **283**:13289-13301.
35. Buttner K, Wenig K, Hopfner KP: **Structural framework for the mechanism of archaeal exosomes in RNA processing.** *Mol Cell* 2005, **20**:461-471.
36. Bycroft M, Hubbard TJ, Proctor M, Freund SM, Murzin AG: **The solution structure of the S1 RNA binding domain: a member of an ancient nucleic acid-binding fold.** *Cell* 1997, **88**:235-242.
37. Salah P, Bisaglia M, Aliprandi P, Uzan M, Sizun C, Bontems F: **Probing the relationship between Gram-negative and Gram-positive S1 proteins by sequence analysis.** *Nucleic Acids Res* 2009, **37**:5578-5588.
38. Schubert M, Edge RE, Lario P, Cook MA, Strynadka NC, Mackie GA, McIntosh LP: **Structural characterization of the RNase E S1 domain and identification of its oligonucleotide-binding and dimerization interfaces.** *J Mol Biol* 2004, **341**:37-54.
39. Bisaglia M, Laalami S, Uzan M, Bontems F: **Activation of the RegB endoribonuclease by the S1 ribosomal protein is due to cooperation between the S1 four C-terminal modules in a substrate-dependant manner.** *J Biol Chem* 2003, **278**:15261-15271.
40. Rajkowitsch L, Schroeder R: **Dissecting RNA chaperone activity.** *RNA* 2007, **13**:2053-2060.
41. Thomas JO, Szer W: **RNA-helix-destabilizing proteins.** *Prog Nucleic Acid Res Mol Biol* 1982, **27**:157-187.

42. Piesiniene L, Truncaite L, Zajanckauskaite A, Nivinskas R: The sequences and activities of RegB endoribonucleases of T4-related bacteriophages. *Nucleic Acids Res* 2004, **32**:5582-5595.

43. Kai T, Selick HE, Yonesaki T: Destabilization of bacteriophage T4 mRNAs by a mutation of gene 61.5. *Genetics* 1996, **144**:7-14.

44. Kanesaki T, Hamada T, Yonesaki T: Opposite roles of the dmd gene in the control of RNase E and RNase LS activities. *Genes Genet Syst* 2005, **80**:241-249.

45. Otsuka Y, Yonesaki T: A novel endoribonuclease, RNase LS, in Escherichia coli. *Genetics* 2005, **169**:13-20.

46. Ueno H, Yonesaki T: Recognition and specific degradation of bacteriophage T4 mRNAs. *Genetics* 2001, **158**:7-17.

47. Ueno H, Yonesaki T: Role of Escherichia coli Hfq in late-gene silencing of bacteriophage T4 dmd mutant. *Genes Genet Syst* 2002, **77**:301-308.

48. Otsuka Y, Ueno H, Yonesaki T: Escherichia coli endoribonucleases involved in cleavage of bacteriophage T4 mRNAs. *J Bacteriol* 2003, **185**:983-990.

49. Otsuka Y, Koga M, Iwamoto A, Yonesaki T: A role of RnlA in the RNase LS activity from Escherichia coli. *Genes Genet Syst* 2007, **82**:291-299.

50. Kai T, Yonesaki T: Multiple mechanisms for degradation of bacteriophage T4 soc mRNA. *Genetics* 2002, **160**:5-12.

51. Yamanishi H, Yonesaki T: RNA cleavage linked with ribosomal action. *Genetics* 2005, **171**:419-425.

52. Iwamoto A, Lemire S, Yonesaki T: Post-transcriptional control of Crp-cAMP by RNase LS in Escherichia coli. *Mol Microbiol* 2008, **70**:1570-1578.

53. Regnier P, Hajnsdorf E: Poly(A)-assisted RNA decay and modulators of RNA stability. *Prog Mol Biol Transl Sci* 2009, **85**:137-185.

54. Hurwitz J, Furth JJ, Anders M, Ortiz PJ, August JT: The enzymatic incorporation of ribonucleotides into RNA and the role of DNA. *Cold Spring Harb Symp Quant Biol* 1961, **26**:91-100.

55. Yonesaki T: Scarce adenylation in bacteriophage T4 mRNAs. *Genes Genet Syst* 2002, **77**:219-225.

56. Khemici V, Toesca I, Poljak L, Vanzo NF, Carpousis AJ: The RNase E of Escherichia coli has at least two binding sites for DEAD-box RNA helicases: functional replacement of RhlB by RhlE. *Mol Microbiol* 2004, **54**:1422-1430.

57. Kido M, Yamanaka K, Mitani T, Niki H, Ogura T, Hiraga S: RNase E polypeptides lacking a carboxyl-terminal half suppress a mukB mutation in Escherichia coli. *J Bacteriol* 1996, **178**:3917-3925.

58. Leroy A, Vanzo NF, Sousa S, Dreyfus M, Carpousis AJ: Function in Escherichia coli of the non-catalytic part of RNase E: role in the degradation of ribosome-free mRNA. *Mol Microbiol* 2002, **45**:1231-1243.

59. Lopez PJ, Marchand I, Joyce SA, Dreyfus M: The C-terminal half of RNase E, which organizes the Escherichia coli degradosome, participates in mRNA degradation but not rRNA processing in vivo. *Mol Microbiol* 1999, **33**:188-199.

60. Celesnik H, Deana A, Belasco JG: Initiation of RNA decay in Escherichia coli by 5' pyrophosphate removal. *Mol Cell* 2007, **27**:79-90.

61. Ueno H, Yonesaki T: Phage-induced change in the stability of mRNAs. *Virology* 2004, **329**:134-141.

62. Williams KP, Kassavetis GA, Herendeen DR, Geiduschek EP: Regulation of late gene expression. In *Molecular Biology of Bacteriophage T4*. Edited by: Karam JD, Drake JD, Kreuzer KN, Mosig G, Hall DW, Eiserling FA, Black LW, Spicer EK, Kutter E, Carlson K, Miller ES. Washington, D.C.: American Society for Microbiology; 1994:161-175.

63. Kolesky S, Ouhammouch M, Brody EN, Geiduschek EP: Sigma competition: the contest between bacteriophage T4 middle and late transcription. *J Mol Biol* 1999, **291**:267-281.

64. Nechaev S, Kamali-Moghaddam M, Andre E, Leonetti JP, Geiduschek EP: The bacteriophage T4 late-transcription coactivator gp33 binds the flap domain of Escherichia coli RNA polymerase. *Proc Natl Acad Sci USA* 2004, **101**:17365-17370.

65. Miller ES, Karam JD, Spicer E: Control of translation initiation: mRNA structure and protein repressors. In *Molecular Biology of Bacteriophage T4*. Edited by: Karam JD, Drake JD, Kreuzer KN, Mosig G, Hall DW, Eiserling FA, Black LW, Spicer EK, Kutter E, Carlson K, Miller ES. Washington, D.C.: American Society for Microbiology; 1994:193-205.

66. Kang C, Chan R, Berger I, Lockshin C, Green L, Gold L, Rich A: Crystal structure of the T4 regA translational regulator protein at 1.9 A resolution. *Science* 1995, **268**:1170-1173.

67. O'Malley SM, Sattar AK, Williams KR, Spicer EK: Mutagenesis of the COOH-terminal region of bacteriophage T4 regA protein. *J Biol Chem* 1995, **270**:5107-5114.

68. Phillips CA, Gordon J, Spicer EK: Bacteriophage T4 regA protein binds RNA as a monomer, overcoming dimer interactions. *Nucleic Acids Res* 1996, **24**:4319-4326.

69. Liu C, Tolic LP, Hofstadler SA, Harms AC, Smith RD, Kang C, Sinha N: Probing RegA/RNA interactions using electrospray ionization-fourier transform ion cyclotron resonance-mass spectrometry. *Anal Biochem* 1998, **262**:67-76.

70. Gordon J, Sengupta TK, Phillips CA, O'Malley SM, Williams KR, Spicer EK: Identification of the RNA binding domain of T4 RegA protein by structure-based mutagenesis. *J Biol Chem* 1999, **274**:32265-32273.

71. Jozwik CE, Miller ES: Regions of bacteriophage T4 and RB69 RegA translational repressor proteins that determine RNA-binding specificity. *Proc Natl Acad Sci USA* 1992, **89**:5053-5057.

72. Sengupta TK, Gordon J, Spicer EK: RegA proteins from phage T4 and RB69 have conserved helix-loop groove RNA binding motifs but different RNA binding specificities. *Nucleic Acids Res* 2001, **29**:1175-1184.

73. Brown D, Brown J, Kang C, Gold L, Allen P: Single-stranded RNA recognition by the bacteriophage T4 translational repressor, regA. *J Biol Chem* 1997, **272**:14969-14974.

74. Allen SV, Miller ES: RNA-binding properties of in vitro expressed histidine-tagged RB69 RegA translational repressor protein. *Anal Biochem* 1999, **269**:32-37.

75. Dean TR, Allen SV, Miller ES: In vitro selection of phage RB69 RegA RNA binding sites yields UAA triplets. *Virology* 2005, **336**:26-36.

76. Williams KR, Shamoo Y, Spicer EK, Coleman JE, Konigsberg WH: Correlating structure to function in proteins: T4 Gp32 as a prototype. In *Molecular Biology of Bacteriophage T4*. Edited by: Karam JD, Drake JD, Kreuzer KN, Mosig G, Hall DW, Eiserling FA, Black LW, Spicer EK, Kutter E, Carlson K, Miller ES. Washington, D.C.: American Society for Microbiology; 1994:301-304.

77. Holland JA, Hansen MR, Du Z, Hoffman DW: An examination of coaxial stacking of helical stems in a pseudoknot motif: the gene 32 messenger RNA pseudoknot of bacteriophage T2. *RNA* 1999, **5**:257-271.

78. Borjac-Natour JM, Petrov VM, Karam JD: Divergence of the mRNA targets for the Ssb proteins of bacteriophages T4 and RB69. *Virol J* 2004, **1**:4.

79. Desplats C, Dez C, Tetart F, Eleaume H, Krisch HM: Snapshot of the genome of the pseudo-T-even bacteriophage RB49. *J Bacteriol* 2002, **184**:2789-2804.

80. Brierley I, Pennell S, Gilbert RJ: Viral RNA pseudoknots: versatile motifs in gene expression and replication. *Nat Rev Microbiol* 2007, **5**:598-610.

81. Gold L, O'Farrell PZ, Russel M: Regulation of gene 32 expression during bacteriophage T4 infection of Escherichia coli. *J Biol Chem* 1976, **251**:7251-7262.

82. Russel M, Gold L, Morrissett H, O'Farrell PZ: Translational, autogenous regulation of gene 32 expression during bacteriophage T4 infection. *J Biol Chem* 1976, **251**:7263-7270.

83. Belin D, Mudd EA, Prentki P, Yi-Yi Y, Krisch HM: Sense and antisense transcription of bacteriophage T4 gene 32. Processing and stability of the mRNAs. *J Mol Biol* 1987, **194**:231-243.

84. Mudd EA, Prentki P, Belin D, Krisch HM: Processing of unstable bacteriophage T4 gene 32 mRNAs into a stable species requires Escherichia coli ribonuclease E. *EMBO J* 1988, **7**:3601-3607.

85. Carpousis AJ, Mudd EA, Krisch HM: Transcription and messenger RNA processing upstream of bacteriophage T4 gene 32. *Mol Gen Genet* 1989, **219**:39-48.

86. Loayza D, Carpousis AJ, Krisch HM: Gene 32 transcription and mRNA processing in T4-related bacteriophages. *Mol Microbiol* 1991, **5**:715-725.

87. Philippe C, Eyermann F, Benard L, Portier C, Ehresmann B, Ehresmann C: Ribosomal protein S15 from Escherichia coli modulates its own translation by trapping the ribosome on the mRNA initiation loading site. *Proc Natl Acad Sci USA* 1993, **90**:4394-4398.

88. Braun F, Le Derout J, Regnier P: Ribosomes inhibit an RNase E cleavage which induces the decay of the rpsO mRNA of Escherichia coli. *EMBO J* 1998, **17**:4790-4797.

89. Pavlov AR, Karam JD: Binding specificity of T4 DNA polymerase to RNA. *J Biol Chem* 1994, **269**:12968-12972.

90. Franklin MC, Wang J, Steitz TA: Structure of the replicating complex of a pol alpha family DNA polymerase. *Cell* 2001, **105**:657-667.

91. Wang CC, Pavlov A, Karam JD: Evolution of RNA-binding specificity in T4 DNA polymerase. *J Biol Chem* 1997, **272**:17703-17710.

92. Petrov VM, Ng SS, Karam JD: Protein determinants of RNA binding by DNA polymerase of the T4-related bacteriophage RB69. *J Biol Chem* 2002, **277**:33041-33048.
93. Petrov VM, Karam JD: RNA determinants of translational operator recognition by the DNA polymerases of bacteriophages T4 and RB69. *Nucleic Acids Res* 2002, **30**:3341-3348.
94. Petrov VM, Karam JD: Diversity of structure and function of DNA polymerase (gp43) of T4-related bacteriophages. *Biochemistry (Mosc)* 2004, **69**:1213-1218.
95. Miller ES, Heidelberg JF, Eisen JA, Nelson WC, Durkin AS, Ciecko A, Feldblyum TV, White O, Paulsen IT, Nierman WC, *et al*: Complete genome sequence of the broad-host-range vibriophage KVP40: comparative genomics of a T4-related bacteriophage. *J Bacteriol* 2003, **185**:5220-5233.
96. Breaker RR: Riboswitches: from ancient gene-control systems to modern drug targets. *Future Microbiol* 2009, **4**:771-773.
97. Dambach MD, Winkler WC: Expanding roles for metabolite-sensing regulatory RNAs. *Curr Opin Microbiol* 2009, **12**:161-169.
98. McPheeters DS, Christensen A, Young ET, Stormo G, Gold L: Translational regulation of expression of the bacteriophage T4 lysozyme gene. *Nucleic Acids Res* 1986, **14**:5813-5826.
99. Kasai T, Bautz EK: Regulation of gene-specific RNA synthesis in bacteriophage T4. *J Mol Biol* 1969, **41**:401-417.
100. Edgell DR, Gibb EA, Belfort M: Mobile DNA elements in T4 and related phages. *Virol J* 2010, **7**:290.
101. Edgell DR, Derbyshire V, Van Roey P, LaBonne S, Stanger MJ, Li Z, Boyd TM, Shub DA, Belfort M: Intron-encoded homing endonuclease I-TevI also functions as a transcriptional autorepressor. *Nat Struct Mol Biol* 2004, **11**:936-944.
102. Gott JM, Zeeh A, Bell-Pedersen D, Ehrenman K, Belfort M, Shub DA: Genes within genes: independent expression of phage T4 intron open reading frames and the genes in which they reside. *Genes Dev* 1988, **2**:1791-1799.
103. Shub DA, Coetzee T, Hall DW, Belfort M: The self-splicing introns of bacteriophage T4. In *Molecular Biology of Bacteriophage T4*. Edited by: Karam J, Drake JW, Kreuzer KN, Mosig G, Hall DH, Eiserling FA, Black LW, Spicer EK, Kutter E, Carlson K, Miller ES. Washington, D.C.: American Society for Microbiology; 1994:186-192.
104. Gibb EA, Edgell DR: Better late than early: delayed translation of intron-encoded endonuclease I-TevI is required for efficient splicing of its host group I intron. *Mol Microbiol* 2010, **78**:35-46.
105. Gibb EA, Edgell DR: Multiple controls regulate the expression of mobE, an HNH homing endonuclease gene embedded within a ribonucleotide reductase gene of phage Aeh1. *J Bacteriol* 2007, **189**:4648-4661.
106. Gibb EA, Edgell DR: An RNA hairpin sequesters the ribosome binding site of the homing endonuclease mobE gene. *J Bacteriol* 2009, **191**:2409-2413.
107. Brok-Volchanskaya VS, Kadyrov FA, Sivogrivov DE, Kolosov PM, Sokolov AS, Shlyapnikov MG, Kryukov VM, Granovsky IE: Phage T4 SegB protein is a homing endonuclease required for the preferred inheritance of T4 tRNA gene region occurring in co-infection with a related phage. *Nucleic Acids Res* 2008, **36**:2094-2105.
108. Nivinskas R, Malys N, Klausa V, Vaiskunaite R, Gineikiene E: Post-transcriptional control of bacteriophage T4 gene 25 expression: mRNA secondary structure that enhances translational initiation. *J Mol Biol* 1999, **288**:291-304.
109. Malys N, Nivinskas R: Non-canonical RNA arrangement in T4-even phages: accommodated ribosome binding site at the gene 26-25 intercistronic junction. *Mol Microbiol* 2009, **73**:1115-1127.
110. Snyder L: Phage-exclusion enzymes: a bonanza of biochemical and cell biology reagents? *Mol Microbiol* 1995, **15**:415-420.
111. Kaufmann G: Anticodon nucleases. *Trends Biochem Sci* 2000, **25**:70-74.
112. Tyndall C, Meister J, Bickle TA: The Escherichia coli prr region encodes a functional type IC DNA restriction system closely integrated with an anticodon nuclease gene. *J Mol Biol* 1994, **237**:266-274.
113. Penner M, Morad I, Snyder L, Kaufmann G: Phage T4-coded Stp: double-edged effector of coupled DNA and tRNA-restriction systems. *J Mol Biol* 1995, **249**:857-868.
114. Amitsur M, Morad I, Chapman-Shimshoni D, Kaufmann G: HSD restriction-modification proteins partake in latent anticodon nuclease. *EMBO J* 1992, **11**:3129-3134.
115. Morad I, Chapman-Shimshoni D, Amitsur M, Kaufmann G: Functional expression and properties of the tRNA(Lys)-specific core anticodon nuclease encoded by Escherichia coli prrC. *J Biol Chem* 1993, **268**:26842-26849.
116. Amitsur M, Benjamin S, Rosner R, Chapman-Shimshoni D, Meidler R, Blanga S, Kaufmann G: Bacteriophage T4-encoded Stp can be replaced as activator of anticodon nuclease by a normal host cell metabolite. *Mol Microbiol* 2003, **50**:129-143.
117. Shterman N, Elroy-Stein O, Morad I, Amitsur M, Kaufmann G: Cleavage of the HIV replication primer tRNALys,3 in human cells expressing bacterial anticodon nuclease. *Nucleic Acids Res* 1995, **23**:1744-1749.
118. Meidler R, Morad I, Amitsur M, Inokuchi H, Kaufmann G: Detection of anticodon nuclease residues involved in tRNALys cleavage specificity. *J Mol Biol* 1999, **287**:499-510.
119. Klaiman D, Amitsur M, Blanga-Kanfi S, Chai M, Davis DR, Kaufmann G: Parallel dimerization of a PrrC-anticodon nuclease region implicated in tRNALys recognition. *Nucleic Acids Res* 2007, **35**:4704-4714.
120. Davidov E, Kaufmann G: RloC: a wobble nucleotide-excising and zinc-responsive bacterial tRNase. *Mol Microbiol* 2008, **69**:1560-1574.
121. Jiang Y, Meidler R, Amitsur M, Kaufmann G: Specific interaction between anticodon nuclease and the tRNA(Lys) wobble base. *J Mol Biol* 2001, **305**:377-388.
122. Jiang Y, Blanga S, Amitsur M, Meidler R, Krivosheyev E, Sundaram M, Bajji AC, Davis DR, Kaufmann G: Structural features of tRNALys favored by anticodon nuclease as inferred from reactivities of anticodon stem and loop substrate analogs. *J Biol Chem* 2002, **277**:3836-3841.
123. Miller WG, Pearson BM, Wells JM, Parker CT, Kapitonov VV, Mandrell RE: Diversity within the Campylobacter jejuni type I restriction-modification loci. *Microbiology* 2005, **151**:337-351.
124. Copeland NA, Kleanthous C: The role of an activating peptide in protease-mediated suicide of Escherichia coli K12. *J Biol Chem* 2005, **280**:112-117.
125. Bingham R, Ekunwe SI, Falk S, Snyder L, Kleanthous C: The major head protein of bacteriophage T4 binds specifically to elongation factor Tu. *J Biol Chem* 2000, **275**:23219-23226.
126. Copeland NA, Bingham R, Georgiou T, Cooper P, Kleanthous C: Identification of essential residues within Lit, a cell death peptidase of Escherichia coli K-12. *Biochemistry* 2004, **43**:7948-7953.
127. Wills NM, O'Connor M, Nelson CC, Rettberg CC, Huang WM, Gesteland RF, Atkins JF: Translational bypassing without peptidyl-tRNA anticodon scanning of coding gap mRNA. *EMBO J* 2008, **27**:2533-2544.
128. Atkins JF, Gesteland RF: mRNA readout at 40. *Nature* 2001, **414**:693.
129. Herr AJ, Gesteland RF, Atkins JF: One protein from two open reading frames: mechanism of a 50 nt translational bypass. *EMBO J* 2000, **19**:2671-2680.
130. Baranov PV, Gesteland RF, Atkins JF: Recoding: translational bifurcations in gene expression. *Gene* 2002, **286**:187-201.
131. Gesteland RF, Atkins JF: Recoding: dynamic reprogramming of translation. *Annu Rev Biochem* 1996, **65**:741-768.
132. Weiss RB, Huang WM, Dunn DM: A nascent peptide is required for ribosomal bypass of the coding gap in bacteriophage T4 gene 60. *Cell* 1990, **62**:117-126.
133. Adamski FM, Atkins JF, Gesteland RF: Ribosomal protein L9 interactions with 23 S rRNA: the use of a translational bypass assay to study the effect of amino acid substitutions. *J Mol Biol* 1996, **261**:357-371.
134. Bucklin DJ, Wills NM, Gesteland RF, Atkins JF: P-site pairing subtleties revealed by the effects of different tRNAs on programmed translational bypassing where anticodon re-pairing to mRNA is separated from dissociation. *J Mol Biol* 2005, **345**:39-49.
135. Herr AJ, Nelson CC, Wills NM, Gesteland RF, Atkins JF: Analysis of the roles of tRNA structure, ribosomal protein L9, and the bacteriophage T4 gene 60 bypassing signals during ribosome slippage on mRNA. *J Mol Biol* 2001, **309**:1029-1048.
136. Herr AJ, Wills NM, Nelson CC, Gesteland RF, Atkins JF: Drop-off during ribosome hopping. *J Mol Biol* 2001, **311**:445-452.
137. Herr AJ, Wills NM, Nelson CC, Gesteland RF, Atkins JF: Factors that influence selection of coding resumption sites in translational bypassing: minimal conventional peptidyl-tRNA:mRNA pairing can suffice. *J Biol Chem* 2004, **279**:11081-11087.
138. Shah AA, Giddings MC, Parvaz JB, Gesteland RF, Atkins JF, Ivanov IP: Computational identification of putative programmed translational frameshift sites. *Bioinformatics* 2002, **18**:1046-1053.

139. Stitt B, Hinton D: **Regulation of middle-mode transcription.** In *Molecular Biology of Bacteriophage T4. Volume American Society for Microbiology.* Edited by: Karam J, Drake JW, Kreuzer KN, Mosig G, Hall DH, Eiserling FA, Black LW, Spicer EK, Kutter E, Carlson K, Miller ES. Washington, D. C; 1994:142-160.

140. Tiemann B, Depping R, Gineikiene E, Kaliniene L, Nivinskas R, Ruger W: **ModA and ModB, two ADP-ribosyltransferases encoded by bacteriophage T4: catalytic properties and mutation analysis.** *J Bacteriol* 2004, **186**:7262-7272.

141. Depping R, Lohaus C, Meyer HE, Ruger W: **The mono-ADP-ribosyltransferases Alt and ModB of bacteriophage T4: target proteins identified.** *Biochem Biophys Res Commun* 2005, **335**:1217-1223.

142. Tiemann B, Depping R, Ruger W: **Overexpression, purification, and partial characterization of ADP-ribosyltransferases modA and modB of bacteriophage T4.** *Gene Expr* 1999, **8**:187-196.

143. Kutter E, Kellenberger E, Carlson K, Eddy S, Neitzel J, Messinger L, North J, Guttman B: **Effects of bacterial growth conditions and physiology on T4 infection.** In *Molecular Biology of Bacteriophage T4.* Edited by: Karam J, Drake JW, Kreuzer KN, Mosig G, Hall DH, Eiserling FA, Black LW, Spicer EK, Kutter E, Carlson K, Miller ES. Washington, D.C.: ASM Press; 1994:406-418.

144. Takagi H, Kakuta Y, Okada T, Yao M, Tanaka I, Kimura M: **Crystal structure of archaeal toxin-antitoxin RelE-RelB complex with implications for toxin activity and antitoxin effects.** *Nat Struct Mol Biol* 2005, **12**:327-331.

145. Kamada K, Hanaoka F: **Conformational change in the catalytic site of the ribonuclease YoeB toxin by YefM antitoxin.** *Mol Cell* 2005, **19**:497-509.

doi:10.1186/1743-422X-7-360
Cite this article as: Uzan and Miller: Post-transcriptional control by bacteriophage T4: mRNA decay and inhibition of translation initiation. *Virology Journal* 2010 7:360.

Kreuzer and Brister *Virology Journal* 2010, **7**:358
http://www.virologyj.com/content/7/1/358

VIROLOGY JOURNAL

Initiation of bacteriophage T4 DNA replication and replication fork dynamics: a review in the Virology Journal series on bacteriophage T4 and its relatives

Kenneth N Kreuzer[1*], J Rodney Brister[2]

Abstract

Bacteriophage T4 initiates DNA replication from specialized structures that form in its genome. Immediately after infection, RNA-DNA hybrids (R-loops) occur on (at least some) replication origins, with the annealed RNA serving as a primer for leading-strand synthesis in one direction. As the infection progresses, replication initiation becomes dependent on recombination proteins in a process called recombination-dependent replication (RDR). RDR occurs when the replication machinery is assembled onto D-loop recombination intermediates, and in this case, the invading 3' DNA end is used as a primer for leading strand synthesis. Over the last 15 years, these two modes of T4 DNA replication initiation have been studied *in vivo* using a variety of approaches, including replication of plasmids with segments of the T4 genome, analysis of replication intermediates by two-dimensional gel electrophoresis, and genomic approaches that measure DNA copy number as the infection progresses. In addition, biochemical approaches have reconstituted replication from origin R-loop structures and have clarified some detailed roles of both replication and recombination proteins in the process of RDR and related pathways. We will also discuss the parallels between T4 DNA replication modes and similar events in cellular and eukaryotic organelle DNA replication, and close with some current questions of interest concerning the mechanisms of replication, recombination and repair in phage T4.

Introduction

Studies during the last 15 years have provided strong evidence that T4 DNA replication initiates from specialized structures, namely R-loops for origin-dependent replication and D-loops for recombination-dependent replication (RDR). The roles of many of the T4 replication and recombination proteins in these processes are now understood in detail, and the transition from origin-dependent replication to RDR has been ascribed to both down-regulation of origin transcripts and activation of the UvsW helicase, which unwinds origin R-loops.

One of the interesting themes that emerged in studies of T4 DNA metabolism is the extensive overlap between different modes of replication initiation and the processes of DNA repair, recombination, and replication fork restart. As discussed in more detail below, the distinction between origin-dependent and recombination-dependent replication is blurred by the involvement of recombination proteins in certain aspects of origin replication. Another example of overlap is the finding that repair of double-strand breaks (DSBs) in phage T4 infections occurs by a mechanism that is very closely related to the process of RDR. The close interconnections between recombination and replication are not unique to phage T4 - it has become obvious that the process of homologous recombination and particular recombination proteins play critical roles in cellular DNA replication and the maintenance of genomic stability [1-4].

* Correspondence: kenneth.kreuzer@duke.edu
[1]Department of Biochemistry, Duke University Medical Center, Durham, NC 27710 USA
Full list of author information is available at the end of the article

BioMed Central

Origin-dependent replication

Most chromosomes that have been studied include defined loci where DNA synthesis is initiated. Such origins of replication have unique physical attributes that contribute to the assembly of processive replisomes, facilitate biochemical transactions by the replisome proteins to initiate DNA synthesis, and serve as key sites for the regulation of replication timing. While the actual determinants of origin activity remain ill defined in many systems, all origins must somehow promote the priming of DNA synthesis. Bacteriophage T4 contains several replication origins that are capable of supporting multiple rounds of DNA synthesis [5,6] and has very well-defined replication proteins [7], making this bacteriophage an ideal model to study origin activation and maintenance.

Localization of T4 origins throughout the genome

Clear evidence for defined T4 origin sequences began to emerge about 30 years ago when the Kozinski and Mosig groups demonstrated that nascent DNA produced early during infection originated from specific regions within the 169 kb phage genome [8-10]. The race was on, and several groups spent the better part of two decades trying to define the T4 origins of replication. These early efforts brought a battery of techniques to bear, including electron microscopy and tritium labeling of nascent viral DNA, localizing origins to particular regions of the genome. The first direct evidence for the DNA sequence elements that constitute a T4 origin emerged from studies of Kreuzer and Alberts [11,12], who isolated small DNA fragments that were capable of driving autonomous replication of plasmids during a T4 infection. Later approaches using two-dimensional gel electrophoresis confirmed that these two origins, *oriF* and *oriG* [also called *ori(uvsY)* and *ori(34)*, respectively], were indeed active in the context of the phage genome [13,14]. All told, at least seven putative origins (termed *oriA* through *oriG*) were identified by these various efforts, yet no strong consensus emerged as whether all seven were *bona fide* origins and how the multiple origins were utilized during infection.

Recent work by Brister and Nossal [5,15] has helped to clarify many issues regarding T4 origin usage. Using an array of PCR fragments, they monitored the accumulation of nascent DNA across the entire viral genome over the course of infection, allowing both the origins and breadth of DNA synthesis to be monitored in real time. This whole-genome approach revealed that at least 5 origins of replication are active early during infection, *oriA, oriC, oriE, oriF,* and *oriG* (see Figure 1). Though all of these origins had been independently identified to some extent in previous studies, this was the first

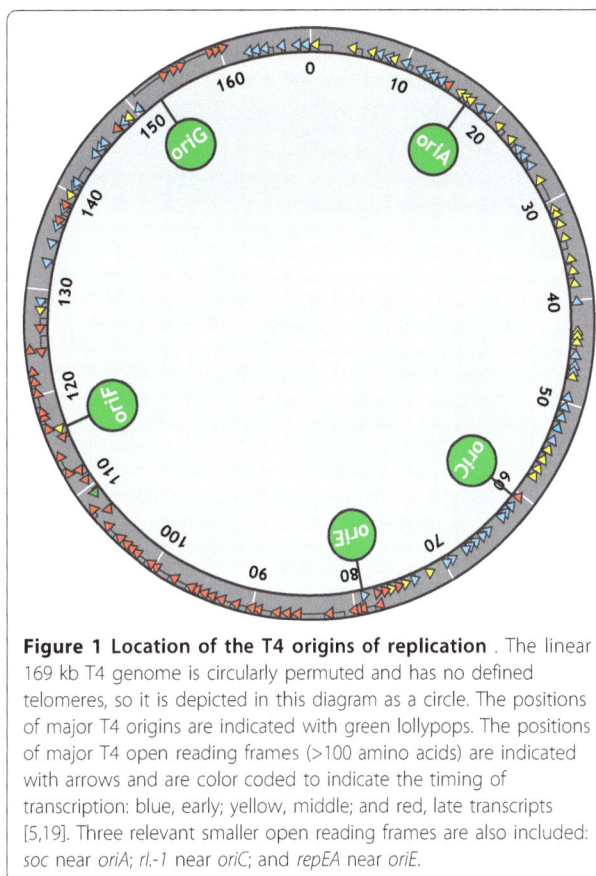

Figure 1 Location of the T4 origins of replication. The linear 169 kb T4 genome is circularly permuted and has no defined telomeres, so it is depicted in this diagram as a circle. The positions of major T4 origins are indicated with green lollypops. The positions of major T4 open reading frames (>100 amino acids) are indicated with arrows and are color coded to indicate the timing of transcription: blue, early; yellow, middle; and red, late transcripts [5,19]. Three relevant smaller open reading frames are also included: *soc* near *oriA*; *rl.-1* near *oriC*; and *repEA* near *oriE*.

observation of concurrent activity from each within a population of infected cells.

There do not appear to be any local sequence motifs shared among all the T4 origins. However, one origin, *oriE*, does include a cluster of evenly spaced, 12-nt direct repeats [16]. Similar "iterons" are also found within syntenic regions of closely related bacteriophage genomes, implying conserved function [17]. Indeed, this arrangement of direct repeats is reminiscent of some plasmid origins, such as the RK6 gamma origin, where replication initiator proteins bind to direct repeats and promote assembly of replisomes [18]. Despite this circumstantial evidence, no association has been established between the T4 iterons and *oriE* replication activity, and to this date their role during T4 infection remains ill defined.

There is some indication that global genome constraints influence the position of T4 origins. Three of the more active T4 origins, *oriE, oriF,* and *oriG* are located near chromosomal regions where the template for viral transcription switches from predominately one strand to predominately the complementary strand [5,19] (see Figure 1). These regions of transcriptional divergence coincide with shifts in nucleotide compositional bias (predominance of particular nucleotides on a particular strand), a hallmark of replication origins in other systems [20]. That said, at least two origins (*oriA* and *oriC*) are well outside regions of intrastrand nucleotide skews and transcriptional divergence, so it is not clear what, if any, physical properties of the T4 chromosome contribute to origin location. Moreover, the T4 genome is circularly permuted with no defined telomeres, so the actual position of a given locus relative to the chromosome ends is variable in a population of replicating virus.

The undulating T4 transcription pattern reflects the modular nature of the viral genome. T4 genes are arranged in functionally related clusters, and diversity among T4-related viruses appears to arise through the horizontal transfer of gene clusters [17,21]. The spacing of T4 origins over the length of the viral genome coincides with some of these clusters and may reflect genome mechanics. Most early T4 DNA synthesis originates from regions within the genome that are dominated by late-mode viral transcription [5,19]. This arrangement suggests an intimate relationship between T4 replication and transcription of late genes, like those encoding viral capsid components. It has been known for some time that late-mode transcription is dependent on gp45 clamp protein, which is a component of both the T4 replisome and late-mode transcription complexes (reviewed by Miller *et al.* [22]), but there is also evidence that the amount of replication directly influences the amount of transcription [23] (Brister, unpublished data).

Molecular mechanism of origin initiation
Though few obvious sequence characteristics are shared between them, all of the T4 origins are thought to facilitate formation of RNA primers used to initiate leading strand DNA synthesis. Most of what is known about the detailed mechanism of T4 replication initiation comes from studies of the two origins (*oriF* and *oriG*) that support autonomous replication of plasmids in T4-infected cells (see above). Origin plasmid replication requires the expected T4-encoded replisome proteins, and like phage genomic DNA replication, is substantially reduced and/or delayed by mutations in the replicative helicase, primase and topoisomerase [24,25].

The DNA sequences required for *oriF* and *oriG* function on recombinant plasmids have been defined by deletion and point mutation studies [26] (Menkens and Kreuzer, unpublished data). A minimal sequence of about 100 bp from each origin was shown to be necessary for autonomous replication, and though there is little homology between *oriF* and *oriG*, both minimal sequences include a middle-mode promoter and an A + T-rich downstream unwinding element (DUE) [26,27]. Middle-mode promoters consist of a binding site for the

viral transcription factor MotA in the -30 region, along with a -10 sequence motif that is indistinguishable from the typical *E. coli* σ70 -10 motif [28,29]. Transcripts initiated from the *oriF* MotA-dependent promoter were shown to form persistent R-loops within the DUE region, leaving the non-template strand hypersensitive to ssDNA cleavage. Formation of these R-loops is not dependent on specific sequences and the endogenous DUE can be substituted with heterologous unwinding elements [13,27].

The *oriF* R-loops are very likely processed by viral RNase H to generate free 3'-OH ends that are used to prime leading strand DNA synthesis [13,27]. Furthermore, the presence of an R-loop presumably holds the origin duplex in an open conformation, giving the gp41/61 primosome complex access to the unpaired non-template strand to allow extensive parental DNA unwinding and priming on the lagging strand. Less is known about replication priming at the other T4 origins [30]. Presumably, *oriG* uses the same mechanism as *oriF* [13,27], and there is some evidence that a transcript from a nearby MotA-dependent promoter is used to initiate replication at *oriA* [30]. Yet, MotA mutations do not fully prevent viral replication [16,31], and other types of viral promoters also appear important to origin function. For example, there are no middle-mode promoters near *oriE*; instead this origin apparently depends on an early-mode promoter, which does not require viral transcription factors for activity [16]. Moreover, mutations that prevent late-mode viral transcription alter replication from T4 *oriC*, without affecting activity from the other origins (Brister, unpublished), raising the possibility that a late-mode promoter is required for activity from this origin.

Discontinuous lagging strand replication is normally primed by the T4-encoded gp61 primase [32-34]. Even though T4 primase is required only for lagging strand synthesis *in vitro*, the *in vivo* results are more complex. First, mutants deficient in primase show a severe DNA-delay phenotype, with very little DNA synthesis occurring early during infection [24,30,35,36]. This implies that primase activity contributes directly to early steps of T4 DNA replication. Either leading strand synthesis at some T4 origins is primed by primase, or normal viral replication requires the coupling of leading strand synthesis with primase-dependent lagging strand synthesis. Second, T4 DNA replication eventually reaches a remarkably vigorous level in primase-deficient infections, even when using a complete primase deletion mutant [24] (also see [37]). One published report suggested that the primase-independent replication was abolished by mutational inactivation of T4 endonuclease VII, leading to a model in which endonuclease VII cleavage of recombination intermediates provides primers for DNA synthesis [38]. However, repetition of this experiment revealed little or no decrease in endonuclease-deficient infections [39], and the strain used in the Mosig study was later found to contain an additional mutation that was contributing to the reduced replication (G. Mosig, personal communication to KNK). The mechanism of extensive DNA replication late in a primase-deficient infection remains unclear, but could possibly result from extensive priming by mRNA transcripts (perhaps in combination with endonuclease cleavage as suggested by Mosig [38]).

In other systems, there are examples of both primase- and transcript-mediated initiation of leading strand DNA synthesis from origins. A transcript is used to prime replication from the ColE1 plasmid origin, as well as mitochondrial DNA origins [40,41], yet primase is used to initiate replication from the major *E. coli* origin, *oriC* [42,43]. Indeed, there are even systems where both mechanisms of initiation are used within a single chromosome. For example, unlike *oriC*, R-loops are apparently used to initiate DNA synthesis at the *oriK* sites in *E. coli* (reviewed in [44]).

The molecular mechanism of T4 replication initiation has been investigated *in vitro* using R-loop substrates constructed by annealing an RNA oligonucleotide to supercoiled *oriF* plasmids [45]. Efficient replication of these preformed R-loop substrates does not require a promoter sequence, but a DUE is necessary. In fact, non-origin plasmids are efficiently replicated *in vitro* by the T4 replisome as long as they have a preformed R-loop within a DUE region, implying that the R-loop itself is the signal for replisome assembly on these substrates. Experiments using radioactively labeled R-loop RNA directly demonstrated that the RNA is used as the primer for DNA synthesis. Several viral proteins are required for significant replication of these R-loop substrates: DNA polymerase (gp43), polymerase clamp (gp45), clamp loader (gp44/62), and single-stranded DNA binding protein (gp32). In addition, without the replicative helicase (gp41), leading-strand synthesis is limited to a relatively short region (about 2.5 kb) and lagging strand synthesis is abolished. While gp41 can load without the helicase loading protein (gp59), the presence of gp59 greatly accelerates the process. Finally, replication on these covalently closed substrates is severely limited when the T4-encoded type II topoisomerase (gp39/52/60) is withheld, as expected due to the accumulation of positive supercoiling ahead of the fork.

Normal viral replication also requires gp59 protein, and though gene *59* mutants make some DNA early, this synthesis is arrested as the infection progresses [5,46,47]. This deficiency was initially thought to reflect a unique requirement for gp59 in recombination-dependent replication (i.e., no requirement in origin-dependent replication). However, gp59 mutations also affect origin activity, reducing the total amount of origin-mediated DNA synthesis, mirroring the *in vitro* studies mentioned above [5]. Further defects are clearly visible at *oriG*, where gene *59* mutations cause problems in the coupling of leading and lagging strand synthesis (but do not prevent replication initiation) [48].

The deleterious effects of gene *59* mutations could reflect several biochemical activities that have been characterized *in vitro*. A major function of gp59 is loading of the replicative helicase gp41 [49]. Gp59 is a branch-specific DNA binding protein with a novel alpha-helical two-domain fold [50]. The gp59 protein is capable of binding a totally duplex fork, but requires a single-stranded gap of more than 5 nucleotides (on the arm corresponding to the lagging strand template) to load gp41 [51]. As expected from this loading activity, gp59 stimulates gp41 helicase activity on branched DNA substrates (e.g. Holliday junction-like molecules). Interestingly, gp59 has another function in the coordination of leading- and lagging-strand synthesis and in this context has been called a "gatekeeper". When gp59 binds to replication fork-like structures in the absence of gp41, it blocks extension by T4 DNA polymerase [45,48,52]. This inhibitory activity of gp59 presumably acts to prevent the generation of excessive single-stranded DNA and allow coordinated and coupled leading and lagging strand synthesis.

Unlike gp59, the viral gp41 helicase is required for extended replication of R-loop substrates *in vitro* (see above) and any appreciable replication during infection [15,45,53]. Yet, some viral replication is observed in gp59-deficient infections (see above), indicating that gp41 helicase can load onto origins at some rate through another means. T4 encodes at least two other helicases, UvsW and Dda, and earlier studies demonstrated that one of them, Dda, stimulates gp41-mediated replication *in vitro* [49]. It was therefore suggested that either gp59 or Dda was sufficient to load gp41 helicase at the T4 origins [49]. Consistent with this notion, *dda* mutants have a DNA delay phenotype and are deficient in early, presumably origin-mediated DNA synthesis, though replication rebounds at later times when it is dependent on viral recombination [15,46]. Moreover, *dda 59* double mutants have a greater defect than either single mutant, essentially showing no replication (either early or late) and indicating a cumulative effect on origin activity [46].

Though there may be some functional overlap between Dda and gp59, DNA replication patterns indicate that each has distinct activities at the T4 origins [15]. Unlike *dda* mutations, which cause a generalized reduction in DNA synthesis that is particularly evident at *oriE*, gene *59* mutations have little effect on replication from this origin [15]. This difference may indicate that *oriE* uses a different mechanism to initiate replication, one less dependent on gp59. This idea has been expressed before and may simply reflect the difference in sequence elements at *oriE* compared to the other origins. One protein in particular, RepEB, has also been implicated in *oriE* activity [16], but *repEB* mutations have a more generalized effect, reducing replication from all origins [15].

Inactivation of origins at late times

The regulation of origin usage has been studied directly for *oriF* and *oriG*, the two origins known to function via an R-loop intermediate. One level of control is exerted by the change in the transcriptional program. The RNA within the *oriF* and *oriG* R-loops are initiated from MotA-dependent middle mode promoters, which are shut off as RNA polymerase is converted into the form for late transcription [28,29]. A second level of control is exerted when the UvsW helicase is expressed from its late promoter [54]. UvsW is a helicase with fairly broad specificity for various branched nucleic acids, including the R-loops that occur at *oriF* and *oriG* [55-57]. Thus, any existing R-loops at these origins are unwound when UvsW is synthesized. While not yet studied directly, R-loops may also occur at one or more other T4 origins (e.g. *oriE*), and thus the mechanisms of regulation could be identical to that of *oriF* and *oriG*. Further work is clearly needed to understand the regulation of other T4 origins.

As will be discussed in more detail below, mutational inactivation of T4 recombination proteins leads to the DNA arrest phenotype, characterized by a paucity of late DNA replication. The additional inactivation of UvsW suppresses this DNA arrest phenotype and allows high levels of DNA synthesis at late times [58-61]. The simplest explanation is that R-loop replication becomes dominant in these double-mutant infections at late times. If true, it seems likely that much of this late replication is initiated at R-loops formed at late promoters, but these "cryptic origin" locations have not yet been experimentally defined.

Recombination-dependent replication

The tight coupling of homologous genetic recombination and DNA replication was first recognized in the phage T4 system when it was found that mutational inactivation of recombination proteins leads to the DNA-arrest phenotype characterized by defective late replication [62]. Based on this and other data, Gisela Mosig proposed that genomic DNA replication can be initiated on the invading 3' ends of D-loop structures generated by the recombination machinery (Figure 2A)

[63]. There is now abundant *in vivo* and *in vitro* evidence supporting this model for phage T4 DNA replication. T4 RDR is an important model for the linkage of recombination and replication, because it has become clear that recombination provides a backup method for restarting DNA replication in both prokaryotes and eukaryotes (see below).

RDR on the phage genome

The infecting T4 DNA is a linear molecule, and early genetic results showed that the (randomly located) DNA ends are preferential sites for homologous genetic recombination [64-66]. When an origin-initiated replication fork reaches one of the DNA ends, one of the two daughter molecules should contain a single-stranded 3' end that is competent for strand invasion and D-loop formation; the other daughter molecule is also presumably competent for strand invasion after processing to generate a 3' end. The complementary sequence that is invaded could be at the other end of the same DNA molecule, since the infecting T4 DNA is terminally redundant, or it may be within the interior region of a co-infecting T4 DNA molecule, since T4 DNA is also circularly permuted. In this way, the process of RDR can in principle initiate soon after an origin-initiated fork reaches a genomic end. As will be described below, RDR or some variant thereof might be needed to continue replication well before origin-initiated forks reach the genome ends. The overall role of RDR in genome replication and the relationship of RDR to the eventual packaging of phage DNA are discussed in detail elsewhere [6,67].

RDR of the phage genome is abolished or greatly reduced by mutational inactivation of most T4-encoded recombination proteins (see [68] for review on the biochemistry of T4 recombination proteins). The strongest DNA arrest phenotypes are caused by inactivation of gp46/47 or gp59, and correspondingly, these are essential proteins. Inactivation of the non-essential UvsX and UvsY proteins eliminate most but not all late DNA replication. These two proteins catalyze the strand invasion reaction that generates D-loops, and so one might expect RDR to be totally abolished. However, a significant amount of T4 genetic recombination still occurs in the absence of UvsX or UvsY, and this has been ascribed to a single-strand annealing pathway [69,70]. Single-strand annealing intermediates may also be used to initiate RDR, which could explain the residual late DNA replication in UvsX or UvsY knockout mutants.

The *uvsW* gene is in the same recombinational repair pathway as *uvsX* and *uvsY* [71]. However, the *uvsW* gene product was not originally implicated in the process of RDR because *uvsW* knockout mutations do not block late DNA replication [71]. This inference was

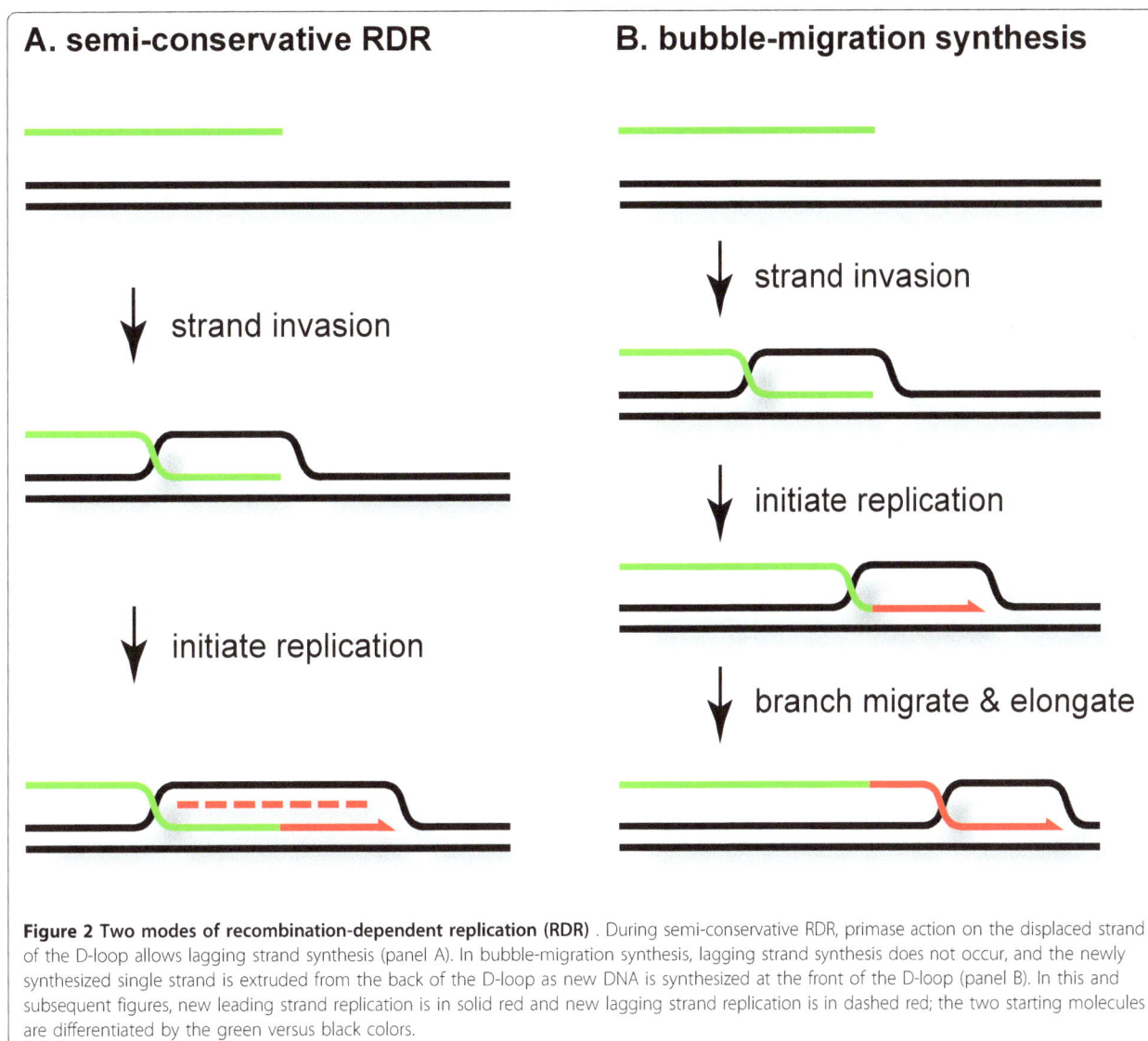

Figure 2 Two modes of recombination-dependent replication (RDR) . During semi-conservative RDR, primase action on the displaced strand of the D-loop allows lagging strand synthesis (panel A). In bubble-migration synthesis, lagging strand synthesis does not occur, and the newly synthesized single strand is extruded from the back of the D-loop as new DNA is synthesized at the front of the D-loop (panel B). In this and subsequent figures, new leading strand replication is in solid red and new lagging strand replication is in dashed red; the two starting molecules are differentiated by the green versus black colors.

probably misleading - as described above, the UvsW helicase apparently unwinds R-loops that could otherwise trigger replication at late times. Thus, inactivation of UvsW could simultaneously reduce or eliminate RDR and activate an R-loop dependent mechanism of late replication, resulting in no net decrease in late DNA replication [54]. Consistent with this model, a *uvsW* mutant has reduced recombination and was shown to be defective in generating phage DNA longer than unit length (in alkaline sucrose gradients) [71]. In addition, UvsW is required for a plasmid-based model for RDR [55] (see below).

The one T4 recombination function that is not required for RDR is endonuclease VII, which resolves Holliday junctions and other branched DNA structures [72,73]. The major function of endonuclease VII during infection is to resolve DNA branches during DNA

packaging [74,75]. Because this is a very late step in genetic recombination, the lack of a role in RDR is unsurprising.

Plasmid model systems for RDR
Plasmid model systems have been productive for analyzing the mechanism of RDR *in vivo*, and have revealed a very close relationship between repair of DSBs and the process of RDR. Plasmids with homology to the T4 genome but no T4 replication origin are replicated during a phage T4 infection, as long as T4-induced host DNA breakdown is prevented [76-78]. This plasmid replication is not dependent on particular T4 sequences, because even plasmid pBR322 can be replicated when the infecting T4 carries an integrated copy of the plasmid [76]. Plasmid replication requires T4 recombination proteins, arguing that it occurs by RDR [77]. The

products of plasmid replication in a T4 infection consist mostly of long plasmid concatamers, arguing that rolling circle replication is induced, but the mechanism of rolling circle formation is unknown [79].

The remarkable discovery of mobile group I introns in T4 [80] led to a simple way to introduce site-specific DSBs during a T4 infection, which has been valuable for *in vivo* studies of T4 RDR. These introns encode site-specific DNA endonucleases, such as the endonuclease I-*Tev*I from the intron of the T4 *td* gene (see below for discussion of the intron mobility/DSB repair events; also see [81]). The recognition site for I-*Tev*I (or another intron-encoded nuclease, SegC) has been introduced into recombinant plasmids and also into ectopic locations in the T4 genome, and in either case, the site is cleaved efficiently during a normal T4 infection when the endonuclease is expressed [76,82-86]. If the regions adjacent to the cut site have a homologous DNA target, either in the T4 genome or another segment of a plasmid residing in the same cell, coupled recombination/replication reactions are efficiently induced [76,79,87].

Using such model systems for RDR, it was shown that T4 recombination proteins UvsX, UvsY, UvsW, gp46/47, and gp59 are required for extensive DSB-directed replication, as are the expected T4 replication fork proteins (gp43, gp44/62, gp45, gp32, gp41, gp61; delayed replication of the plasmid occurs in the gp61-deficient infection, similar to the delayed replication of chromosomal DNA) [24,55,77]. In addition, by limiting the homology to just one side of the break, a single double-strand end was shown to be sufficient to induce RDR, as predicted by the Mosig model [76,86].

Molecular mechanism of RDR

The heart of the RDR process is the strand-invasion reaction that creates D-loops, which is described in more detail in the review on T4 recombination [68]. Briefly, DNA ends are prepared for strand invasion by the gp46/47 helicase/nuclease complex, transient regions of ssDNA are coated by the single-strand binding protein gp32, UvsY acts as a mediator protein in loading UvsX onto gp32-coated ssDNA, and UvsX is the strand-invasion protein (RecA and Rad51 homolog). Recent evidence argues that the UvsW helicase also plays a direct role in strand invasion, promoting 3-strand branch migration to stabilize the D-loop [88].

As described in more detail by Kreuzer and Morrical [6], early reconstitution of a T4 RDR reaction *in vitro* generated a conservative replication reaction called bubble-migration synthesis [89]. In bubble-migration synthesis, the 3' invading end in the D-loop is extended by DNA polymerase as the junction at the back of the D-loop undergoes branch migration in the same direction (Figure 2B). The net result is that a newly synthesized single-strand copy is created and then quickly extruded from its template, and lagging-strand synthesis does not occur within the D-loop.

In the RDR reactions analyzed by Formosa and Alberts [90], the T4 DNA polymerase holoenzyme complex (polymerase gp43, clamp gp45 and clamp loader gp44/62) catalyzed synthesis in reactions containing only UvsX and gp32. Interestingly, synthesis did not occur if the host RecA protein was substituted for UvsX (even if host SSB protein was added), suggesting that the T4 polymerase complex has specific interactions with the phage-encoded strand-exchange protein. The extent of synthesis was limited unless a helicase was added to facilitate parental DNA unwinding - Dda was used in these initial experiments and allowed extensive bubble-migration synthesis [90].

Since the publication of Molecular Biology of Bacteriophage T4 in 1994 [91], much progress has been made in understanding the mechanism of loading of the helicase/primase complex onto D-loops. When T4 RDR reactions are supplemented with gp59, gp41 and gp61, lagging-strand synthesis is efficiently reconstituted on the displaced strand of the D-loop, and a conventional semiconservative replication fork is established (Figure 2A) (see [6]). As described above, gp59 is a branch-specific DNA binding protein that loads gp41, and gp59 interacts specifically with both gp41 and gp32 in the loading reaction [50,51,92-97]. Jones et al. [94] showed that gp59 can load helicase onto a structure that closely resembles a D-loop, reflecting its role in RDR. Once the replicative helicase is loaded onto the displaced strand of the D-loop (which becomes the lagging-strand template), leading strand synthesis by T4 DNA polymerase (gp43) is activated. Because the T4 primase gp61 binds to and functions with gp41 (see [7]), loading of gp41 is critical to begin lagging-strand synthesis as well.

Overlap between origin- and recombination-dependent mechanisms

The transition between origin- and recombination-dependent replication is not entirely clear cut during T4 infection, and there is significant interplay between the two replication modes. Moreover, the relationship between origin- and recombination-dependent replication is dynamic, which is clearly seen in experiments with varying multiplicities of infection. In singly infected cells, there is a prolonged period early during infection when the recombination protein UvsX is not required for replication. Yet, when cells are infected with an average of five viruses, the timing changes, and even very early replication is dependent on UvsX [5]. Though the mechanism of this regulation is not clear, it is evident that the infection program can somehow sense the amount of infecting viral DNA and switch replication modes under conditions where there are ample templates for RDR.

Recombination proteins also appear to be more important to replication from some origins compared to others. As mentioned earlier, genetic requirements vary among the multiple T4 replication origins that are active within a single population of infected cells. At least one origin, *oriA*, appears more active later during infection, when replication is dependent on the viral recombination machinery. Moreover, replication from this origin is significantly reduced when the viral recombination protein UvsX is mutated [5]. Though these observations underscore a role for T4 recombination machinery at *oriA*, it is not clear whether RDR is preferentially initiated near *oriA* or if normal *oriA*-mediated replication is partially dependent on UvsX.

One hint to the role of UvsX during origin-mediated replication comes from the apparently slow movement of replication forks across the T4 chromosome. Once initiated, T4 replication forks do not simply progress from an origin to the ends of the chromosome at the 30-45 kb per minute rate observed *in vitro* [5]. Rather, replication forks appear to move more slowly than expected, resulting in the accumulation of sub-genomic length DNAs early during infection. Only later are these short DNAs efficiently elongated into full-length genomes. This behavior was initially noticed by Cunningham and Berger [58], who analyzed the length of newly replicated single-stranded DNA using alkaline sucrose gradients. They also showed that efficient maturation of nascent DNAs into full genome length products requires the viral replication proteins UvsX or UvsY. A similar effect was observed during array studies where the elongation of nascent DNAs was greatly delayed in *uvsX* mutant infections compared to normal infections [5].

So why is there a delay in the elongation of T4 nascent DNAs? One possibility is that physical factors (e.g. tightly bound proteins) impede the progress of the replication forks across the T4 chromosome, causing replisomes to stall or disassociate from the DNA template. Rescue of model stalled forks *in vitro* can be catalyzed by UvsX and either gp41 helicase (with gp59) or Dda helicase [98]. Thus, one model is that UvsX is required *in vivo* to restart origin-initiated forks that have stalled before completing replication, and so the elongation of nascent DNAs is compromised during *uvsX* mutant infections.

Several factors may impede the progress of replication forks (also see below). T4 replication occurs concurrently with transcription during infection [19] (Brister, unpublished results), so replisomes must compete with the transcriptional machinery for template. Head-on collisions with RNA polymerase cause pausing of T4 replisomes *in vitro* [99], and undulating patterns of T4

transcription imply that replication forks must eventually pass through regions of head-on transcription. Furthermore, if multiple origins are active on a single chromosome, then replication forks initiated at different origins would speed towards one another, plowing through the duplex template. In this scenario intervening sequences would be wound into impassable torsion springs, and T4 topoisomerase (gp39/52/60) would be necessary to relax the duplex and allow progression. Indeed, gene *52* mutants produce shorter than normal DNA replication products early during infection, similar to *uvsX* mutants [100].

Interrelationship between replication, recombination and repair

Studies in many different biological systems have uncovered key roles of recombination proteins in the replication of damaged DNA [1-4]. One major set of pathways involves the repair of DSBs and broken replication forks. In addition, recombination proteins are involved in multiple pathways proposed for replication fork restart after blockage by non-coding lesions, some pathways coupled to repair of the DNA damage and others that result in bypass of the damage. Here, we briefly review unique contributions to this field that emerged from the phage T4 system.

Tight linkage of DSB repair and RDR

As indicated above, DSB repair in phage T4 is closely related to the process of RDR. Studies of DSB repair were greatly accelerated by the discovery of the mobile group I introns and their associated endonucleases. Intron mobility involves the generation of a DSB within the recipient (initially intron-free) DNA by an intron endonuclease, followed by a DSB repair reaction that introduces a copy of the intron from the donor DNA, such that both recipient and donor end up with a copy of the intron [80,81,101].

A variety of approaches have been used to study the detailed mechanism of DSB repair *in vivo* using intron endonuclease-mediated DSBs. One series of studies using a plasmid model system indicated that the DSBs are repaired by a pathway called synthesis-dependent strand annealing (SDSA), in which the induced DNA replication is limited to the region near the DSB (Figure 3A) [102,103]. The SDSA repair mechanism is closely related to the bubble-migration reaction described above, and has been implicated in DSB repair in eukaryotic systems such as Drosophila [104,105]. Other studies, however, argue that the DSB leads to the generation of fully functional replication forks in a process that is very closely related to the RDR pathway that occurs in the phage genome [79,85,87,106]. This so-called extensive chromosomal replication (ECR) model leads to *bona fide* DSB

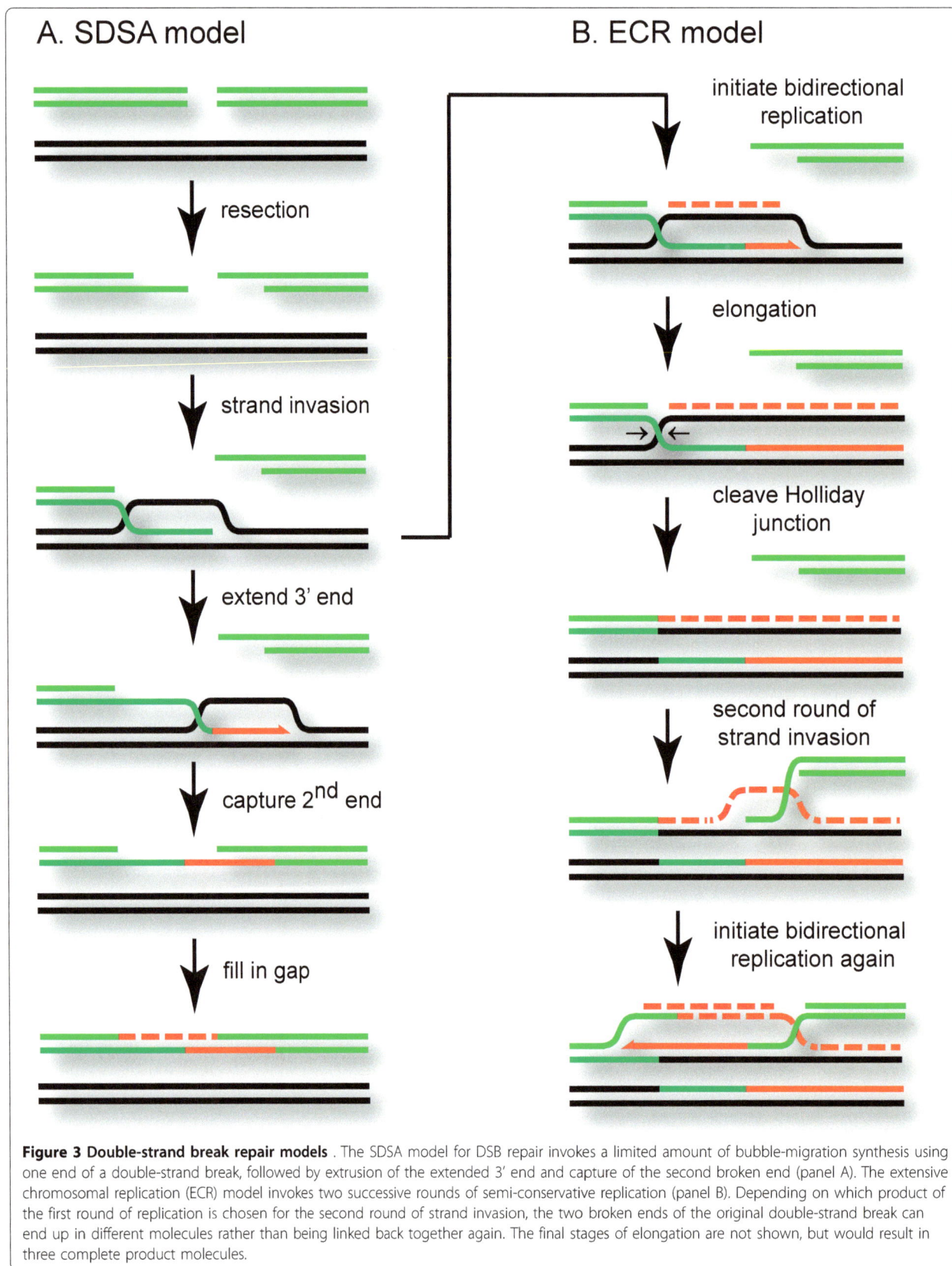

Figure 3 Double-strand break repair models . The SDSA model for DSB repair invokes a limited amount of bubble-migration synthesis using one end of a double-strand break, followed by extrusion of the extended 3' end and capture of the second broken end (panel A). The extensive chromosomal replication (ECR) model invokes two successive rounds of semi-conservative replication (panel B). Depending on which product of the first round of replication is chosen for the second round of strand invasion, the two broken ends of the original double-strand break can end up in different molecules rather than being linked back together again. The final stages of elongation are not shown, but would result in three complete product molecules.

repair, even though the two broken ends of the DSB can end up in different molecules (Figure 3B).

A major difference between the SDSA and ECR models for DSB repair is that SDSA does not involve primase (gp61)-dependent lagging strand synthesis, while the ECR model does. Perhaps either repair model can occur when a DSB occurs on the phage genome, but the choice of pathways depends on whether the helicase/primase complex is successfully loaded onto the displaced strand of the initial D-loop. Considering that gp59 efficiently inhibits polymerase and loads helicase/primase (see above), it is difficult to see how the bubble-migration pathway and SDSA could occur *in vivo*, unless there is some additional level of regulation that has not yet been uncovered. Shcherbakov et al. [106] have presented additional evidence that DSBs trigger normal replication like that postulated in the ECR model, and provided arguments against a major role for the SDSA pathway during wild-type T4 infections.

If a DNA end can trigger a new replication fork by invading homologous DNA, there would seem to be no need to coordinate the processing of the two ends of a DSB - each could simply start a new replication fork on any homologous DNA molecule. Indeed, if the two DNA segments flanking a DSB are homologous to two different plasmid molecules, the DSB is repaired by inducing replication of both plasmids [86]. While this result clearly shows that the two ends can act independently when forced to do so, other experiments demonstrate that the two broken ends of a DSB are often repaired in a coordinated fashion, using the same template molecule [86,106]. Moreover, Shcherbakov et al. [106] presented striking evidence that the end coordination is dependent on the gp46/47 complex. The eukaryotic homolog, Rad50/Mre11, has also been implicated in end coordination in DSB repair by a mechanism involving tethering of the two ends via a protein bridge [107,108]. How does end tethering relate to the extensive replication triggered by the broken ends? The simplest explanation is that one end of the DSB triggers a new replication fork on a homolog, and then the second broken end invades one of the two newly-replicated products from that first replication event and triggers a second replication fork in the opposite direction, as diagrammed in Figure 3B[86,106].

Replication fork blockage and restart

Replication forks can be blocked or stalled by template lesions, lack of nucleotide substrates, or problems with the replication apparatus. In addition to the natural blockage that appears to occur in normal infections (see above), the consequences of fork blockage and possible pathways for fork restart have been studied using two different inhibitors. First, hydroxyurea (HU) inhibits the reduction of ribonucleotides to deoxyribonucleotides and thereby depletes the nucleotide precursors for replication [109]. Second, the topoisomerase inhibitor 4'-(9-acridinylamino)-methanesulfon-*m*-anisidide (*m*-AMSA) stabilizes covalent topoisomerase-DNA complexes and thereby physically blocks T4 replication forks [110].

Wild-type T4 induces breakdown of host DNA, providing a significant source of deoxynucleotide precursors for phage replication and thereby making the phage relatively resistant to HU. One class of HU hypersensitive mutants consists of those defective in the breakdown of host DNA (e.g., *denA* which encodes DNA endonuclease II) [111,112]. A second well-studied HU hypersensitive mutant class consists of those with knockouts of the *uvsW* gene [71,113]. These mutants are not defective in host DNA breakdown, and the HU hypersensitivity of *uvsW* mutants was shown to result from a different genetic pathway than that of *denA* mutants. We will suggest below that the UvsW protein plays a special role in processing blocked replication forks, namely that it catalyzes a process called replication fork regression. We also suggest that fork regression might somehow lead to efficient replication fork restart, although the details are unclear. Interestingly, the HU hypersensitivity of *uvsW* knockout mutants can be eliminated by additional knockout of *uvsX* or *uvsY* [58]. This result suggests that the UvsXY homologous recombination system creates some kind of toxic intermediate/product from stalled replication forks when the UvsW protein is unavailable - the nature of this toxic structure is currently unknown.

The phage T4 type II DNA topoisomerase is sensitive to anticancer agents, including *m*-AMSA, that inhibit mammalian type II topoisomerases [114]. For both enzymes, the drugs stabilize an otherwise transient intermediate in which the enzyme is covalently attached to DNA with a latent enzyme-induced DNA break at the site of linkage. Treatment of phage T4 infections with *m*-AMSA thereby leads to replication fork blockage at the sites of topoisomerase action [110]. Interestingly, the blocked replication fork does not immediately resume synthesis when the topoisomerase dissociates from its site of action (and reseals the latent DNA break in the process). This result strongly suggests that key components of the replisome had been disassembled upon fork blockage, so that a fork restart pathway must be used to resume DNA replication.

Mutations in genes *46/47*, *59*, *uvsX*, *uvsY*, and *uvsW* each lead to hypersensitivity to *m*-AMSA, arguing that the RDR pathway or some close variant is required to survive damage caused by *m*-AMSA [115,116]. Consistent with this model, continued replication of an origin-containing plasmid in the presence of the drug (but not in its absence) was shown to be inhibited in a *46 uvsX* double knockout mutant [110]. The simplest interpretation is that the T4 RDR system allows the restart of replication after the fork blockage event. One plausible scenario is that the blocked replication forks are especially prone to cleavage, for example by a recombination nuclease such as endonuclease VII (gp49), and that the RDR pathway provides a mechanism to restart the broken forks. Evidence supporting this view was obtained when it was found that endonuclease VII can indeed cleave blocked replication forks *in vitro*, and that blocked forks accumulate to a higher level during infections with a gene *49* knockout mutant [117]. These latter results led the authors to propose a "collateral damage" model, in which cytotoxic DNA damage from these anticancer agents results from endonuclease-mediated cleavage of stalled replication forks. It should be noted that the processing of forks stalled by HU and *m*-AMSA must differ significantly, because inactivation of *uvsX* or *uvsY* causes hypersensitivity to *m*-AMSA but not to HU, suggesting that only *m*-AMSA leads to high levels of broken forks.

In an attempt to further study the restart pathway(s) of blocked replication forks in T4 infections, Long and Kreuzer [118,119] analyzed the fork-shaped intermediates ("origin forks") that accumulate at *oriG* after one replication fork has left the origin region. Novel intermediates were detected by two-dimensional gel electrophoresis at a relatively low abundance in wild-type infections, and these were ascribed to replication fork regression [118]. Replication fork regression is a process in which the two newly synthesized strands of a replication fork are unwound from their complementary partners and rewound together, backing up (regressing) the location of the fork along the DNA. Many years ago, Higgins et al. [120] proposed this general model as a step in the accurate replication of damaged DNA in mammalian cells (Figure 4).

What is the significance of the fork regression at *oriG*? A clue was uncovered when the amount of regressed fork was found to be substantially increased when either gp46/47 or gp49 (endonuclease VII) was mutationally inactivated [118]. The authors therefore proposed that gp46/47 normally processes the extruded duplex of the regressed fork, and that endonuclease VII can cleave the regressed fork (which resembles a Holliday junction). Either of these steps could initiate a fork restart pathway, and it is possible that they normally function together as a single fork reactivation pathway (see [118]). One possible model for the fork restart pathway is that the extruded duplex in the regressed fork undergoes a strand invasion reaction ahead of the position of the fork, and thereby initiates replication by an RDR reaction (also see above).

In a subsequent study, the UvsW helicase was shown to be required for detection of the regressed forks *in*

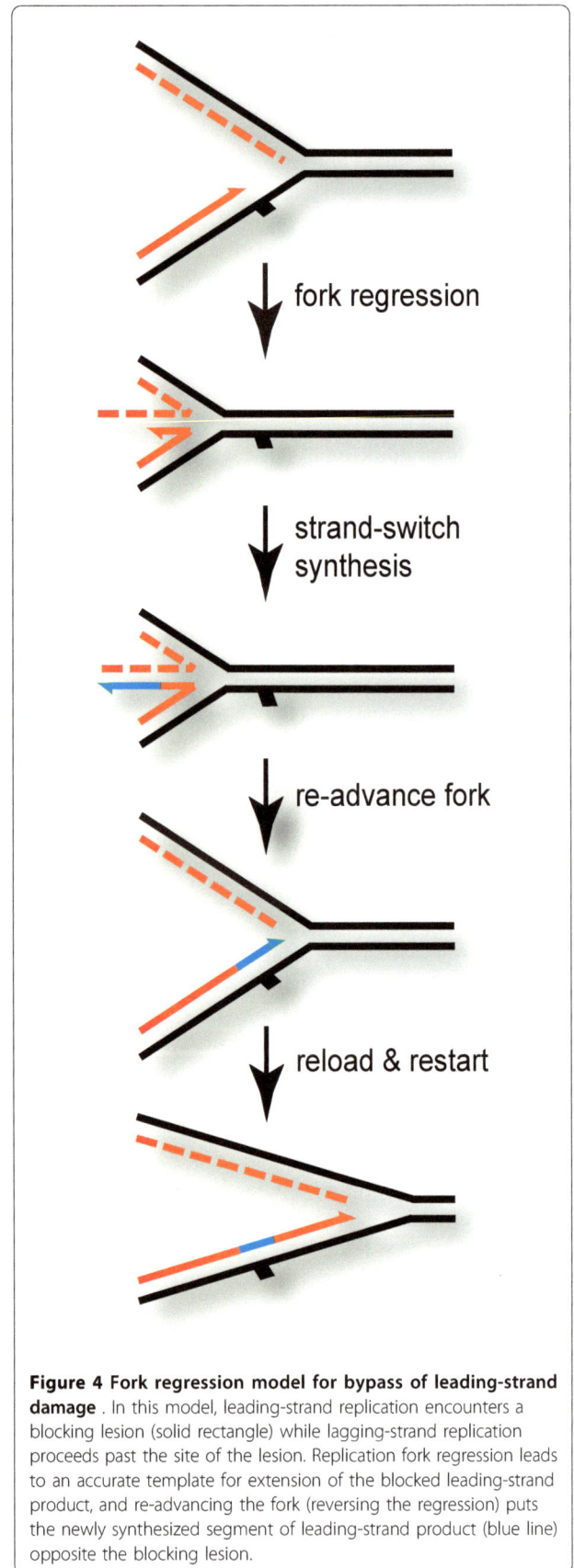

Figure 4 Fork regression model for bypass of leading-strand damage . In this model, leading-strand replication encounters a blocking lesion (solid rectangle) while lagging-strand replication proceeds past the site of the lesion. Replication fork regression leads to an accurate template for extension of the blocked leading-strand product, and re-advancing the fork (reversing the regression) puts the newly synthesized segment of leading-strand product (blue line) opposite the blocking lesion.

vivo, and also that purified UvsW helicase can catalyze fork regression *in vitro* with forked DNA isolated from a T4 infection [119]. These results strongly suggest that fork regression is an active process that contributes to survival after DNA damage, since *uvsW* knockout mutants are hypersensitive to DNA damaging agents [71,113,116].

Is fork regression required for restart of stalled forks (or other fork-shaped structures) in a T4 infection? The above experiments do not provide a clear answer to this question, because the regressed fork intermediates were only a relatively small subset of the blocked forks and because some of the RDR proteins could themselves be involved in loading a replisome onto a simple fork structure. Indeed, gp59 can bind to simple fork structures *in vitro* and load the gp41 helicase (see above), supporting the possibility of a simple direct loading pathway. Perhaps multiple processing pathways compete for access to blocked/stalled forks in T4, and the different pathways have unique capabilities to resolve different kinds of problems (e.g., different forms of DNA damage).

Replication of damaged DNA by strand switching

The general involvement of recombination in the successful replication of damaged DNA was first uncovered in the pioneering experiments of Luria [121], who discovered the phenomenon of multiplicity reactivation (MR). In MR, co-infection with multiple phages, each of which has extensive DNA damage, results in viable progeny, while single infections with the same phage particles result in no burst. Subsequent studies clearly showed the involvement of both replication and recombination functions in MR (reviewed by [122,123]). A favored model to explain MR involves DNA polymerase strand switching upon encounter with DNA damage, but the molecular details of MR have yet to be elucidated (see [123] for discussion of this and other models).

While further experiments are needed to test the strand-switching model for MR, studies over the last 15 years have provided direct evidence for strand switching in the T4 system. Strand switching may also play important roles in the process of post-replication recombination repair (PRRR) and a pathway called replication repair (see [123]).

Strand switching events can promote the accurate replication of damaged DNA when the second template is a *bona fide* homolog, either from the opposite daughter duplex behind a replication fork or from another homologous DNA molecule. An *in vitro* model for this process was established by Kadyrov and Drake [98,124], who engineered replication-fork like substrates with a blocking lesion in the leading strand and a pre-existing

lagging strand product that extended past the site of blockage. They were able to demonstrate that the leading-strand product can be extended past the site of blockage by a strand switching event that allows extension using the longer lagging-strand product as template. The simplest way to model the strand switching event is by replication fork regression, followed by polymerase extension on the extruded duplex [98] (Figure 4). To complete the error-free bypass of the DNA damage, a second strand switching event is needed, and this can occur by reversal of the fork regression process. This event was also detected in the studies of Kadyrov and Drake [98].

The *in vitro* strand switching analyzed by Kadyrov and Drake had several properties that resemble the replication of damaged DNA during T4 infections. Certain alleles of gene *32* and *41* compromise a process called replication repair *in vivo*, and these same alleles greatly reduced the strand switching process *in vitro* [124]. Furthermore, the UvsX recombinase is centrally important in survival after DNA damage *in vivo*, and greatly stimulated the strand switching reaction *in vitro* [98]. The Dda helicase was also shown to stimulate the *in vitro* strand switching, but Kadyrov and Drake [98] suggested that the UvsW helicase was more likely to promote this role *in vivo* based on the phenotypes of *dda* and *uvsW* mutants. Consistent with this suggestion, the UvsW protein was subsequently shown to catalyze fork regression (see above), and very recent evidence has directly demonstrated *in vitro* strand switching promoted by UvsW [125].

Even earlier evidence for strand switching *in vitro* came from reactions in which the polymerase changed templates, presumably at inverted repeat sequences [37,126]. In this reaction, the 3' end of a newly replicated strand base pairs with a short complementary sequence that happens to be on the same strand, resulting in a replication event in which the same strand is used as both template and primer (also see [127]). This reaction is genetically aberrant and would create genome rearrangements rather than assist in the replication of damaged DNA. Indeed, Schultz et al. [128] presented evidence that a similar kind of aberrant strand switching can lead to "templated" mutations during a T4 infection. These mutations apparently arise from sequential strand switching events in which DNA polymerase copies an imperfect repeat elsewhere in the template and then returns to the correct initial location on the template. Interestingly, these templated mutations became more frequent with certain mutations in genes *32*, *41* and *uvsX*, arguing that these proteins normally help to accurately direct the template switching events, e.g. to the opposite daughter strand rather than to ectopic locations elsewhere in the genome.

Conclusions and perspectives

Phage T4 has continued to provide an important model system for studies of the mechanisms of DNA replication, recombination and repair, and in several cases has led the way in illuminating the interconnections between these processes. A major example is RDR, a process that was first studied in detail in phage T4, that was originally thought to be an odd peculiarity of the phage's life cycle, but that is now appreciated as central in the completion of cellular genomic replication and the repair of DSB's in prokaryotic and eukaryotic chromosomes (for reviews, see [1-4]). Recombination-related pathways, including RDR, strand-switching and replication fork regression, are now appreciated to be critical in the maintenance of genome stability in mammalian systems and thereby important in cancer biology. There seems to be particularly strong parallels between DNA metabolism in phage T4 and in eukaryotic mitochondrial DNA and mitochondrial plasmid DNA. In these systems, evidence has been obtained for both R-loop-mediated replication and RDR, as well as EM data showing branched concatameric DNA similar to that of intracellular replicating T4 DNA [129-135].

While we have learned much about how T4 initiates replication at both origins and from recombination structures, many important questions remain to be answered. We will close with a few of the most interesting, which suggest that important principles and lessons remain to be uncovered using the T4 model system:

(i) What are the rules governing R-loop formation at *oriF* and *oriG*?

(ii) Do other T4 origins use an R-loop mechanism or some other initiation process?

(iii) What are the factors that govern origin usage and change the pattern of origin function?

(iv) What are the precise roles of recombination proteins, gp59 and Dda in origin usage?

(v) Why do replication forks often fail to complete replication, or move very slowly, at early times of infection?

(vi) Does T4 use a "direct restart" pathway *in vivo*, in which the replisome is loaded directly onto a fork structure?

(vii) What are the detailed roles of the gp46/47 complex, the homolog of eukaryotic Mre11/Rad50 complex?

(viii) How does replication fork regression contribute to replication fork restart?

(ix) How frequently does T4 use strand switching mechanisms *in vivo*, which proteins are required, and how is the process regulated?

Author details
[1]Department of Biochemistry, Duke University Medical Center, Durham, NC 27710 USA. [2]National Center for Biotechnology Information, National Library of Medicine, National Institutes of Health, Bethesda, MD 20894 USA.

Authors' contributions
KK wrote the first drafts of the "Introduction", the section on "Recombination-dependent replication", the section on "Interrelationship between replication, recombination and repair", and the "Conclusions and perspectives"; JRB wrote the first draft of the section on "Overlap between origin- and recombination-dependent mechanisms"; both authors contributed to the first draft of the section on "Origin-dependent replication"; both authors revised all sections and read and approved the final draft.

Competing interests
The authors declare that they have no competing interests.

Received: 17 August 2010 Accepted: 3 December 2010
Published: 3 December 2010

References
1. Aguilera A, Gomez-Gonzalez B: **Genome instability: a mechanistic view of its causes and consequences.** *Nature Reviews Genetics* 2008, **9**:204-217.
2. Barbour L, Xiao W: **Regulation of alternative replication bypass pathways at stalled replication forks and its effects on genome stability: a yeast model.** *Mutation Research* 2003, **532**:137-155.
3. Kreuzer KN: **Interplay between DNA replication and recombination in prokaryotes.** *Annual Review of Microbiology* 2005, **59**:43-67.
4. Lambert S, Froget B, Carr AM: **Arrested replication fork processing: interplay between checkpoints and recombination.** *DNA Repair (Amst)* 2007, **6**:1042-1061.
5. Brister JR, Nossal NG: **Multiple origins of replication contribute to a discontinuous pattern of DNA synthesis across the T4 genome during infection.** *Journal of Molecular Biology* 2007, **368**:336-348.
6. Kreuzer KN, Morrical SW: **Initiation of DNA replication.** In *Molecular Biology of Bacteriophage T4.* Edited by: Karam JD. Washington, DC: ASM Press; 1994:28-42.
7. Nossal NG: **The bacteriophage T4 DNA replication fork.** In *Molecular Biology of Bacteriophage T4.* Edited by: Karam JD. Washington, DC: ASM Press; 1994:43-53.
8. Halpern ME, Mattson T, Kozinski AW: **Origins of phage T4 DNA replication as revealed by hybridization to cloned genes.** *Proceedings of the National Academy of Sciences of the United States of America* 1979, **76**:6137-6141.
9. Kozinski AW, Ling SK, Hutchinson N, Halpern ME, Mattson T: **Differential amplification of specific areas of phage T4 genome as revealed by hybridization to cloned genetic segments.** *Proceedings of the National Academy of Sciences of the United States of America* 1980, **77**:5064-5068.
10. Macdonald PM, Seaby RM, Brown W, Mosig G: **Initiator DNA from a primary origin and induction of a secondary origin of bacteriophage T4 DNA replication.** In *Microbiology - 1983. Volume 1.* Edited by: Schlessinger D. Washington, D.C.: American Society for Microbiology; 1983:111-116.
11. Kreuzer KN, Alberts BM: **A defective phage system reveals bacteriophage T4 replication origins that coincide with recombination hot spots.** *Proceedings of the National Academy of Sciences of the United States of America* 1985, **82**:3345-3349.
12. Kreuzer KN, Alberts BM: **Characterization of a defective phage system for the analysis of bacteriophage T4 DNA replication origins.** *Journal of Molecular Biology* 1986, **188**:185-198.
13. Belanger KG, Kreuzer KN: **Bacteriophage T4 initiates bidirectional DNA replication through a two-step process.** *Molecular Cell* 1998, **2**:693-701.
14. Doan PL, Belanger KG, Kreuzer KN: **Two types of recombination hotspots in bacteriophage T4: one requires DNA damage and a replication origin and the other does not.** *Genetics* 2001, **157**:1077-1087.
15. Brister JR: **Origin activation requires both replicative and accessory helicases during T4 infection.** *Journal of Molecular Biology* 2008, **377**:1304-1313.
16. Vaiskunaite R, Miller A, Davenport L, Mosig G: **Two new early bacteriophage T4 genes, repEA and repEB, that are important for DNA replication initiated from origin E.** *Journal of Bacteriology* 1999, **181**:7115-7125.

17. Petrov VM, Nolan JM, Bertrand C, Levy D, Desplats C, Krisch HM, Karam JD: Plasticity of the gene functions for DNA replication in the T4-like phages. *Journal of Molecular Biology* 2006, **361**:46-68.
18. Filutowicz M, Rakowski SA: Regulatory implications of protein assemblies at the gamma origin of plasmid. *Gene* 1998, **223**:195-204.
19. Luke K, Radek A, Liu XP, Campbell J, Uzan M, Haselkorn R, Kogan Y: Microarray analysis of gene expression during bacteriophage T4 infection. *Virology* 2002, **299**:182-191.
20. Kano-Sueoka T, Lobry JR, Sueoka N: Intra-strand biases in bacteriophage T4 genome. *Gene* 1999, **238**:59-64.
21. Petrov VM, Ratnayaka S, Nolan JM, Miller ES, Karam JD: Genomes of the T4-related bacteriophages as windows on microbial genome evolution. *Virol J* 2010, **7**:292.
22. Miller ES, Kutter E, Mosig G, Arisaka F, Kunisawa T, Ruger W: Bacteriophage T4 genome. *Microbiology and Molecular Biology Reviews* 2003, **67**:86-156.
23. Riva S, Cascino A, Geiduschek EP: Coupling of late transcription to viral DNA replication in bacteriophage T4 development. *Journal of Molecular Biology* 1970, **54**:85-102.
24. Benson KH, Kreuzer KN: Plasmid models for bacteriophage T4 DNA replication: Requirements for fork proteins. *Journal of Virology* 1992, **66**:6960-6968.
25. Kreuzer KN, Engman HW, Yap WY: Tertiary initiation of replication in bacteriophage T4. Deletion of the overlapping uvsY promoter/replication origin from the phage genome. *J Biol Chem* 1988, **263**:11348-11357.
26. Menkens AE, Kreuzer KN: Deletion analysis of bacteriophage T4 tertiary origins. A promoter sequence is required for a rifampicin-resistant replication origin. *J Biol Chem* 1988, **263**:11358-11365.
27. Carles-Kinch K, Kreuzer KN: RNA-DNA hybrid formation at a bacteriophage T4 replication origin. *Journal of Molecular Biology* 1997, **266**:915-926.
28. Stitt B, Hinton D: Regulation of middle gene transcription. In *Molecular Biology of Bacteriophage T4*. Edited by: Karam J. Washington, D.C.: ASM; 1994:142-160.
29. Hinton DM: Transcriptional control in the prereplicative phase of T4 development. *Virol J* 2010, **7**:289.
30. Mosig G, Colowick NE, Gruidl ME, Chang A, Harvey A: Multiple initiation mechanisms adapt phage T4 DNA replication to physiological changes during T4's development. *FEMS Microbiological Reviews* 1995, **17**:83-98.
31. Benson KH, Kreuzer KN: Role of MotA transcription factor in bacteriophage T4 DNA replication. *Journal of Molecular Biology* 1992, **228**:88-100.
32. Hinton DM, Nossal NG: Bacteriophage T4 DNA primase-helicase. *J Biol Chem* 1987, **262**:10873-10878.
33. Liu CC, Alberts BM: Pentaribonucleotides of mixed sequence are synthesized and efficiently prime de novo DNA chain starts in the T4-bacteriophage DNA-replication system. *Proceedings of the National Academy of Sciences of the United States of America* 1980, **77**:5698-5702.
34. Nossal NG: RNA priming of DNA replication by bacteriophage T4 proteins. *J Biol Chem* 1980, **255**:2176-2182.
35. Edgar RS, Wood WB: Morphogenesis of bacteriophage T4 in extracts of mutant-infected cells. *Proceedings of the National Academy of Sciences of the United States of America* 1966, **55**:498-505.
36. Yegian CD, Mueller M, Selzer G, Russo V, Stahl FW: Properties of DNA-delay mutants of bacteriophage T4. *Virology* 1971, **46**:900-919.
37. Belanger KG, Mirzayan C, Kreuzer HE, Alberts BM, Kreuzer KN: Two-dimensional gel analysis of rolling circle replication in the presence and absence of bacteriophage T4 primase. *Nucleic Acids Research* 1996, **24**:2166-2175.
38. Mosig G, Luder A, Ernst A, Canan N: Bypass of a primase requirement for bacteriophage T4 DNA replication in vivo by a recombination enzyme, endonuclease VII. *The New Biologist* 1991, **3**:1-11.
39. Belanger KG: Origin-dependent DNA replication in bacteriophage T4. *PhD Thesis, Duke University* 1997.
40. Itoh T, Tomizawa J: Formation of an RNA primer for initiation of replication of ColE1 DNA by ribonuclease H. *Proceedings of the National Academy of Sciences of the United States of America* 1980, **77**:2450-2454.
41. Xu BJ, Clayton DA: RNA-DNA hybrid formation at the human mitochondrial heavy- strand origin ceases at replication start sites: An implication for RNA-DNA hybrids serving as primers. *EMBO Journal* 1996, **15**:3135-3143.

42. Baker TA, Sekimizu K, Funnell BE, Kornberg A: Extensive unwinding of the plasmid template during staged enzymatic initiation of DNA replication from the origin of the Escherichia coli chromosome. *Cell* 1986, **45**:53-64.
43. Van der Ende A, Baker TA, Ogawa T, Kornberg A: Initiation of enzymatic replication at the origin of the Escherichia coli chromosome - Primase as the sole priming enzyme. *Proceedings of the National Academy of Sciences of the United States of America* 1985, **82**:3954-3958.
44. Asai T, Kogoma T: D-loops and R-loops: alternative mechanisms for the initiation of chromosome replication in E. coli. *Journal of Bacteriology* 1994, **176**:1807-1812.
45. Nossal NG, Dudas KC, Kreuzer KN: Bacteriophage T4 proteins replicate plasmids with a preformed R loop at the T4 ori(uvsY) replication origin in vitro. *Molecular Cell* 2001, **7**:31-41.
46. Gauss P, Park K, Spencer TE, Hacker KJ: DNA helicase requirements for DNA replication during bacteriophage T4 infection. *Journal of Bacteriology* 1994, **176**:1667-1672.
47. Shah DB: Replication and Recombination of Gene 59 Mutant of Bacteriophage T4D. *Journal of Virology* 1976, **17**:175-182.
48. Dudas KC, Kreuzer KN: Bacteriophage T4 helicase loader protein gp59 functions as gatekeeper in origin-dependent replication in vivo. *J Biol Chem* 2005, **280**:21561-21569.
49. Barry J, Alberts B: Purification and characterization of bacteriophage T4 gene 59 protein. A DNA helicase assembly protein involved in DNA replication. *J Biol Chem* 1994, **269**:33049-33062.
50. Mueser TC, Jones CE, Nossal NG, Hyde CC: Bacteriophage T4 gene 59 helicase assembly protein binds replication fork DNA. The 1.45 A resolution crystal structure reveals a novel a-helical two-domain fold. *Journal of Molecular Biology* 2000, **296**:597-612.
51. Jones CE, Mueser TC, Dudas KC, Kreuzer KN, Nossal NG: Bacteriophage T4 gene 41 helicase and gene 59 helicase-loading protein: a versatile couple with roles in replication and recombination. *Proceedings of the National Academy of Sciences of the United States of America* 2001, **98**:8312-8318.
52. Xi J, Zhuang ZH, Zhang ZQ, Selzer T, Spiering MM, Hammes GG, Benkovic SJ: The interaction between the T4 helicase loading protein (gp59) and the DNA polymerase (gp43): a locking mechanism to delay replication during replisome assembly. *Biochemistry-Us* 2005, **44**:12264-12264.
53. Epstein RH, Bolle A, Steinberg CM, Kellenberger E, Boy de la Tour E, Chevalley R, Edgar RS, Susman M, Denhardt GH, Lielausis A: Physiological studies of conditional lethal mutants of bacteriophage T4D. *Cold Spring Harbor Symposium of Quantitative Biology* 1963, **28**:375-392.
54. Derr LK, Kreuzer KN: Expression and function of the uvsW gene of bacteriophage T4. *Journal of Molecular Biology* 1990, **214**:643-656.
55. Carles-Kinch K, George JW, Kreuzer KN: Bacteriophage T4 UvsW protein is a helicase involved in recombination, repair, and the regulation of DNA replication origins. *EMBO Journal* 1997, **16**:4142-4151.
56. Dudas KC, Kreuzer KN: UvsW protein regulates bacteriophage T4 origin-dependent replication by unwinding R-loops. *Molecular and Cellular Biology* 2001, **21**:2706-2715.
57. Nelson SW, Benkovic SJ: The T4 phage UvsW protein contains both DNA unwinding and strand annealing activities. *J Biol Chem* 2007, **282**:407-416.
58. Cunningham RP, Berger H: Mutations affecting genetic recombination in bacteriophage T4D. I. Pathway analysis. *Virology* 1977, **80**:67-82.
59. Wu JR, Yeh YC: New late gene, dar, involved in DNA replication of bacteriophage T4 I. Isolation, characterization, and genetic location. *Journal of Virology* 1975, **15**:1096-1106.
60. Wu JR, Yeh YC, Ebisuzaki K: Genetic analysis of dar, uvsW, and uvsY in bacteriophage T4: dar and uvsW are alleles. *Journal of Virology* 1984, **52**:1028-1031.
61. Yonesaki T, Minagawa T: Studies on the recombination genes of bacteriophage T4: Suppression of uvsX and uvsY mutations by uvsW mutations. *Genetics* 1987, **115**:219-227.
62. Luder A, Mosig G: Two alternative mechanisms for initiation of DNA replication forks in bacteriophage T4: Priming by RNA polymerase and by recombination. *Proceedings of the National Academy of Sciences of the United States of America* 1982, **79**:1101-1105.
63. Mosig G: Relationship of T4 DNA replication and recombination. In *Bacteriophage T4. Volume 1.* Edited by: Mathews CK, Kutter EM, Mosig G, Berget PB. Washington, D.C.: American Society for Microbiology; 1983:120-130.

64. Doermann AH, Boehner L: **An experimental analysis of bacteriophage T4 heterozygotes.1. Mottled plaques from crosses involving six rII loci.** *Virology* 1963, **21**:551-567.

65. Mosig G: **Genetic recombination in bacteriophage T4 during replication of DNA fragments.** *Cold Spring Harbor Symposium of Quantitative Biology* 1963, **28**:35-41.

66. Womack FC: **An analysis of single-burst progeny of bacteria singly infected with a bacteriophage heterozygote.** *Virol* 1963, **21**:232-241.

67. Kreuzer KN: **Recombination-dependent DNA replication in phage T4.** *Trends in Biochemical Sciences* 2000, **25**:165-173.

68. Liu J, Morrical SW: **Assembly and dynamics of the bacteriophage T4 homologous recombination machinery.** *Virol J* 2010, **7**:357.

69. Mosig G: **Homologous recombination.** In *Molecular Biology of Bacteriophage T4.* Edited by: Karam JD. Washington, DC: ASM Press; 1994:54-82.

70. Tomso DJ, Kreuzer KN: **Double strand break repair in tandem repeats during bacteriophage T4 infection.** *Genetics* 2000, **155**:1493-1504.

71. Hamlett NV, Berger H: **Mutations altering genetic recombination and repair of DNA in bacteriophage T4.** *Virology* 1975, **63**:539-567.

72. Jensch F, Kemper B: **Endonuclease VII resolves Y-junctions in branched DNA in vitro.** *EMBO Journal* 1986, **5**:181-189.

73. Mizuuchi K, Kemper B, Hays J, Weisberg RA: **T4 endonuclease VII cleaves Holliday structures.** *Cell* 1982, **29**:357-365.

74. Frankel FR, Batchele ML, Clark CK: **Role of gene 49 in DNA replication and head morphogenesis in bacteriophage T4.** *Journal of Molecular Biology* 1971, **62**:439-463.

75. Golz S, Kemper B: **Association of Holliday-structure resolving endonuclease VII with gp20 from the packaging machine of phage T4.** *Journal of Molecular Biology* 1999, **285**:1131-1144.

76. Kreuzer KN, Saunders M, Weislo LJ, Kreuzer HWE: **Recombination-dependent DNA replication stimulated by double-strand breaks in bacteriophage T4.** *Journal of Bacteriology* 1995, **177**:6844-6853.

77. Kreuzer KN, Yap WY, Menkens AE, Engman HW: **Recombination-dependent replication of plasmids during bacteriophage T4 infection.** *J Biol Chem* 1988, **263**:11366-11373.

78. Mattson T, Van Houwe G, Bolle A, Epstein R: **Fate of cloned bacteriophage T4 DNA after phage T4 infection of clone-bearing cells.** *Journal of Molecular Biology* 1983, **169**:343-355.

79. George JW, Kreuzer KN: **Repair of double-strand breaks in bacteriophage T4 by a mechanism that involves extensive DNA replication.** *Genetics* 1996, **143**:1507-1520.

80. Clyman J, Quirk S, Belfort M: **Mobile introns in the T-even phages.** In *Molecular Biology of Bacteriophage T4.* Edited by: Karam JD. Washington, DC: ASM Press; 1994:83-88.

81. Edgell DR, Gibb EA, Belfort M: **Mobile DNA elements in T4 and related phages.** *Virol J* 2010, **7**:290.

82. Bell-Pedersen D, Quirk SM, Aubrey M, Belfort M: **A site-specific endonuclease and co-conversion of flanking exons associated with the mobile td intron of phage T4.** *Gene* 1989, **82**:119-126.

83. Clyman J, Belfort M: **Trans and cis requirements for intron mobility in a prokaryotic system.** *Genes & Development* 1992, **6**:1269-1279.

84. Mueller JE, Clyman T, Huang YJ, Parker MM, Belfort M: **Intron mobility in phage T4 occurs in the context of recombination-dependent DNA replication by way of multiple pathways.** *Genes & Development* 1996, **10**:351-364.

85. Shcherbakov V, Granovsky I, Plugina L, Shcherbakova T, Sizova S, Pyatkov K, Shlyapnikov M, Shubina O: **Focused genetic recombination of bacteriophage T4 initiated by double-strand breaks.** *Genetics* 2002, **162**:543-556.

86. Stohr BA, Kreuzer KN: **Coordination of DNA ends during double-strand-break repair in bacteriophage T4.** *Genetics* 2002, **162**:1019-1030.

87. George JW, Stohr BA, Tomso DJ, Kreuzer KN: **The tight linkage between DNA replication and double-strand break repair in bacteriophage T4.** *Proceedings of the National Academy of Sciences of the United States of America* 2001, **98**:8290-8297.

88. Gajewski S, Webb MR, Galkin V, Egelman EH, Kreuzer KN, White SW: **Crystal structure of the phage T4 recombinase UvsX and its functional interaction with the T4 SF2 helicase UvsW.** *J Mol Biol* .

89. Formosa T, Alberts BM: **Purification and characterization of the T4 bacteriophage uvsX protein.** *J Biol Chem* 1986, **261**:6107-6118.

90. Formosa T, Alberts BM: **DNA synthesis dependent on genetic recombination: Characterization of a reaction catalyzed by purified T4 proteins.** *Cell* 1986, **47**:793-806.

91. Karam JD: *Molecular Biology of Bacteriophage T4* Washington: ASM Press; 1994.

92. Barry J, Alberts B: **A role for two DNA helicases in the replication of T4 bacteriophage DNA.** *J Biol Chem* 1994, **269**:33063-33068.

93. Jones CE, Mueser TC, Nossal NG: **Bacteriophage T4 32 protein is required for helicase-dependent leading strand synthesis when the helicase is loaded by the T4 59 helicase-loading protein.** *J Biol Chem* 2004, **279**:12067-12075.

94. Jones CF, Mueser TC, Nossal NG: **Interaction of the bacteriophage T4 gene 59 helicase loading protein and gene 41 helicase with each other, and with fork, flap, and cruciform DNA.** *J Biol Chem* 2000, **275**:27145-27154.

95. Ma Y, Wang T, Villemain JL, Giedroc DP, Morrical SW: **Dual functions of single-stranded DNA-binding protein in helicase loading at the bacteriophage T4 DNA replication fork.** *J Biol Chem* 2004, **279**:19035-19045.

96. Morrical SW, Beernink HTH, Dash A, Hempstead K: **The gene 59 protein of bacteriophage T4. Characterization of protein-protein interactions with gene 32 protein, the T4 single-stranded DNA binding protein.** *J Biol Chem* 1996, **271**:20198-20207.

97. Nelson SW, Yang JS, Benkovic SJ: **Site-directed mutations of T4 helicase loading protein (gp59) reveal multiple modes of DNA polymerase inhibition and the mechanism of unlocking by gp41 helicase.** *J Biol Chem* 2006, **281**:8697-8706.

98. Kadyrov FA, Drake JW: **UvsX recombinase and Dda helicase rescue stalled bacteriophage T4 DNA replication forks in vitro.** *J Biol Chem* 2004, **279**:35735-35740.

99. Liu B, Alberts BM: **Head-on collision between a DNA replication apparatus and RNA polymerase transcription complex.** *Science* 1995, **267**:1131-1137.

100. Naot Y, Shalitin C: **Defective concatemer formation in cells infected with deoxyribonucleic acid-delay mutants of bacteriophage T4.** *Journal of Virology* 1972, **10**:858-862.

101. Belfort M, Perlman PS: **Mechanisms of intron mobility.** *J Biol Chem* 1995, **270**:30237-30240.

102. Huang YJ, Parker MM, Belfort M: **Role of exonucleolytic degradation in group I intron homing in phage T4.** *Genetics* 1999, **153**:1501-1512.

103. Mueller JE, Smith D, Belfort M: **Exon coconversion biases accompanying intron homing: Battle of the nucleases.** *Genes & Development* 1996, **10**:2158-2166.

104. McVey M, Adams M, Staeva-Vieira E, Sekelsky JJ: **Evidence for multiple cycles of strand invasion during repair of double-strand gaps in Drosophila.** *Genetics* 2004, **167**:699-705.

105. Shinohara A, Ogawa T: **Homologous recombination and the roles of double-strand breaks.** *Trends in Biochemical Sciences* 1995, **20**:387-391.

106. Shcherbakov VP, Plugina L, Shcherbakova T, Sizova S, Kudryashova E: **Double-strand break repair in bacteriophage T4: Coordination of DNA ends and effects of mutations in recombinational genes.** *DNA Repair* 2006, **5**:773-787.

107. Connelly JC, Leach DR: **Tethering on the brink: the evolutionarily conserved Mre11-Rad50 complex.** *Trends in Biochemical Sciences* 2002, **27**:410-418.

108. Hopfner KP, Craig L, Moncalian G, Zinkel RA, Usui T, Owen BAL, Karcher A, Henderson B, Bodmer JL, McMurray CT, *et al*: **The Rad50 zinc-hook is a structure joining Mre11 complexes in DNA recombination and repair.** *Nature* 2002, **418**:562-566.

109. Warner HR, Hobbs MD: **Effect of hydroxyurea on replication of bacteriophage T4 in Escherichia coli.** *Journal of Virology* 1969, **3**:331-336.

110. Hong G, Kreuzer KN: **An antitumor drug-induced topoisomerase cleavage complex blocks a bacteriophage T4 replication fork in vivo.** *Molecular and Cellular Biology* 2000, **20**:594-603.

111. Hercules K: **Mutants in a nonessential gene of bacteriophage T4 which are defective in degradation of Escherichia coli deoxyribonucleic acid.** *Journal of Virology* 1971, **7**:95-105.

112. Warner HR, Snustad DP, Jorgensen SE, Koerner JF: **Isolation of bacteriophage T4 mutants defective in the ability to degrade host deoxyribonucleic acid.** *Journal of Virology* 1970, **5**:700-708.

113. Derr LK, Drake JW: **Isolation and genetic characterization of new uvsW alleles of bacteriophage T4.** *Molecular and General Genetics* 1990, **222**:257-264.
114. Kreuzer KN: **Bacteriophage T4, a model system for understanding the mechanism of type II topoisomerase inhibitors.** *Biochimica et Biophysica Acta: Gene Structure and Expression* 1998, **1400**:339-347.
115. Neece SH, Carles-Kinch K, Tomso DJ, Kreuzer KN: **Role of recombinational repair in sensitivity to an antitumor agent that inhibits bacteriophage T4 type II DNA topoisomerase.** *Molecular Microbiology* 1996, **20**:1145-1154.
116. Woodworth DL, Kreuzer KN: **Bacteriophage T4 mutants hypersensitive to an antitumor agent that induces topoisomerase-DNA cleavage complexes.** *Genetics* 1996, **143**:1081-1090.
117. Hong G, Kreuzer KN: **Endonuclease cleavage of blocked replication forks: An indirect pathway of DNA damage from antitumor drug-topoisomerase complexes.** *Proceedings of the National Academy of Sciences of the United States of America* 2003, **100**:5046-5051.
118. Long DT, Kreuzer KN: **Regression supports two mechanisms of fork processing in phage T4.** *Proceedings of the National Academy of Sciences of the United States of America* 2008, **105**:6852-6857.
119. Long DT, Kreuzer KN: **Fork regression is an active helicase-driven pathway in bacteriophage T4.** *EMBO Reports* 2009, **10**:394-399.
120. Higgins NP, Kato K, Strauss B: **A model for replication repair in mammalian cells.** *Journal of Molecular Biology* 1976, **101**:417-425.
121. Luria S: **Reactivation of irradiated bacteriophage by transfer of self-reproducing units.** *Proceedings of the National Academy of Sciences of the United States of America* 1947, **33**:253-264.
122. Bernstein C, Wallace SS: **DNA repair.** In *Bacteriophage T4. Volume 1.* Edited by: Mathews CK, Kutter EM, Mosig G, Berget PB. Washington, D.C.: American Society for Microbiology; 1983:138-151.
123. Kreuzer KN, Drake JW: **Repair of lethal DNA damage.** In *Molecular Biology of Bacteriophage T4.* Edited by: Karam JD. Washington, DC: ASM Press; 1994:89-97.
124. Kadyrov FA, Drake JW: **Properties of bacteriophage T4 proteins deficient in replication repair.** *J Biol Chem* 2003, **278**:25247-25255.
125. Nelson SW, Benkovic SJ: **Response of the bacteriophage T4 replisome to noncoding lesions and regression of a stalled replication fork.** *J Mol Biol* 2010, **401**:743-756.
126. Morrical SW, Wong ML, Alberts BM: **Amplification of snap-back DNA synthesis reactions by the uvsX recombinase of bacteriophage T4.** *J Biol Chem* 1991, **266**:14031-14038.
127. Englund PT: **The initial step of in vitro synthesis of deoxyribonucleic acid by T4 deoxyribonucleic acid polymerase.** *J Biol Chem* 1971, **246**:5684-5687.
128. Schultz GE, Carver GT, Drake JW: **A role for replication repair in the genesis of templated mutations.** *Journal of Molecular Biology* 2006, **358**:963-973.
129. Backert S, Borner T: **Phage T4-like intermediates of DNA replication and recombination in the mitochondria of the higher plant Chenopodium album (L.).** *Current Genetics* 2000, **37**:304-314.
130. Backert S, Dorfel P, Lurz R, Borner T: **Rolling-circle replication of mitochondrial DNA in the higher plant Chenopodium album (L).** *Molecular and Cellular Biology* 1996, **16**:6285-6294.
131. Bendich AJ: **Reaching for the ring: the study of mitochondrial genome structure.** *Current Genetics* 1993, **24**:279-290.
132. Lee DY, Clayton DA: **Initiation of mitochondrial DNA replication by transcription and R-loop processing.** *J Biol Chem* 1998, **273**:30614-30621.
133. Preiser PR, Wilson RJM, Moore PW, McCready S, Hajibagheri MAN, Blight KJ, Strath M, Williamson DH: **Recombination associated with replication of malarial mitochondrial DNA.** *EMBO Journal* 1996, **15**:684-693.
134. Shadel GS, Clayton DA: **Mitochondrial DNA maintenance in vertebrates.** *Annual Review of Biochemistry* 1997, **66**:409-435.
135. Gerhold JM, Aun A, Sedman T, Joers P, Sedman J: **Strand invasion structures in the inverted repeat of *Candida albicans* mitochondrial DNA reveal a role for homologous recombination in replication.** *Molecular Cell* 2010, **39**:851-861.

doi:10.1186/1743-422X-7-358
Cite this article as: Kreuzer and Brister: Initiation of bacteriophage T4 DNA replication and replication fork dynamics: a review in the Virology Journal series on bacteriophage T4 and its relatives. *Virology Journal* 2010 **7**:358.

Liu and Morrical *Virology Journal* 2010, **7**:357
http://www.virologyj.com/content/7/1/357

VIROLOGY JOURNAL

REVIEW   Open Access

Assembly and dynamics of the bacteriophage T4 homologous recombination machinery

Jie Liu[1], Scott W Morrical[2*]

Abstract

Homologous recombination (HR), a process involving the physical exchange of strands between homologous or nearly homologous DNA molecules, is critical for maintaining the genetic diversity and genome stability of species. Bacteriophage T4 is one of the classic systems for studies of homologous recombination. T4 uses HR for high-frequency genetic exchanges, for homology-directed DNA repair (HDR) processes including DNA double-strand break repair, and for the initiation of DNA replication (RDR). T4 recombination proteins are expressed at high levels during T4 infection in *E. coli*, and share strong sequence, structural, and/or functional conservation with their counterparts in cellular organisms. Biochemical studies of T4 recombination have provided key insights on DNA strand exchange mechanisms, on the structure and function of recombination proteins, and on the coordination of recombination and DNA synthesis activities during RDR and HDR. Recent years have seen the development of detailed biochemical models for the assembly and dynamics of presynaptic filaments in the T4 recombination system, for the atomic structure of T4 UvsX recombinase, and for the roles of DNA helicases in T4 recombination. The goal of this chapter is to review these recent advances and their implications for HR and HDR mechanisms in all organisms.

Introduction

Homologous recombination (HR) is a conserved biological process in which DNA strands are physically exchanged between DNA molecules of identical or nearly identical sequence (Figure 1). The DNA strand exchange mechanism in HR allows gene conversion events to occur, which is important for maintaining genetic diversity within populations of organisms. The DNA strand exchange mechanism in HR is also essential for the high-fidelity repair of DNA double-strand breaks (DSBs) and daughter-strand gaps, which is important for maintaining genome stability [1-3]. These homology-directed DNA repair (HDR) processes require the coordination of activities between HR and DNA replication machineries.

Homologous recombination in bacteriophage T4

The bacteriophage T4 recombination system provides an important model for understanding recombination transactions including DNA strand exchange, recombination-

dependent replication (RDR), and homology-directed DNA repair (HDR) [4-6]. The relatively simple, but functionally conserved, core recombination machinery of T4 facilitates detailed mechanistic studies of DNA strand exchange reactions and intermediates. The T4 paradigm for presynaptic filament assembly is widely used as a basis for studying presynaptic filaments in many cellular organisms including humans. At the same time, because of the close linkages between its DNA recombination, replication, and repair pathways, bacteriophage T4 has yielded novel insights on the cross-talk that occurs between recombination and replication proteins. This is especially true in the case of T4 DNA helicases, which are seen as critical for the channeling of recombination intermediates into RDR and HDR pathways.

Single-stranded DNA and presynaptic filaments

The generation of single-stranded DNA is a common early step of HR pathways [7,8]. ssDNA production typically occurs as a result of nucleolytic resection of DSBs (Figure 1), or due to replication fork stalling or collapse. In T4 recombination, exonuclease activities of a Gp46/Gp47 complex (orthologous to eukaryotic Mre11/Rad50) appear to be critical for DSB resection [9].

* Correspondence: smorrica@uvm.edu
[2]Department of Biochemistry, University of Vermont College of Medicine, Burlington, VT 05405 USA
Full list of author information is available at the end of the article

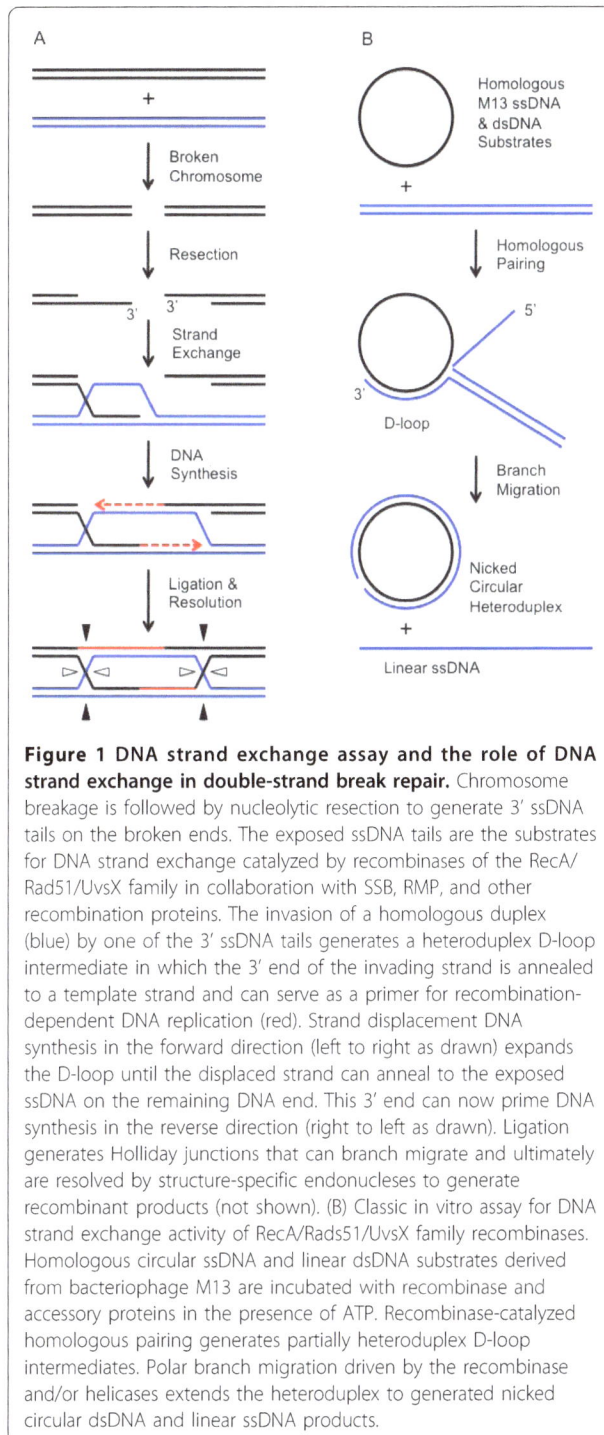

Figure 1 DNA strand exchange assay and the role of DNA strand exchange in double-strand break repair. Chromosome breakage is followed by nucleolytic resection to generate 3' ssDNA tails on the broken ends. The exposed ssDNA tails are the substrates for DNA strand exchange catalyzed by recombinases of the RecA/Rad51/UvsX family in collaboration with SSB, RMP, and other recombination proteins. The invasion of a homologous duplex (blue) by one of the 3' ssDNA tails generates a heteroduplex D-loop intermediate in which the 3' end of the invading strand is annealed to a template strand and can serve as a primer for recombination-dependent DNA replication (red). Strand displacement DNA synthesis in the forward direction (left to right as drawn) expands the D-loop until the displaced strand can anneal to the exposed ssDNA on the remaining DNA end. This 3' end can now prime DNA synthesis in the reverse direction (right to left as drawn). Ligation generates Holliday junctions that can branch migrate and ultimately are resolved by structure-specific endonucleases to generate recombinant products (not shown). (B) Classic in vitro assay for DNA strand exchange activity of RecA/Rads51/UvsX family recombinases. Homologous circular ssDNA and linear dsDNA substrates derived from bacteriophage M13 are incubated with recombinase and accessory proteins in the presence of ATP. Recombinase-catalyzed homologous pairing generates partially heteroduplex D-loop intermediates. Polar branch migration driven by the recombinase and/or helicases extends the heteroduplex to generated nicked circular dsDNA and linear ssDNA products.

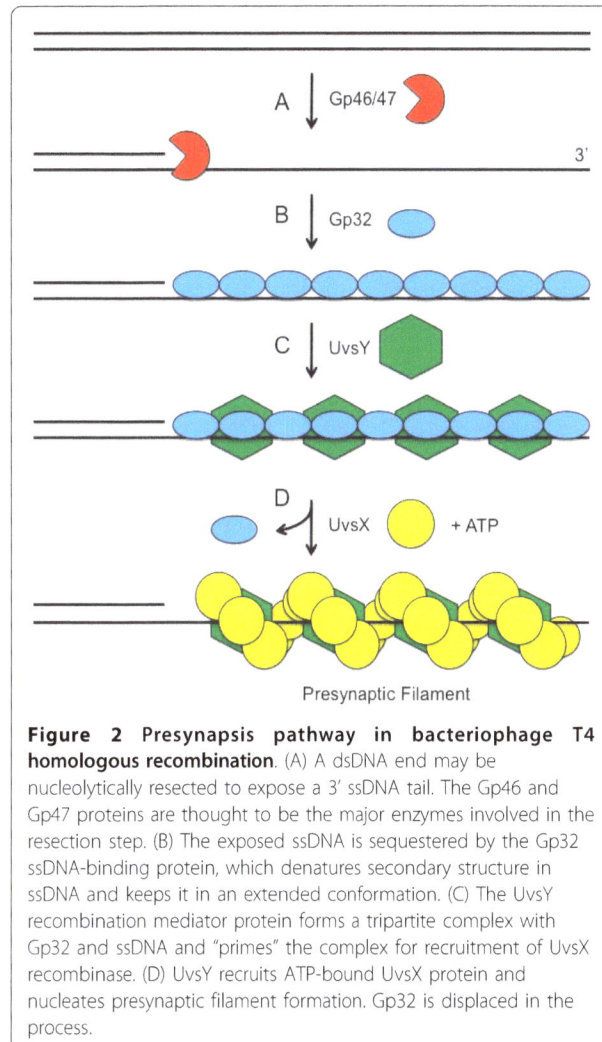

Figure 2 Presynapsis pathway in bacteriophage T4 homologous recombination. (A) A dsDNA end may be nucleolytically resected to expose a 3' ssDNA tail. The Gp46 and Gp47 proteins are thought to be the major enzymes involved in the resection step. (B) The exposed ssDNA is sequestered by the Gp32 ssDNA-binding protein, which denatures secondary structure in ssDNA and keeps it in an extended conformation. (C) The UvsY recombination mediator protein forms a tripartite complex with Gp32 and ssDNA and "primes" the complex for recruitment of UvsX recombinase. (D) UvsY recruits ATP-bound UvsX protein and nucleates presynaptic filament formation. Gp32 is displaced in the process.

In addition to DNA damage-linked production of ssDNA, bacteriophage T4 routinely generates ssDNA during replication of its linear chromosome ends. The production of ssDNA tails or gaps in otherwise duplex DNA allows the assembly of core recombination machinery including *presynaptic filaments* on ssDNA. Presynaptic filaments are helical nucleoprotein filaments

consisting of a recombinase enzyme and its accessory proteins bound cooperatively to ssDNA (Figure 2). Presynaptic filament assembly activates the enzymatic activities of the recombinase including ATPase and DNA strand exchange activities. Filament dynamics controls DNA strand exchange and its coupling to the downstream, replicative steps of HDR. These processes require the timely assembly of presynaptic filaments on recombinagenic ssDNA. Equally important is the coordinated disassembly or translocation of filaments, which appears to be required to make way for the assembly of replication enzymes on recombination intermediates [10,11].

The transition from recombination to DNA replication and repair

The transition from recombination intermediate to replication fork occurs very efficiently in bacteriophage T4, which has evolved to use this as a major mode of DNA replication initiation. The transition likely involves not only the built-in dynamics of the presynaptic filament but also the coordinated activities of DNA helicases. In the following sections of this chapter, we will review what is known about presynaptic filament dynamics in the T4 system, as well as what is known about the influences of DNA helicases on recombination, and how these two ATP-driven machines may cooperate with each other to successfully couple HR to recombination-dependent replication and repair.

Properties of the T4 Core Recombination Machinery

Although relatively simple, the core activities of the T4 recombination system are highly conserved. Three core protein components are required for T4 presynaptic filament assembly and for DNA strand exchange under physiological conditions: UvsX, the phage recombinase (orthologous to bacterial RecA and eukaryotic Rad51); Gp32, the phage ssDNA-binding protein (equivalent to bacterial SSB and eukaryotic RPA); and UvsY, the phage recombination mediator protein (equivalent to bacterial RecOR, eukaryotic Rad52, Brca2, and others) [4,5]. The DNA binding properties of UvsX, Gp32, and UvsY are presented below in context with their physical and enzymatic properties.

UvsX recombinase

UvsX protein (44 kDa) is a member of the RecA/Rad51 recombinase family and shares 28% sequence identity and 51% sequence similarity with the catalytic core domain of *E. coli* RecA [12]. UvsX catalyzes DNA strand exchange reactions that play central roles in T4 HR, RDR, and HDR pathways [4,6]. UvsX binds sequence-non-specifically to both ssDNA and dsDNA and can bind to both lattices simultaneously via two different binding sites (Maher, R.L. and S.W. Morrical: Coordinated binding of ssDNA and dsDNA substrates by UvsX recombinase and its regulation by ATP, unpublished). UvsX has higher affinity for dsDNA in the absence of other factors, but simultaneous ssDNA binding lowers UvsX-dsDNA binding affinity unless the duplex sequence is homologous to the bound ssDNA (Maher, R.L. and S.W. Morrical: Coordinated binding of ssDNA and dsDNA substrates by UvsX recombinase and its regulation by ATP, unpublished). At the same time, UvsX-ssDNA interactions are selectively stabilized by nucleoside triphosphates ATP, dATP, or their non-hydrolyzable analogs, and by UvsY protein [13,14]. These combined factors help to target UvsX filament assembly onto recombinagenic ssDNA even in the presence of excess dsDNA as would normally be found in the T4-infected cell. Binding of UvsX to ssDNA, not dsDNA, specifically activates catalysis by UvsX including ATPase and DNA strand exchange activities.

Quantitative binding studies established the intrinsic ssDNA-binding parameters of UvsX [13]. Its average binding site size on ssDNA is 4 nucleotide residues per protomer. UvsX exhibits moderate affinity and cooperativity for ssDNA with $K_{obs} = K\omega \approx 10^6$ M^{-1} at physiological ionic strength, where the cooperativity parameter $\omega \approx 100$ [13]. The observed cooperativity of UvsX is consistent with the formation of long filaments on ssDNA at high binding density.

The ATPase activity of UvsX is strongly ssDNA-dependent under normal solution conditions [15], although very high salt concentrations can also stimulate ATP hydrolysis by UvsX in the absence of ssDNA. Double-stranded DNA does not activate UvsX ATPase activity. UvsX ATPase activity is also highly unusual in that it generates both ADP and AMP as products [15,16]. The two products appear to be generated independently by two different classes of active sites within UvsX-ssDNA presynaptic filaments, as indicated by results of steady-state kinetics studies [16]. These sites have different K_m and k_{cat}/K_m values for the ATP and ssDNA substrates. One type of active site appears to produce ADP exclusively, while the other appears to generate AMP via a sequential mechanism (ATP → ADP → AMP) without releasing the ADP intermediate from the active site [16]. Thus UvsX presynaptic filaments exhibit active site asymmetry (Figure 2). This asymmetry may be important for UvsX-catalyzed DNA strand exchange reactions, since increases in ADP/AMP product ratio observed in UvsX site-directed mutants correlate inversely with strand exchange activity [16]. Active site asymmetry may be a general property of presynaptic filaments in many species, since evidence exists for two classes of active sites in filaments of *E. coli* RecA and *S. cerevisiae* Rad51 recombinases [17,18].

UvsX-ssDNA filaments rapidly search for homology in dsDNA substrates, leading to efficient homologous pairing and strand exchange. ATP binding (not hydrolysis) is required for homologous pairing, however ATP hydrolysis is needed to drive extensive polar (5' → 3') branch migration during strand exchange [19-21]. There is a strong requirement for Gp32 to stimulate UvsX-catalyzed strand exchange at normal concentrations of the recombinase [15,22,23]. In vitro, this Gp32 requirement can be circumvented by raising the UvsX concentration to super-saturating levels with respect to ssDNA binding sites. Stimulation of strand exchange by Gp32 requires the correct order of protein addition: Adding Gp32 to ssDNA prior to the addition of UvsX typically inhibits strand exchange. This ssDNA-binding protein/recombinase order of addition effect is a characteristic of all well-characterized recombination systems [24], and is reflective of the competition between the two proteins for binding sites on ssDNA. Similar inhibition of

UvsX-catalyzed strand exchange is seen at high concentrations of Gp32 and/or at elevated salt concentrations, i.e. conditions that favor Gp32-ssDNA over UvsX-ssDNA interactions. Under conditions such as these there is an absolute requirement for the UvsY recombination mediator protein for strand exchange reactions in vitro [23,25]. This mimics the in vivo situation in which T4 recombination transactions are equally dependent on UvsX and UvsY [26-28].

Branched networks of single- and double-stranded DNA are the major products of UvsX-catalyzed DNA strand exchange, indicating that each DNA substrate molecule participates in many homologous pairing events [15,29]. One plausible explanation for this behavior is that UvsX appears to catalyze homologous pairing much more rapidly than branch migration. Therefore it is possible for different regions of one long ssDNA substrate to pair with homologous regions of different dsDNA substrates before any of the resulting D-loop intermediates can be completely extended into heteroduplex DNA. Rapid homologous pairing by UvsX may be an evolutionary adaptation for efficiently capturing 3' ssDNA tails and using them to prime recombination-dependent replication. Furthermore, branch migration appears to be dependent on T4-encoded DNA helicases, as we discuss in a later section.

Gp32 ssDNA-binding protein

Gp32 (34 kDa) is the prototype ssDNA-binding protein and a key component of the T4 replisome. Gp32 also plays important roles in homologous recombination and DNA repair. The biochemical properties of Gp32 have been thoroughly characterized [30-45], and the atomic structure of its central DNA-binding domain (DBD) has been solved [32]. The DBD contains an oligonucleotide/oligosaccharide-binding (OB)-fold motif plus a structural Zn^{++} atom. An N-terminal domain (so-called *basic* or "B-domain") is required for self-association and cooperativity, whereas a C-terminal domain (so-called *acidic* or "A-domain") is the site for protein-protein interactions with various recombination and replication enzymes including UvsX and UvsY.

Gp32 binds sequence-non-specifically to polynucleotides, with the highest observed affinity for ssDNA ($K_{obs} \approx 10^9$ M^{-1} at physiological ionic strength), moderate affinity for single-stranded RNA, and very low affinity for dsDNA. The binding site size of Gp32 on ssDNA is approximately 7 nucleotide residues. Binding to ssDNA is highly cooperative ($\omega \approx 1000$), meaning that Gp32 exists almost exclusively in clusters or long filaments on ssDNA at protein concentrations normally encountered in in vitro DNA strand exchange assays as well as in vivo.

Gp32 affects both pre- and post-synaptic steps of UvsX-catalyzed DNA strand exchange reactions [15,22,23,25,46,47]. An important function of Gp32 in presynapsis is to denature secondary structure in the ssDNA substrate, which *eventually* allows UvsX to saturate the ssDNA by forming long presynaptic filaments. Paradoxically, the immediate effect of Gp32 on UvsX-ssDNA filament formation is negative under physiological conditions, because Gp32 competes effectively with UvsX for binding sites [13]. Overcoming Gp32 inhibition requires either pre-incubation of UvsX with ssDNA in the presence of ATP (the previously mentioned order of addition effect), or the inclusion of UvsY in reaction mixtures (see below) [4,24]. Gp32 has also been shown to play a post-synaptic role in strand exchange, stimulating the reaction by sequestering the outgoing ssDNA strand that is displaced during D-loop formation and subsequent branch migration [47].

UvsY recombination mediator protein

UvsY is the prototype recombination mediator protein or RMP [24]. By definition, RMPs are proteins that load recombinases of the RecA/Rad51 family onto ssDNA molecules that are pre-saturated with cognate ssDNA-binding protein. UvsY is absolutely required for UvsX-catalyzed DNA strand exchange in the presence of Gp32 under physiological or high-salt conditions [22,48,49]. In vivo, UvsY is also absolutely required for UvsX-dependent recombination since mutations knocking out either gene product have equivalent recombination-deficient phenotypes including the small-plaque phenotype associated with defective RDR [26-28]. UvsY is the only member of the core T4 recombination machinery that forms a discreet oligomeric structure: It exists as a stable hexamer of identical 15.8 kDa subunits in solution, and binds to ssDNA in this form [50].

UvsY binds to both ssDNA and dsDNA, but has a much higher affinity for the former under relaxed DNA conditions [51]. The preference of UvsY for ssDNA may be an important factor in directing UvsX filament assembly onto ssDNA in the presence of excess dsDNA, since UvsX itself has a relatively high affinity for non-homologous dsDNA (Maher, R.L. and S.W. Morrical: Coordinated binding of ssDNA and dsDNA substrates by UvsX recombinase and its regulation by ATP, unpublished). UvsY has a binding site size on ssDNA of 4 nucleotide residues per protomer, or 24 nucleotide residues per hexamer [52]. The protomeric binding site sizes of UvsY and UvsX are identical. UvsY binds to ssDNA with high affinity ($K_{obs} \approx 10^7$ M^{-1} at physiological ionic strength), but with little or no cooperativity ($\omega \approx 1$). Therefore UvsY has higher *intrinsic* affinity, but lower cooperativity, for ssDNA than either UvsX or Gp32 under conditions that are relevant for strand exchange reactions in vitro and in vivo. UvsY-ssDNA interactions are weakened by mutations at residues Lys-58 and Arg-60, which form part of a conserved LKARLDY motif (so-called 'KARL' motif) found in the N-terminal domain of UvsY, which is thought to comprise part of its DNA binding surface [14,48,51,53,54]. The KARL motif is also found in certain DNA helicases, however no helicase activity has ever been associated with UvsY, which lacks a motor domain. The C-terminal domain of UvsY is essential for hexamerization. Deletion of this domain drastically reduces the affinity of UvsY-ssDNA interactions, demonstrating the importance of UvsY hexamers as the relevant ssDNA-binding unit [55].

Several lines of evidence indicate that UvsY hexamers have the ability to wrap ssDNA strands around themselves, and that wrapping is responsible for the high affinity of UvsY-ssDNA interactions. Evidence includes the observation that a C-terminally deleted, monomeric form of UvsY has 10^4-fold lower affinity for ssDNA than wild-type [55]. The wrapping hypothesis is supported by the finding that mutiple subunits within each UvsY hexamer are in contact with ssDNA [51]. Other evidence comes from results of single-molecule DNA stretching studies, which showed that the ssDNA that is created by the treatment of individual stretched dsDNA molecules with glyoxal is strongly wrapped by UvsY [54]. Wrapping of ssDNA occurs at low stretching forces where the DNA is relatively relaxed. At high stretching forces, where the DNA is under tension, wrapping is suppressed. The tension-dependent suppression of wrapping leads to the loss of preferential binding to ssDNA as shown by the fact that UvsY binds tighter to stretched dsDNA than to stretched ssDNA [54]. This contrasts with the observation that UvsY has ~1000-fold higher affinity for ssDNA than for dsDNA under relaxed conditions [51]. Therefore high-affinity binding of UvsY to ssDNA requires wrapping, which also imposes a preference for binding to ssDNA over dsDNA. Presumably UvsY cannot wrap dsDNA because its persistence length is much higher than that of ssDNA [56]. The surprising observation that UvsY binds tightly to stretched dsDNA could have important implications for presynaptic filament assembly. The binding of Gp32 to ssDNA creates an extended or "stiff" DNA conformation that might be recognized by UvsY in an unwrapped mode similar to its interaction with stretched dsDNA. Converting this extended ssDNA structure into a wrapped one might be an important step in the recruitment of UvsX recombinase, as we discuss in a later section.

UvsY is absolutely required for UvsX-catalyzed DNA strand exchange assays performed under physiological conditions of Gp32 and salt [4,24], consistent with the co-dependency of recombination on UvsX and UvsY in vivo [26-28]. In vitro, UvsY lowers the critical concentration of

UvsX for RDR and other recombination reactions [46,57]. UvsY stimulates the ssDNA-dependent ATPase activity of UvsX, possibly by acting as a nucleotide exchange factor for the recombinase [58]. The greatest stimulation of ATPase activity is seen when UvsY and Gp32 act together synergistically on the reaction [23,49]. UvsY stimulates the catalytic activities of UvsX mainly by promoting presynaptic filament assembly. The mechanism of UvsY's recombination mediator activity will be explored in greater detail below.

Assembly and Dynamics of the T4 Presynaptic Filament

Regulation of UvsX-ssDNA interactions by the ATPase cycle

Like all RecA/Rad51 recombinases, UvsX is a member of the AAA$^+$ ATPase super-family and its interactions with ssDNA are regulated by ATP binding and hydrolysis. The analog ATPγS, which is tightly bound but slowly hydrolyzed by UvsX, induces a stable, high-affinity ssDNA binding state of the enzyme [13,14]. ATP itself transiently induces high-affinity ssDNA binding by UvsX until it is hydrolyzed to ADP or AMP [15,16]. Both of these hydrolytic products are associated with decreased ssDNA-binding affinity states of UvsX under steady-state conditions [16].

Regulation of protein-ssDNA interactions by UvsY

Most evidence indicates that UvsX and Gp32 undergo mutually exclusive binding to ssDNA [48,59,60]. On the other hand there is overwhelming evidence that UvsY can co-occupy ssDNA binding sites simultaneously with either UvsX or Gp32 [14,19,25,60-62]. The interaction of UvsY with either Gp32-ssDNA or UvsX-ssDNA complexes alters the properties of both in ways that favor presynaptic filament formation and the activation of UvsX catalytic activities.

UvsY forms a stable tripartite complex with Gp32 and ssDNA at physiologically relevant salt conditions [61]. These complexes contain stoichiometric amounts of both UvsY and Gp32 with respect to their normal binding site sizes on ssDNA (Figure 2). Gp32-ssDNA interactions are destabilized within the UvsY-Gp32-ssDNA complex as shown by their increased sensitivity to disruption by salt compared to Gp32-ssDNA complexes in the absence of UvsY [61]. Results of single-molecule DNA stretching studies confirm that UvsY destabilizes Gp32-DNA interactions [54]. It has been proposed that, since cooperativity is such a large component of K_{obs} for Gp32-ssDNA interactions, UvsY could destabilize Gp32-ssDNA by lowering Gp32's cooperativity parameter [61]. This is probably the major pathway for destabilizing Gp32-ssDNA under physiological or high-salt conditions. It has also been proposed, based on results of single-molecule DNA stretching experiments, that UvsY directly displaces Gp32 from ssDNA under low-salt conditions [54]. In either case, the destabilization of Gp32-ssDNA interactions by UvsY lowers the energy barrier necessary for UvsX to displace Gp32 from ssDNA, which is necessary for nucleation and propagation of presynaptic filaments on ssDNA that is pre-saturated with Gp32 (as is likely to be the case in vivo).

Biochemical studies demonstrate that UvsY stabilizes UvsX-ssDNA interactions [14]. UvsY, UvsX, and ssDNA form a tripartite complex with a stoichiometry of ~1 UvsY hexamer per 6 UvsX protomers, consistent with their equivalent binding site sizes (4 nucleotide residues/protomer). The increased stability of UvsX-ssDNA interactions within these complexes is demonstrated by their higher resistance to salt compared to filaments formed in the absence of UvsY. The most stable complex is formed when UvsY and ATPγS are both present, indicating that the RMP and nucleoside triphosphate act synergistically to stabilize UvsX-ssDNA [14]. UvsY also stabilizes UvsX-ssDNA in the presence of ADP or no nucleotide, so its effects are global. Results of recent kinetics studies are consistent with the idea that UvsY acts as a nucleotide exchange factor for UvsX, promoting the release of hydrolytic products so that new ATP substrate can bind to the active sites [58]. It is postulated that UvsY-enhanced nucleotide exchange allows UvsX to remain longer in its ATP-bound form with higher affinity for ssDNA, which would tend to stabilize presynaptic filaments and increase their catalytic activites activity. Through its dual activities in destabilizing Gp32-ssDNA and stabilizing UvsX-ssDNA interactions, UvsY allows UvsX filaments to nucleate and propagate on Gp32-covered ssDNA (Figure 2).

ssDNA hand-offs govern filament assembly

UvsX and UvsY interact specifically with the C-terminal "A-domain" of Gp32, and with each other [35,36,49,60]. Protein-protein interactions play a significant role in the overall DNA strand exchange reaction. Nevertheless, studies of UvsY have shown that its ability to destabilize Gp32-ssDNA complexes is independent of UvsY-Gp32 interactions [54,61], indicating that the ssDNA-binding activity of UvsY is responsible for destabilizing Gp32-ssDNA interactions. Results of in vitro complementation assays between UvsX and UvsY mutants further suggest that UvsY-ssDNA interactions create an optimal ssDNA conformation for high-affinity binding by UvsX [58]. Studies showed that UvsY KARL-motif mutants K58A and K58A/R60A have reduced affinities for ssDNA compared to wild-type [53]. Similarly UvsX missense mutants H195Q and H195A exhibit reduced affinities for ssDNA as well as altered enzymatic activities compared to wild-type [16]. Unlike wild-type UvsX, the

ssDNA-dependent ATPase activities of UvsX-H195Q/A are strongly inhibited by wild-type UvsY at both low and high concentrations of the mediator. The UvsY KARL-motif mutants partially relieve this inhibition [58]. Furthermore the UvsX-H195Q mutant has weak DNA strand exchange activity that is inhibited by wild-type UvsY, but stimulated by the UvsY KARL-motif mutants [58]. These and other results support a mechanism in which presynaptic filament assembly involves a hand-off of ssDNA from UvsY to UvsX, with the efficiency of the hand-off controlled by the relative ssDNA-binding affinities of the two proteins.

Evidence increasingly supports the notion that DNA and RNA pathways channel their substrates through series of hand-off transactions in which intermediate nucleic acid structures are passed directly from one protein in the pathway to the next [63]. This strategy avoids potential cytotoxic effects of the free nucleic acid structure and protects it from unprogrammed side reactions or degradation. The available data suggest that T4 presynaptic filament assembly is also governed by a sequence of hand-off events involving intermediate ssDNA structures generated by Gp32 and UvsY (Figure 3). Initially, Gp32 binding converts ssDNA into an extended conformation that resembles the mechanically stretched DNA created in force-spectroscopy experiments. In the first hand-off event, a UvsY hexamer binds to the extended ssDNA and converts it into a wrapped conformation that destabilizes Gp32-ssDNA interactions. The wrapped UvsY-ssDNA complex is thought to be in equilibrium between "closed" and "open" states. The "closed" state destabilizes Gp32-ssDNA interactions but is inaccessible to UvsX, whereas the "open" state favors high-affinity UvsX-ssDNA interactions. In the second hand-off event, ATP-bound UvsX binds to the "open" form of the wrapped UvsY-ssDNA structure, allowing nucleation of a UvsX-ssDNA filament while displacing Gp32 from the ssDNA. Other ssDNA hand-off transactions may occur as the filament transitions from the nucleation to the propagation phase, or as UvsY performs its nucleotide exchange factor function. In addition, the linkage of the UvsX ATPase cycle to the sequential hand-off mechanism creates opportunities for dynamic instability in presynaptic filaments, which we will address in a later section.

UvsX-Gp32 exchanges on ssDNA

Gp32F is a fluorescein-conjugated form of Gp32 that is useful as a fluorescence probe for Gp32 displacement from ssDNA and to study the kinetics of presynaptic filament assembly in real time [48]. As UvsX filaments assemble on Gp32F-covered ssDNA, Gp32F is displaced and the fluorescence of its fluorescein moiety decreases. This assay was used to study presynaptic filament assembly both in the absence of UvsY (low-salt

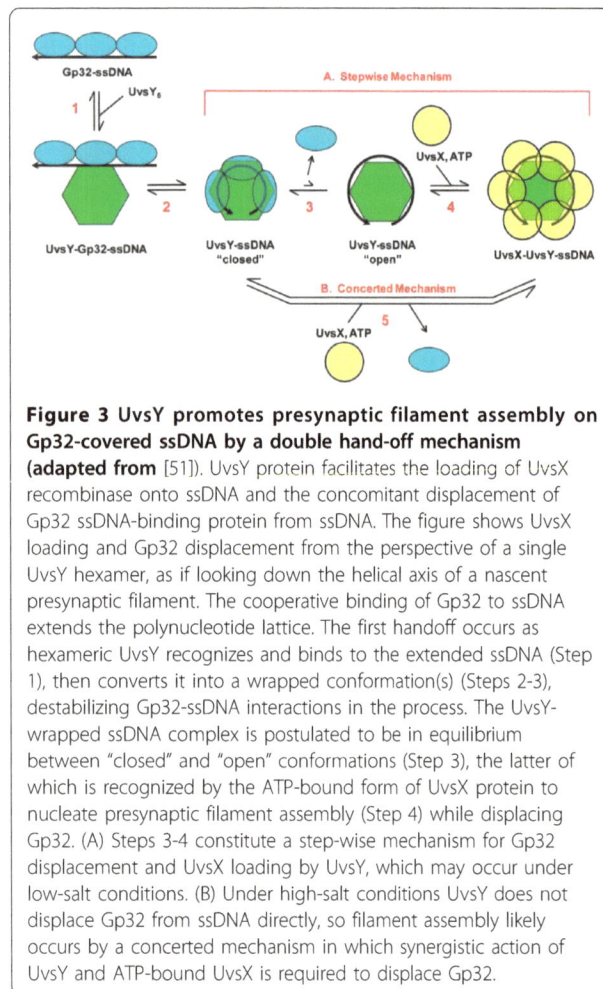

Figure 3 UvsY promotes presynaptic filament assembly on Gp32-covered ssDNA by a double hand-off mechanism (adapted from [51]). UvsY protein facilitates the loading of UvsX recombinase onto ssDNA and the concomitant displacement of Gp32 ssDNA-binding protein from ssDNA. The figure shows UvsX loading and Gp32 displacement from the perspective of a single UvsY hexamer, as if looking down the helical axis of a nascent presynaptic filament. The cooperative binding of Gp32 to ssDNA extends the polynucleotide lattice. The first handoff occurs as hexameric UvsY recognizes and binds to the extended ssDNA (Step 1), then converts it into a wrapped conformation(s) (Steps 2-3), destabilizing Gp32-ssDNA interactions in the process. The UvsY-wrapped ssDNA complex is postulated to be in equilibrium between "closed" and "open" conformations (Step 3), the latter of which is recognized by the ATP-bound form of UvsX protein to nucleate presynaptic filament assembly (Step 4) while displacing Gp32. (A) Steps 3-4 constitute a step-wise mechanism for Gp32 displacement and UvsX loading by UvsY, which may occur under low-salt conditions. (B) Under high-salt conditions UvsY does not displace Gp32 from ssDNA directly, so filament assembly likely occurs by a concerted mechanism in which synergistic action of UvsY and ATP-bound UvsX is required to displace Gp32.

conditions only) and in the presence of UvsY (physiological or high-salt conditions). The salt-dependence of the UvsY requirement for Gp32 displacement is a consequence of differential salt effects on the intrinsic association constants (K parameters) of UvsX and Gp32 for ssDNA [13,41,44,45,64]. Under low-salt conditions (≤ 50 mM NaCl), the ATP or ATPγS-bound forms of UvsX possess sufficient affinity for ssDNA to compete with Gp32 and displace it from the lattice, causing a time-dependent decrease in the fluorescence of the Gp32F probe [48]. ADP-bound, AMP-bound, or *apo* forms of UvsX cannot displace Gp32 from ssDNA under any conditions. At higher, more physiologically relevant salt concentrations, all forms of UvsX lack the ability to displace Gp32 from the ssDNA. Under these conditions, the addition of UvsY restores UvsX-ssDNA filament formation and Gp32 displacement, as measured by the decrease in Gp32F fluorescence [48]. The UvsY-dependent reactions still require ATP or ATPγS as a prerequisite for filament assembly; ADP-, AMP-, and *apo*-UvsX conditions do not support Gp32 displacement. This observation is consistent with the previous finding that UvsY and ATPγS-binding stabilize UvsX-ssDNA filaments synergistically [14], which implies the cooperation of these two factors during filament nucleation and/or propagation steps.

Following timecourses of Gp32F displacement from ssDNA allows detailed analyses of the kinetics of presynaptic filament assembly in a fully-reconstituted *in vitro* T4 recombination system (UvsX, UvsY, and Gp32). This has led to important new discoveries about filament dynamics and about the mechanism of UvsY in recombination mediation (Liu, J., C. Berger, and S.W. Morrical: Kinetics of Presynaptic Filament Assembly in the Presence of SSB and Mediator Proteins, unpublished). Under low-salt conditions, the ATP-dependent, UvsY-independent nucleation of UvsX filaments on Gp32F-covered ssDNA is highly salt-sensitive. Nevertheless nucleation rates are faster than propagation rates, suggesting that UvsX nucleates rapidly at many different sites. Under high-salt conditions, UvsY appears to specifically enhance the nucleation step to overcome the salt-sensitivity of UvsX filament assembly (Liu, J., C. Berger, and S.W. Morrical: Kinetics of Presynaptic Filament Assembly in the Presence of SSB and Mediator Proteins, unpublished). Rapid, salt-sensitive nucleation may be a general property of recombinase-DNA interactions, since similar behavior is observed for human Rad51 filament assembly on dsDNA [65]. It will be interesting to learn whether human RMPs such as Rad52, Brca2, or Rad51 paralogs also work by decreasing the salt-sensitivity of Rad51 filament nucleation.

A simplified kinetic scheme for T4 presynaptic filament assembly is shown in Figure 4, based on data derived from analysis of Gp32F displacement timecourses (Liu, J., C. Berger, and S.W. Morrical: Kinetics of Presynaptic Filament Assembly in the Presence of SSB and Mediator Proteins, unpublished). Results are consistent with a two-phase model, nucleation and propagation, both of which include a fast and reversible binding step (K_1 or K_3) followed by a slow isomerization step (k_2 or k_4) that is essentially irreversible under pre-steady-state conditions. We found that UvsY specifically enhances K_1, thereby stabilizing the product of the reversible binding step during the filament nucleation phase. This product may be thought of as a "pre-nucleation complex". Therefore UvsY overcomes the salt-sensitivity of filament nucleation by stabilizing the pre-nucleation complex at high salt concentrations. We also found that k_4, the rate constant for the isomerization step of filament propagation, is rate-limiting under all conditions (Liu, J., C. Berger, and S.W. Morrical: Kinetics of Presynaptic Filament Assembly in the Presence of SSB and Mediator Proteins, unpublished). This suggests that long presynaptic filaments are likely to be assembled from many shorter filaments that arise at multiple nucleation centers. In accord with this idea,

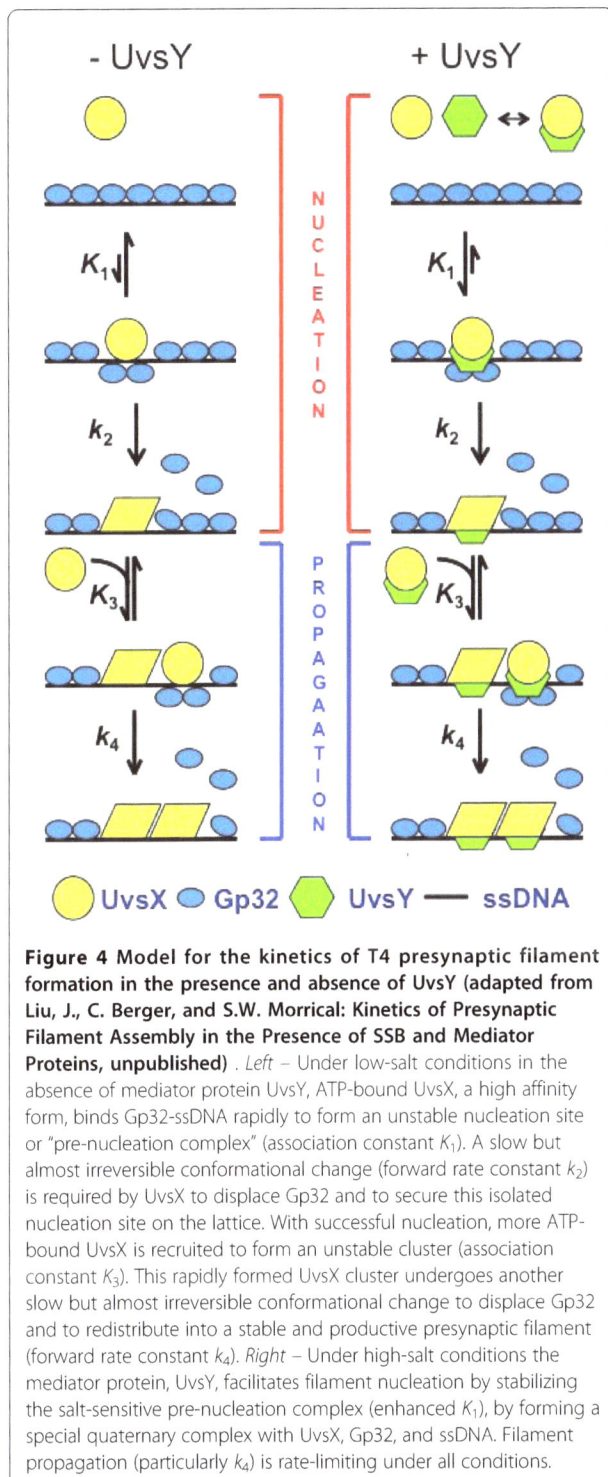

Figure 4 Model for the kinetics of T4 presynaptic filament formation in the presence and absence of UvsY (adapted from Liu, J., C. Berger, and S.W. Morrical: Kinetics of Presynaptic Filament Assembly in the Presence of SSB and Mediator Proteins, unpublished) . *Left* – Under low-salt conditions in the absence of mediator protein UvsY, ATP-bound UvsX, a high affinity form, binds Gp32-ssDNA rapidly to form an unstable nucleation site or "pre-nucleation complex" (association constant K_1). A slow but almost irreversible conformational change (forward rate constant k_2) is required by UvsX to displace Gp32 and to secure this isolated nucleation site on the lattice. With successful nucleation, more ATP-bound UvsX is recruited to form an unstable cluster (association constant K_3). This rapidly formed UvsX cluster undergoes another slow but almost irreversible conformational change to displace Gp32 and to redistribute into a stable and productive presynaptic filament (forward rate constant k_4). *Right* – Under high-salt conditions the mediator protein, UvsY, facilitates filament nucleation by stabilizing the salt-sensitive pre-nucleation complex (enhanced K_1), by forming a special quaternary complex with UvsX, Gp32, and ssDNA. Filament propagation (particularly k_4) is rate-limiting under all conditions.

human Rad51 assembles on dsDNA from many rapidly-formed nucleation sites and the cluster growth from each site is limited in length [65]. The requirement for many filament nucleation events may explain the observation that an apparent 1:1 stoichiometry between UvsX and

UvsY has to be maintained for optimal recombination activity [22,46,60].

Dynamic instability in presynaptic filaments

Presynaptic filaments are predicted to exhibit dynamic instability, or vectorial growth and collapse, due to the coupling of the recombinase ATPase cycle to changes in ssDNA binding affinity [15,19,47,60]. The Gp32F probe provides an indirect readout of the dynamic instability of UvsX-ssDNA filaments [49]. Results demonstrate that the dynamic instability of T4 presynaptic filaments depends not only on UvsX-catalyzed ATP hydrolysis, but also on competition between UvsX and Gp32 for binding sites on ssDNA (Figure 5). Experiments were designed in which UvsX and Gp32 undergo a pre-steady-state competition for a limited number of binding sites on ssDNA at physiological ionic strength [48]. The order of addition is controlled so that ssDNA is added to a pre-existing mixture of recombination proteins, which mimics the most likely pathway for filament assembly/disassembly in vivo. Filament assembly/disassembly is then monitored by following Gp32F dissociation/association using fluorescence. The data show that presynaptic filaments formed in the presence of Gp32 undergo constant assembly and collapse that is closely linked to the ATPase cycle of UvsX [48]. The reactions occur in three sequential phases (Figure 5): *Phase 1–preparing the lattice*. Gp32 rapidly binds and saturates all of the available ssDNA (rapid Gp32F fluorescence increase). *Phase 2–filament growth*. ATP-bound UvsX is loaded by UvsY and gradually displaces Gp32 (slow Gp32F fluorescence decrease). There is a stringent requirement for UvsY and either ATP or ATPγS in this phase, and the rate is optimal when UvsY stoichiometry is 1:1 with respect to UvsX and ssDNA binding sites. *Phase 3–filament collapse*. Depletion of ATP allows Gp32 to slowly re-occupy the ssDNA and drive off UvsX, which is now mainly in the low-affinity ADP/AMP forms [16,48] (slow Gp32F fluorescence increase). This collapse phase is sensitive to the nucleotide substrate/product ratio and does not occur if ATP is regenerated or if ATPγS is substituted. These observations are consistent with a dynamically unstable T4 presynaptic filament. Dynamic instability could take the form of treadmilling as shown in Figure 5, in which UvsX-ssDNA filaments simultaneously grow at an ATP-capped end and contract at an ADP- or AMP-capped end. The vectorial motion would be reinforced by Gp32 which would out-compete UvsX for ssDNA binding sites preferentially at the ADP/AMP-capped filament end.

Atomic Structure of T4 UvsX Recombinase
A recently solved, high-resolution UvsX crystal structure provides important new information on the mechanism

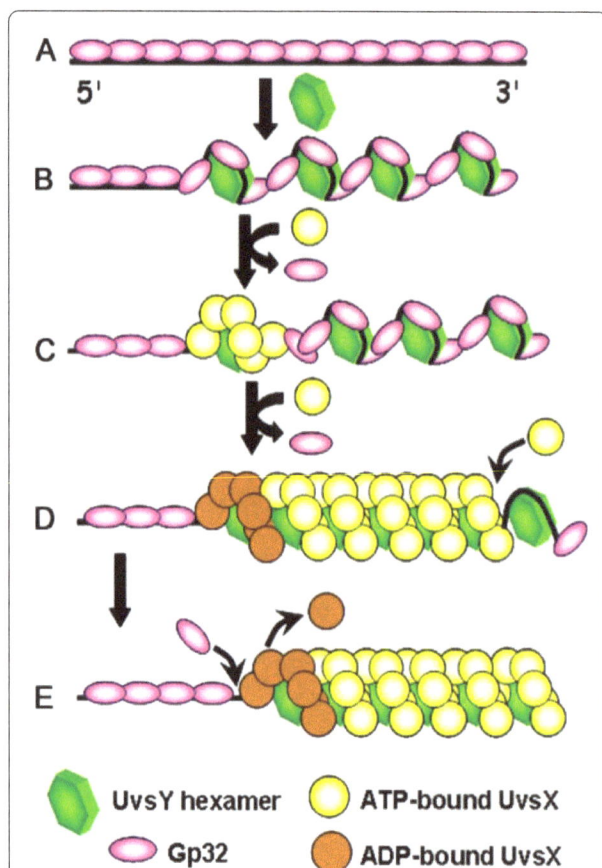

Figure 5 Dynamic instability in T4 presynaptic filaments is coupled to the UvsX ATPase cycle and to UvsX/Gp32 competition for binding sites (adapted from [48]). A, Gp32 covers free ssDNA rapidly to protect it from nuclease digestion and to remove secondary structure. B. Hexameric UvsY protein weakens Gp32-ssDNA interactions by binding to the complex and wrapping the ssDNA lattice. C. ATP-bound UvsX is recruited to the tripartite UvsY-Gp32-ssDNA intermediate. ATP and UvsY both contribute to a synergistic increase in UvsX-ssDNA binding affinity that allows the recombinase to locally displace Gp32 from the lattice. D. Propagation occurs in the 5′ → 3′ direction as ATP-bound UvsX subunits slowly add to the 3′ filament end, displacing more Gp32 subunits in the process. E. The first UvsX subunits to bind are the first to hydrolyze ATP, generating a relatively aged, ADP-capped 5′ filament end. The ADP-bound UvsX subunits are now vulnerable to displacement by Gp32. Differential competitive effects between Gp32 and the ATP- vs. ADP-capped filament ends creates dynamic instability in the complex, which could lead to filament treadmilling.

of the T4 recombinase [66]. The crystal was obtained from a truncation mutant UvsX$_{30-358}$ (full-length UvsX = 391 amino acid residues), which lacks the N-terminal protein-protein association domain and the extreme C-terminal region. The crystal has a P6$_1$ space group and the asymmetric unit is composed of dimer of identical subunits with a two-fold axis. In the crystal lattice these dimers are arranged as a right-handed helical filament, with one subunit of each dimer forming the

filament while the opposite subunit in each dimer decorates the surface of the filament without interacting with its symmetry partners. The dimer interface in the asymmetric unit occludes the ATP binding site, therefore no bound ATP is observed in the structure. The DNA binding loops L1 and L2 of UvsX are disordered as is the case for all RecA family proteins crystallized in the absence of DNA.

As expected, UvsX shares high similarity with *E. coli* RecA protein in overall architecture and protein folding, in spite of the remote sequence homology [67]. Compared to RecA, UvsX contains a larger N-terminal α/β motif, and a smaller C-terminal domain filled with helices and a small three-stranded β-sheet. The α/β ATPase core is highly conserved between UvsX and RecA in terms of structural motifs, locations, and amino acid compositions. The two nucleotide-binding motifs of UvsX, the Walker A and Walker B boxes, are located at similar positions compared to RecA structures. For example, the aromatic ring of Tyr99 in UvsX stacks with the adenine ring of ATP, similar to Tyr103 in RecA [66].

Docking of the UvsX structure into models of extended and compressed filament forms reconstituted from EM studies revealed additional details about the active site (Figure 6) [66]. Docking into the high-pitch "active" filament (ADP-AlF$_4$ form) indicated that the ATPase site spans the filament interface, as is the case for high-pitch filaments of *E. coli* RecA and *S. cerevisiae* Rad51 [17,68,69]. Conserved residue Glu92 is positioned to activate a water molecule for nucleophilic attack on ATP γ-phosphate. Significantly, residues Lys246′ and Arg248′ reach across the filament interface and form salt bridges with the phosphates of ATP and with Glu92. These residues are structurally equivalent to the Lys248′ and Lys250′ bridges and to catalytic residue Glu96 in *E. coli* RecA. The lysine bridges are thought to promote catalysis by stabilizing the transition state during ATP hydrolysis [69]. This strategy is apparently conserved between RecA and UvsX. Interestingly, eukaryotic Rad51 and Dmc1 recombinases lack the entire motif containing the basic bridge residues, and no other basic residues take their places in the Rad51 crystal structures [17,68]. Thus there is a divergence of active site structure and function between the prokaryotic and eukaryotic recombinases, with UvsX more closely aligned to the prokaryotic mechanism.

Docking of the UvsX structure into the low-pitch "inactive" filament (ADP form) indicates that residues Lys246′ to Lys254′ move by about 4 Å so that the ATP binding site no longer spans the filament interface. These observations indicate that changes in filament pitch observed at different stages of the ATPase cycle are accompanied by extensive remodeling of the active

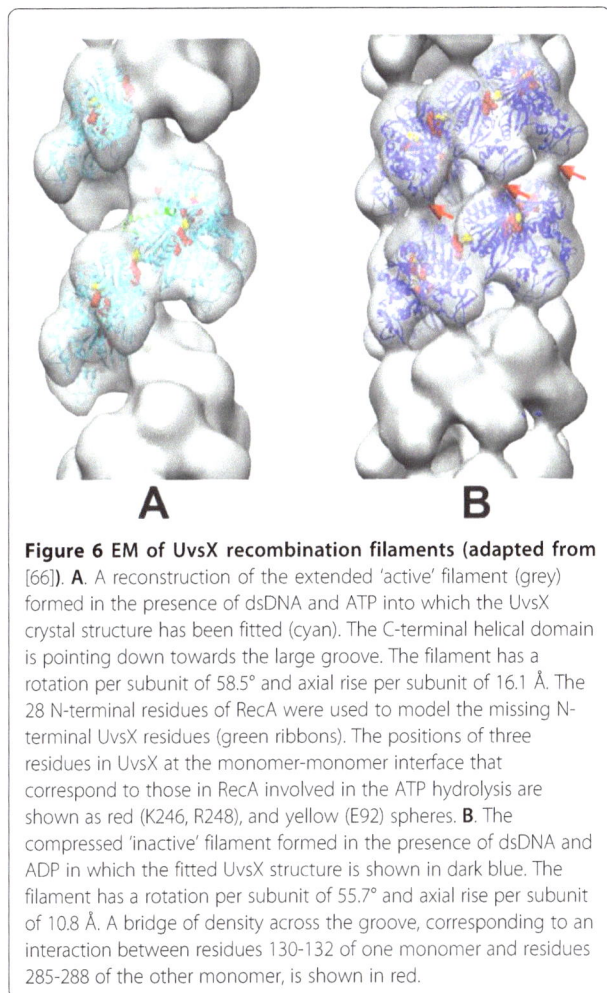

Figure 6 EM of UvsX recombination filaments (adapted from [66]). **A**. A reconstruction of the extended 'active' filament (grey) formed in the presence of dsDNA and ATP into which the UvsX crystal structure has been fitted (cyan). The C-terminal helical domain is pointing down towards the large groove. The filament has a rotation per subunit of 58.5° and axial rise per subunit of 16.1 Å. The 28 N-terminal residues of RecA were used to model the missing N-terminal UvsX residues (green ribbons). The positions of three residues in UvsX at the monomer-monomer interface that correspond to those in RecA involved in the ATP hydrolysis are shown as red (K246, R248), and yellow (E92) spheres. **B**. The compressed 'inactive' filament formed in the presence of dsDNA and ADP in which the fitted UvsX structure is shown in dark blue. The filament has a rotation per subunit of 55.7° and axial rise per subunit of 10.8 Å. A bridge of density across the groove, corresponding to an interaction between residues 130-132 of one monomer and residues 285-288 of the other monomer, is shown in red.

site itself. Overall, the high-resolution structure of UvsX [66] provides exciting new opportunities to investigate its catalytic and allosteric mechanisms.

Actions of Helicases in DNA Strand Exchange Reactions

The bacteriophage T4 recombination system provided one of the earliest demonstrations that a DNA helicase, Dda protein, can stimulate a recombinase-catalyzed DNA strand exchange reaction [70]. Subsequent work has shown that at least three T4-encoded helicases (Dda, Gp41, and UvsW) are capable of influencing recombination and/or recombination-dependent replication transactions in vitro, and probably in vivo as well. In this section we will focus on the impacts of Dda, Gp41, and UvsW on reconstituted strand exchange reactions *in vitro*.

Helicase processing of recombination intermediates

After a UvsX-catalyzed homology search and strand pairing, a joint molecule is formed between the invading

3' single-stranded DNA (ssDNA) tail and the homologous double-stranded DNA (dsDNA) template in the form of a displacement-loop (D-loop) (Figure 1). ssDNA regions of the D-loop are potential targets for helicase assembly. Depending on which strand the helicase translocates on, and on the polarity of the helicase, processing of the D-loop could have three different outcomes: extension of the heteroduplex by branch migration, unwinding of the heteroduplex by branch or bubble migration, or conversion of the D-loop into a nascent replication fork. In addition, certain helicases may use their translocase activity to remove presynaptic filaments from ssDNA. It appears likely that all four of these processes occur at some point during T4 DNA metabolism. It has been shown that all three T4 helicases, Dda, Gp41, and UvsW, are capable of catalyzing branch migration *in vitro* [29,70,71]. However, the biological functions of these helicases are distinctive, in spite of the overlapping branch migration activities.

Dda helicase

Dda is a unique helicase compared to Gp41 and UvsW, since it may regulate recombination both positively and negatively at two different stages: presynaptic filament formation and branch migration. *E. coli* UvrD and yeast Srs2 proteins are two translocases/helicases functioning to remove recombinases from ssDNA and to prevent improper presynaptic filament formation and illegitimate recombination events [72-74]. To date, no T4 helicase has been identified as a direct functional homolog of UvrD or Srs2. Dda may share some properties of these helicases though, since the phenotypes of certain *dda* mutants are consistent with a role in anti-recombination [75], and since Dda inhibits UvsX-mediated homologous strand pairing reactions in vitro [76]. It is speculated that destabilizing UvsX-ssDNA filaments through its translocase activity is one factor contributing to the observed inhibition of homologous pairing. Similarly, Dda might apply this translocation activity to DNA replication by allowing the fork to bypass DNA-bound proteins on the template in vitro [77-79]. If Dda protein does disrupt presynaptic filaments then its mechanism must differ somewhat from Srs2 and UvrD, since the latter two have 3' to 5' polarity while Dda has 5' to 3' polarity [80-82].

The strand exchange assay routinely uses a circular M13 ssDNA and a linearized M13 dsDNA as substrates. The extent of branch migration after initial synapsis can be monitored by the restriction endonuclease digestion pattern of the end-radiolabeled dsDNA [70]. This nicely-designed assay system allowed Kodadek and Alberts to monitor and measure the rate of branch migration of UvsX-catalyzed strand exchange in the presence and absence of Dda. The late addition of Dda after synapsis stimulates the rate of branch migration more than four-fold, from ~15 bp/sec to ~70 bp/sec [70]. Dda was the first helicase documented to stimulate strand exchange reactions by stimulating branch migration, on the premise that it is added late into the reconstituted reaction after synapsis has occurred. Furthermore, the specific protein-protein interaction between Dda and UvsX might be important for this stimulation, since Dda cannot stimulate RecA-catalyzed strand exchange reactions.

In vitro, Dda's inhibition of homologous pairing and stimulation of branch migration can be separated by manipulating the addition sequence of Dda into the reconstituted reaction, either simultaneously with UvsX during presynapsis, or after the initiation of synapsis. How Dda balances these opposite activities and cooperates with UvsX *in vivo* remains largely unknown, however. It is observed that UvsX and Dda act synergistically in template switching to allow DNA lesion bypass and to rescue stalled replication forks [4,83]. Furthermore, protein-protein interactions between Dda and the C-terminal domain of Gp32 are required for the DNA replication activities of Dda [37]. These observations suggest that interactions with UvsX or with Gp32 could recruit Dda onto different nucleoprotein intermediates at different stages of the strand exchange process, perhaps regulating the recombination vs. anti-recombination functions of Dda.

Gp41 helicase and Gp59 helicase loading protein

Gp41, the essential replicative helicase in T4, facilitates both leading strand DNA synthesis catalyzed by the T4 DNA polymerase holoenzyme (Gp43, Gp44/Gp62, and Gp45 proteins), and lagging strand DNA synthesis by recruiting primase Gp61 to reconstitute the T4 primosome [4]. The Gp41 helicase translocates processively on the displaced strand in a 5' → 3' direction, as an asymmetric hexagonal ring on the DNA [84,85].

Gp59 has been classified as a replication mediator protein or helicase loading protein, based on the observation that it is required to load Gp41 onto Gp32-covered ssDNA [4,38,77,86]. Gp59 acts as an adapter protein by interacting with Gp32 at the N-terminus and with Gp41 at the C-terminus [86-88]. It is the key factor for the strand-specific recruitment of primosome onto the displaced strand of a D-loop to covert it into a replication fork during RDR, and to initiate new lagging-strand DNA synthesis during RDR. Gp41 cannot stimulate UvsX-dependent strand exchange unless Gp59 is present, and this stimulation occurs through branch migration [70]. UvsY stimulates homologous pairing, but strongly inhibits branch migration. The branch migration activity can only be recovered by adding Gp41 and Gp59. The protein-protein interaction

between Gp59 and the C-terminal acidic domain of Gp32 is important for this rescue [70].

Interestingly, the formation and stability of Gp32-ssDNA clusters is a key factor for strand- and structure-specific loading of Gp41 helicase by Gp59. Gp59 targets Gp41 helicase assembly onto Gp32-ssDNA clusters [4,37,38]. The interplay between Gp32 and Gp59 is complicated. The formation of a tripartite Gp59-Gp32-ssDNA complex decreases the stability of Gp32-ssDNA interaction, but Gp32 also helps modulate the strand specificity of Gp59 [4,38]. Gp59-mediated primosome assembly is precluded from ssDNA that is saturated with UvsX and UvsY, but allowed when a few Gp32 clusters interrupt the presynaptic filament. In DNA strand exchange, the invading strand is typically saturated with UvsX and UvsY and therefore resistant to Gp41/Gp59 loading. However, Gp32 rapidly sequesters the displaced strand of the D-loop [19,47], forming a target for Gp41/Gp59. Thus UvsX/UvsY and Gp32/Gp59 enforce strand specific loading of Gp41 onto the displaced strand, where it is poised to catalyze branch migration using its 5' to 3' helicase activity (Figure 7). UvsX/UvsY prevent D-loop resolution (anti-recombination) by Gp41/Gp59 by preventing their assembly on the invading ssDNA strand. An identical partitioning mechanism is used during RDR to ensure primosome assembly on the displaced strand of the D-loop, assuring complete reconstitution of semi-conservative DNA synthesis beginning with a recombination event [4].

In the absence of UvsX and UvsY, the sole presence of excessive amount of Gp32 can produce joint molecules from M13 dsDNA with a 3' single-stranded termini of about 100 nucleotides and a circular M13 ssDNA [89]. The initial binding of Gp32 onto the single-stranded tail is probably sufficient to destabilize the double-stranded helix, starting from the junction point, and to promote spontaneous joint molecule formation. When coupled with Gp59 and Gp41, the polar branch migration mediated by Gp41 can drive the formation of nicked circle, the final product of standard three-strand exchange reactions [89]. This synergism between Gp32 and Gp41/Gp59 is also crucial for extensive strand displacement synthesis by the T4 DNA polymerase holoenzyme [39,90].

UvsW helicase

UvsW plays a central role in T4 recombination and in the transition from origin to recombination-dependent replication. UvsW mutations cause hypersensitivity to UV and hydroxyurea, and a decreased frequency of recombination [91,92]. UvsW is a 3' to 5' RNA/DNA and DNA/DNA helicase with specificity for branched-DNA substrates such as X-shaped Holliday junctions

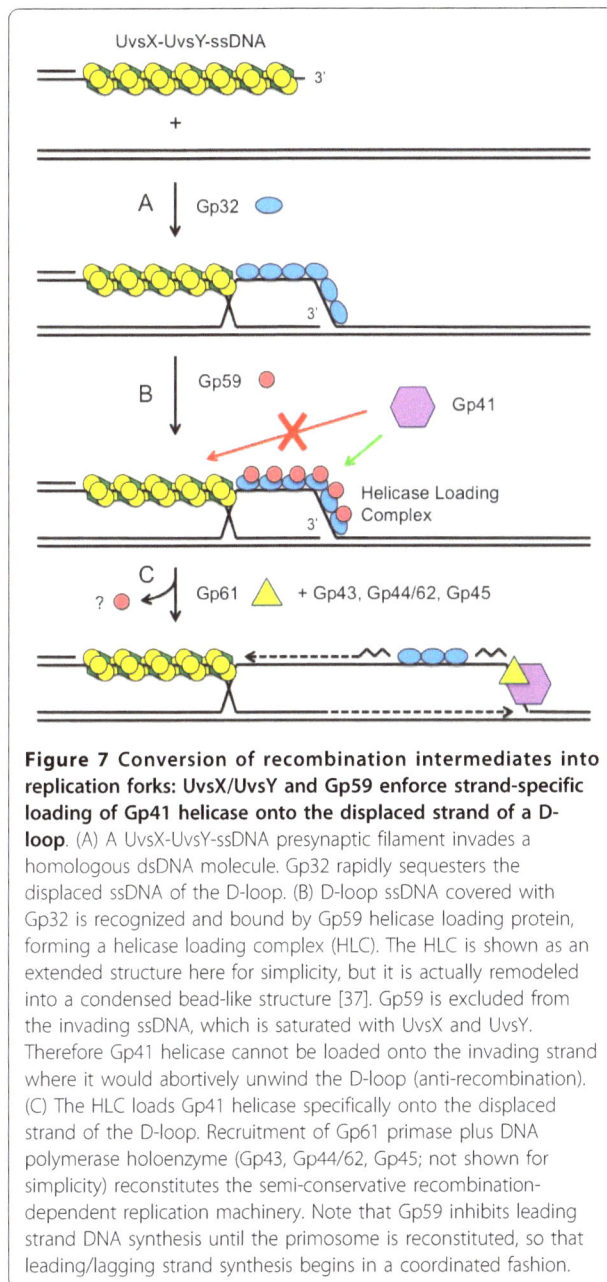

Figure 7 Conversion of recombination intermediates into replication forks: UvsX/UvsY and Gp59 enforce strand-specific loading of Gp41 helicase onto the displaced strand of a D-loop. (A) A UvsX-UvsY-ssDNA presynaptic filament invades a homologous dsDNA molecule. Gp32 rapidly sequesters the displaced ssDNA of the D-loop. (B) D-loop ssDNA covered with Gp32 is recognized and bound by Gp59 helicase loading protein, forming a helicase loading complex (HLC). The HLC is shown as an extended structure here for simplicity, but it is actually remodeled into a condensed bead-like structure [37]. Gp59 is excluded from the invading ssDNA, which is saturated with UvsX and UvsY. Therefore Gp41 helicase cannot be loaded onto the invading strand where it would abortively unwind the D-loop (anti-recombination). (C) The HLC loads Gp41 helicase specifically onto the displaced strand of the D-loop. Recruitment of Gp61 primase plus DNA polymerase holoenzyme (Gp43, Gp44/62, Gp45; not shown for simplicity) reconstitutes the semi-conservative recombination-dependent replication machinery. Note that Gp59 inhibits leading strand DNA synthesis until the primosome is reconstituted, so that leading/lagging strand synthesis begins in a coordinated fashion.

and Y-shaped replication forks [71,93,94]. It does not unwind linear duplex substrates with either blunt ends or single-stranded tails. Substrate recognition may occur through a small but highly electropositive N-terminal domain and an arginine/aromatic-rich loop, as revealed by its crystal structure [95]. The mutant phenotype and substrate specificity lead to the hypothesis that UvsW might drive branch migration to resolve recombination intermediates during strand invasion and transfer. Indeed, purified UvsW protein can catalyze Holliday junction branch migration through more than 1 kb of

DNA sequence, using a plasmid-based Holliday junction-containing substrate [71]. Recent data show that UvsW promotes branch migration in UvsX-catalyzed DNA strand exchange reactions [66]. In the classic three-strand exchange reaction with M13 circular ssDNA and linear dsDNA substrates, UvsW promotes resolution of the branched ssDNA/dsDNA networks formed by UvsX, leading to the robust generation of nicked circular heteroduplex product. Reactions occur in the presence of Gp32 and in either the presence or absence of UvsY. Thus UvsW appears to provide a "missing link" in the biochemistry of T4 recombination, since it can provide physiologically reasonable mechanisms for generating extensive heteroduplex DNA, involving the translocation of either 3- or 4-strand junctions.

In summary, Dda, Gp41, and UvsW are three helicases all capable of stimulating branch migration, but with clearly different biological roles in T4 recombination. Dda may act as a negative regulator of homologous pairing, but may also be used to accelerate branch migration or to couple recombination to bubble migration DNA synthesis [70,75,76,96]. The major role of Gp41/Gp59 in recombination is likely to be the channeling of recombination intermediates into structures that can support RDR, and then launching lagging strand synthesis in the semi-conservative RDR mechanism [4]. UvsW on the other hand optimizes strand exchange and the formation of long heteroduplex DNA [66]. Complex interplays between the three different helicase activities are likely to modulate many aspects of T4 recombination metabolism.

Conclusions

Studies of the T4 recombination system have provided insights on recombination mechanisms that are highly relevant to HR and HDR processes in cellular organisms including eukaryotes. Work with T4 UvsY protein has helped to define the roles that recombination mediator proteins play in promoting presynaptic filament assembly and in the trafficking of recombination proteins (SSB, RMP, and recombinase) on ssDNA that occurs during the early stages of recombination and homology-directed DNA repair processes. It is clear that the UvsY model for assembly of recombinase filaments on ssDNA covered with ssDNA-binding protein is highly conserved [24], including in human beings where at least three classes of proteins with UvsY-like mediator activity participate in genome stability pathways. These include Rad52, the human Rad51 paralogs Rad51B, Rad51C, Rad51D, Xrcc2, and Xrcc3, and breast cancer susceptibility gene Brca2 [97-100]. Details of T4 presynaptic filament assembly and dynamics, such as ssDNA hand-offs and dynamic instability, suggest mechanisms that may be used by recombination machineries in many organisms to capture recombinagenic ssDNA, perform strand exchange, and pass the intermediates on to other repair enzymes such as the replicative components of HDR pathways.

Recent biochemical and structural studies of UvsX recombinase shed light on its mechanism and relationship to other recombinases of the RecA/Rad51 superfamily. The observation that ssDNA-binding by UvsX allosterically regulates the enzyme's affinity for homologous vs. non-homologous dsDNA at a second site is an important breakthrough [66]. The sensitive fluorescence assay developed for this study represents an excellent opportunity to explore how micro-heterology affects homologous pairing, as well as the similarities and differences between pairing mechanisms used by recombinases from various organisms. The X-ray crystal structure of UvsX and its modeling in EM filament structures shows that UvsX shares the same extended filament structure in its active form as do *E. coli* and yeast filament structures (Gajewski, S., M.R. Webb, V. Galkin, E.H. Egelman, K.N. Kreuzer, and S.W. White: Crystal structure of the phage T4 recombinase UvsX and its functional interation with the T4 SF2 helicase UvsW, unpublished). The observation that UvsX appears to share the lysine bridges found at the active site of *E. coli* RecA-DNA places UvsX mechanistically closer to prokaryotic than to eukaryotic recombinases, at least in this detail. Opportunities for structure-driven mutagenesis and mechanistic studies, as well as for evolutionary studies, of UvsX will surely follow from this important structure.

The T4 field pioneered studies of helicases in recombination, which are now known to be pervasive regulators of recombination and HDR metabolism in all organisms [100]. The biochemistry of T4 helicases demonstrates the diverse ways that these enzymes can influence recombination outcomes, including both positive and negative regulation of homologous pairing and strand exchange. It is noteworthy that T4 encodes threes different helicases on its phage genome that appear to have both unique and overlapping functions in recombination. Of particular relevance is the role of helicases in channeling strand exchange reactions toward the formation of intermediates that can serve as initiators of recombination-dependent DNA replication [4,6,96]. T4 RDR requires either Dda (for bubble-migration DNA synthesis) or Gp41/Gp59 (for semi-conservative DNA synthesis) to initiate replication via a recombination event. The biochemical role of UvsW in the RDR machine remains to be elucidated but is likely to be central given its ability to promote extensive branch migration. The coupling of recombination to replication is fundamental for DNA repair and genome stability in all organisms. Eukaryotic DNA helicases/

translocases such as Rad54, Srs2 and others are known to play important roles in processing recombination intermediates, either for regulatory purposes or to facilitate access of downstream DNA replication and repair enzymes to the products of strand exchange [10,11,72-74,100]. The T4 helicases offer an excellent opportunity to study more about the mechanism of recombination/replication coupling, the findings of which will directly inform studies of genome stability mechanisms in cellular organisms including humans.

Abbreviations
HR: homologous recombination; HDR: homology-directed repair; RDR: recombination-dependent replication; DSB: double-strand break; ssDNA: single-stranded DNA; dsDNA: double-stranded DNA; SSB: single-stranded DNA binding protein; RMP: recombination mediator protein; ATPγS: adenosine 5'-O-(3-thio)triphosphate; Gp32F: fluorescein-labeled bacteriophage T4 gene 32 protein (Gp32).

Author details
¹Section of Microbiology, University of California-Davis, Davis, CA 95616 USA. ²Department of Biochemistry, University of Vermont College of Medicine, Burlington, VT 05405 USA.

Authors' contributions
JL and SM made equal intellectual contributions to this review and participated equally in writing the manuscript.

Competing interests
The authors declare that they have no competing interests.

Received: 1 November 2010 Accepted: 3 December 2010
Published: 3 December 2010

References
1. Jasin M: **Homologous repair of DNA damage and tumorigenesis: the BRCA connection.** *Oncogene* 2002, **21**:8981-8993.
2. Symington LS: **Role of RAD52 epistasis group genes in homologous recombination and double-strand break repair.** *Microbiol Mol Biol Rev* 2002, **66**:630-670.
3. Pierce AJ, Stark JM, Araujo FD, Moynahan ME, Berwick M, Jasin M: **Double-strand breaks and tumorigenesis.** *Trends Cell Biol* 2001, **11**:S52-59.
4. Bleuit JS, Xu H, Ma Y, Wang T, Liu J, Morrical SW: **Mediator proteins orchestrate enzyme-ssDNA assembly during T4 recombination-dependent DNA replication and repair.** *Proc Natl Acad Sci USA* 2001, **98**:8298-8305.
5. Mosig G: **Homologous recombination.** In *Molecular Biology of Bacteriophage T4.* Edited by: Karam JD. ASM Press, Washington, DC; 1994:54-82.
6. Kreuzer KN: **Recombination-dependent DNA replication in phage T4.** *Trends Biochem Sci* 2000, **25**:165-173.
7. Sun H, Treco D, Szostak JW: **Extensive 3'-overhanging, single-stranded DNA associated with the meiosis-specific double-strand breaks at the ARG4 recombination initiation site.** *Cell* 1991, **64**:1155-1161.
8. Haber JE: **In vivo biochemistry: physical monitoring of recombination induced by site-specific endonucleases.** *BioEssays* 1995, **17**:609-620.
9. Mickelson C, Wiberg JS: **Membrane-associated DNase activity controlled by genes 46 and 47 of bacteriophage T4D and elevated DNase activity associated with the T4 das mutation.** *J Virol* 1981, **40**:65-77.
10. Li X, Heyer WD: **RAD54 controls access to the invading 3'-OH end after RAD51-mediated DNA strand invasion in homologous recombination in Saccharomyces cerevisiae.** *Nucleic Acids Res* 2009, **37**:638-646.
11. Macris MA, Sung P: **Multifaceted role of the Saccharomyces cerevisiae Srs2 helicase in homologous recombination regulation.** *Biochem Soc Trans* 2005, **33**:1447-1450.
12. Bianco PR, Tracy RB, Kowalczykowski SC: **DNA strand exchange proteins: a biochemical and physical comparison.** *Front Biosci* 1998, **3**:570-603.

13. Ando RA, Morrical SW: Single-stranded DNA binding properties of the UvsX recombinase of bacteriophage T4: binding parameters and effects of nucleotides. *J Mol Biol* 1998, 283:785-796.
14. Liu J, Bond JP, Morrical SW: Mechanism of presynaptic filament stabilization by the bacteriophage T4 UvsY recombination mediator protein. *Biochemistry* 2006, 45:5493-5502.
15. Formosa T, Alberts BM: Purification and characterization of the T4 bacteriophage uvsX protein. *J Biol Chem* 1986, 261:6107-6118.
16. Farb JN, Morrical SW: Role of allosteric switch residue histidine 195 in maintaining active-site asymmetry in presynaptic filaments of bacteriophage T4 UvsX recombinase. *J Mol Biol* 2009, 385:393-404.
17. Conway AB, Lynch TW, Zhang Y, Fortin GS, Fung CW, Symington LS, Rice PA: Crystal structure of a Rad51 filament. *Nat Struct Mol Biol* 2004, 11:791-796.
18. Lauder SD, Kowalczykowski SC: Asymmetry in the recA protein-DNA filament. *J Biol Chem* 1991, 266:5450-5458.
19. Kodadek T, Wong ML, Alberts BM: The mechanism of homologous DNA strand exchange catalyzed by the bacteriophage T4 uvsX and gene 32 proteins. *J Biol Chem* 1988, 263:9427-9436.
20. Riddles PW, Lehman IR: The formation of plectonemic joints by the recA protein of Escherichia coli. Requirement for ATP hydrolysis. *J Biol Chem* 1985, 260:170-173.
21. Kowalczykowski SC, Krupp RA: DNA-strand exchange promoted by RecA protein in the absence of ATP: implications for the mechanism of energy transduction in protein-promoted nucleic acid transactions. *Proc Natl Acad Sci USA* 1995, 92:3478-3482.
22. Harris LD, Griffith JD: UvsY protein of bacteriophage T4 is an accessory protein for in vitro catalysis of strand exchange. *J Mol Biol* 1989, 206:19-27.
23. Yonesaki T, Minagawa T: Synergistic action of three recombination gene products of bacteriophage T4, uvsX, uvsY, and gene 32 proteins. *J Biol Chem* 1989, 264:7814-7820.
24. Beernink HT, Morrical SW: RMPs: recombination/replication mediator proteins. *Trends Biochem Sci* 1999, 24:385-389.
25. Kodadek T, Gan DC, Stemke-Hale K: The phage T4 uvsY recombination protein stabilizes presynaptic filaments. *J Biol Chem* 1989, 264:16451-16457.
26. Melamede RJ, Wallace SS: Properties of the nonlethal recombinational repair x and y mutants of bacteriophage T4. II. DNA synthesis. *J Virol* 1977, 24:28-40.
27. Melamede RJ, Wallace SS: Properties of the nonlethal recombinational repair deficient mutants of bacteriophage T4. III. DNA replicative intermediates and T4w. *Mol Gen Genet* 1980, 177:501-509.
28. Kreuzer KN, Morrical SW: Initiation of DNA replication. In *Molecular Biology of Bacteriophage T4*. Edited by: Karam JD. ASM Press, Washington, DC; 1994:28-42.
29. Salinas F, Kodadek T: Phage T4 homologous strand exchange: a DNA helicase, not the strand transferase, drives polar branch migration. *Cell* 1995, 82:111-119.
30. Chase JW, Williams KR: Single-stranded DNA binding proteins required for DNA replication. *Annu Rev Biochem* 1986, 55:103-136.
31. Karpel RL: T4 bacteriophage gene 32 protein. In *The Biology of nonspecific DNA-protein interactions*. Edited by: Revzin A. CRC Press; Boca Raton, FL; 1990:103-130.
32. Shamoo Y, Friedman AM, Parsons MR, Konigsberg WH, Steitz TA: Crystal structure of a replication fork single-stranded DNA binding protein (T4 gp32) complexed to DNA. *Nature* 1995, 376:362-366.
33. Williams KR, Shamoo Y, Spicer EK, Coleman JE, Konigsberg WH: Correlating structure to function in proteins: T4 Gp32 as a prototype. In *Molecular Biology of Bacteriophage T4*. Edited by: Karam JD. ASM Press, Washington, DC; 1994:301-304.
34. Giedroc DP, Khan R, Barnhart K: Overexpression, purification, and characterization of recombinant T4 gene 32 protein22-301 (g32P-B). *J Biol Chem* 1990, 265:11444-11455.
35. Hurley JM, Chervitz SA, Jarvis TC, Singer BS, Gold L: Assembly of the bacteriophage T4 replication machine requires the acidic carboxy terminus of gene 32 protein. *J Mol Biol* 1993, 229:398-418.
36. Jiang H, Giedroc D, Kodadek T: The role of protein-protein interactions in the assembly of the presynaptic filament for T4 homologous recombination. *J Biol Chem* 1993, 268:7904-7911.

37. Ma Y, Wang T, Villemain JL, Giedroc DP, Morrical SW: Dual functions of single-stranded DNA-binding protein in helicase loading at the bacteriophage T4 DNA replication fork. *J Biol Chem* 2004, 279:19035-19045.
38. Morrical SW, Beernink HT, Dash A, Hempstead K: The gene 59 protein of bacteriophage T4. Characterization of protein-protein interactions with gene 32 protein, the T4 single-stranded DNA binding protein. *J Biol Chem* 1996, 271:20198-20207.
39. Xu H, Wang Y, Bleuit JS, Morrical SW: Helicase assembly protein Gp59 of bacteriophage T4: fluorescence anisotropy and sedimentation studies of complexes formed with derivatives of Gp32, the phage ssDNA binding protein. *Biochemistry* 2001, 40:7651-7661.
40. Kowalczykowski SC: Thermodynamic data for protein-nucleic acid interactions.Edited by: Saenger W. Berlin: Springer-Verlag; 1990:244-263, Landolt-Bornstein: Numerical Data and Functional Relationships in Science and Technology (New Series) Group VII: Biophysics, Nucleic Acids 1d.
41. Kowalczykowski SC, Lonberg N, Newport JW, von Hippel PH: Interactions of bacteriophage T4-coded gene 32 protein with nucleic acids. I. Characterization of the binding interactions. *J Mol Biol* 1981, 145:75-104.
42. Newport JW, Lonberg N, Kowalczykowski SC, von Hippel PH: Interactions of bacteriophage T4-coded gene 32 protein with nucleic acids. II. Specificity of binding to DNA and RNA. *J Mol Biol* 1981, 145:105-121.
43. Pant K, Karpel RL, Rouzina I, Williams MC: Salt dependent binding of T4 gene 32 protein to single and double-stranded DNA: single molecule force spectroscopy measurements. *J Mol Biol* 2005, 349:317-330.
44. Rouzina I, Pant K, Karpel RL, Williams MC: Theory of electrostatically regulated binding of T4 gene 32 protein to single- and double-stranded DNA. *Biophys J* 2005, 89:1941-1956.
45. Shokri L, Rouzina I, Williams MC: Interaction of bacteriophage T4 and T7 single-stranded DNA-binding proteins with DNA. *Phys Biol* 2009, 6:15096-15103.
46. Morrical SW, Alberts BM: The UvsY protein of bacteriophage T4 modulates recombination-dependent DNA synthesis in vitro. *J Biol Chem* 1990, 265:15096-15103.
47. Kodadek T: The role of the bacteriophage T4 gene 32 protein in homologous pairing. *J Biol Chem* 1990, 265:20966-20969.
48. Liu J, Qian N, Morrical SW: Dynamics of bacteriophage T4 presynaptic filament assembly from extrinsic fluorescence measurements of Gp32-single-stranded DNA interactions. *J Biol Chem* 2006, 281:26308-26319.
49. Yassa DS, Chou KM, Morrical SW: Characterization of an amino-terminal fragment of the bacteriophage T4 uvsY recombination protein. *Biochimie* 1997, 79:275-285.
50. Beernink HT, Morrical SW: The uvsY recombination protein of bacteriophage T4 forms hexamers in the presence and absence of single-stranded DNA. *Biochemistry* 1998, 37:5673-5681.
51. Xu H, Beernink HT, Morrical SW: DNA-binding properties of T4 UvsY recombination mediator protein: polynucleotide wrapping promotes high-affinity binding to single-stranded DNA. *Nucleic Acids Res* 2010, 38:4821-4833.
52. Sweezy MA, Morrical SW: Single-stranded DNA binding properties of the uvsY recombination protein of bacteriophage T4. *J Mol Biol* 1997, 266:927-938.
53. Bleuit JS, Ma Y, Munro J, Morrical SW: Mutations in a conserved motif inhibit single-stranded DNA binding and recombination mediator activities of bacteriophage T4 UvsY protein. *J Biol Chem* 2004, 279:6077-6086.
54. Pant K, Shokri L, Karpel RL, Morrical SW, Williams MC: Modulation of T4 gene 32 protein DNA binding activity by the recombination mediator protein UvsY. *J Mol Biol* 2008, 380:799-811.
55. Ando RA, Morrical SW: Relationship between hexamerization and ssDNA binding affinity in the uvsY recombination protein of bacteriophage T4. *Biochemistry* 1999, 38:16589-16598.
56. McGhee JD: Theoretical calculations of the helix-coil transition of DNA in the presence of large, cooperatively binding ligands. *Biopolymers* 1976, 15:1345-1375.
57. Morrical SW, Wong ML, Alberts BM: Amplification of snap-back DNA synthesis reactions by the uvsX recombinase of bacteriophage T4. *J Biol Chem* 1991, 266:14031-14038.
58. Farb JN, Morrical SW: Functional complementation of UvsX and UvsY mutations in the mediation of T4 homologous recombination. *Nucleic Acids Res* 2009, 37:2336-2345.

59. Griffith J, Formosa T: The uvsX protein of bacteriophage T4 arranges single-stranded and double-stranded DNA into similar helical nucleoprotein filaments. *J Biol Chem* 1985, **260**:4484-4491.

60. Kodadek T: Functional interactions between phage T4 and E. coli DNA-binding proteins during the presynapsis phase of homologous recombination. *Biochem Biophys Res Commun* 1990, **172**:804-810.

61. Sweezy MA, Morrical SW: Biochemical interactions within a ternary complex of the bacteriophage T4 recombination proteins uvsY and gp32 bound to single-stranded DNA. *Biochemistry* 1999, **38**:936-944.

62. Hashimoto K, Yonesaki T: The characterization of a complex of three bacteriophage T4 recombination proteins, uvsX protein, uvsY protein, and gene 32 protein, on single-stranded DNA. *J Biol Chem* 1991, **266**:4883-4888.

63. Echols H: Multiple DNA-protein interactions governing high-precision DNA transactions. *Science* 1986, **233**:1050-1056.

64. Lohman TM, Kowalczykowski SC: Kinetics and mechanism of the association of the bacteriophage T4 gene 32 (helix destabilizing) protein with single-stranded nucleic acids. Evidence for protein translocation. *J Mol Biol* 1981, **152**:67-109.

65. Hilario J, Amitani I, Baskin RJ, Kowalczykowski SC: Direct imaging of human Rad51 nucleoprotein dynamics on individual DNA molecules. *Proc Natl Acad Sci USA* 2009, **106**:361-368.

66. Gajewski S, Webb MR, Galkin V, Egelman EH, Kreuzer KN, White SW: Crystal Structure of the Phage T4 Recombinase UvsX and Its Functional Interaction with the T4 SF2 Helicase UvsW. *J Mol Biol* 2010, [Epub ahead of print].

67. Fujisawa H, Yonesaki T, Minagawa T: Sequence of the T4 recombination gene, uvsX, and its comparison with that of the recA gene of Escherichia coli. *Nucleic Acids Res* 1985, **13**:7473-7481.

68. Chen J, Villanueva N, Rould MA, Morrical SW: Insights into the mechanism of Rad51 recombinase from the structure and properties of a filament interface mutant. *Nucleic Acids Res* 2010, **38**:4889-4906.

69. Chen Z, Yang H, Pavletich NP: Mechanism of homologous recombination from the RecA-ssDNA/dsDNA structures. *Nature* 2008, **453**:489-484.

70. Kodadek T, Alberts BM: Stimulation of protein-directed strand exchange by a DNA helicase. *Nature* 1987, **326**:312-314.

71. Webb MR, Plank JL, Long DT, Hsieh TS, Kreuzer KN: The phage T4 protein UvsW drives Holliday junction branch migration. *J Biol Chem* 2007, **282**:34401-34411.

72. Krejci L, Van Komen S, Li Y, Villemain J, Reddy MS, Klein H, Ellenberger T, Sung P: DNA helicase Srs2 disrupts the Rad51 presynaptic filament. *Nature* 2003, **423**:305-309.

73. Veaute X, Delmas S, Selva M, Jeusset J, Le Cam E, Matic I, Fabre F, Petit MA: UvrD helicase, unlike Rep helicase, dismantles RecA nucleoprotein filaments in Escherichia coli. *Embo J* 2005, **24**:180-189.

74. Veaute X, Jeusset J, Soustelle C, Kowalczykowski SC, Le Cam E, Fabre F: The Srs2 helicase prevents recombination by disrupting Rad51 nucleoprotein filaments. *Nature* 2003, **423**:309-312.

75. Mosig G: Recombination and recombination-dependent DNA replication in bacteriophage T4. *Annu Rev Genet* 1998, **32**:379-413.

76. Kodadek T: Inhibition of protein-mediated homologous pairing by a DNA helicase. *J Biol Chem* 1991, **266**:9712-9718.

77. Barry J, Alberts B: A role for two DNA helicases in the replication of T4 bacteriophage DNA. *J Biol Chem* 1994, **269**:33063-33068.

78. Bedinger P, Hochstrasser M, Jongeneel CV, Alberts BM: Properties of the T4 bacteriophage DNA replication apparatus: the T4 dda DNA helicase is required to pass a bound RNA polymerase molecule. *Cell* 1983, **34**:115-123.

79. Gauss P, Park K, Spencer TE, Hacker KJ: DNA helicase requirements for DNA replication during bacteriophage T4 infection. *J Bacteriol* 1994, **176**:1667-1672.

80. Jongeneel CV, Formosa T, Alberts BM: Purification and characterization of the bacteriophage T4 dda protein. A DNA helicase that associates with the viral helix-destabilizing protein. *J Biol Chem* 1984, **259**:12925-12932.

81. Nanduri B, Byrd AK, Eoff RL, Tackett AJ, Raney KD: Pre-steady-state DNA unwinding by bacteriophage T4 Dda helicase reveals a monomeric molecular motor. *Proc Natl Acad Sci USA* 2002, **99**:14722-14727.

82. Raney KD, Benkovic SJ: Bacteriophage T4 Dda helicase translocates in a unidirectional fashion on single-stranded DNA. *J Biol Chem* 1995, **270**:22236-22242.

83. Kadyrov FA, Drake JW: UvsX recombinase and Dda helicase rescue stalled bacteriophage T4 DNA replication forks in vitro. *J Biol Chem* 2004, **279**:35735-35740.

84. Dong F, Gogol EP, von Hippel PH: The phage T4-coded DNA replication helicase (gp41) forms a hexamer upon activation by nucleoside triphosphate. *J Biol Chem* 1995, **270**:7462-7473.

85. Norcum MT, Warrington JA, Spiering MM, Ishmael FT, Trakselis MA, Benkovic SJ: Architecture of the bacteriophage T4 primosome: electron microscopy studies of helicase (gp41) and primase (gp61). *Proc Natl Acad Sci USA* 2005, **102**:3623-3626.

86. Morrical SW, Hempstead K, Morrical MD: The gene 59 protein of bacteriophage T4 modulates the intrinsic and single-stranded DNA-stimulated ATPase activities of gene 41 protein, the T4 replicative DNA helicase. *J Biol Chem* 1994, **269**:33069-33081.

87. Delagoutte E, von Hippel PH: Mechanistic studies of the T4 DNA (gp41) replication helicase: functional interactions of the C-terminal Tails of the helicase subunits with the T4 (gp59) helicase loader protein. *J Mol Biol* 2005, **347**:257-275.

88. Ishmael FT, Alley SC, Benkovic SJ: Assembly of the bacteriophage T4 helicase: architecture and stoichiometry of the gp41-gp59 complex. *J Biol Chem* 2002, **277**:20555-20562.

89. Kong D, Nossal NG, Richardson CC: Role of the bacteriophage T7 and T4 single-stranded DNA-binding proteins in the formation of joint molecules and DNA helicase-catalyzed polar branch migration. *J Biol Chem* 1997, **272**:8380-8387.

90. Nossal NG: The bacteriophage T4 replication fork. In *Molecular Biology of Bacteriophage T4*. Edited by: Karam JD. ASM Press, Washington, DC; 1994:43-53.

91. Derr LK, Drake JW: Isolation and genetic characterization of new uvsW alleles of bacteriophage T4. *Mol Gen Genet* 1990, **222**:257-264.

92. Derr LK, Kreuzer KN: Expression and function of the uvsW gene of bacteriophage T4. *J Mol Biol* 1990, **214**:643-656.

93. Carles-Kinch K, George JW, Kreuzer KN: Bacteriophage T4 UvsW protein is a helicase involved in recombination, repair and the regulation of DNA replication origins. *Embo J* 1997, **16**:4142-4151.

94. Dudas KC, Kreuzer KN: UvsW protein regulates bacteriophage T4 origin-dependent replication by unwinding R-loops. *Mol Cell Biol* 2001, **21**:2706-2715.

95. Sickmier EA, Kreuzer KN, White SW: The crystal structure of the UvsW helicase from bacteriophage T4. *Structure* 2004, **12**:583-592.

96. Formosa T, Alberts BM: DNA synthesis dependent on genetic recombination: characterization of a reaction catalyzed by purified bacteriophage T4 proteins. *Cell* 1986, **47**:793-806.

97. Liu J, Doty T, Gibson B, Heyer WD: Human BRCA2 protein promotes RAD51 filament formation on RPA-covered single-stranded DNA. *Nat Struct Mol Biol* 2010, **17**:1260-1262.

98. Jensen RB, Carreira A, Kowalczykowski SC: Purified human BRCA2 stimulates RAD51-mediated recombination. *Nature* 2010, **467**:678-683.

99. San Filippo J, Sung P, Klein H: Mechanism of eukaryotic homologous recombination. *Annu Rev Biochem* 2008, **77**:229-257.

100. Sung P, Klein H: Mechanism of homologous recombination: mediators and helicases take on regulatory functions. *Nat Rev Mol Cell Biol* 2006, **7**:739-750.

doi:10.1186/1743-422X-7-357
Cite this article as: Liu and Morrical: Assembly and dynamics of the bacteriophage T4 homologous recombination machinery. *Virology Journal* 2010 **7**:357.

Mueser *et al. Virology Journal* 2010, **7**:359
http://www.virologyj.com/content/7/1/359

VIROLOGY JOURNAL

REVIEW

Structural analysis of bacteriophage T4 DNA replication: a review in the Virology Journal series on *bacteriophage T4 and its relatives*

Timothy C Mueser[1]*, Jennifer M Hinerman[2], Juliette M Devos[3], Ryan A Boyer[4], Kandace J Williams[5]

Abstract

The bacteriophage T4 encodes 10 proteins, known collectively as the replisome, that are responsible for the replication of the phage genome. The replisomal proteins can be subdivided into three activities; the replicase, responsible for duplicating DNA, the primosomal proteins, responsible for unwinding and Okazaki fragment initiation, and the Okazaki repair proteins. The replicase includes the gp43 DNA polymerase, the gp45 processivity clamp, the gp44/62 clamp loader complex, and the gp32 single-stranded DNA binding protein. The primosomal proteins include the gp41 hexameric helicase, the gp61 primase, and the gp59 helicase loading protein. The RNaseH, a 5' to 3' exonuclease and T4 DNA ligase comprise the activities necessary for Okazaki repair. The T4 provides a model system for DNA replication. As a consequence, significant effort has been put forth to solve the crystallographic structures of these replisomal proteins. In this review, we discuss the structures that are available and provide comparison to related proteins when the T4 structures are unavailable. Three of the ten full-length T4 replisomal proteins have been determined; the gp59 helicase loading protein, the RNase H, and the gp45 processivity clamp. The core of T4 gp32 and two proteins from the T4 related phage RB69, the gp43 polymerase and the gp45 clamp are also solved. The T4 gp44/62 clamp loader has not been crystallized but a comparison to the *E. coli* gamma complex is provided. The structures of T4 gp41 helicase, gp61 primase, and T4 DNA ligase are unknown, structures from bacteriophage T7 proteins are discussed instead. To better understand the functionality of T4 DNA replication, in depth structural analysis will require complexes between proteins and DNA substrates. A DNA primer template bound by gp43 polymerase, a fork DNA substrate bound by RNase H, gp43 polymerase bound to gp32 protein, and RNase H bound to gp32 have been crystallographically determined. The preparation and crystallization of complexes is a significant challenge. We discuss alternate approaches, such as small angle X-ray and neutron scattering to generate molecular envelopes for modeling macromolecular assemblies.

Bacteriophage T4 DNA Replication

The semi-conservative, semi-discontinuous process of DNA replication is conserved in all life forms. The parental anti-parallel DNA strands are separated and copied following hydrogen bonding rules for the keto form of each base as proposed by Watson and Crick [1]. Progeny cells therefore inherit one parental strand and one newly synthesized strand comprising a new duplex DNA genome. Protection of the integrity of genomic DNA is vital to the survival of all organisms. In a masterful dichotomy, the genome encodes proteins that are also the caretakers of the genome. RNA can be viewed as the evolutionary center of this juxtaposition of DNA and protein. Viruses have also played an intriguing role in the evolutionary process, perhaps from the inception of DNA in primordial times to modern day lateral gene transfer. Simply defined, viruses are encapsulated genomic information. Possibly an ancient encapsulated virus became the nucleus of an ancient prokaryote, a symbiotic relationship comparable to mitochondria, as some have recently proposed [2-4]. This early relationship has evolved into highly complex eukaryotic cellular processes of replication, recombination and repair requiring multiple signaling pathways to coordinate activities required for the processing of complex genomes. Throughout evolution, these processes have become

* Correspondence: timothy.mueser@utoledo.edu
[1]Department of Chemistry, University of Toledo, Toledo OH, USA
Full list of author information is available at the end of the article

increasing complicated with protein architecture becoming larger and more complex. Our interest, as structural biologists, is to visualize these proteins as they orchestrate their functions, posing them in sequential steps to examine functional mechanisms. Efforts to crystallize proteins and protein:DNA complexes are hampered for multiple reasons, from limited solubility and sample heterogeneity to the fundamental lack of crystallizability due to the absence of complimentary surface contacts required to form an ordered lattice. For crystallographers, the simpler organisms provide smaller proteins with greater order which have a greater propensity to crystallize. Since the early days of structural biology, viral and prokaryotic proteins were successfully utilized as model systems for visualizing biological processes. In this review, we discuss our current progress to complete a structural view of DNA replication using the viral proteins encoded by bacteriophage T4 or its relatives.

DNA replication initiation is best exemplified by interaction of the *E. coli* DnaA protein with the *OriC* sequence which promotes DNA unwinding and the subsequent bi-directional loading of DnaB, the replicative helicase [5]. Assembly of the replication complex and synthesis of an RNA primer by DnaG initiates the synthesis of complimentary DNA polymers, comprising the elongation phase. The bacteriophage T4 encodes all of the proteins essential for its DNA replication. Table 1 lists these proteins, their functions and corresponding T4 genes. Through the pioneering work of Nossal, Alberts, Konigsberg, and others, the T4 DNA replication proteins have all been isolated, analyzed, cloned, expressed, and purified to homogeneity. The replication process has been reconstituted, using purified recombinant proteins, with velocity and accuracy comparable to *in vivo* reactions [6]. Initiation of phage DNA replication within the T4-infected cell is more complicated than for the *E. coli* chromosome, as the multiple circularly permuted linear copies of the phage genome appear as concatemers with homologous recombination events initiating strand synthesis during middle and late stages of infection ([7], see Kreuzer and Brister this series).

The bacteriophage T4 replisome can be subdivided into two components, the DNA replicase and the primosome. The DNA replicase is composed of the gene 43-encoded DNA polymerase (gp43), the gene 45 sliding clamp (gp45), the gene 44 and 62 encoded ATP-dependent clamp loader complex (gp44/62), and the gene 32 encoded single-stranded DNA binding protein (gp32) [6]. The gp45 protein is a trimeric, circular molecular clamp that is equivalent to the eukaryotic processivity factor, proliferating cell nuclear antigen (PCNA) [8]. The gp44/62 protein is an accessory protein required for gp45 loading onto DNA [9]. The gp32 protein assists in the unwinding of DNA and the gp43 DNA polymerase extends the invading strand primer into the next genome, likely co-opting the *E. coli* gyrase (topo II) to reduced positive supercoiling ahead of the polymerase [10]. The early stages of elongation involves replication of the leading strand template in which gp43 DNA polymerase can continuously synthesize a daughter strand in a 5' to 3' direction. The lagging strand requires segmental synthesis of Okazaki fragments which are initiated by the second component of the replication complex, the primosome. This T4 replicative complex is composed of the gp41 helicase and the gp61 primase, a DNA directed RNA polymerase [11]. The gp41 helicase is a homohexameric protein that encompasses the lagging strand and traverses in the 5' to 3' direction, hydrolyzing ATP as it unwinds the duplex in front of the replisome [12]. Yonesaki and Alberts demonstrated that gp41 helicase cannot load onto replication forks protected by the gp32 protein single-stranded DNA binding protein [13,14]. The T4 gp59 protein is a helicase loading protein comparable to *E. coli* DnaC and is required for the loading of gp41 helicase if DNA is preincubated with the gp32 single-stranded DNA binding protein [15]. We have shown that

Table 1 DNA Replication Proteins Encoded by Bacteriophage T4

Protein	Function
Replicase	
gp43 DNA polymerase	DNA directed 5' to 3' DNA polymerase
gp45 protein	Polymerase clamp enhances processivity of gp43 polymerase and RNase H
gp44/62 protein	clamp loader utilizes ATP to open and load the gp45 clamp
gp32 protein	cooperative single stranded DNA binding protein assists in unwinding duplex
Primosome	
gp41 helicase	processive 5' to 3' replicative helicase
gp61 primase	DNA dependent RNA polymerase generates lagging strand RNA pentamer primers in concert with gp41 helicase
gp59 protein	helicase assembly protein required for loading the gp41 helicase in the presence of gp32 protein
Lagging strand repair	
RNase H	5' to 3' exonuclease cleaves Okazaki RNA primers
gp30 DNA ligase	ATP-dependent ligation of nicks after lagging strand gap repair

the gp59 protein preferentially recognizes branched DNA and Holliday junction architectures and can recruit gp32 single-strand DNA binding protein to the 5' arm of a short fork of DNA [16,17]. The gp59 helicase loading protein also delays progression of the leading strand polymerase, allowing for the assembly and coordination of lagging strand synthesis. Once gp41 helicase is assembled onto the replication fork by gp59 protein, the gp61 primase synthesizes an RNA pentaprimer to initiate lagging strand Okazaki fragment synthesis. It is unlikely that the short RNA primer, in an A-form hybrid duplex with template DNA, would remain annealed in the absence of protein, so a hand-off from primase to either gp32 protein or gp43 polymerase is probably necessary [18].

Both the leading and lagging strands of DNA are synthesized by the gp43 DNA polymerase simultaneously, similar to most prokaryotes. Okazaki fragments are initiated stochastically every few thousand bases in prokaryotes (eukaryotes have slower pace polymerases with primase activity every few hundred bases) [19]. The lagging strand gp43 DNA polymerase is physically coupled to the leading strand gp43 DNA polymerase. This juxtaposition coordinates synthesis while limiting the generation of single-stranded DNA[20]. As synthesis progresses, the lagging strand duplex extrude from the complex creating a loop, or as Alberts proposed, a trombone shape (Figure 1) [21]. Upon arrival at the previous Okazaki primer, the lagging strand gp43 DNA polymerase halts, releases the newly synthesized duplex, and rebinds to a new gp61 generated primer. The RNA primers are removed from the lagging strands by the T4 *rnh* gene encoded RNase H, assisted by gp32 single-strand binding protein if the polymerase has yet to arrive or by gp45 clamp protein if gp43 DNA polymerase has reached the primer prior to processing [22-24]. For this latter circumstance, the gap created by RNase H can be filled either by reloading of gp43 DNA polymerase or by *E. coli* Pol I [25]. The *rnh⁻* phage are viable indicating that *E. coli* Pol I 5' to 3' exonuclease activity can substitute for RNase H [25]. Repair of the gap leaves a single-strand nick with a 3' OH and a 5' monophosphate, repaired by the gp30 ATP-dependent DNA ligase; better known as T4 ligase [26]. Coordination of each step involves molecular interactions between both DNA and the proteins discussed above. Elucidation of the structures of DNA replication proteins reveals the protein folds and active sites as well as insight into molecular recognition between the various proteins as they mediate transient interactions.

Crystal Structures of the T4 DNA Replication Proteins

In the field of protein crystallography, approximately one protein in six will form useful crystals. However, the odds frequently appear to be inversely proportional to overall interest in obtaining the structure. Our first encounter with T4 DNA replication proteins was a draft of Nancy Nossal's review "The Bacteriophage T4 DNA Replication Fork" subsequently published as Chapter 5 in the 1994 edition of "Molecular Biology of Bacteriophage T4" [6]. At the beginning of our collaboration (NN with TCM), the recombinant T4 replication system had been reconstituted and all 10 proteins listed in Table 1 were available [27]. Realizing the low odds for successful crystallization, all 10 proteins were purified and screened. Crystals were observed for 4 of the 10 proteins; gp43 DNA polymerase, gp45 clamp, RNase H, and gp59 helicase loading protein. We initially focused our efforts on solving the RNase H crystal structure, a protein first described by Hollingsworth and Nossal [24] and subsequently determined to be more structurally similar to the FEN-1 5' to 3' exonuclease family, rather than RNase H proteins [28]. The second crystal we observed was of the gp59 helicase loading protein first described by Yonesaki and Alberts [13,14]. To date, T4 RNase H, gp59 helicase loading protein, and gp45 clamp are the only full length T4 DNA replication proteins for which structures are available [17,28,29]. When proteins do not crystallize, there are several approaches to take. One avenue is to search for homologous organisms, such as the T4 related genome sequences ([30]; Petrov et al. this series) in which the protein function is the same but the surface residues may have diverged sufficiently to provide compatible lattice interactions in crystals. For example, the Steitz group has solved two structures from a related bacteriophage, the RB69 gp43 DNA polymerase and gp45 sliding clamp [31,32]. Our efforts with a more distant relative, the vibriophage KVP40, unfortunately yielded insoluble proteins. Another approach is to cleave flexible regions of proteins using either limited proteolysis or mass spectrometry fragmentation. The stable fragments are sequenced using mass spectrometry and molecular cloning is used to prepare core proteins for crystal trials. Again, the Steitz group successfully used proteolysis to solve the crystal structure of the core fragment of T4 gp32 single-stranded DNA binding protein (ssb) [33]. This accomplishment has brought the total to five complete or partial structures of the ten DNA replication proteins from T4 or related bacteriophage. To complete the picture, we must rely on other model systems, the bacteriophage T7 and *E. coli* (Figure 2). We provide here a summary of our collaborative efforts with the late Dr. Nossal, and also the work of many others, that, in total, has created a pictorial view of prokaryotic DNA replication. A list of proteins of the DNA replication fork along with the relevant protein data bank (PDB) numbers is provided in Table 2.

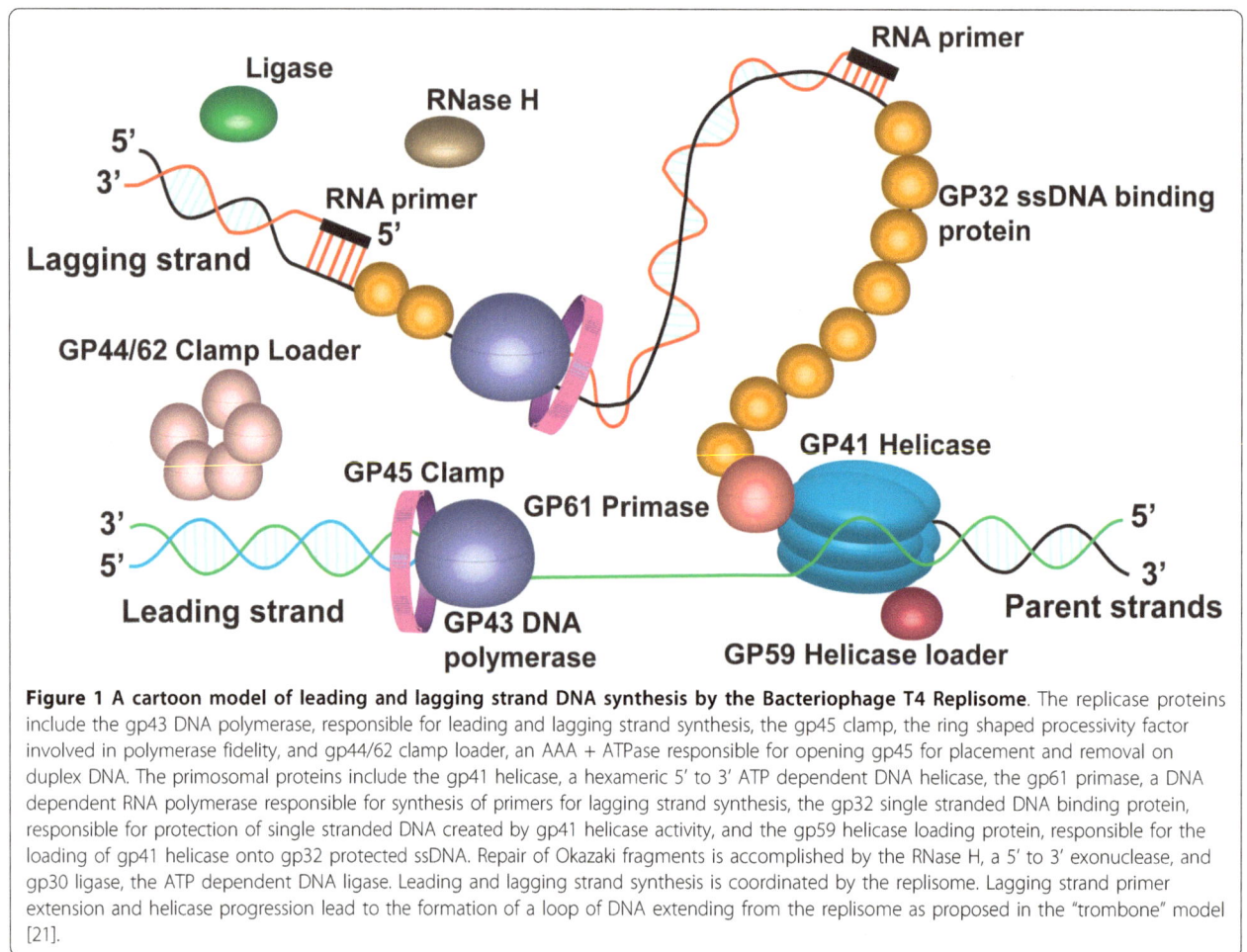

Figure 1 A cartoon model of leading and lagging strand DNA synthesis by the Bacteriophage T4 Replisome. The replicase proteins include the gp43 DNA polymerase, responsible for leading and lagging strand synthesis, the gp45 clamp, the ring shaped processivity factor involved in polymerase fidelity, and gp44/62 clamp loader, an AAA + ATPase responsible for opening gp45 for placement and removal on duplex DNA. The primosomal proteins include the gp41 helicase, a hexameric 5' to 3' ATP dependent DNA helicase, the gp61 primase, a DNA dependent RNA polymerase responsible for synthesis of primers for lagging strand synthesis, the gp32 single stranded DNA binding protein, responsible for protection of single stranded DNA created by gp41 helicase activity, and the gp59 helicase loading protein, responsible for the loading of gp41 helicase onto gp32 protected ssDNA. Repair of Okazaki fragments is accomplished by the RNase H, a 5' to 3' exonuclease, and gp30 ligase, the ATP dependent DNA ligase. Leading and lagging strand synthesis is coordinated by the replisome. Lagging strand primer extension and helicase progression lead to the formation of a loop of DNA extending from the replisome as proposed in the "trombone" model [21].

Replicase Proteins
Gene 43 DNA Polymerase
The T4 gp43 DNA polymerase (gi:118854, NP_049662), an 898 amino acid residue protein related to the Pol B family, is used in both leading and lagging strand DNA synthesis. The Pol B family includes eukaryotic pol α, δ, and ε. The full length T4 enzyme and the exo⁻ mutant (D219A) have been cloned, expressed and purified [34,35]. While the structure of the T4 gp43 DNA polymerase has yet to be solved, the enzyme from the RB69 bacteriophage has been solved individually (PDB 1waj) and in complex with a primer template DNA duplex (PDB 1ig9, Figure 3A) [32,36]. The primary sequence alignment reveals that the T4 gp43 DNA polymerase is 62% identical and 74% similar to RB69 gp43 DNA polymerase, a 903 residue protein [37,38].

E. coli Pol I, the first DNA polymerase discovered by Kornberg, has three domains, an N-terminal 5' to 3' exonuclease (cleaved to create the Klenow fragment), a 3' to 5' editing exonuclease domain, and a C-terminal polymerase domain [5]. The structure of the *E. coli* Pol I Klenow fragment was described through anthropomorphic terminology of fingers, palm, and thumb domains [39,40]. The RB69 gp43 DNA polymerase has two active sites, the 3' to 5' exonuclease (residues 103 - 339) and the polymerase domain (residues 381 - 903), comparable to Klenow fragment domains [41]. The gp43 DNA polymerase also has an N-terminal domain (residues 1 - 102 and 340 - 380) and a C-terminal tail containing a PCNA interacting peptide (PIP box) motif (residues 883 - 903) that interacts with the 45 sliding clamp protein. The polymerase domain contains a fingers subunit (residues 472 - 571) involved in template display (Ser 565, Lys 560, amd Leu 561) and NTP binding (Asn 564) and a palm domain (residues 381 - 471 and 572 - 699) which contains the active site, a cluster of aspartate residues (Asp 411, 621, 622, 684, and 686) that coordinates the two divalent active site metals (Figure 3B). The T4 gp43 DNA polymerase appears to be active in a monomeric form, however it has been suggested that polymerase dimerization is necessary to coordinate leading and lagging strand synthesis [6,20].

Gene 45 Clamp
The gene 45 protein (gi:5354263, NP_049666), a 228 residue protein, is the polymerase-associated processivity

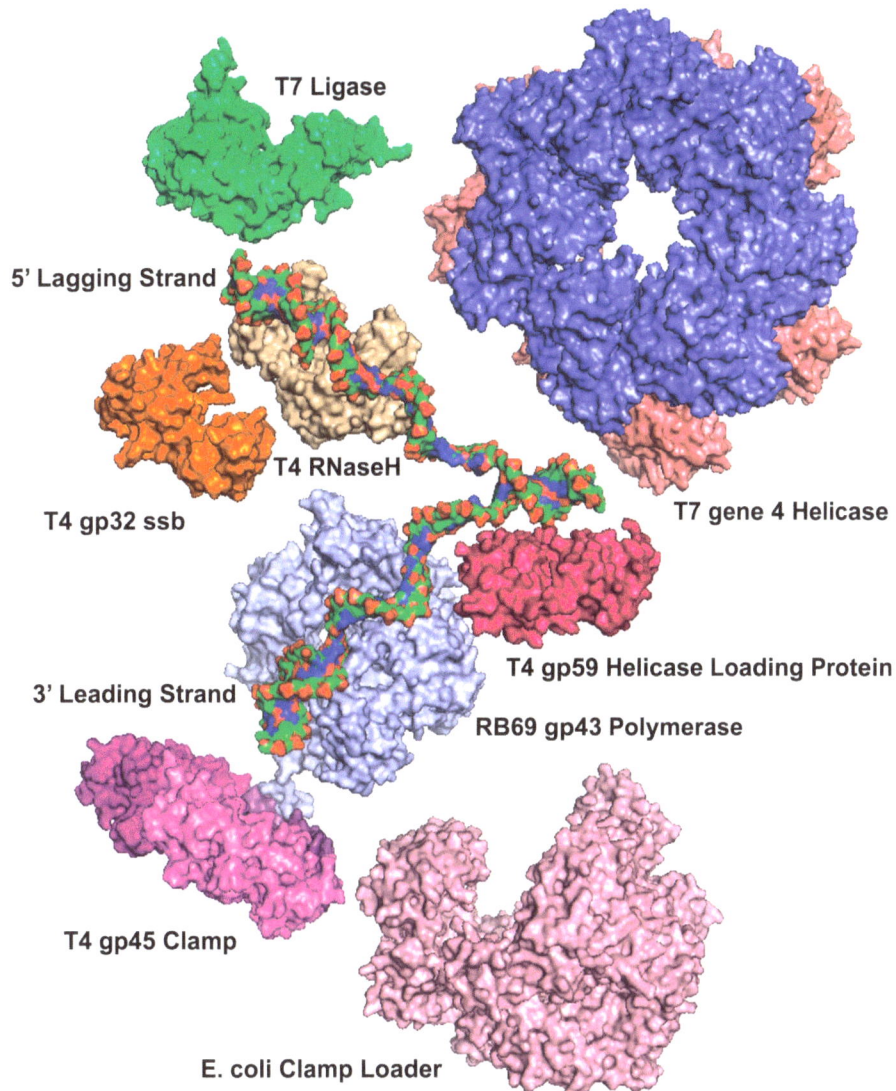

Figure 2 The molecular models, rendered to scale, of a DNA replication fork. Structures of four of ten T4 proteins are known; the RNase H (tan), the gp59 helicase loading protein (rose), the gp45 clamp (magenta), and the gp32 ssb (orange). Two additional structures from RB69, a T4 related phage, have also been completed; the RB69 gp43 polymerase (light blue) and the gp45 clamp (not shown). The E. coli clamp loader (γ complex) (pink) is used here in place of the T4 gp44/62 clamp loader, and two proteins from bacteriophage T7, T7 ligase (green) and T7 gene 4 helicase-primase (blue/salmon) are used instead of T4 ligase, and gp41/gp61, respectively.

clamp, and is a functional analog to the β subunit of *E. coli* Pol III holoenzyme and the eukaryotic proliferating cell nuclear antigen (PCNA) [8]. All proteins in this family, both dimeric (*E. co*li β) and trimeric (gp45, PCNA), form a closed ring represented here by the structure of the T4 gp45 (PDB 1czd, Figure 4A) [29]. The diameter of the central opening of all known clamp rings is slightly larger than duplex B-form DNA. When these clamps encircle DNA, basic residues lining the rings (T4 gp45 residues Lys 5 and 12, Arg 124, 128, and 131) interact with backbone phosphates. The clamps have an α/β structure with α-helices creating the inner wall of the ring. The anti-parallel β-sandwich fold forms

the outer scaffolding. While most organisms utilize a polymerase clamp, some exceptions are known. For example, bacteriophage T7 gene 5 polymerase sequesters *E. coli* thioredoxin for use as a processivity factor [42].

The gp45 related PCNA clamp proteins participate in many protein/DNA interactions including DNA replication, repair, and repair signaling proteins. A multitude of different proteins have been identified that contain a PCNA interaction protein box (PIP box) motif Qxxhxxaa where x is any residue, h is L, I or M, and a is aromatic [43]. In T4, PIP box sequences have been identified in the C-terminal domain of gp43 DNA

Table 2 Proteins of the DNA Replication Fork and Protein Database (pdb) reference numbers

	T4 and Related Phage	T7 Phage	*E. coli*	Eukaryotes
Replicative DNA polymerase	gp43 polymerase (*pdb* 1ig9, 1clq, 1noy, 1ih7)	Gene 5 Polymerase (*pdb* 1t7p, 1skr, 1t8e, 1x9m)	Pol III (αεθ) (*pdb* 2hnh, 1ido)	Pol δ (subunits p125, p50, p66, p12) (*pdb* 3e0j)
Sliding clamp	gp45 protein (*pdb* 1czd, 1b8h)	*E. coli* Thioredoxin (*pdb* 1t7p, 1skr)	β (*pdb* 2pol)	PCNA (*pdb* 1plr, 1axc, 1ul1)
Clamp loader	gp44/62 protein		γ complex (γδδ'ψχ) (*pdb* 1jr3, 1jqj, 3glf)	RF-C (subunits p140, p40, p38, p37, p36) (*pdb* 1sxj)
ssDNA binding protein	gp32 protein (*pdb* 1gpc)	Gene 2.5 protein (*pdb* 1je5)	SSB (*pdb* 1qvc)	RP-A (subunits p14, p32, p70) (*pdb* 1fgu, 2b29, 1jmc, 1l1o, 2pi2)
Replicative helicase	gp41 helicase	Gene 4 helicase (*pdb* 1e0j, 1e0k, 1q57)	DnaB (*pdb* 1b79, 2r6a)	MCM (*pdb* 3f9v, 1ltl)
helicase assembly protein	gp59 protein (*pdb* 1c1k)		DnaC, PriA (*pdb* 3ec2)	
primase	gp61 primase	Gene 4 primase (*pdb* 1nui, 1q57)	DnaG (*pdb* 1dd9, 3b39, 2r6c)	Pol α/primase (*pdb* 3flo)
5' to 3' Exonuclease	RNase H (*pdb* 1tfr, 2ihn)		Pol I N-domain	FEN-1, RNase H1 (*pdb* 1ul1, 2qk9)
DNA ligase 1	T4 ligase (gp30)	T7 ligase (*pdb* 1a0i)	DNA ligase (pdb 2owo)	DNA ligase I

polymerase, mentioned above, and in the N-terminal domain of RNase H, discussed below. The C-terminal PIP box peptide from RB69 gp43 DNA polymerase has been co-crystallized with RB69 gp45 clamp protein (PDB 1b8h, Figures 3A and 3C) and allows modeling of the gp45 clamp and gp43 DNA polymerase complex (Figure 3A) [31]. The gp45 clamp trails behind the 43 DNA polymerase, coupled through the gp43 C-terminal PIP box bound to a pocket on the outer surface of the gp45 clamp protein. Within RB69 gp45 clamp protein, the binding pocket is primarily hydrophobic (residues Tyr 39, Ile 107, Phe 109, Trp 199, and Val 217) with two basic residues (Arg 32 and Lys 204) interacting with the acidic groups in the PIP box motif. The rate of DNA synthesis, in the presence and absence of gp45 clamp protein, is approximately 400 nucleotides per second, indicating that the accessory gp45 clamp protein does not affect the enzymatic activity of the gp43 DNA polymerase [6]. More discussion about the interactions between T4 gp43 polymerase and T4 gp45 clamp can be found in Geiduschek and Kassavetis, this series. While the gp45 clamp is considered to be a processivity factor, this function may be most prevalent when misincorporation occurs. When a mismatch is introduced, the template strand releases, activating the 3' to 5' exonuclease activity of the gp43 DNA polymerase. During the switch, gp45 clamp maintains the interaction between the replicase and DNA.

Gene 44/62 Clamp Loader

The mechanism for loading of the ring shaped PCNA clamps onto duplex DNA is a conundrum; imagine a magician's linking rings taken apart and reassembled without an obvious point for opening. The clamp loaders, the magicians opening the PCNA rings, belong to the AAA + ATPase family which include the *E. coli*

gamma (γ) complex and eukaryotic replication factor C (RF-C) [44,45]. The clamp loaders bind to the sliding clamps, open the rings through ATP hydrolysis, and then close the sliding clamps around DNA, delivering these ring proteins to initiating replisomes or to sites of DNA repair. The gp44 clamp loader protein (gi:5354262, NP_049665) is a 319 residue, two-domain, homotetrameric protein. The N-domain of gp44 clamp loader protein has a Walker A p-loop motif (residues 45-52, **GTRGVGKT**) [38]. The gp62 clamp loader protein (gi:5354306, NP_049664) at 187 residues, is half the size of gp44 clamp loader protein and must be co-expressed with gp44 protein to form an active recombinant complex [46].

The T4 gp44/62 clamp loader complex is analogous to the *E. coli* heteropentameric γ complex (γ₃δ'δ) and yeast RF-C despite an almost complete lack of sequence homology with these clamp loaders [46]. The yeast p36, p37, and p40 subunits of RF-C are equivalent to the *E. coli* γ, yeast p38 subunit is equivalent to δ', and yeast p140 subunit is equivalent to δ[47]. The T4 homotetrameric gp44 clamp loader protein is equivalent to the *E. coli* γ₃δ' and T4 gp62 clamp loader is equivalent to the *E. coli* δ. The first architectural view of clamp loaders came from the collaborative efforts of John Kuriyan and Mike O'Donnell who have completed crystal structures of several components of the *E. coli* Pol III holoenzyme including the ψ-χ complex (PDB 1em8), the β-δ complex (PDB 1jqj) and the full γ complex γ₃δ'δ (PDB 1jr3, Figure 4B) [48-50]. More recently, the yeast RF-C complex has been solved (PDB 1sxj) [47]. Mechanisms of all clamp loaders are likely very similar, therefore comparison of T4 gp44/62 clamp loader protein with the *E. coli* model system is most appropriate. The *E. coli* γ₃δ', referred to as the motor/stator (equivalent to T4

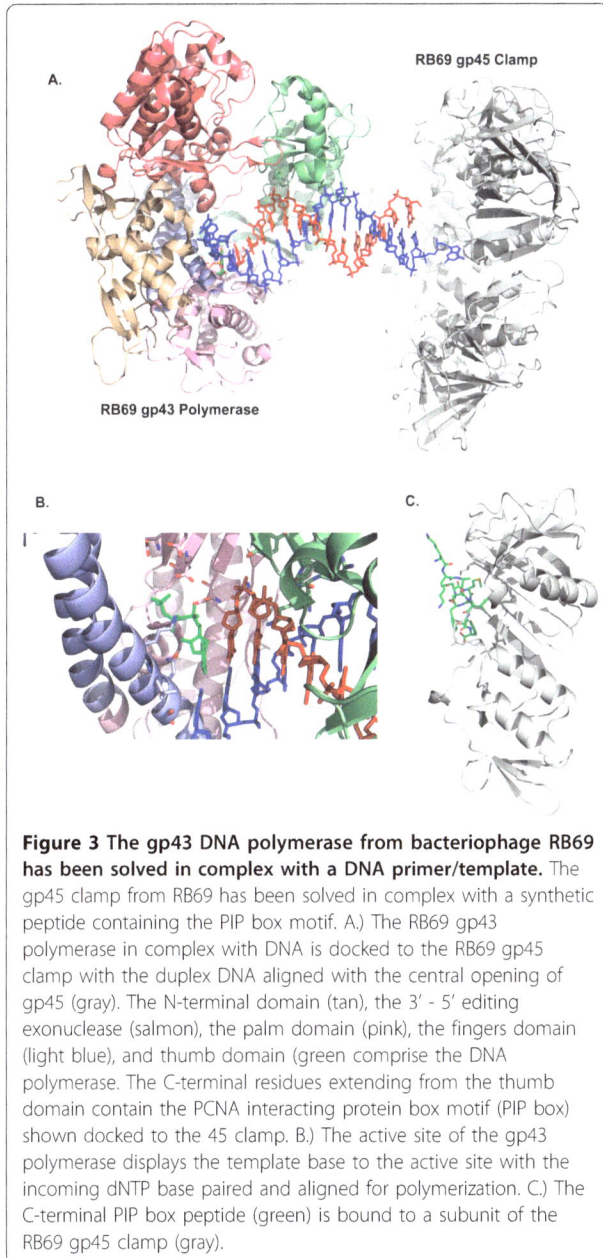

Figure 3 The gp43 DNA polymerase from bacteriophage RB69 has been solved in complex with a DNA primer/template. The gp45 clamp from RB69 has been solved in complex with a synthetic peptide containing the PIP box motif. A.) The RB69 gp43 polymerase in complex with DNA is docked to the RB69 gp45 clamp with the duplex DNA aligned with the central opening of gp45 (gray). The N-terminal domain (tan), the 3' - 5' editing exonuclease (salmon), the palm domain (pink), the fingers domain (light blue), and thumb domain (green comprise the DNA polymerase. The C-terminal residues extending from the thumb domain contain the PCNA interacting protein box motif (PIP box) shown docked to the 45 clamp. B.) The active site of the gp43 polymerase displays the template base to the active site with the incoming dNTP base paired and aligned for polymerization. C.) The C-terminal PIP box peptide (green) is bound to a subunit of the RB69 gp45 clamp (gray).

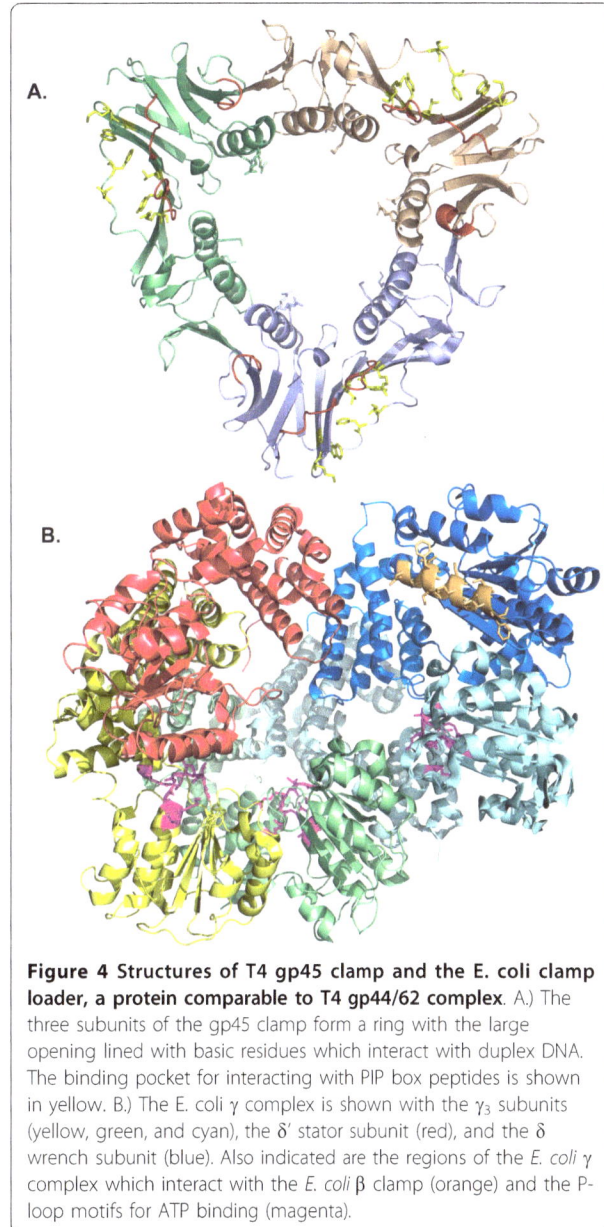

Figure 4 Structures of T4 gp45 clamp and the E. coli clamp loader, a protein comparable to T4 gp44/62 complex. A.) The three subunits of the gp45 clamp form a ring with the large opening lined with basic residues which interact with duplex DNA. The binding pocket for interacting with PIP box peptides is shown in yellow. B.) The E. coli γ complex is shown with the γ₃ subunits (yellow, green, and cyan), the δ' stator subunit (red), and the δ wrench subunit (blue). Also indicated are the regions of the *E. coli* γ complex which interact with the *E. coli* β clamp (orange) and the P-loop motifs for ATP binding (magenta).

gp44 clamp loader protein), binds and hydrolyzes ATP, while the δ subunit, known as the wrench (equivalent to T4 gp62 clamp loader protein), binds to the β clamp (T4 gp45 clamp protein). The *E. coli* γ complex is comparable in size to the *E. coli* β clamp and the two proteins interact face to face, with one side of the β clamp dimer interface bound to the δ (wrench) subunit, and the other positioned against the δ' (stator). Upon hydrolysis of ATP, the γ (motor) domains rotate, the δ subunit pulls on one side of a β clamp interface as the δ' subunit pushes against the other side of the β clamp, resulting in ring opening. For the T4 system, interaction with DNA and the presence of the gp43 DNA

polymerase releases the gp45 clamp from the gp44/62 clamp loader. In the absence of gp43 DNA polymerase, the gp44/62 clamp loader complex becomes a clamp unloader[6]. Current models of the *E. coli* Pol III holoenzyme have leading and lagging strand synthesis coordinated with a single clamp loader coupled to two DNA polymerases through the τ subunit and to single-stranded DNA binding protein through the χ subunit [51]. There are no T4 encoded proteins that are comparable to *E. coli* τ or χ.

Gene 32 Single-Stranded DNA Binding Protein
Single-stranded DNA binding proteins have an oligonucleotide-oligosaccharide binding fold (OB fold), an open

curved antiparallel β-sheet [52,53]. The aromatic residues within the OB fold stack with bases, thereby reducing the rate of spontaneous deamination of single-stranded DNA [54]. The OB fold is typically lined with basic residues for interaction with the phosphate backbone to increase stability of the interaction. Cooperative binding of ssb proteins assists in unwinding the DNA duplex at replication forks, recombination intermediates, and origins of replication. The T4 gp32 single-stranded DNA binding protein (gi:5354247, NP_049854) is a 301 residue protein consisting of three domain. The N-terminal basic B-domain (residues 1 - 21) is involved in cooperative interactions, likely through two conformations[55]. In the absence of DNA, the unstructured N-terminal domain interferes with the protein multimerization. In the presence of DNA, the lysine residues within the N-terminal peptide presumably interact with the phosphate backbone of DNA. Organization of the gp32 N-terminus by DNA creates the cooperative binding site for assembly of gp32 ssb filaments [56].

The crystal structure of the core domain of T4 gp32 ssb protein (residues 22 - 239) containing the single OB fold has been solved (Figure 5A) [33]. Two extended and two short antiparallel β-strands form the open cavity of the OB fold for nucleotide interaction. Two helical regions stabilize the β-strands, the smaller of which, located at the N-terminus of the core, has a structural zinc finger motif (residues His 64, and Cys 77, 87, and 90). The C-terminal acidic domain A-domain (residues 240 - 301) is involved in protein assembly, interacting with other T4 proteins, including gp61 primase, gp59 helicase assembly protein, and RNase H [57]. We have successfully crystallized the gp32(-B) construct (residues 21 - 301), but have found the A-domain disordered in the crystals with only the gp32 ssb core visible in the electron density maps (Hinerman, unpublished data). The analogous protein in eukaryotes is the heterotrimeric replication protein A (RPA) [58]. Several structures of Archaeal and Eukaryotic RPAs have been reported including the crystal structure of a core fragment of human RPA70 [59,60]. The RPA70 protein is the largest of the three proteins in the RPA complex and has two OB fold motifs with 9 bases of single-stranded DNA bound (PDB 1jmc). The *E. coli* ssb contains four OB fold motifs and functions as a homotetramer. A structure of the full length version of *E. coli* ssb (PDB 1sru) presents evidence that the C terminus (equivalent to the T4 32 A domain) is also disordered [61].

Primosomal Proteins

Gene 41 Helicase

The replicative helicase family of enzymes, which includes bacteriophages T4 gp41 helicase and T7 gene 4 helicase, *E. coli* DnaB, and the eukaryotic MCM proteins, are responsible for the unwinding of duplex DNA in

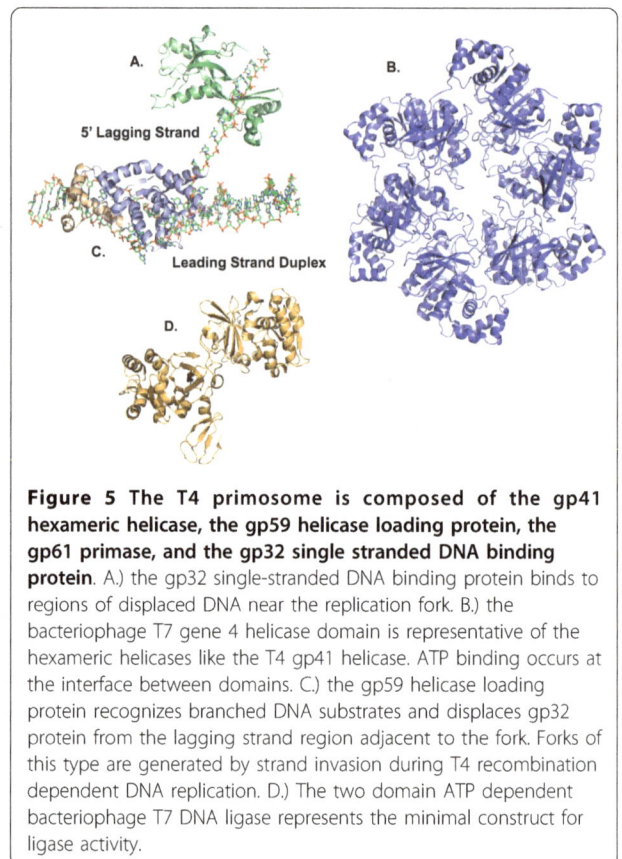

Figure 5 The T4 primosome is composed of the gp41 hexameric helicase, the gp59 helicase loading protein, the gp61 primase, and the gp32 single stranded DNA binding protein. A.) the gp32 single-stranded DNA binding protein binds to regions of displaced DNA near the replication fork. B.) the bacteriophage T7 gene 4 helicase domain is representative of the hexameric helicases like the T4 gp41 helicase. ATP binding occurs at the interface between domains. C.) the gp59 helicase loading protein recognizes branched DNA substrates and displaces gp32 protein from the lagging strand region adjacent to the fork. Forks of this type are generated by strand invasion during T4 recombination dependent DNA replication. D.) The two domain ATP dependent bacteriophage T7 DNA ligase represents the minimal construct for ligase activity.

front of the leading strand replisome [62]. The T4 gp41 protein (gi:9632635, NP_049654) is the 475 residue helicase subunit of the primase(gp61)-helicase(gp41) complex and a member of the p-loop NTPase family of proteins [63]. Similar to other replicative helicases, the gp41 helicase assembles by surrounding the lagging strand and excluding the leading strand of DNA. ATP hydrolysis translocates the enzyme 5' to 3' along the lagging DNA strand, thereby unwinding the DNA duplex approximately one base pair per hydrolyzed ATP molecule. Efforts to crystallize full length or truncated gp41 helicase individually, in complex with nucleotide analogs, or in complex with other T4 replication proteins have not been successful in part due to the limited solubility of this protein. In addition, the protein is a heterogeneous mixture of dimers, trimers and hexamers, according to dynamic light scattering measurements. The solubility of T4 41 helicase can be improved to greater than 40 mg/ml of homogenous hexamers by eliminating salt and using buffer alone (10 mM TAPS pH 8.5) [64]. However, the low ionic strength crystal screen does not produce crystals [65]. To understand the T4 gp41 helicase, we must therefore look to related model systems.

Like T4 41 helicase, efforts to crystallize *E. coli* DnaB have met with minimal success. Thus far only a fragment of the non-hexameric N-terminal domain (PDB 1b79) has been crystallized successfully for structural determinations [66]. More recently, thermal stable eubacteria (*Bacillus* and *Geobacillus stearothermophilis*) have been utilized by the Steitz lab to yield more complete structures of the helicase-primase complex (PDB 2r6c and 2r6a, respectively) [67]. A large central opening in the hexamer appears to be the appropriate size for enveloping single-stranded DNA, as it is too small for duplex DNA. Collaborative efforts between the Wigley and Ellenberger groups revealed the hexameric structure of T7 gene 4 helicase domain alone (residues 261 - 549, PDB 1eOk) and in complex with a non-hydrolyzable ATP analog (PDB 1e0h) [68]. Interestingly, the central opening in the T7 gene 4 helicase hexamer is smaller than other comparable helicase, suggesting that a fairly large rearrangement is necessary to accomplish DNA binding. A more complete structure from the Ellenberger lab of T7 gene 4 helicase that includes a large segment of the N-terminal primase domain (residues 64 - 566) reveals a heptameric complex with a larger central opening (Figure 5B) [69]. Both the eubacterial and bacteriophage helicase have similar α/β folds. The C-terminal Rec A like domain follows 6-fold symmetry and has nucleotide binding sites at each interface. In the eubacterial structures, the helical N-domains alternate orientation and follow a three-fold symmetry with domain swapping. The T4 gp41 helicase is a hexameric two-domain protein with Walker A p-loop motif (residues 197 - 204, **GVNVGKS**) located at the beginning of the conserved NTPase domain (residues 170 - 380), likely near the protein:protein interfaces, similar to the T7 helicase structure.

Gene 59 Helicase Assembly Protein

The progression of the DNA replisome is restricted in the absence of either gp32 ssb protein or the gp41 helicase [6]. In the presence of gp32 ssb protein, loading of the gp41 helicase is inhibited. In the absence of gp32 ssb protein, the addition of gp41 helicase improves the rate of DNA synthesis but displays a significant lag prior to reaching maximal DNA synthesis [13]. The gp59 helicase loading protein (gi:5354296, NP_049856) is a 217 residue protein that alleviates the lag phase of gp41 helicase [13,14]. In the presence gp32 ssb protein, the loading of gp41 helicase requires gp59 helicase loading protein. This activity is similar to the *E. coli* DnaC loading of DnaB helicase [70,71]. Initially, 59 helicase loading protein was thought to be a single-stranded DNA binding protein that competes with 32 ssb protein on the lagging strand [13,72]. In that model, the presence of gp59 protein within the gp32 filament presumably created a docking site for gp41 helicase. However, the gp59 helicase loading protein is currently known to have more specific binding affinity for branched and

Holliday junctions [16,17]. This activity is comparable to the *E. coli* replication rescue protein, PriA, which was first described as the PAS recognition protein (n' protein) in ϕX174 phage replication [73]. Using short pseudo-Y junction DNA substrates, gp59 helicase loading protein has been shown to recruit gp32 ssb protein to the 5' (lagging strand) arm, a scenario relevant to replication fork assembly [74].

The high-resolution crystal structure of 59 helicase loading protein reveals a two-domain, α-helical structure that has no obvious cleft for DNA binding [17]. The *E. coli* helicase loader, DnaC, is also a two domain protein. However, the C-terminal domain of DnaC is an AAA + ATPase related to DnaA, as revealed by the structure of a truncated DnaC from *Aquifex aeolicus* (pdb 3ec2) [75]. The DnaC N-domain interacts with the hexameric DnaB in a one-to-one ratio forming a second hexameric ring. Sequence alignments of gp59 helicase loading protein reveal an "ORFaned" (orphaned open reading frame) protein; a protein that is unique to the T-even and other related bacteriophages [4,17]. Interestingly, searches for structural alignments of the gp59 protein, using both Dali [76] and combinatorial extension [77], have revealed partial homology with the eukaryotic high mobility group 1A (HMG1A) protein, a nuclear protein involved in chromatin remodeling [78]. Using the HMG1A:DNA structure as a guide, we have successfully modeled gp59 helicase assembly protein bound to a branched DNA substrate which suggests a possible mode of cooperative interaction with 32 ssb protein (Figure 5C) [17]. Attempts to co-crystallize gp59 protein with DNA, or with gp41 helicase, or with gp32 ssb constructs have all been unsuccessful. The 59 helicase assembly protein combined with 32(-B) ssb protein yields a homogenous solution of heterodimers, amenable for small angle X-ray scattering analysis (Hinerman, unpublished data).

Gene 61 Primase

The gp61 DNA dependent RNA polymerase (gi:5354295, NP_049648) is a 348 residue enzyme that is responsible for the synthesis of short RNA primers used to initiate lagging strand DNA synthesis. In the absence of gp41 helicase and gp32 ssb proteins, the gp61 primase synthesizes ppp(Pu)pC dimers that are not recognized by DNA polymerase [79,80]. A monomer of gp61 primase and a hexamer of gp41 helicase are essential components of the initiating primosome [63,81]. Each subunit of the hexameric gp41 helicase has the ability to bind a gp61 primase. Higher occupancies of association have been reported but physiological relevance is unclear [82,83]. When associated with gp41 helicase, the gp61 primase synthesizes pentaprimers that begin with 5'-pppApC onto template 3'-TG; a very short primer that does not remain annealed in the absence of protein [79]. An interaction between gp32 ssb protein and gp61 primase likely coordinates the handoff of the RNA primer to the gp43 DNA polymerase, establishing a synergy between leading strand progression and lagging strand synthesis [84]. The gp32 ssb protein will bind to single-stranded DNA unwound by gp41 helicase. This activity inhibits the majority of 3'-TG template sites for gp61 primase and therefore increases the size of Okazaki fragments [6]. Activity of gp61 primase is obligate to the activity of the gp41 helicase. The polymerase accessory proteins, gp45 clamp and gp44/62 clamp loader, are essential for primer synthesis when DNA is covered by gp32 ssb protein [85]. Truncation of 20 amino acids from the C-terminus of gp41 helicase protein retains interaction with gp61 primase but eliminates gp45 clamp and gp44/62 clamp loader stimulation of primase activity [86].

The gp61 primase contains an N-terminal zinc finger DNA binding domain (residues cys 37, 40, 65, and 68) and a central toprim catalytic core domain (residues 179 - 208) [87,88]. Crystallization trials of full length gp61 primase and complexes with gp41 helicase have been unsuccessful. Publication of a preliminary crystallization report of gp61 primase C-terminal domain (residues 192 - 342) was limited in resolution, and a crystal structure has not yet been published [89]. A structure of the toprim core fragment of E. coli DnaG primase (residues 110 to 433 of 582) has been solved, concurrently by the Berger and Kuriyan labs (PDB 1dd9, [90]) (PDB 1eqn, [91]). To accomplish this, the N-terminal Zn finger and the C-terminal DnaB interacting domain were removed. More recently, this same DnaG fragment has been resolved in complex with single-stranded DNA revealing a binding track adjacent to the toprim domain (PDB 3b39, [92]). Other known primase structures include the Stearothermophilis enzymes solved in complex with helicase (discussed above) and the primase domain of T7 gene 4 primase (PDB 1nui) (Figure 5D) [69]. The primase domain of T7 gene 4 is comprised of the N-terminal Zn finger (residues 1 - 62) and toprim domain (residues 63 - 255). This structure is actually a primase-helicase fusion protein.

Okazaki Repair Proteins
RNase H, 5' to 3' exonuclease
RNase H activity of the bacteriophage T4 rnh gene product (gi:5354347, NP_049859) was first reported by Hollingsworth and Nossal [24]. The structure of the 305 residue enzyme with two metals bound in the active site was completed in collaboration with the Nossal laboratory (PDB 1tfr) (Figure 6A) [28]. Mutations of highly conserved residues which abrogate activity are associated with the two hydrated magnesium ions [93]. The site I metal is coordinated by four highly conserved aspartate residues (D19, D71, D132, and D155) and

mutation of any one to asparagines eliminates nuclease activity. The site II metal is fully hydrated and hydrogen bonded to three aspartates (D132, D157, and D200) and to the imino nitrogen of an arginine, R79. T4 RNase H has 5' to 3' exonuclease activity on RNA/DNA, DNA/DNA 3'overhang, and nicked substrate, with 5' to 3' endonuclease activity on 5' fork and flap DNA substrates. The crystal structure of T4 RNase H in complex with a pseudo Y junction DNA substrate has been solved (PDB 2ihn, Figure 6B) [94]. To obtain this structure, it was necessary to use an active site mutant (D132N); Asp132 is the only residue in RNase H that is inner sphere coordinated to the active site metals [28].

The processivity of RNase H exonuclease activity is enhanced by the gp32 ssb protein. Protein interactions can be abrogated by mutations in the C-terminal domain of RNase H [22] and within the core domain of

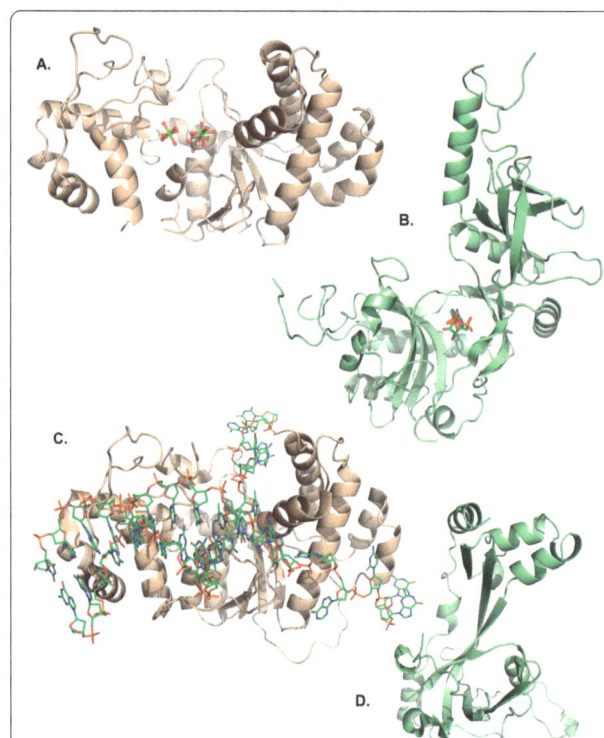

Figure 6 Lagging strand DNA synthesis requires repair of the Okazaki fragments. A.) The T4 RNase H, shown with two hydrated magnesium ions (green) in the active site, is a member of the rad2/FEN-1 family of 5' - 3' exonucleases. The enzyme is responsible for the removal of lagging strand RNA primers and several bases of DNA adjacent to the RNA primer which are synthesized with low fidelity by the gp43 DNA polymerase. B.) The T4 DNA ligase, shown with ATP bound in the active site, repairs nicks present after primer removal and gap synthesis by the DNA polymerase. C.) The T4 RNase H structure has been solved with a pseudo-Y junction DNA substrate. D.) The gp32 single stranded binding protein increases the processivity of the RNase H. The two proteins interact between the C-terminal domain of RNase H and the core domain of gp32 on the 3' arm of the replication fork.

gp32 ssb protein (Mueser, unpublished data). Full length gp32 ssb protein and RNase H do not interact in the absence of DNA substrate. Removal of the N-terminal peptide of gp32 ssb protein (gp32(-B)), responsible for gp32 ssb cooperativity, yields a protein that has high affinity for RNase H. It is likely that the reorganization of gp32 B-domain when bound to DNA reveals a binding site for RNase H and therefore helps to coordinate 5'-3' primer removal after extension by the DNA polymerase. This is compatible with the model proposed for the cooperative self assembly of gp32 protein. The structure of RNase H in complex with gp32(-B) has been solved using X-ray crystallography and small angle X-ray scattering (Mueser, unpublished data) (Figure 6C). The gp45 clamp protein enhances the processivity of RNase H on nicked and flap DNA substrates [23]. Removal of the N-terminal peptide of RNase H eliminates the interaction between RNase H and gp45 clamp protein and decreases processivity of RNase H. The structure of the N-terminal peptide of RNase H in complex with gp45 clamp protein reveals that binding occurs within the gp45 clamp PIP-box motif of RNase H (Devos, unpublished data).

Sequence alignment of T4 RNase H reveals membership to a highly conserved family of nucleases that includes yeast rad27, rad2, human FEN-1, and xeroderma pigmentosa group G (XPG) proteins. The domain structure of both FEN-1 and XPG proteins is designated N, I, and C [95]. The yeast rad2 and human XPG proteins are much larger than the yeast rad27 and human FEN-1 proteins. This is due to a large insertion in the middle of rad2 and XPG proteins between the N and I domains. The N and I domains are not separable in the T4 RNase H protein as the N-domain forms part of the α/β structure responsible for fork binding and half of the active site. The I domain is connected to the N-domain by a bridge region above the active site which is unstructured in the presence of active site metals and DNA substrate. It is this region that corresponds with the position of the large insertions of rad2 and XPG. Curiously, this bridge region of T4 RNase H becomes a highly ordered a-helical structure in the absence of metals. Arg and Lys residues are interdigitated between the active site Asp groups within the highly ordered structure (Mueser, unpublished data). The I domain encompasses the remainder of the larger α/β subdomain and the α-helical H3TH motif responsible for duplex binding. The C-domain is truncated at the helical cap that interacts with gp32 ssb and the PIP motif is located in the N-terminus of T4 RNase H. In the FEN-1 family of proteins, the C-domain, located opposite the H3TH domain, contains a helical cap and an unstructured C-terminal PIP-box motif for interaction with a PCNA clamp.

Gene 30 DNA Ligase

The T4 gp30 protein (gi: 5354233, NP_049813) is best known as T4 DNA ligase, a 487 residue ATP-dependent ligase. DNA ligases repair nicks in double-stranded DNA containing 3' OH and 5' phosphate ends. Ligases are activated by the covalent modification of a conserved lysine with AMP donated by NADH or ATP. The conserved lysine and the nucleotide binding site reside in the adenylation domain (NTPase domain) of ligases. Sequence alignment of the DNA ligase family Motif 1 (**KXDGXR**) within the adenylation domain identifies Lys 159 in T4 DNA ligase (159 **K**ADGAR 164) as the moiety for covalent modification [96]. The bacterial ligases are NADH-dependent, while all eukaryotic enzymes are ATP-dependent [97]. Curiously, T4 phage, whose existence is confined within a prokaryote, encodes an ATP-dependent ligase. During repair, the AMP group from the activated ligase is transferred to the 5' phosphate of the DNA nick. This activates the position for condensation with the 3' OH, releasing AMP in the reaction. The T4 ligase has been cloned, expressed, and purified but attempts to crystallize T4 ligase, with and without cofactor, have not been successful. The structure of the bacteriophage T7 ATP-dependent ligase has been solved (PDB 1a0i, Figure 6C) [98,99], which has a similar fold to T4 DNA ligase [100]. The minimal two-domain structure of the 359 residue T7 ligase has a large central cleft, with the larger N-terminal adenylation domain containing the cofactor binding site and a C-terminal OB domain. In contrast, the larger 671 residue E. coli DNA ligase has five domains; the N-terminal adenylation and OB fold domains, similar to T7 and T4 ligase, including a Zn finger, HtH and BRCT domains present in the C-terminal half of the protein [97]. Sequence alignment of DNA ligases indicate that the highly conserved ligase signature motifs reside in the central DNA binding cleft, the active site lysine, and the nucleotide binding site [98]. Recently, the structure of NAD-dependent *E. coli* DNA ligase has been solved in complex with nicked DNA containing an adenylated 5' PO_4 (pdb 2owo) [101]. This flexible, multidomain ligase encompasses the duplex DNA with the adenylation domain binding to the nick; a binding mode also found in the human DNA ligase 1 bound to nicked DNA (pdb 1x9n) [102]. T4 DNA ligase is used routinely in molecular cloning for repairing both sticky and blunt ends. The smaller two-domain structure of T4 DNA ligase has lower affinity for DNA than the multidomain ligases. The lack of additional domains to encompass the duplex DNA likely explains the sensitivity of T4 ligase activity to salt concentration.

Conclusion and Future Directions of Structural Analysis

The bacteriophage T4 model system has been an invaluable resource for investigating fundamental aspects of

DNA replication. The phage DNA replication system has been reconstituted for both structural and enzymatic studies. For example, the in vitro rates and fidelity of DNA synthesis are equivalent to those measured in vivo. These small, compact proteins define the minimal requirements for enzymatic activity and are the most amenable to structural studies. The T4 DNA replication protein structures reveal the basic molecular requirements for DNA synthesis. These structures, combined with those from other systems, allow us to create a visual image of the complex process of DNA replication.

Macromolecular crystallography is a biophysical technique that is now available to any biochemistry enabled laboratory. Dedicated crystallographers are no longer essential; a consequence of advances in technology. Instead, biologists and biochemists utilize the technique to compliment their primary research. In the past, the bottleneck to determining X-ray structures was data collection and analysis. Over the past two decades, multiple wavelength anomalous dispersion phasing (MAD phasing) has been accompanied by the adaptation of charge-coupled device (CCD) cameras for rapid data collection, and the construction of dedicated, tunable X-ray sources at the National Laboratory facilities such as the National Synchrotron Light Source (NSLS) at Brookhaven National Labs (BNL), the Advanced Light Source (ALS) at Lawrence Berkeley National Labs (LBNL), and the Advanced Photon Source (APS) at Argonne National Labs (ANL). These advances have transformed crystallography to a fairly routine experimental procedure. Today, many of these national facilities provide mail-in service with robotic capability for remote data collection, eliminating the need for expensive in-house equipment. The current bottle neck for protein crystallography has shifted into the realm of molecular cloning and protein purification of macromolecules amenable to crystallization. Even this aspect of crystallography has been commandeered by high throughput methods as structural biology centers attempt to fill "fold space".

A small investment in crystallization tools, by an individual biochemistry research lab, can take advantage of the techniques of macromolecular crystallography. Dedicated suppliers (e.g. Hampton Research) sell crystal screens and other tools for the preparation, handling, and cryogenic preservation of crystals, along with web-based advice. The computational aspects of crystallography are simplified and can operate on laptop computers using open access programs. Data collection and reduction software are typically provided by the beam lines. Suites of programs such as CCP4 [103] and PHENIX [104,105] provide data processing, phasing, and model refinement. Visualization software has been dominated in recent years by the Python [106] based programs COOT [107] for model building and PYMOL, developed by the late Warren DeLano, for the presentation of models for publication. In all, a modest investment in time and resources can convert any biochemistry lab into a structural biology lab.

What should independent structural biology research labs focus on, in the face of competition from high throughput centers? A promising frontier is the visualization of complexes, exemplified by the many protein: DNA complexes with known structures. A multitude of transient interactions occur during DNA replication and repair, a few of these have been visualized in the phage-encoded DNA replication system. The RB69 gp43 polymerase has been crystallized in complex with DNA, and with gp32 ssb as a fusion protein [36,108]. The gp45 clamp bound with PIP box motif peptides have been used to model the gp43:gp45 interaction [31]. The bacteriophage T4 RNase has been solved in complex with a fork DNA substrate and in complex with gp32 for modeling of the RNaseH:gp32:DNA ternary complex. These few successes required investigation of multiple constructs to obtain a stable, homogeneous complex, therefore indicating that the probability for successful crystallization of protein:DNA constructs can be significantly lower than for solitary protein domains.

Small angle X-ray and Neutron scattering

Thankfully, the inability to crystallize complexes does not preclude structure determination. Multiple angle and dynamic light scattering techniques (MALS and DLS, respectively) use wavelengths of light longer than the particle size. This allows the determination of the size and shape of macromolecular complex. Higher energy light with wavelengths significantly shorter than the particle size provides sufficient information to generate a molecular envelope comparable to those manifested from cryoelectron microscopy image reconstruction. Small angle scattering techniques including X-ray (SAXS) and neutron (SANS) are useful for characterizing proteins and protein complexes in solution. These low-resolution techniques provide information about protein conformation (folded, partially folded and unfolded), aggregation, flexibility, and assembly of higher-ordered protein oligomers and/or complexes [109]. The scattering intensity of biological macromolecules in solution is equivalent to momentum transfer $q = [4\pi \sin \theta/\lambda]$, where 2θ is the scattering angle and λ is the wavelength of the incident X-ray beam. Larger proteins will have a higher scattering intensity (at small angles) compared to smaller proteins or buffer alone. Small angle neutron scattering is useful for contrast variation studies of protein-DNA and protein-RNA complexes (using deuterated components) [110]. The contrast variation method uses the neutron scattering differences between hydrogen isotopes. For specific ratios of D_2O to H_2O in the solvent, the scattering contribution from DNA, RNA, or perdeuterated protein becomes negligible. This allows for the determination of spatial arrangement of components within the macromolecular complex [111]. There are dedicated SAXS beamlines available at NSLS and LBNL. Neutron studies, almost non-existent in the US in the 1990's, have made a comeback with the recent commissioning of the Spallation Neutron Source (SNS) and the High Flux Isotope Reactor (HFIR) at Oak Ridge National Laboratory (ORNL) to compliment the existing facility at the National Institute of Standards and Technology (NIST). The bombardment by neutrons is harmless to biological molecules, unlike high energy X-rays that induce significant damage to molecules in solution.

To conduct a scattering experiment, the protein samples should be monodisperse and measurements at different concentrations used to detect concentration-dependent aggregation. The scattering intensity from buffer components is subtracted from the scattering intensity of the protein sample, producing a 1-D scattering curve that is used for data analysis. These corrected scattering curves are evaluated using programs such as GNOM and PRIMUS, components of the ATSAS program suite [112]. Each program allows the determination of the radius of gyration (R_G), maximum particle distance, and molecular weight of the species in solution as well as the protein conformation. The 1-D scattering profiles are utilized to generate 3-D models. There are several methods of generating molecular envelopes including *ab initio* reconstruction (GASBOR, DAMMIN, GA_STRUCT), models based on known atomic structure (SASREF, MASSHA, CRYSOL), and a combination of *ab initio*/atomic structure models (CREDO, CHADD, GLOOPY). The *ab initio* programs use simulated annealing and dummy atoms or dummy atom chains to generate molecular envelopes, while structural-based modeling programs, like SASREF, use rigid-body modeling to orient the known X-ray structures into the experimental scattering intensities (verified by comparing experimental scattering curves to theoretical scattering curves). We have used these programs to generate molecular envelopes for the RNaseH:gp32(-B) complex and for the gp59:gp32(-B) complexes. The high resolution crystal structures of the components can be placed into the envelopes to model the complex.

Abbreviations

ALS: Advanced Light Source; ANL: Argonne National Labs; APS: Advanced Photon Source; BNL: Brookhaven National Labs; CCD: Charge coupled device; DLS: Dynamic light scattering; HFIR: High Flux Isotope Reactor; LBNL: Lawrence Berkeley National Labs; MAD: Multiple wavelength anomalous dispersion; MALS: Multiple angle light scattering; NIST: National Institute for Standards and Technology; NSLS: National Synchrotron Light Source; OB fold: Oligonucleotide-oligosaccharide binding fold; ORNL: Oak Ridge National Laboratory; PCNA: Proliferating cell nuclear antigen; PIP box: PCNA interaction protein box; RF-C: Replication factor - C; SAXS: Small angle X-ray scattering; SANS: Small angle neutron scattering; SNS: Spallation Neutron Source; ssb: single-stranded DNA binding; Toprim: topoisomerase-primase.

Acknowledgements
The authors wish to thank Dr. Leif Hanson for helpful suggestions. We also wish to dedicate this review to the memory of Dr. Nancy G. Nossal.

Author details
[1]Department of Chemistry, University of Toledo, Toledo OH, USA. [2]Department of Molecular Genetics, Biochemistry & Microbiology, University of Cincinnati College of Medicine, Cincinnati, OH, USA. [3]European Molecular Biology Laboratory, Grenoble Outstation, Grenoble, France. [4]Texas State University, San Marcos, TX, USA. [5]Department of Biochemistry and Cancer Biology, University of Toledo College of Medicine, Toledo OH, USA.

Authors' contributions
TM was the primary author of this manuscript and created the final constructions of tables and figures. JH contributed the review of scattering methods and assisted in drafting the manuscript. JD created Figures 1 and 2 and assisted in outlining the manuscript. RB created the movies for the supplemental information. KW assisted in drafting the manuscript, provided the expertise in eukaryotic DNA replication and repair, and contributed the majority of editorial assistance. All authors have read and approved the final manuscript.

Competing interests
The authors declare that they have no competing interests.

Received: 8 October 2010 Accepted: 3 December 2010
Published: 3 December 2010

References
1. Watson JD, Crick FH: **The structure of DNA**. *Cold Spring Harb Symp Quant Biol* 1953, **18**:123-131.
2. Bell PJ: **Viral eukaryogenesis: was the ancestor of the nucleus a complex DNA virus?** *J Mol Evol* 2001, **53**:251-256.
3. Bell PJ: **The viral eukaryogenesis hypothesis: a key role for viruses in the emergence of eukaryotes from a prokaryotic world environment**. *Ann N Y Acad Sci* 2009, **1178**:91-105.
4. Koonin EV, Senkevich TG, Dolja VV: **The ancient Virus World and evolution of cells**. *Biol Direct* 2006, **1**:29.
5. Kornberg A, Baker TA: *DNA Replication*. Second edition. New York: W.H. Freeman and Company; 1992.
6. Nossal NG: *The Bacteriophage T4 DNA Replication Fork* Washington, D.C.: American Society for Microbiology; 1994.
7. Mosig G: **Recombination and recombination-dependent DNA replication in bacteriophage T4**. *Annu Rev Genet* 1998, **32**:379-413.
8. Yao N, Turner J, Kelman Z, Stukenberg PT, Dean F, Shechter D, Pan ZQ, Hurwitz J, O'Donnell M: **Clamp loading, unloading and intrinsic stability of the PCNA, beta and gp45 sliding clamps of human, E. coli and T4 replicases**. *Genes Cells* 1996, **1**:101-113.
9. Venkatesan M, Nossal NG: **Bacteriophage T4 gene 44/62 and gene 45 polymerase accessory proteins stimulate hydrolysis of duplex DNA by T4 DNA polymerase**. *J Biol Chem* 1982, **257**:12435-12443.
10. Karam JD, Konigsberg WH: **DNA polymerase of the T4-related bacteriophages**. *Prog Nucleic Acid Res Mol Biol* 2000, **64**:65-96.
11. Venkatesan M, Silver LL, Nossal NG: **Bacteriophage T4 gene 41 protein, required for the synthesis of RNA primers, is also a DNA helicase**. *J Biol Chem* 1982, **257**:12426-12434.
12. Young MC, Schultz DE, Ring D, von Hippel PH: **Kinetic parameters of the translocation of bacteriophage T4 gene 41 protein helicase on single-stranded DNA**. *J Mol Biol* 1994, **235**:1447-1458.
13. Barry J, Alberts B: **Purification and characterization of bacteriophage T4 gene 59 protein. A DNA helicase assembly protein involved in DNA replication**. *J Biol Chem* 1994, **269**:33049-33062.
14. Yonesaki T: **The purification and characterization of gene 59 protein from bacteriophage T4**. *J Biol Chem* 1994, **269**:1284-1289.
15. Benkovic SJ, Valentine AM, Salinas F: **Replisome-mediated DNA replication**. *Annu Rev Biochem* 2001, **70**:181-208.

16. Jones CE, Mueser TC, Dudas KC, Kreuzer KN, Nossal NG: **Bacteriophage T4 gene 41 helicase and gene 59 helicase-loading protein: a versatile couple with roles in replication and recombination.** *Proc Natl Acad Sci USA* 2001, **98**:8312-8318.

17. Mueser TC, Jones CE, Nossal NG, Hyde CC: **Bacteriophage T4 gene 59 helicase assembly protein binds replication fork DNA. The 1.45 A resolution crystal structure reveals a novel alpha-helical two-domain fold.** *J Mol Biol* 2000, **296**:597-612.

18. Nelson SW, Kumar R, Benkovic SJ: **RNA primer handoff in bacteriophage T4 DNA replication: the role of single-stranded DNA-binding protein and polymerase accessory proteins.** *J Biol Chem* 2008, **283**:22838-22846.

19. Chastain PD, Makhov AM, Nossal NG, Griffith JD: **Analysis of the Okazaki fragment distributions along single long DNAs replicated by the bacteriophage T4 proteins.** *Mol Cell* 2000, **6**:803-814.

20. Salinas F, Benkovic SJ: **Characterization of bacteriophage T4-coordinated leading- and lagging-strand synthesis on a minicircle substrate.** *Proc Natl Acad Sci USA* 2000, **97**:7196-7201.

21. Sinha NK, Morris CF, Alberts BM: **Efficient in vitro replication of double-stranded DNA templates by a purified T4 bacteriophage replication system.** *J Biol Chem* 1980, **255**:4290-4293.

22. Gangisetty O, Jones CE, Bhagwat M, Nossal NG: **Maturation of bacteriophage T4 lagging strand fragments depends on interaction of T4 RNase H with T4 32 protein rather than the T4 gene 45 clamp.** *J Biol Chem* 2005, **280**:12876-12887.

23. Bhagwat M, Hobbs LJ, Nossal NG: **The 5′-exonuclease activity of bacteriophage T4 RNase H is stimulated by the T4 gene 32 single-stranded DNA-binding protein, but its flap endonuclease is inhibited.** *J Biol Chem* 1997, **272**:28523-28530.

24. Hollingsworth HC, Nossal NG: **Bacteriophage T4 encodes an RNase H which removes RNA primers made by the T4 DNA replication system in vitro.** *J Biol Chem* 1991, **266**:1888-1897.

25. Hobbs LJ, Nossal NG: **Either bacteriophage T4 RNase H or Escherichia coli DNA polymerase I is essential for phage replication.** *J Bacteriol* 1996, **178**:6772-6777.

26. Weiss B, Richardson CC: **Enzymatic breakage and joining of deoxyribonucleic acid, I. Repair of single-strand breaks in DNA by an enzyme system from Escherichia coli infected with T4 bacteriophage.** *Proc Natl Acad Sci USA* 1967, **57**:1021-1028.

27. Nossal NG, Hinton DM, Hobbs LJ, Spacciapoli P: **Purification of bacteriophage T4 DNA replication proteins.** *Methods Enzymol* 1995, **262**:560-584.

28. Mueser TC, Nossal NG, Hyde CC: **Structure of bacteriophage T4 RNase H, a 5′ to 3′ RNA-DNA and DNA-DNA exonuclease with sequence similarity to the RAD2 family of eukaryotic proteins.** *Cell* 1996, **85**:1101-1112.

29. Moarefi I, Jeruzalmi D, Turner J, O'Donnell M, Kuriyan J: **Crystal structure of the DNA polymerase processivity factor of T4 bacteriophage.** *J Mol Biol* 2000, **296**:1215-1223.

30. Nolan JM, Petrov V, Bertrand C, Krisch HM, Karam JD: **Genetic diversity among five T4-like bacteriophages.** *Virol J* 2006, **3**:30.

31. Shamoo Y, Steitz TA: **Building a replisome from interacting pieces: sliding clamp complexed to a peptide from DNA polymerase and a polymerase editing complex.** *Cell* 1999, **99**:155-166.

32. Wang J, Sattar AK, Wang CC, Karam JD, Konigsberg WH, Steitz TA: **Crystal structure of a pol alpha family replication DNA polymerase from bacteriophage RB69.** *Cell* 1997, **89**:1087-1099.

33. Shamoo Y, Friedman AM, Parsons MR, Konigsberg WH, Steitz TA: **Crystal structure of a replication fork single-stranded DNA binding protein (T4 gp32) complexed to DNA.** *Nature* 1995, **376**:362-366.

34. Frey MW, Nossal NG, Capson TL, Benkovic SJ: **Construction and characterization of a bacteriophage T4 DNA polymerase deficient in 3′->5′ exonuclease activity.** *Proc Natl Acad Sci USA* 1993, **90**:2579-2583.

35. Nossal NG: **A new look at old mutants of T4 DNA polymerase.** *Genetics* 1998, **148**:1535-1538.

36. Franklin MC, Wang J, Steitz TA: **Structure of the replicating complex of a pol alpha family DNA polymerase.** *Cell* 2001, **105**:657-667.

37. Spicer EK, Rush J, Fung C, Reha-Krantz LJ, Karam JD, Konigsberg WH: **Primary structure of T4 DNA polymerase. Evolutionary relatedness to eucaryotic and other procaryotic DNA polymerases.** *J Biol Chem* 1988, **263**:7478-7486.

38. Yeh LS, Hsu T, Karam JD: **Divergence of a DNA replication gene cluster in the T4-related bacteriophage RB69.** *J Bacteriol* 1998, **180**:2005-2013.

39. Freemont PS, Friedman JM, Beese LS, Sanderson MR, Steitz TA: **Cocrystal structure of an editing complex of Klenow fragment with DNA.** *Proc Natl Acad Sci USA* 1988, **85**:8924-8928.

40. Steitz TA, Beese L, Freemont PS, Friedman JM, Sanderson MR: **Structural studies of Klenow fragment: an enzyme with two active sites.** *Cold Spring Harb Symp Quant Biol* 1987, **52**:465-471.

41. Wang CC, Yeh LS, Karam JD: **Modular organization of T4 DNA polymerase. Evidence from phylogenetics.** *J Biol Chem* 1995, **270**:26558-26564.

42. Tabor S, Huber HE, Richardson CC: **Escherichia coli thioredoxin confers processivity on the DNA polymerase activity of the gene 5 protein of bacteriophage T7.** *J Biol Chem* 1987, **262**:16212-16223.

43. Moldovan GL, Pfander B, Jentsch S: **PCNA, the maestro of the replication fork.** *Cell* 2007, **129**:665-679.

44. Davey MJ, Jeruzalmi D, Kuriyan J, O'Donnell M: **Motors and switches: AAA + machines within the replisome.** *Nat Rev Mol Cell Biol* 2002, **3**:826-835.

45. Jeruzalmi D, O'Donnell M, Kuriyan J: **Clamp loaders and sliding clamps.** *Curr Opin Struct Biol* 2002, **12**:217-224.

46. Janzen DM, Torgov MY, Reddy MK: **In vitro reconstitution of the bacteriophage T4 clamp loader complex (gp44/62).** *J Biol Chem* 1999, **274**:35938-35943.

47. Bowman GD, O'Donnell M, Kuriyan J: **Structural analysis of a eukaryotic sliding DNA clamp-clamp loader complex.** *Nature* 2004, **429**:724-730.

48. Gulbis JM, Kazmirski SL, Finkelstein J, Kelman Z, O'Donnell M, Kuriyan J: **Crystal structure of the chi:psi sub-assembly of the Escherichia coli DNA polymerase clamp-loader complex.** *Eur J Biochem* 2004, **271**:439-449.

49. Jeruzalmi D, O'Donnell M, Kuriyan J: **Crystal structure of the processivity clamp loader gamma (gamma) complex of E. coli DNA polymerase III.** *Cell* 2001, **106**:429-441.

50. Jeruzalmi D, Yurieva O, Zhao Y, Young M, Stewart J, Hingorani M, O'Donnell M, Kuriyan J: **Mechanism of processivity clamp opening by the delta subunit wrench of the clamp loader complex of E. coli DNA polymerase III.** *Cell* 2001, **106**:417-428.

51. O'Donnell M, Jeruzalmi D, Kuriyan J: **Clamp loader structure predicts the architecture of DNA polymerase III holoenzyme and RFC.** *Curr Biol* 2001, **11**:R935-946.

52. Bochkarev A, Bochkareva E: **From RPA to BRCA2: lessons from single-stranded DNA binding by the OB-fold.** *Curr Opin Struct Biol* 2004, **14**:36-42.

53. Theobald DL, Mitton-Fry RM, Wuttke DS: **Nucleic acid recognition by OB-fold proteins.** *Annu Rev Biophys Biomol Struct* 2003, **32**:115-133.

54. Lindahl T: **Instability and decay of the primary structure of DNA.** *Nature* 1993, **362**:709-715.

55. Villemain JL, Giedroc DP: **The N-terminal B-domain of T4 gene 32 protein modulates the lifetime of cooperatively bound Gp32-ss nucleic acid complexes.** *Biochemistry* 1996, **35**:14395-14404.

56. Villemain JL, Ma Y, Giedroc DP, Morrical SW: **Mutations in the N-terminal cooperativity domain of gene 32 protein alter properties of the T4 DNA replication and recombination systems.** *J Biol Chem* 2000, **275**:31496-31504.

57. Krassa KB, Green LS, Gold L: **Protein-protein interactions with the acidic COOH terminus of the single-stranded DNA-binding protein of the bacteriophage T4.** *Proc Natl Acad Sci USA* 1991, **88**:4010-4014.

58. Wold MS: **Replication protein A: a heterotrimeric, single-stranded DNA-binding protein required for eukaryotic DNA metabolism.** *Annu Rev Biochem* 1997, **66**:61-92.

59. Bochkarev A, Pfuetzner RA, Edwards AM, Frappier L: **Structure of the single-stranded-DNA-binding domain of replication protein A bound to DNA.** *Nature* 1997, **385**:176-181.

60. Pfuetzner RA, Bochkarev A, Frappier L, Edwards AM: **Replication protein A. Characterization and crystallization of the DNA binding domain.** *J Biol Chem* 1997, **272**:430-434.

61. Savvides SN, Raghunathan S, Futterer K, Kozlov AG, Lohman TM, Waksman G: **The C-terminal domain of full-length E. coli SSB is disordered even when bound to DNA.** *Protein Sci* 2004, **13**:1942-1947.

62. von Hippel PH, Delagoutte E: **A general model for nucleic acid helicases and their "coupling" within macromolecular machines.** *Cell* 2001, **104**:177-190.

63. Dong F, von Hippel PH: **The ATP-activated hexameric helicase of bacteriophage T4 (gp41) forms a stable primosome with a single subunit of T4-coded primase (gp61).** *J Biol Chem* 1996, **271**:19625-19631.

64. Collins BK, Tomanicek SJ, Lyamicheva N, Kaiser MW, Mueser TC: **A preliminary solubility screen used to improve crystallization trials: crystallization and preliminary X-ray structure determination of Aeropyrum pernix flap endonuclease-1.** *Acta Crystallogr D Biol Crystallogr* 2004, **60**:1674-1678.

65. Izaac A, Schall CA, Mueser TC: **Assessment of a preliminary solubility screen to improve crystallization trials: uncoupling crystal condition searches.** *Acta Crystallogr D Biol Crystallogr* 2006, **62**:833-842.

66. Fass D, Bogden CE, Berger JM: **Crystal structure of the N-terminal domain of the DnaB hexameric helicase.** *Structure* 1999, **7**:691-698.

67. Bailey S, Eliason WK, Steitz TA: **Structure of hexameric DnaB helicase and its complex with a domain of DnaG primase.** *Science* 2007, **318**:459-463.

68. Singleton MR, Sawaya MR, Ellenberger T, Wigley DB: **Crystal structure of T7 gene 4 ring helicase indicates a mechanism for sequential hydrolysis of nucleotides.** *Cell* 2000, **101**:589-600.

69. Toth EA, Li Y, Sawaya MR, Cheng Y, Ellenberger T: **The crystal structure of the bifunctional primase-helicase of bacteriophage T7.** *Mol Cell* 2003, **12**:1113-1123.

70. Wahle E, Lasken RS, Kornberg A: **The dnaB-dnaC replication protein complex of Escherichia coli. II. Role of the complex in mobilizing dnaB functions.** *J Biol Chem* 1989, **264**:2469-2475.

71. Wahle E, Lasken RS, Kornberg A: **The dnaB-dnaC replication protein complex of Escherichia coli. I. Formation and properties.** *J Biol Chem* 1989, **264**:2463-2468.

72. Lefebvre SD, Wong ML, Morrical SW: **Simultaneous interactions of bacteriophage T4 DNA replication proteins gp59 and gp32 with single-stranded (ss) DNA. Co-modulation of ssDNA binding activities in a DNA helicase assembly intermediate.** *J Biol Chem* 1999, **274**:22830-22838.

73. Jones JM, Nakai H: **PriA and phage T4 gp59: factors that promote DNA replication on forked DNA substrates microreview.** *Mol Microbiol* 2000, **36**:519-527.

74. Jones CE, Mueser TC, Nossal NG: **Interaction of the bacteriophage T4 gene 59 helicase loading protein and gene 41 helicase with each other and with fork, flap, and cruciform DNA.** *J Biol Chem* 2000, **275**:27145-27154.

75. Mott ML, Erzberger JP, Coons MM, Berger JM: **Structural synergy and molecular crosstalk between bacterial helicase loaders and replication initiators.** *Cell* 2008, **135**:623-634.

76. Holm L, Sander C: **Dali: a network tool for protein structure comparison.** *Trends Biochem Sci* 1995, **20**:478-480.

77. Shindyalov IN, Bourne PE: **A database and tools for 3-D protein structure comparison and alignment using the Combinatorial Extension (CE) algorithm.** *Nucleic Acids Res* 2001, **29**:228-229.

78. Bianchi ME, Agresti A: **HMG proteins: dynamic players in gene regulation and differentiation.** *Curr Opin Genet Dev* 2005, **15**:496-506.

79. Nossal NG, Hinton DM: **Bacteriophage T4 DNA primase-helicase. Characterization of the DNA synthesis primed by T4 61 protein in the absence of T4 41 protein.** *J Biol Chem* 1987, **262**:10879-10885.

80. Hinton DM, Nossal NG: **Bacteriophage T4 DNA primase-helicase. Characterization of oligomer synthesis by T4 61 protein alone and in conjunction with T4 41 protein.** *J Biol Chem* 1987, **262**:10873-10878.

81. Jing DH, Dong F, Latham GJ, von Hippel PH: **Interactions of bacteriophage T4-coded primase (gp61) with the T4 replication helicase (gp41) and DNA in primosome formation.** *J Biol Chem* 1999, **274**:27287-27298.

82. Valentine AM, Ishmael FT, Shier VK, Benkovic SJ: **A zinc ribbon protein in DNA replication: primer synthesis and macromolecular interactions by the bacteriophage T4 primase.** *Biochemistry* 2001, **40**:15074-15085.

83. Norcum MT, Warrington JA, Spiering MM, Ishmael FT, Trakselis MA, Benkovic SJ: **Architecture of the bacteriophage T4 primosome: electron microscopy studies of helicase (gp41) and primase (gp61).** *Proc Natl Acad Sci USA* 2005, **102**:3623-3626.

84. Cha TA, Alberts BM: **Effects of the bacteriophage T4 gene 41 and gene 32 proteins on RNA primer synthesis: coupling of leading- and lagging-strand DNA synthesis at a replication fork.** *Biochemistry* 1990, **29**:1791-1798.

85. Richardson RW, Nossal NG: **Characterization of the bacteriophage T4 gene 41 DNA helicase.** *J Biol Chem* 1989, **264**:4725-4731.

86. Richardson RW, Nossal NG: **Trypsin cleavage in the COOH terminus of the bacteriophage T4 gene 41 DNA helicase alters the primase-helicase activities of the T4 replication complex in vitro.** *J Biol Chem* 1989, **264**:4732-4739.

87. Aravind L, Leipe DD, Koonin EV: **Toprim-a conserved catalytic domain in type IA and II topoisomerases, DnaG-type primases, OLD family nucleases and RecR proteins.** *Nucleic Acids Res* 1998, **26**:4205-4213.

88. Ilyina TV, Gorbalenya AE, Koonin EV: **Organization and evolution of bacterial and bacteriophage primase-helicase systems.** *J Mol Evol* 1992, **34**:351-357.

89. Korndorfer IP, Salerno J, Jing D, Matthews BW: **Crystallization and preliminary X-ray analysis of a bacteriophage T4 primase fragment.** *Acta Crystallogr D Biol Crystallogr* 2000, **56**:95-97.

90. Keck JL, Roche DD, Lynch AS, Berger JM: **Structure of the RNA polymerase domain of E. coli primase.** *Science* 2000, **287**:2482-2486.

91. Podobnik M, McInerney P, O'Donnell M, Kuriyan J: **A TOPRIM domain in the crystal structure of the catalytic core of Escherichia coli primase confirms a structural link to DNA topoisomerases.** *J Mol Biol* 2000, **300**:353-362.

92. Corn JE, Pelton JG, Berger JM: **Identification of a DNA primase template tracking site redefines the geometry of primer synthesis.** *Nat Struct Mol Biol* 2008, **15**:163-169.

93. Bhagwat M, Meara D, Nossal NG: **Identification of residues of T4 RNase H required for catalysis and DNA binding.** *J Biol Chem* 1997, **272**:28531-28538.

94. Devos JM, Tomanicek SJ, Jones CE, Nossal NG, Mueser TC: **Crystal structure of bacteriophage T4 5' nuclease in complex with a branched DNA reveals how flap endonuclease-1 family nucleases bind their substrates.** *J Biol Chem* 2007, **282**:31713-31724.

95. Harrington JJ, Lieber MR: **Functional domains within FEN-1 and RAD2 define a family of structure-specific endonucleases: implications for nucleotide excision repair.** *Genes Dev* 1994, **8**:1344-1355.

96. Lindahl T, Barnes DE: **Mammalian DNA ligases.** *Annu Rev Biochem* 1992, **61**:251-281.

97. Shuman S: **DNA ligases: progress and prospects.** *J Biol Chem* 2009, **284**:17365-17369.

98. Subramanya HS, Doherty AJ, Ashford SR, Wigley DB: **Crystal structure of an ATP-dependent DNA ligase from bacteriophage T7.** *Cell* 1996, **85**:607-615.

99. Doherty AJ, Ashford SR, Subramanya HS, Wigley DB: **Bacteriophage T7 DNA ligase. Overexpression, purification, crystallization, and characterization.** *J Biol Chem* 1996, **271**:11083-11089.

100. Shuman S, Schwer B: **RNA capping enzyme and DNA ligase: a superfamily of covalent nucleotidyl transferases.** *Mol Microbiol* 1995, **17**:405-410.

101. Nandakumar J, Nair PA, Shuman S: **Last stop on the road to repair: structure of E. coli DNA ligase bound to nicked DNA-adenylate.** *Mol Cell* 2007, **26**:257-271.

102. Pascal JM, O'Brien PJ, Tomkinson AE, Ellenberger T: **Human DNA ligase I completely encircles and partially unwinds nicked DNA.** *Nature* 2004, **432**:473-478.

103. Collaborative Computational Project N: **The CCP4 suite: programs for protein crystallography.** *Acta Crystallogr D Biol Crystallogr* 1994, **50**:760-763.

104. Zwart PH, Afonine PV, Grosse-Kunstleve RW, Hung LW, Ioerger TR, McCoy AJ, McKee E, Moriarty NW, Read RJ, Sacchettini JC, *et al*: **Automated structure solution with the PHENIX suite.** *Methods Mol Biol* 2008, **426**:419-435.

105. Adams PD, Grosse-Kunstleve RW, Hung LW, Ioerger TR, McCoy AJ, Moriarty NW, Read RJ, Sacchettini JC, Sauter NK, Terwilliger TC: **PHENIX: building new software for automated crystallographic structure determination.** *Acta Crystallogr D Biol Crystallogr* 2002, **58**:1948-1954.

106. Sanner MF: **Python: a programming language for software integration and development.** *J Mol Graph Model* 1999, **17**:57-61.

107. Emsley P, Cowtan K: **Coot: model-building tools for molecular graphics.** *Acta Crystallogr D Biol Crystallogr* 2004, **60**:2126-2132.

108. Sun S, Geng L, Shamoo Y: **Structure and enzymatic properties of a chimeric bacteriophage RB69 DNA polymerase and single-stranded DNA binding protein with increased processivity.** *Proteins* 2006, **65**:231-238.

109. Svergun DI, Petoukhov MV, Koch MH: **Determination of domain structure of proteins from X-ray solution scattering.** *Biophys J* 2001, **80**:2946-2953.

110. Niimura N: **Neutrons expand the field of structural biology.** *Curr Opin Struct Biol* 1999, **9**:602-608.

111. Koch MH, Vachette P, Svergun DI: **Small-angle scattering: a view on the properties, structures and structural changes of biological macromolecules in solution.** *Q Rev Biophys* 2003, **36**:147-227.

112. Konarev PV, Petoukhov MV, Volkov VV, Svergun DI: **ATSAS 2.1, a program package for small-angle scattering data analysis.** *Journal of Applied Crystallography* 2006, **39**:277-286.

doi:10.1186/1743-422X-7-359
Cite this article as: Mueser *et al.:* Structural analysis of bacteriophage T4 DNA replication: a review in the Virology Journal series on *bacteriophage T4 and its relatives. Virology Journal* 2010 **7**:359.

Rao and Black *Virology Journal* 2010, **7**:356
http://www.virologyj.com/content/7/1/356

VIROLOGY JOURNAL

Structure and assembly of bacteriophage T4 head

Venigalla B Rao[1]*, Lindsay W Black[2]

Abstract

The bacteriophage T4 capsid is an elongated icosahedron, 120 nm long and 86 nm wide, and is built with three essential proteins; gp23*, which forms the hexagonal capsid lattice, gp24*, which forms pentamers at eleven of the twelve vertices, and gp20, which forms the unique dodecameric portal vertex through which DNA enters during packaging and exits during infection. The past twenty years of research has greatly elevated the understanding of phage T4 head assembly and DNA packaging. The atomic structure of gp24 has been determined. A structural model built for gp23 using its similarity to gp24 showed that the phage T4 major capsid protein has the same fold as that found in phage HK97 and several other icosahedral bacteriophages. Folding of gp23 requires the assistance of two chaperones, the *E. coli* chaperone GroEL and the phage coded gp23-specific chaperone, gp31. The capsid also contains two non-essential outer capsid proteins, Hoc and Soc, which decorate the capsid surface. The structure of Soc shows two capsid binding sites which, through binding to adjacent gp23 subunits, reinforce the capsid structure. Hoc and Soc have been extensively used in bipartite peptide display libraries and to display pathogen antigens including those from HIV, *Neisseria meningitides*, *Bacillus anthracis*, and FMDV. The structure of Ip1*, one of the components of the core, has been determined, which provided insights on how IPs protect T4 genome against the *E. coli* nucleases that degrade hydroxymethylated and glycosylated T4 DNA. Extensive mutagenesis combined with the atomic structures of the DNA packaging/terminase proteins gp16 and gp17 elucidated the ATPase and nuclease functional motifs involved in DNA translocation and headful DNA cutting. Cryo-EM structure of the T4 packaging machine showed a pentameric motor assembled with gp17 subunits on the portal vertex. Single molecule optical tweezers and fluorescence studies showed that the T4 motor packages DNA at a rate of up to 2000 bp/sec, the fastest reported to date of any packaging motor. FRET-FCS studies indicate that the DNA gets compressed during the translocation process. The current evidence suggests a mechanism in which electrostatic forces generated by ATP hydrolysis drive the DNA translocation by alternating the motor between tensed and relaxed states.

Introduction

The T4-type bacteriophages are ubiquitously distributed in nature and occupy environmental niches ranging from mammalian gut to soil, sewage, and oceans. More than 130 such viruses that show similar morphological features as phage T4 have been described; from the T4 superfamily ~1400 major capsid protein sequences have been correlated to its 3D structure [1-3]. The features include large elongated (prolate) head, contractile tail, and a complex baseplate with six long, kinked tail fibers radially emanating from it. Phage T4 historically has served as an excellent model to elucidate the mechanisms of head assembly of not only T-even phages but of large icosahedral viruses in general, including the widely distributed eukaryotic viruses such as the herpes viruses. This review will focus on the advances in the past twenty years on the basic understanding of phage T4 head structure and assembly and the mechanism of DNA packaging. Application of some of this knowledge to develop phage T4 as a surface display and vaccine platform will also be discussed. The reader is referred to the comprehensive review by Black et al [4], for the early work on T4 head assembly.

* Correspondence: rao@cua.edu
[1]Department of Biology, The Catholic University of America, Washington, DC, USA
Full list of author information is available at the end of the article

Structure of phage T4 capsid

The overall architecture of the phage T4 head determined earlier by negative stain electron microscopy of the procapsid, capsid, and polyhead, including the positions of the dispensable Hoc and Soc proteins, has basically not changed as a result of cryo-electron microscopic structure determination of isometric capsids [5]. However, the dimensions of the phage T4 capsid and its inferred protein copy numbers have been slightly altered on the basis of the higher resolution cryo-electron microscopy structure. The width and length of the elongated prolate icosahedron [5] are T_{end} = 13 laevo and T_{mid} = 20 (86 nm wide and 120 nm long), and the copy numbers of gp23, Hoc and Soc are 960, 155, and 870, respectively (Figure 1).

The most significant advance was the crystal structure of the vertex protein, gp24, and by inference the structure of its close relative, the major capsid protein gp23 [6]. This ~0.3 nm resolution structure permits rationalization of head length mutations in the major capsid protein as well as of mutations allowing bypass of the vertex protein. The former map to the capsomer's periphery and the latter within the capsomer. It is likely that the special gp24 vertex protein of phage T4 is a relatively recent evolutionary addition as judged by the ease with which it can be bypassed. Cryo-electron microscopy showed that in the bypass mutants that substitute pentamers of the major capsid protein at the vertex, additional Soc decoration protein subunits surround these gp23* molecules, which does not occur in the gp23*-gp24* interfaces of the wild-type capsid [7].

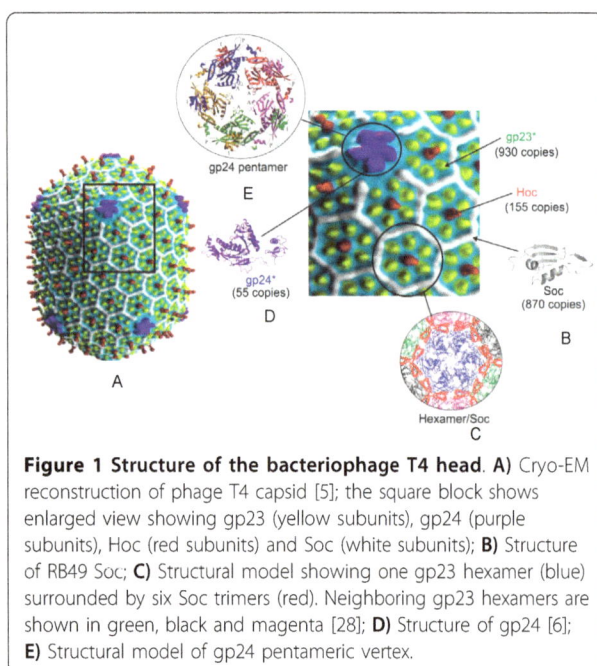

Figure 1 Structure of the bacteriophage T4 head. A) Cryo-EM reconstruction of phage T4 capsid [5]; the square block shows enlarged view showing gp23 (yellow subunits), gp24 (purple subunits), Hoc (red subunits) and Soc (white subunits); **B)** Structure of RB49 Soc; **C)** Structural model showing one gp23 hexamer (blue) surrounded by six Soc trimers (red). Neighboring gp23 hexamers are shown in green, black and magenta [28]; **D)** Structure of gp24 [6]; **E)** Structural model of gp24 pentameric vertex.

Nevertheless, despite the rationalization of major capsid protein affecting head size mutations, it should be noted that these divert only a relatively small fraction of the capsids to altered and variable sizes. The primary determinant of the normally invariant prohead shape is thought to be its scaffolding core, which grows concurrently with the shell [4]. However, little progress has been made in establishing the basic mechanism of size determination or in determining the structure of the scaffolding core.

The gp24 and inferred gp23 structures are closely related to the structure of the major capsid protein of bacteriophage HK97, most probably also the same protein fold as the majority of tailed dsDNA bacteriophage major capsid proteins [8]. Interesting material bearing on the T-even head size determination mechanism is provided by "recent" T-even relatives of increased and apparently invariant capsid size, unlike the T4 capsid size mutations that do not precisely determine size (e.g. KVP40, 254 kb, apparently has a single T_{mid} greater than the 170 kb T4 T_{mid} = 20) [9]. However, few if any in depth studies have been carried out on these phages to determine whether the major capsid protein, the morphogenetic core, or other factors are responsible for the different and precisely determined volumes of their capsids.

Folding of the major capsid protein gp23

Folding and assembly of the phage T4 major capsid protein gp23 into the prohead requires a special utilization of the GroEL chaperonin system and an essential phage co-chaperonin gp31. gp31 replaces the GroES co-chaperonin that is utilized for folding the 10-15% of *E. coli* proteins that require folding by the GroEL folding chamber. Although T4 gp31 and the closely related RB49 co-chaperonin CocO have been demonstrated to replace the GroES function for all essential *E. coli* protein folding, the GroES-gp31 relationship is not reciprocal; i.e. GroES cannot replace gp31 to fold gp23 because of special folding requirements of the latter protein [10,11]. The N-terminus of gp23 appears to strongly target associated fusion proteins to the GroEL chaperonin [12-14]. Binding of gp23 to the GroEL folding cage shows features that are distinct from those of most bound *E. coli* proteins. Unlike substrates such as RUBISCO, gp23 occupies both chambers of the GroEL folding cage, and only gp31 is able to promote efficient capped single "cis" chamber folding, apparently by creating a larger folding chamber [15]. On the basis of the gp24 inferred structure of gp23, and the structures of the GroES and gp31 complexed GroEL folding chambers, support for a critical increased chamber size to accommodate gp23 has been advanced as the explanation for the gp31 specificity [14]. However, since comparable size T-even phage gp31 homologs display preference for folding their own gp23s, more subtle features of the various T-even phage structured folding cages may also determine specificity.

Structure of the packaged components of the phage T4 head

Packaged phage T4 DNA shares a number of general features with other tailed dsDNA phages: 2.5 nm side to side packing of predominantly B-form duplex DNA condensed to ~500 mg/ml. However, other features differ among phages; e.g. T4 DNA is packed in an orientation that is parallel to the head tail axis together with ~1000 molecules of imbedded and mobile internal proteins, unlike the DNA arrangement that traverses head-tail axis and is arranged around an internal protein core as seen in phage T7 [16]. Use of the capsid targeting sequence of the internal proteins allows encapsidation of foreign proteins such as GFP and staphylococcal nuclease within the DNA of active virus [17,18]. Digestion by the latter nuclease upon addition of calcium yields a pattern of short DNA fragments, predominantly a 160 bp repeat [19]. This pattern supports a discontinuous pattern of DNA packing such as in the icosahedral-bend or spiral-fold models. A number of proposed models (Figure 2) and experimental evidence bearing on these are summarized in [17].

In addition to the uncertain arrangement at the nucleotide level of packaged phage DNA, the structure of other internal components is poorly understood in comparison to surface capsid proteins. The internal

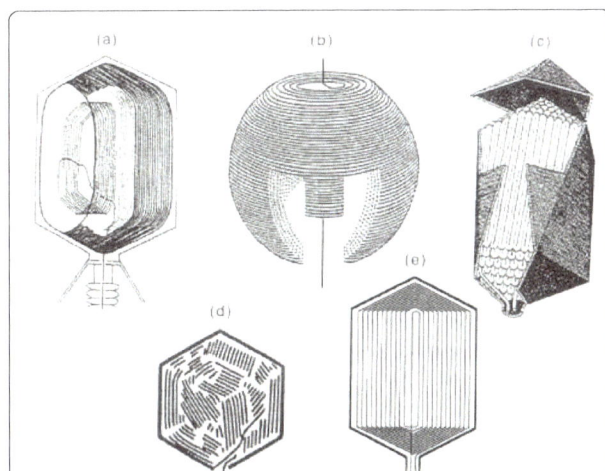

Figure 3 Structure and function of T4 internal protein I*. The NMR structure of IP1*, a highly specific inhibitor of the two-subunit CT (gmrS/gmrD) glucosyl-hmC DNA directed restriction endonuclease (right panel); shown are DNA modifications blocking such enzymes. The IPI* structure is compact with an asymmetric charge distribution on the faces (blue are basic residues) that may allow rapid DNA bound ejection through the portal and tail without unfolding-refolding.

protein I* (IPI*) of phage T4 is injected to protect the DNA from a two subunit gmrS + gmrD glucose modified restriction endonuclease of a pathogenic *E. coli* that digests glucosylated hydroxymethylcytosine DNA of T-even phages [20,21]. The 76-residue proteolyzed mature form of the protein has a novel compact protein fold consisting of two beta sheets flanked with N- and C-terminal alpha helices, a structure that is required for its inhibitor activity that is apparently due to binding the gmrS/gmrD proteins (Figure 3) [22]. A single chain gmrS/gmrD homolog enzyme with 90% identity in its sequence to the two subunit enzyme has evolved IPI* inhibitor immunity. It thus appears that the phage T-evens have co-evolved with their hosts, a diverse and highly specific set of internal proteins to counter the hmC modification dependent restriction endonucleases. Consequently the internal protein components of the T-even phages are a highly diverse set of defense proteins against diverse attack enzymes with only a conserved capsid targeting sequence (CTS) to encapsidate the proteins into the precursor scaffolding core [23].

Genes 2 and 4 of phage T4 likely are associated in function and gp2 was previously shown by Goldberg and co-workers to be able to protect the ends of mature T4 DNA from the recBCD exonuclease V, likely by binding to the DNA termini. The gp2 protein has not been identified within the phage head because of its low abundance but evidence for its presence in the head comes from the fact that gp2 can be added to gp2

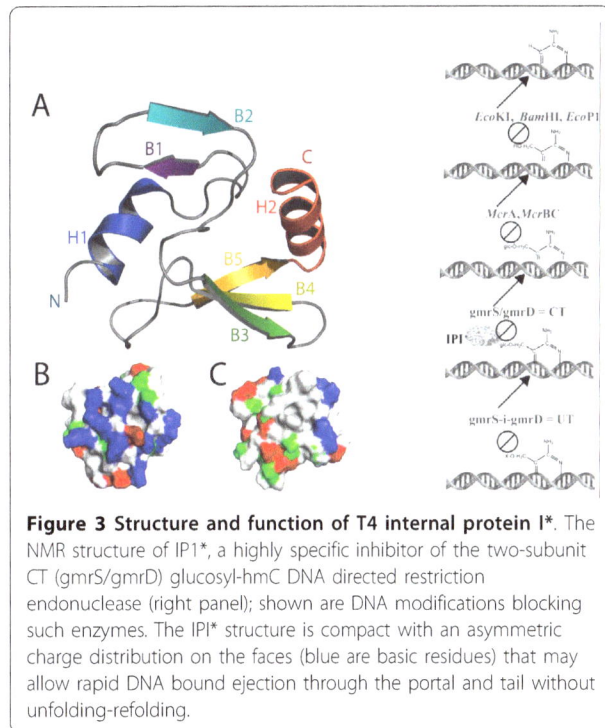

Figure 2 Models of packaged DNA structure. **a)** T4 DNA is packed longitudinally to the head-tail axis [91], unlike the transverse packaging in T7 capsids [16]**(b)**. Other models shown include spiral fold **(c)**, liquid-crystal **(d)**, and icosahedral-bend **(e)**. Both packaged T4 DNA ends are located in the portal [79]. For references and evidence bearing on packaged models see [19].

deficient full heads to confer exonuclease V protection. Thus gp2 affects head-tail joining as well as protecting the DNA ends likely with as few as two copies per particle binding the two DNA ends [24].

Solid state NMR analysis of the phage T4 particle shows the DNA is largely B form and allows its electrostatic interactions to be tabulated [25]. This study reveals high resolution interactions bearing on the internal structure of the phage T4 head. The DNA phosphate negative charge is balanced among lysyl amines, polyamines, and mono and divalent cations. Interestingly, among positively charged amino acids, only lysine residues of the internal proteins were seen to be in contact with the DNA phosphates, arguing for specific internal protein DNA structures. Electrostatic contributions from internal proteins and polyamines' interactions with DNA entering the prohead to the packaging motor were proposed to account for the higher packaging rates achieved by the phage T4 packaging machine when compared to that of Phi29 and lambda phages.

Display on capsid

In addition to the essential capsid proteins, gp23, gp24, and gp20, the T4 capsid is decorated with two nonessential outer capsid proteins: Hoc (highly antigenic outer capsid protein), a dumbbell shaped monomer at the center of each gp23 hexon, up to 155 copies per capsid (39 kDa; red subunits); and Soc (small outer capsid protein), a rod-shaped molecule that binds between gp23 hexons, up to 870 copies per capsid (9 kDa; white subunits) (Figure 1). Both Hoc and Soc are dispensable, and bind to the capsid after the completion of capsid assembly [26,27]. Null (amber or deletion) mutations in either or both the genes do not affect phage production, viability, or infectivity.

The structure of Soc has recently been determined [28]. It is a tadpole shaped molecule with two binding sites for gp23*. Interaction of Soc to the two gp23 molecules glues adjacent hexons. Trimerization of the bound Soc molecules results in clamping of three hexons, and 270 such clamps form a cage reinforcing the capsid structure. Soc assembly thus provides great stability to phage T4 to survive under hostile environments such as extreme pH (pH 11), high temperature (60°C), osmotic shock, and a host of denaturing agents. Soc-minus phage lose viability at pH10.6 and addition of Soc enhances its survival by $\sim10^4$-fold. On the other hand, Hoc does not provide significant additional stability. With its Ig-like domains exposed on the outer surface, Hoc may interact with certain components of the bacterial surface, providing additional survival advantage (Sathaliyawala and Rao, unpublished results).

The above properties of Hoc and Soc are uniquely suited to engineer the T4 capsid surface by arraying

pathogen antigens. Ren et al and Jiang et al developed recombinant vectors that allowed fusion of pathogen antigens to the N- or C-termini of Hoc and Soc [29-32]. The fusion proteins were expressed in *E. coli* and upon infection with *hoc⁻soc⁻* phage, the fusion proteins assembled on the capsid. The phages purified from the infected extracts are decorated with the pathogen antigens. Alternatively, the fused gene can be transferred into T4 genome by recombinational marker rescue and infection with the recombinant phage expresses and assembles the fusion protein on the capsid as part of the infection process. Short peptides or protein domains from a variety of pathogens, *Neisseria meningitides* [32], polio virus [29], HIV [29,33], swine fever virus [34], and foot and mouth disease virus [35], have been displayed on T4 capsid using this approach.

The T4 system can be adapted to prepare bipartite libraries of randomized short peptides displayed on T4 capsid Hoc and Soc and use these libraries to "fish out" peptides that interact with the protein of interest [36]. Biopanning of libraries by the T4 large packaging protein gp17 selected peptides that matches with the sequences of proteins that are thought to interact with p17. Of particular interest was the selection of a peptide that matched with the T4 late sigma factor, gp55. The gp55 deficient extracts packaged concatemeric DNA about 100-fold less efficiently suggesting that the gp17

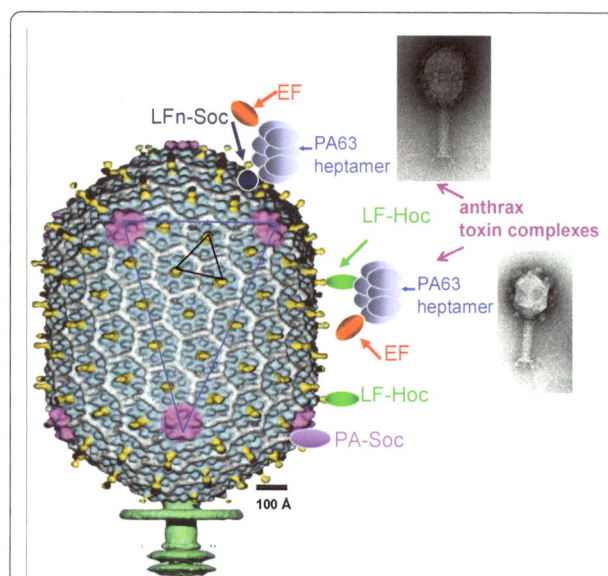

Figure 4 In vitro display of antigens on bacteriophage T4 capsid. Schematic representation of the T4 capsid decorated with large antigens, PA (83 kDa) and LF (89 kDa), or hetero-oligomeric anthrax toxin complexes through either Hoc or Soc binding [39,41]. See text for details. The insets show electron micrographs of T4 phage with the anthrax toxin complexes displayed through Soc (top) or Hoc (bottom). Note the copy number of the complexes is lower with the Hoc display than with the Soc display.

interaction with gp55 helps loading the packaging terminase onto the viral genome [36,37].

An in vitro display system has been developed taking advantage of the high affinity interactions between Hoc or Soc and the capsid (Figure 4) [38,39]. In this system, the pathogen antigen fused to Hoc or Soc with a hexahistidine tag was overexpressed in *E. coli* and purified. The purified protein was assembled on *hoc⁻soc⁻* phage by simply mixing the purified components. This system has certain advantages over the in vivo display: i) a functionally well characterized and conformationally homogeneous antigen is displayed on the capsid; ii) the copy number of displayed antigen can be controlled by altering the ratio of antigen to capsid binding sites; and iii) multiple antigens can be displayed on the same capsid. This system was used to display full-length antigens from HIV [33] and anthrax [38,39] that are as large as 90 kDa.

All 155 Hoc binding sites can be filled with anthrax toxin antigens, protective antigen (PA, 83 kDa), lethal factor (LF, 89 kDa), or edema factor (EF, 90 kDa) [36,40]. Fusion to the N-terminus of Hoc did not affect the apparent binding constant (K_d) or the copy number per capsid (B_{max}), but fusion to the C-terminus reduced the K_d by 500-fold [32,40]. All 870 copies of Soc binding sites can be filled with Soc-fused antigens but the size of the fused antigen must be ~30 kDa or less; otherwise, the copy number is significantly reduced [39]. For example, the 20-kDa PA domain-4 and the 30 kDa LFn domain fused to Soc can be displayed to full capacity. An insoluble Soc-HIV gp120 V3 loop domain fusion protein with a 43 aa C-terminal addition could be refolded and bound with ~100% occupancy to mature phage head type-polyheads [29]. Large 90 kDa anthrax toxins can also be displayed but the B_{max} is reduced to about 300 presumably due to steric constraints. Antigens can be fused to either the N- or C-terminus, or both the termini of Soc simultaneously, without significantly affecting the K_d or B_{max}. Thus, as many as 1895 antigen molecules or domains can be attached to each capsid using both Hoc and Soc [39].

The in vitro system offers novel avenues to display macromolecular complexes through specific interactions with the already attached antigens [41]. Sequential assembly was performed by first attaching LF-Hoc and/or LFn-Soc to *hoc⁻soc⁻* phage and exposing the N-domain of LF on the surface. Heptamers of PA were then assembled through interactions between the LFn domain and the N-domain of cleaved PA (domain 1' of PA63). EF was then attached to the PA63 heptamers, completing the assembly of the ~700 kDa anthrax toxin complex on phage T4 capsid (Figure 4). CryoEM reconstruction shows that native PA63$_{(7)}$-LFn$_{(3)}$ complexes are assembled in which three adjacent capsid-bound

LFn "legs" support the PA63 heptamers [42]. Additional layers of proteins can be built on the capsid through interactions with the respective partners.

One of the main applications of the T4-antigen particles is their potential use in vaccine delivery. A number of independent studies showed that the T4-displayed particulate antigens without any added adjuvant elicit strong antibody responses, and to a lesser extent cellular responses [28,32]. The 43 aa V3 loop of HIV gp120 fused to Soc displayed on T4 phage was highly immunogenic in mice and induced anti-gp120 antibodies; so was the Soc-displayed IgG anti-EWL [29]. The Hoc fused 183 aa N-terminal portion of HIV CD4 receptor protein is displayed in active form. Strong anthrax lethal-toxin neutralization titers were elicited upon immunization of mice and rabbits with phage T4-displayed PA either through Hoc or Soc ([38,40], Rao, unpublished data). When multiple anthrax antigens were displayed, immune responses against all the displayed antigens were elicited [40]. The T4 particles displaying PA and LF, or those displaying the major antigenic determinant cluster mE2 (123 aa) and the primary antigen E2 (371 aa) of the classical swine fever virus elicited strong antibody titers [34]. Furthermore, mice immunized with the Soc displayed foot and mouth disease virus (FMDV) capsid precursor polyprotein (P1, 755 aa) and proteinase 3C (213 aa) were completely protected upon challenge with a lethal dose of FMDV [34,35]. Pigs immunized with a mixture of T4-P1 and T4-3C particles were also protected when these animals were co-housed with FMDV infected pigs. In another type of application, T4-displayed mouse Flt4 tumor antigen elicited anti-Flt4 antibodies and broke immune tolerance to self-antigens. These antibodies provided antitumor and anti-metastasis immunity in mice [43].

The above studies provide abundant evidence that the phage T4 nanoparticle platform has the potential to engineer human as well as veterinary vaccines.

DNA packaging

Two nonstructural terminase proteins, gp16 (18 kDa) and gp17 (70 kDa), link head assembly and genome processing [44-46]. These proteins are thought to form a hetero-oligomeric complex, which recognizes the concatemeric DNA and makes an endonucleolytic cut (hence the name "terminase"). The terminase-DNA complex docks on the prohead through gp17 interactions with the special portal vertex formed by the dodecameric gp20, thus assembling a DNA packaging machine. The gp49 EndoVII Holliday structure resolvase also specifically associates with the portal dodecamer thereby positioning this enzyme to repair packaging-arrested branched-structure-containing concatemers [47]. The ATP-fueled machine translocates DNA into the capsid until the head is full, equivalent to about 1.02 times the genome length (171 kb). The terminase dissociates from the packaged head, makes a second cut to terminate DNA packaging and attaches the concatemeric DNA to another empty head to continue translocation in a processive fashion. Structural and functional analyses of the key parts of the machine - gp16, gp17, and gp20 - as described below, led to models for the packaging mechanism.

gp16

gp16, the 18 kDa small terminase subunit, is dispensable for packaging linear DNA in vitro but it is essential in vivo; amber mutations in gene 16 accumulate empty proheads resulting in null phenotype [37,48].

Mutational and biochemical analyses suggest that gp16 is involved in the recognition of viral DNA [49,50] and regulation of gp17 functions [51]. gp16 is predicted to contain three domains, a central domain that is important for oligomerization, and N- and C-terminal domains that are important for DNA binding, ATP binding, and/or gp17-ATPase stimulation [51,52] (Figure 5). gp16 forms oligomeric single and side-by-side double rings, each ring having a diameter of ~8 nm with ~2 nm central channel [49,52]. Recent mass spectrometry determination shows that the single and double rings are 11-mers and 22-mers respectively [53]. A number of *pac* site phages produce comparable small terminase subunit multimeric ring structures. Sequence analyses predict 2-3 coiled coil motifs in gp16 [48]. All the T4 family gp16s as well as other phage small terminases consist of one or more coiled coil motifs, consistent with their propensity to form stable oligomers. Oligomerization presumably occurs through parallel coiled-coil interactions between neighboring subunits. Mutations in the long central α-helix of T4 gp16 that perturb coiled coil interactions lose the ability to oligomerize [48].

gp16 appears to oligomerize following interaction with viral DNA concatemer, forming a platform for the assembly of the large terminase gp17. A predicted helix-turn-helix in the N-terminal domain is thought to be involved in DNA-binding [49,52]. The corresponding motif in the phage lambda small terminase protein, gpNu1, has been well characterized and demonstrated to bind the DNA. In vivo genetic studies and in vitro DNA binding studies show that a 200 bp 3'-end sequence of gene 16 is a preferred "*pac*" site for gp16 interaction [49,50]. It was proposed that the stable gp16 double rings were two turn lock washers that constituted the structural basis for synapsis of two *pac* site DNAs. This could promote the gp16 dependent gene amplifications observed around the *pac* site that can be selected in *alt-* mutants that package more DNA; such

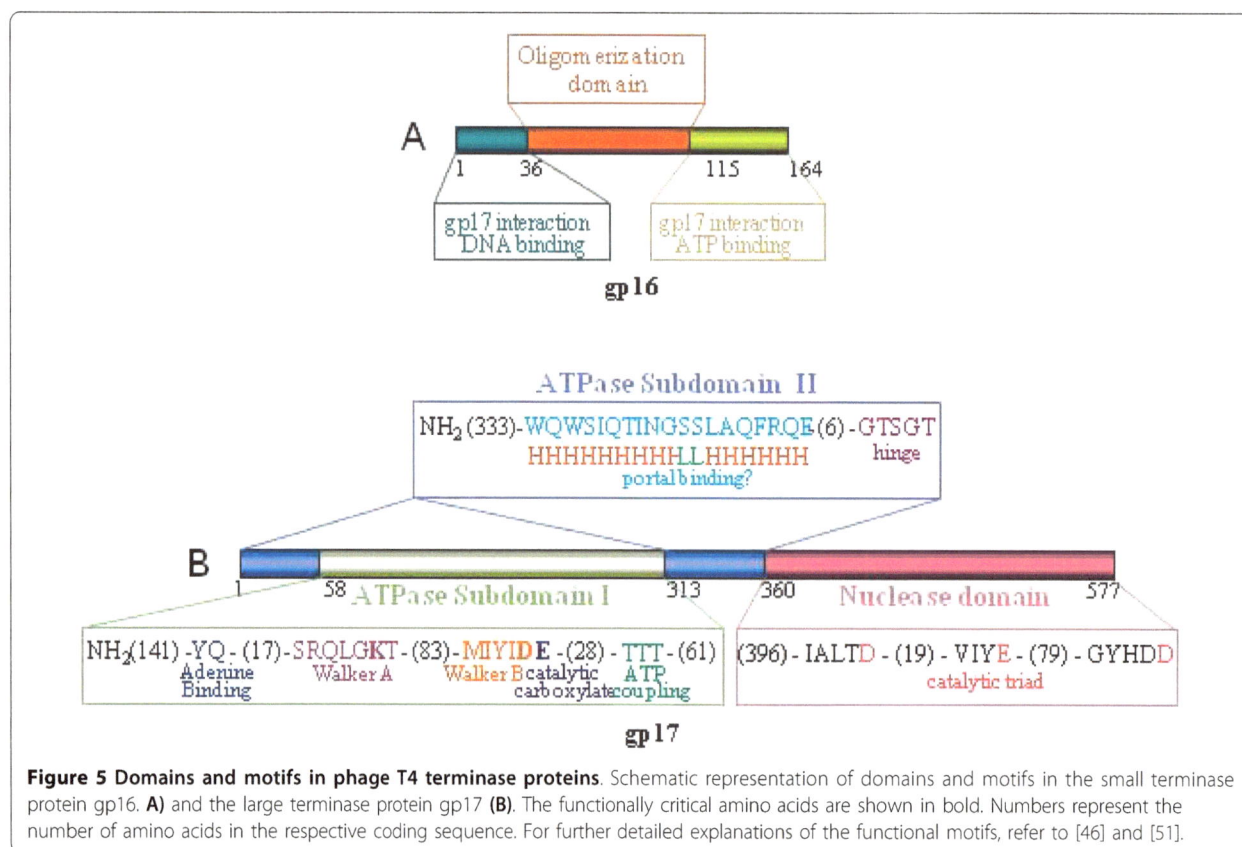

Figure 5 Domains and motifs in phage T4 terminase proteins. Schematic representation of domains and motifs in the small terminase protein gp16. **A)** and the large terminase protein gp17 **(B)**. The functionally critical amino acids are shown in bold. Numbers represent the number of amino acids in the respective coding sequence. For further detailed explanations of the functional motifs, refer to [46] and [51].

synapsis could function as a gauge of DNA concatemer maturation [54-56].

gp16 stimulates the gp17-ATPase activity by > 50-fold [57,58]. Stimulation is likely via oligomerization of gp17 which does not require gp16 association [58]. gp16 also stimulates in vitro DNA packaging activity in the crude system where phage infected extracts containing all the DNA replication/transcription/recombination proteins are present [57,59], but inhibits the packaging activity in the defined system where only two purified components, proheads and gp17, are present [37,60]. It stimulates gp17-nuclease activity when T4 transcription factors are also present but inhibits the nuclease in a pure system [51]. gp16 also inhibits gp17's binding to DNA [61]. Both the N- and C-domains are required for ATPase stimulation or nuclease inhibition [51]. Maximum effects were observed at a ratio of approximately 8 gp16 molecules to 1 gp17 molecule suggesting that in the holoterminase complex one gp16 oligomer interacts with one gp17 monomer [62].

gp16 contains an ATP binding site with broad nucleotide specificity [49,51], however it lacks the canonical nucleotide binding signatures such as Walker A and Walker B [52]. No correlation was evident between nucleotide binding and gp17-ATPase stimulation or gp17-nuclease inhibition. Thus it is unclear what the role of ATP binding plays in gp16 function.

The evidence thus far suggests that gp16 is a regulator of the DNA packaging machine, modulating the ATPase, translocase, and nuclease activities of gp17. Although the regulatory functions can be dispensable for in vitro DNA packaging, these are essential in vivo to coordinate the packaging process and produce an infectious virus particle [51].

gp17

gp17 is the 70 kDa large subunit of the terminase holoenzyme and the motor protein of the DNA packaging machine. gp17 consists of two functional domains (Figure 5); an N-terminal ATPase domain having the classic ATPase signatures such as Walker A, Walker B, and catalytic carboxylate, and a C-terminal nuclease domain having a catalytic metal cluster with conserved aspartic and glutamic acid residues coordinating with Mg [62].

gp17 alone is sufficient to package DNA in vitro. gp17 exhibits a weak ATPase activity (K_{cat} = ~1-2 ATPs hydrolyzed per gp17 molecule/min), which is stimulated by > 50-fold by the small terminase protein gp16 [57,58]. Any mutation in the predicted catalytic residues of the N-terminal ATPase center results in a loss of stimulated ATPase and DNA packaging activities [63]. Even subtle conservative substitutions such as aspartic acid to glutamic acid and *vice versa* in the Walker B motif resulted in complete loss of DNA packaging

suggesting that this ATPase provides energy for DNA translocation [64,65].

The ATPase domain also exhibits DNA binding activity, which may be involved in the DNA cutting and translocation functions of the packaging motor. There is genetic evidence that gp17 may interact with gp32 [66,67], but highly purified preparations of gp17 do not show appreciable affinity for ss or ds DNA. There seem to be complex interactions between the terminase proteins, the concatemeric DNA, and the DNA replication/recombination/repair and transcription proteins that transition the DNA metabolism into the packaging phase [37].

One of the ATPase mutants, the DE-ED mutant in which the sequence of Walker B and catalytic carboxylate was reversed, showed tighter binding to ATP than the wild-type gp17 but failed to hydrolyze ATP [64]. Unlike the wild-type gp17 or the ATPase domain which failed to crystallize, the ATPase domain with the ED mutation crystallized readily, probably because it trapped the ATPase in an ATP-bound conformation. The X-ray structure of the ATPase domain was determined up to 1.8 Å resolution in different bound states; apo, ATP-bound, and ADP-bound [68]. It is a flat structure consisting of two subdomains; a large subdomain I (NsubI) and a smaller subdomain II (NsubII) forming a cleft in which ATP binds (Figure 6A). The NsubI consists of the classic nucleotide binding fold (Rossmann fold), a parallel β-sheet of six β-strands interspersed with helices. The structure showed that the predicted catalytic residues are oriented into the ATP pocket, forming a network of interactions with bound ATP. These also include an arginine finger that is proposed to trigger βγ-phosphoanhydride bond cleavage. In addition, the structure showed the movement of a loop near the adenine binding motif in response to ATP hydrolysis,

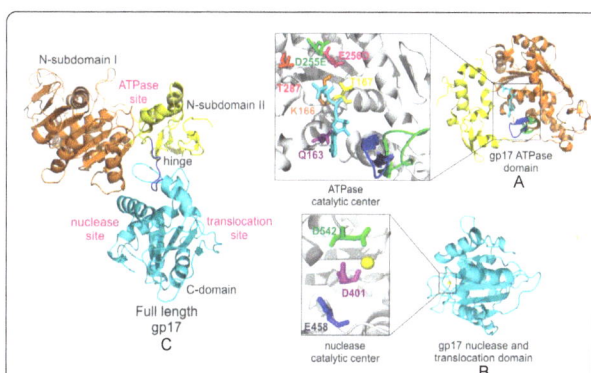

Figure 6 Structures of the T4 packaging motor protein, gp17.
Structures of the ATPase domain: **A)** nuclease/translocation domain; **B)**, and full-length gp17; **C)**. Various functional sites and critical catalytic residues are labeled. See references [68] and [74] for further details.

which may be important for transduction of ATP energy into mechanical motion.

gp17 exhibits a sequence nonspecific endonuclease activity [69,70]. Random mutagenesis of gene 17 and selection of mutants that lost nuclease activity identified a histidine-rich site in the C-terminal domain being critical for DNA cleavage [71]. Extensive site-directed mutagenesis of this region combined with the sequence alignments identified a cluster of conserved aspartic acid and glutamic acid residues that are essential for DNA cleavage [72]. Unlike the ATPase mutants, these mutants retained the gp16-stimulated ATPase activity as well as the DNA packaging activity as long as the substrate is a linear molecule. However these mutants fail to package circular DNA as they are defective in cutting DNA that is required for packaging initiation.

The structure of the C-terminal nuclease domain from a T4-family phage, RB49, which has 72% sequence identity to the T4 C-domain, was determined to 1.16Å resolution [73] (Figure 6B). It has a globular structure consisting mostly of anti-parallel β-strands forming an RNase H fold that is found in resolvases, RNase Hs and integrases. As predicted from the mutagenesis studies, the structures showed that the residues D401, E458 and D542 form a catalytic triad coordinating with Mg ion. In addition the structure showed the presence of a DNA binding groove lined with a number of basic residues. The acidic catalytic metal center is buried at one end of this groove. Together, these form the nuclease cleavage site of gp17.

The crystal structure of the full-length T4 gp17 (ED mutant) was determined to 2.8Å resolution (Figure 6C) [74]. The N- and C-domain structures of the full-length gp17 superimpose with those solved using individually crystallized domains with only minor deviations. The full-length structure however has additional features that are relevant to the mechanism. A flexible "hinge" or "linker" connects the ATPase and nuclease domains. Previous biochemical studies showed that splitting gp17 into two domains at the linker retained the respective ATPase and nuclease functions but DNA translocation activity was completely lost [62]. Second, the N- and C-domains have a > 1000 square Å complementary surface area consisting of an array of five charged pairs and hydrophobic patches [74]. Third, the gp17 has a bound phosphate ion in the crystal structure. Docking of B-form DNA guided by shape and charge complementarity with one of the DNA phosphates superimposed on the bound phosphate aligns a number of basic residues, lining what appears to be a shallow translocation groove. Thus the C-domain appears to have two DNA grooves on different faces of the structure, one that aligns with the nuclease catalytic site and the second that aligns with the translocating DNA (Figure 6). Mutation of one

of the groove residues (R406) showed a novel phenotype; loss of DNA translocation activity but the ATPase and nuclease activities are retained.

Motor

A functional DNA packaging machine could be assembled by mixing proheads and purified gp17. gp17 assembles into a packaging motor through specific interactions with the portal vertex [75] and such complexes can package the 171 kb phage T4 DNA, or any linear DNA [37,60]. If short DNA molecules are added as the DNA substrate, the motor keeps packaging DNA until the head is full [76].

Packaging can be studied in real time either by fluorescence correlation spectroscopy [77] or by optical tweezers [78]. The translocation kinetics of rhodamine (R6G) labeled 100 bp DNA was measured by determining the decrease in diffusion coefficient as the DNA gets confined inside the capsid. Fluorescence resonance energy transfer between the green fluorescent protein labeled proteins within the prohead interior and the translocated rhodamine-labeled DNA confirmed the ATP-powered movement of DNA into the capsid and the packaging of multiple segments per procapsid [77]. Analysis of FRET dye pair end labeled DNA substrates showed that upon packaging the two ends of the packaged DNA were held 8-9 nm apart in the procapsid, likely fixed in the portal channel and crown, and suggesting that a loop rather than an end of DNA is translocated following initiation at an end [79].

In the optical tweezers system, the prohead-gp17 complexes were tethered to a microsphere coated with capsid protein antibody, and the biotinylated DNA is tethered to another microsphere coated with streptavidine. The microspheres are brought together into near contact, allowing the motor to capture the DNA. Single packaging events were monitored and the dynamics of the T4 packaging process were quantified [78]. The T4 motor, like the Phi29 DNA packaging motor, generates forces as high as ~60 pN, which is ~20-25 times that of myosin ATPase and a rate as high as ~2000 bp/sec, the highest recorded to date. Slips and pauses occur but these are relatively short and rare and the motor recovers and recaptures DNA continuing translocation. The high rate of translocation is in keeping with the need to package the 171 kb size T4 genome in about 5 minutes. The T4 motor generates enormous power; when an external load of 40 pN was applied, the T4 motor translocates at a speed of ~380 bp/sec. When scaled up to a macromotor, the T4 motor is approximately twice as powerful as a typical automobile engine.

CryoEM reconstruction of the packaging machine showed two rings of density at the portal vertex [74] (Figure 7). The upper ring is flat, resembling the ATPase domain structure and the lower ring is spherical,

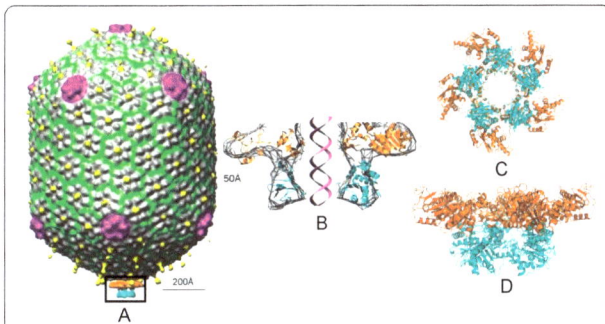

Figure 7 Structure of the T4 DNA packaging machine. **A)** Cryo-EM reconstruction of the phage T4 DNA packaging machine showing the pentameric motor assembled at the special portal vertex. **B-D)** Cross section, top and side views of the pentameric motor respectively, by fitting the X-ray structures of the gp17 ATPase and nuclease/translocation domains into the cryo-EM density.

resembling the C-domain structure. This was confirmed by docking of the X-ray structures of the domains into the cryoEM density. The motor has pentamer stoichiometry, with the ATP binding surface facing the portal and interacting with it. It has an open central channel that is in line with the portal channel and the translocation groove of the C-domain faces the channel. There are minimal contacts between the adjacent subunits suggesting that the ATPases may fire relatively independently during translocation.

Unlike the cryoEM structure where the two lobes (domains) of the motor are separated ("relaxed" state), the domains in the full-length gp17 are in close contact ("tensed" state) [74]. In the tensed state, the subdomain II of ATPase is rotated by 6° degrees and the C-domain is pulled upwards by 7Å, equivalent to 2 bp. The "arginine finger" located between subI and NsubII is positioned towards the βγ phosphates of ATP and the ion pairs are aligned.

Mechanism

Of many models proposed to explain the mechanism of viral DNA translocation, the portal rotation model attracted the most attention. According to the original and subsequent rotation models, the portal and DNA are locked like a nut and bolt [80,81]. The symmetry mismatch between the 5-fold capsid and 12-fold portal means that only one portal subunit aligns with one capsid subunit at any given time, causing the associated terminase-ATPase to fire causing the portal, the nut, to rotate, allowing the DNA, the bolt, to move into the capsid. Indeed, the overall structure of the dodecameric portal is well conserved in numerous bacteriophages and even in HSV, despite no significant sequence similarity. However, the X-ray structures of Phi29 and SPP1 portals did not show any rigid groove-like features that are complementary to the DNA structure [81-83]. The

structures are nevertheless consistent with the proposed portal rotation and newer, more specific, models such as the rotation-compression-relaxation [81], electrostatic gripping [82], and molecular lever [83], have been proposed.

Protein fusions to either the N or C terminal end of the portal protein could be incorporated into up to ~one-half of the dodecamer positions without loss of prohead function. As compared to wild-type, portals containing C-terminal GFP fusions lock the proheads into the unexpanded conformation unless terminase packages DNA, suggesting that the portal plays a central role in controlling prohead expansion. Expansion is required to protect the packaged DNA from nuclease but not for packaging itself as measured by FCS [84]. Moreover retention of DNA packaging function of such portals argues against the portal rotation model, since rotation would require that the bulky C-terminal GFP fusion proteins within the capsid rotate through the densely packaged DNA. A more direct test tethered the portal to the capsid through Hoc interactions [85]. Hoc is a nonessential T4 outer capsid protein that binds as a monomer at the center of the major capsid protein hexon (see above; Figure 1). Hoc binding sites are not present in the unexpanded proheads but are exposed following capsid expansion. To tether the portal, unexpanded proheads were first prepared with 1 to 6 of the 12 portal subunits replaced by the N-terminal Hoc-portal fusion proteins. The proheads were then expanded in vitro to expose Hoc binding sites. The Hoc portion of the portal fusion would bind to the center of the nearest hexon, tethering 1 to 5 portal subunits to the capsid. The Hoc-capsid interaction is thought to be irreversible and thus should prevent the rotation of the portal. If portal rotation were to be central to DNA packaging, the tethered expanded proheads should show very little or no packaging activity. However, the efficiency and rate of packaging of tethered proheads were comparable to those of wild-type proheads, suggesting that portal rotation is not an obligatory requirement for packaging [85]. This was more recently confirmed by single molecule fluorescence spectroscopy of actively packaging Phi29 packaging complexes [86].

In the second class of models, the terminase not only provides the energy but also actively translocates DNA [87]. Conformational changes in the terminase domains cause changes in the DNA binding affinity resulting in binding and releasing DNA, reminiscent of the inchworm-type translocation by helicases. gp17 and numerous large terminases possess an ATPase coupling motif that is commonly present in helicases and translocases [87]. Mutations in the coupling motif present at the junction of NSubI and NSubII result in loss of ATPase and DNA packaging activities.

The cryoEM and X-ray structures (Figure 7) combined with the mutational analyses led to the postulation of a terminase-driven packaging mechanism [74]. The pentameric T4 packaging motor can be considered to be analogous to a five cylinder engine. It consists of an ATPase center in NsubI, which is the engine that provides energy. The C-domain has a translocation groove, which is the wheel that moves DNA. The smaller NsubII is the transmission domain, coupling the engine to the wheel via a flexible hinge. The arginine finger is a spark plug that fires ATPase when the motor is locked in the firing mode. Charged pairs generate electrostatic force by alternating between relaxed and tensed states (Figure 8). The nuclease groove faces away from translocating DNA and is activated when packaging is completed.

In the relaxed conformational state (cryoEM structure), the hinge is extended (Figure 8). Binding of DNA to the translocation groove and of ATP to NsubI locks the motor in translocation mode (A) and brings the arginine finger into position, firing ATP hydrolysis (B). The repulsion between the negatively charged ADP(3-) and Pi(3-) drive them apart, causing NsubII to rotate by 6° (C), aligning the charge pairs between the N- and C-

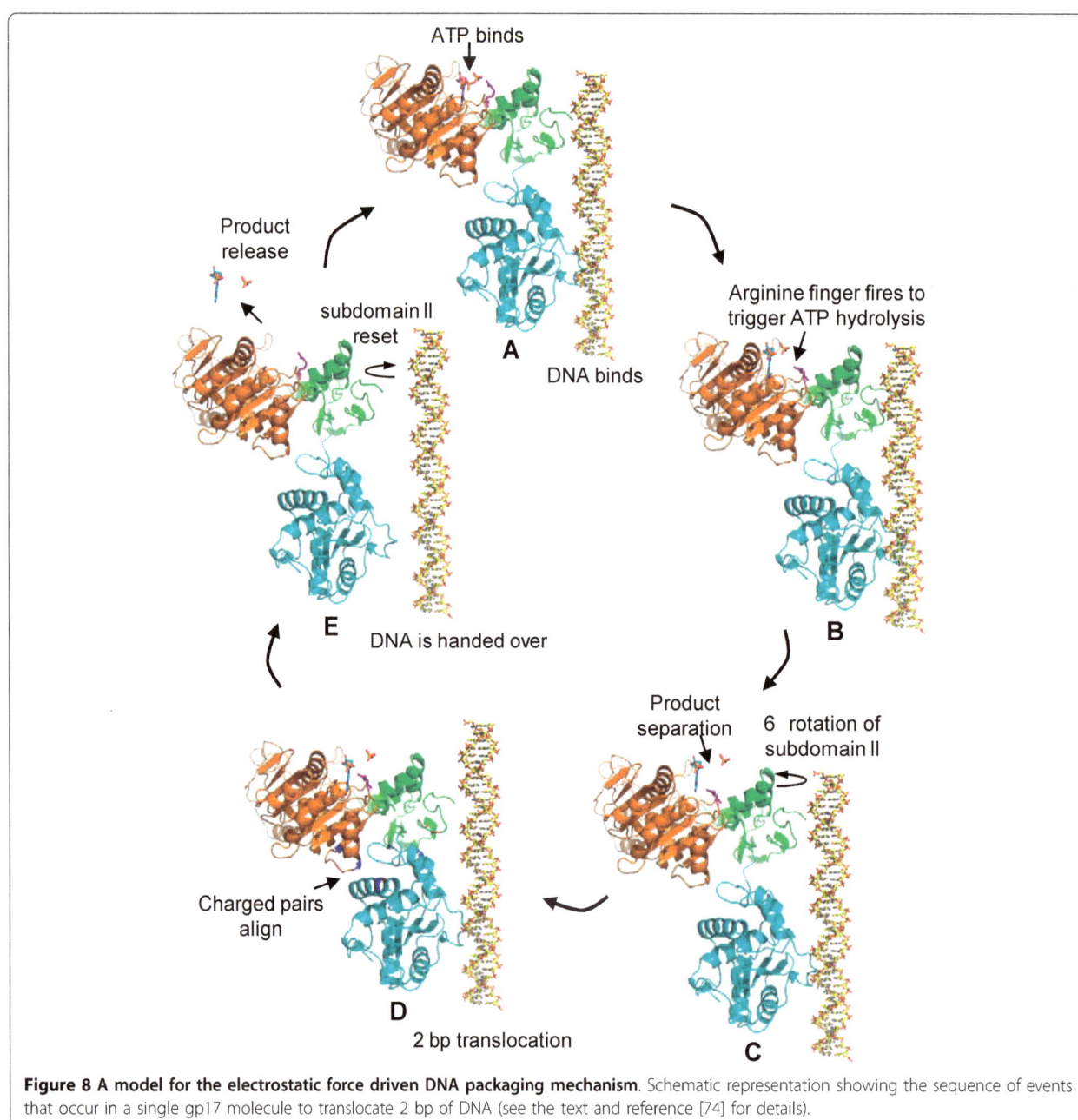

Figure 8 A model for the electrostatic force driven DNA packaging mechanism. Schematic representation showing the sequence of events that occur in a single gp17 molecule to translocate 2 bp of DNA (see the text and reference [74] for details).

domains. This generates electrostatic force, attracting the C-domain-DNA complex and causing 7Å upward movement, the tensed conformational state (X-ray structure) (D). Thus 2 bp of DNA is translocated into the capsid in one cycle. Product release and loss of 6 negative charges causes NsubII to rotate back to original position, misaligning the ion pairs and returning the C-domain to the relaxed state (E).

Translocation of 2 bp would bring the translocation groove of the adjacent subunit into alignment with the backbone phosphates. DNA is then handed over to the next subunit, by the matching motor and DNA symmetries. Thus, ATPase catalysis causes conformational changes which generate electrostatic force, which is then converted to mechanical force. The pentameric motor translocates 10 bp (one turn of the helix) when all five gp17 subunits fire in succession, bringing the first gp17 subunit once again in alignment with the DNA phosphates. Synchronized orchestration of the motor's movements translocates DNA up to ~2000 bp/sec.

Short (< 200 bp) DNA substrate translocation by gp17 is blocked by nicks, gaps, hairpin ends, RNA-containing duplexes, 20-base mismatches and D-loops, but not by 10-base internal mismatches [88]. Packaging of DNAs as short as 20 bp and initiation at almost any type DNA end suggests translocation rather than initiation deficiency of these short centrally nicked or gapped DNAs. Release from the motor of 100 bp nicked DNA segments supported a torsional compression portal-DNA-grip-and-release mechanism, where the portal grips the DNA while the gp17 imparts a linear force that may be stored in the DNA as compression or dissipated by a nick (Figure 9). Use of a DNA leader joined to a Y-DNA structure showed packaging of the leader segment; the Y-junction was arrested in proximity to a prohead portal containing GFP fusions, allowing FRET transfer between the Y-junction located dye molecule and the portal GFPs [89] (Figure 9D). Comparable stalled Y-DNA substrates containing FRET-pair dyes in the Y-stem showed that the motor compresses the stem held in the portal channel by 22-24% (Figure 9E. This finding supports the proposal that torsional compression of B DNA by the terminase motor by a portal-DNA-grip-and-release mechanism helps to drive translocation [88]. Attaching a longer DNA leader to the Y-DNA allows such abnormal structure substrates to be anchored in the procapsid for successful translocation, most likely by multiple motor cycles [89]. Differences in DNA substrate size may at least in part account for much less stringent DNA structural requirements measured in the Phi29 packaging system [90].

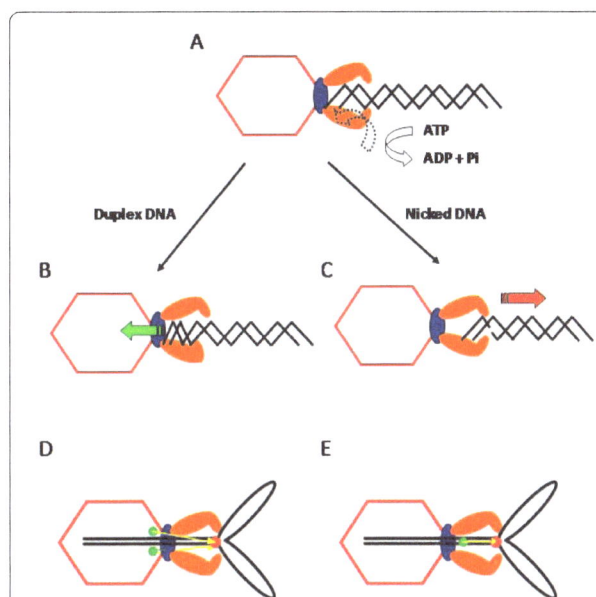

Figure 9 A model for the torsional compression portal-DNA-grip-and-release packaging mechanism. A-C) Short nicked or other abnormal structure containing DNA substrates are released from the motor. **D)** Leader containing Y-DNA substrates are retained by the motor and are anchored in the procapsid in proximity to portal GFP fusions; and **E)** compression of the Y-stem B segment in the stalled complex is observed by FRET [88,89]

Conclusions

It is clear from the above discussion that major advances have been made in recent years on the understanding of the phage T4 capsid structure and mechanism of DNA packaging. These advances, by combining genetics and biochemistry with structure and biophysics, set the stage to probe the packaging mechanism with even greater depth and precision. It is reasonable to hope that this would lead to the elucidation of catalytic cycle, mechanistic details, and motor dynamics to near atomic resolution. The accumulated and emerging basic knowledge should also lead to medical applications such as the development of vaccines and phage therapy.

List of abbreviations

EF: edema factor; EM: electron microscopy; FCS: fluorescence correlation spectroscopy; FMDV: foot and mouth disease virus; FRET: fluorescence resonance energy transfer; gp: gene product; HIV: human immunodeficiency virus; Hoc: highly antigenic outer capsid protein; IP: internal protein; LF: lethal factor; PA: protective antigen; Soc: small outer capsid protein;

Acknowledgements

The authors thank Dr. Bonnie Draper, and Ms. Alice Kuaban, for preparing the figures, references and proof reading. The research in the authors' laboratories has been funded by National Science Foundation (VBR: MCB-0923873) and National Institutes of Health (VBR: NIAID-AI081726; LWB: NIAID-AI011676). Special thanks to our present and former lab members for their contributions over the years.

Author details
[1]Department of Biology, The Catholic University of America, Washington, DC,
USA. [2]Department of Biochemistry and Molecular Biology, University of
Maryland Medical School, Baltimore, MD, USA.

Authors' contributions
VR and LWB made equal contributions to drafts of this review. Both authors
revised all sections of the article and read and approved the final
manuscript.

Competing interests
The authors declare that they have no competing interests.

Received: 6 August 2010 Accepted: 3 December 2010
Published: 3 December 2010

References
1. Comeau AM, Krisch HM: **The capsid of the T4 phage superfamily: the evolution, diversity, and structure of some of the most prevalent proteins in the biosphere.** *Mol Biol Evol* 2008, **25**(7):1321-32.
2. Krisch HM, Comeau AM: **The immense journey of bacteriophage T4–from d'Herelle to Delbruck and then to Darwin and beyond.** *Res Microbiol* 2008, **159**:314-324.
3. Tetart F, Desplats C, Krisch HM: **Genome plasticity in the distal tail fiber locus of the T-even bacteriophage: recombination between conserved motifs swaps adhesin specificity.** *J Mol Biol* 1998, **282**:543-556.
4. Black LW, Showe MK, Steven AC: **Morphogenesis of the T4 head.** 1994, 218-258.
5. Fokine A, Chipman PR, Leiman PG, Mesyanzhinov VV, Rao VB, Rossmann MG: **Molecular architecture of the prolate head of bacteriophage T4.** *Proc Natl Acad Sci USA* 2004, **101**:6003-6008.
6. Fokine A, Leiman PG, Shneider MM, Ahvazi B, Boeshans KM, Steven AC, Black LW, Mesyanzhinov VV, Rossmann MG: **Structural and functional similarities between the capsid proteins of bacteriophages T4 and HK97 point to a common ancestry.** *Proc Natl Acad Sci USA* 2005, **102**:7163-7168.
7. Fokine A, Battisti AJ, Kostyuchenko VA, Black LW, Rossmann MG: **Cryo-EM structure of a bacteriophage T4 gp24 bypass mutant: the evolution of pentameric vertex proteins in icosahedral viruses.** *J Struct Biol* 2006, **154**:255-259.
8. Wikoff WR, Liljas L, Duda RL, Tsuruta H, Hendrix RW, Johnson JE: **Topologically linked protein rings in the bacteriophage HK97 capsid.** *Science* 2000, **289**:2129-2133.
9. Miller ES, Kutter E, Mosig G, Arisaka F, Kunisawa T, Ruger W: **Bacteriophage T4 genome.** *Microbiol Mol Biol Rev* 2003, **67**:86-156, table of contents.
10. Keppel F, Rychner M, Georgopoulos C: **Bacteriophage-encoded cochaperonins can substitute for Escherichia coli's essential GroES protein.** *EMBO Rep* 2002, **3**:893-898.
11. Andreadis JD, Black LW: **Substrate mutations that bypass a specific Cpn10 chaperonin requirement for protein folding.** *J Biol Chem* 1998, **273**:34075-34086.
12. Snyder L, Tarkowski HJ: **The N terminus of the head protein of T4 bacteriophage directs proteins to the GroEL chaperonin.** *J Mol Biol* 2005, **345**:375-386.
13. Bakkes PJ, Faber BW, van Heerikhuizen H, van der Vies SM: **The T4-encoded cochaperonin, gp31, has unique properties that explain its requirement for the folding of the T4 major capsid protein.** *Proc Natl Acad Sci USA* 2005, **102**:8144-8149.
14. Clare DK, Bakkes PJ, van Heerikhuizen H, van der Vies SM, Saibil HR: **An expanded protein folding cage in the GroEL-gp31 complex.** *J Mol Biol* 2006, **358**:905-911.
15. Clare DK, Bakkes PJ, van Heerikhuizen H, van der Vies SM, Saibil HR: **Chaperonin complex with a newly folded protein encapsulated in the folding chamber.** *Nature* 2009, **457**:107-110.
16. Cerritelli ME, Cheng N, Rosenberg AH, McPherson CE, Booy FP, Steven AC: **Encapsidated conformation of bacteriophage T7 DNA.** *Cell* 1997, **91**:271-280.
17. Mullaney JM, Thompson RB, Gryczynski Z, Black LW: **Green fluorescent protein as a probe of rotational mobility within bacteriophage T4.** *J Virol Methods* 2000, **88**:35-40.
18. Mullaney JM, Black LW: **Capsid targeting sequence targets foreign proteins into bacteriophage T4 and permits proteolytic processing.** *J Mol Biol* 1996, **261**:372-385.
19. Mullaney JM, Black LW: **Activity of foreign proteins targeted within the bacteriophage T4 head and prohead: implications for packaged DNA structure.** *J Mol Biol* 1998, **283**:913-929.
20. Bair CL, Rifat D, Black LW: **Exclusion of glucosyl-hydroxymethylcytosine DNA containing bacteriophages is overcome by the injected protein inhibitor IPI*.** *J Mol Biol* 2007, **366**:779-789.
21. Bair CL, Black LW: **A type IV modification dependent restriction nuclease that targets glucosylated hydroxymethyl cytosine modified DNAs.** *J Mol Biol* 2007, **366**:768-778.
22. Rifat D, Wright NT, Varney KM, Weber DJ, Black LW: **Restriction endonuclease inhibitor IPI* of bacteriophage T4: a novel structure for a dedicated target.** *J Mol Biol* 2008, **375**:720-734.
23. Repoila F, Tetart F, Bouet JY, Krisch HM: **Genomic polymorphism in the T-even bacteriophages.** *EMBO J* 1994, **13**:4181-4192.
24. Wang GR, Vianelli A, Goldberg EB: **Bacteriophage T4 self-assembly: in vitro reconstitution of recombinant gp2 into infectious phage.** *J Bacteriol* 2000, **182**:672-679.
25. Yu TY, Schaefer J: **REDOR NMR characterization of DNA packaging in bacteriophage T4.** *J Mol Biol* 2008, **382**:1031-1042.
26. Ishii T, Yanagida M: **The two dispensable structural proteins (soc and hoc) of the T4 phage capsid; their purification and properties, isolation and characterization of the defective mutants, and their binding with the defective heads in vitro.** *J Mol Biol* 1977, **109**:487-514.
27. Ishii T, Yamaguchi Y, Yanagida M: **Binding of the structural protein soc to the head shell of bacteriophage T4.** *J Mol Biol* 1978, **120**:533-544.
28. Qin L, Fokine A, O'Donnell E, Rao VB, Rossmann MG: **Structure of the Small Outer Capsid Protein, Soc: A Clamp for Stabilizing Capsids of T4-like Phages.** *J Mol Biol* 2010, **29;395**(4):728-41.
29. Ren ZJ, Lewis GK, Wingfield PT, Locke EG, Steven AC, Black LW: **Phage display of intact domains at high copy number: a system based on SOC, the small outer capsid protein of bacteriophage T4.** *Protein Sci* 1996, **5**:1833-1843.
30. Ren ZJ, Baumann RG, Black LW: **Cloning of linear DNAs in vivo by overexpressed T4 DNA ligase: construction of a T4 phage hoc gene display vector.** *Gene* 1997, **195**:303-311.
31. Ren Z, Black LW: **Phage T4 SOC and HOC display of biologically active, full-length proteins on the viral capsid.** *Gene* 1998, **215**:439-444.
32. Jiang J, Abu-Shilbayeh L, Rao VB: **Display of a PorA peptide from Neisseria meningitidis on the bacteriophage T4 capsid surface.** *Infect Immun* 1997, **65**:4770-4777.
33. Sathaliyawala T, Rao M, Maclean DM, Birx DL, Alving CR, Rao VB: **Assembly of human immunodeficiency virus (HIV) antigens on bacteriophage T4: a novel in vitro approach to construct multicomponent HIV vaccines.** *J Virol* 2006, **80**:7688-7698.
34. Wu J, Tu C, Yu X, Zhang M, Zhang N, Zhao M, Nie W, Ren Z: **Bacteriophage T4 nanoparticle capsid surface SOC and HOC bipartite display with enhanced classical swine fever virus immunogenicity: a powerful immunological approach.** *J Virol Methods* 2007, **139**:50-60.
35. Ren ZJ, Tian CJ, Zhu QS, Zhao MY, Xin AG, Nie WX, Ling SR, Zhu MW, Wu JY, Lan HY, *et al*: **Orally delivered foot-and-mouth disease virus capsid protomer vaccine displayed on T4 bacteriophage surface: 100% protection from potency challenge in mice.** *Vaccine* 2008, **26**:1471-1481.
36. Malys N, Chang DY, Baumann RG, Xie D, Black LW: **A bipartite bacteriophage T4 SOC and HOC randomized peptide display library: detection and analysis of phage T4 terminase (gp17) and late sigma factor (gp55) interaction.** *J Mol Biol* 2002, **319**:289-304.
37. Black LW, Peng G: **Mechanistic coupling of bacteriophage T4 DNA packaging to components of the replication-dependent late transcription machinery.** *J Biol Chem* 2006, **281**:25635-25643.
38. Shivachandra SB, Rao M, Janosi L, Sathaliyawala T, Matyas GR, Alving CR, Leppla SH, Rao VB: **In vitro binding of anthrax protective antigen on bacteriophage T4 capsid surface through Hoc-capsid interactions: a strategy for efficient display of large full-length proteins.** *Virology* 2006, **345**:190-198.
39. Li Q, Shivachandra SB, Zhang Z, Rao VB: **Assembly of the small outer capsid protein, Soc, on bacteriophage T4: a novel system for high**

density display of multiple large anthrax toxins and foreign proteins on phage capsid. *J Mol Biol* 2007, **370**:1006-1019.

40. Shivachandra SB, Li Q, Peachman KK, Matyas GR, Leppla SH, Alving CR, Rao M, Rao VB: Multicomponent anthrax toxin display and delivery using bacteriophage T4. *Vaccine* 2007, **25**:1225-1235.

41. Li Q, Shivachandra SB, Leppla SH, Rao VB: Bacteriophage T4 capsid: a unique platform for efficient surface assembly of macromolecular complexes. *J Mol Biol* 2006, **363**:577-588.

42. Fokine A, Bowman VD, Battisti AJ, Li Q, Chipman PR, Rao VB, Rossmann MG: Cryo-electron microscopy study of bacteriophage T4 displaying anthrax toxin proteins. *Virology* 2007, **367**:422-427.

43. Ren SX, Ren ZJ, Zhao MY, Wang XB, Zuo SG, Yu F: Antitumor activity of endogenous mFlt4 displayed on a T4 phage nanoparticle surface. *Acta Pharmacol Sin* 2009, **30**:637-645.

44. Rao VB, Black LW: DNA packaging in Bacteriophage T4. In *"Viral Genome Packaging Machines: Genetics, Structure, and Mechanism"*. Edited by: Catalano CE. Eurekah.com and Kluwer Academic/Plenum Publishers; 2006:40-58.

45. Black LW: DNA packaging in dsDNA bacteriophages. *Annu Rev Microbiol* 1989, **43**:267-292.

46. Rao VB, Feiss M: The bacteriophage DNA packaging motor. *Annu Rev Genet* 2008, **42**:647-681.

47. Golz S, Kemper B: Association of holliday-structure resolving endonuclease VII with gp20 from the packaging machine of phage T4. *J Mol Biol* 1999, **285**:1131-1144.

48. Kondabagil KR, Rao VB: A critical coiled coil motif in the small terminase, gp16, from bacteriophage T4: insights into DNA packaging initiation and assembly of packaging motor. *J Mol Biol* 2006, **358**:67-82.

49. Lin H, Simon MN, Black LW: Purification and characterization of the small subunit of phage T4 terminase, gp16, required for DNA packaging. *J Biol Chem* 1997, **272**:3495-3501.

50. Lin H, Black LW: DNA requirements in vivo for phage T4 packaging. *Virology* 1998, **242**:118-127.

51. Al-Zahrani AS, Kondabagil K, Gao S, Kelly N, Ghosh-Kumar M, Rao VB: The small terminase, gp16, of bacteriophage T4 is a regulator of the DNA packaging motor. *J Biol Chem* 2009, **284**:24490-24500.

52. Mitchell MS, Matsuzaki S, Imai S, Rao VB: Sequence analysis of bacteriophage T4 DNA packaging/terminase genes 16 and 17 reveals a common ATPase center in the large subunit of viral terminases. *Nucleic Acids Res* 2002, **30**:4009-4021.

53. Duijn EV: Current Limitations in Native Mass Spectrometry Based Structural Biology. *Journal of the American Chemical Society* 2010, **21(6)**:971-8.

54. Wu CH, Lin H, Black LW: Bacteriophage T4 gene 17 amplification mutants: evidence for initiation by the T4 terminase subunit gp16. *J Mol Biol* 1995, **247**:523-528.

55. Black LW: DNA packaging and cutting by phage terminases: control in phage T4 by a synaptic mechanism. *Bioessays* 1995, **17**:1025-1030.

56. Wu CH, Black LW: Mutational analysis of the sequence-specific recombination box for amplification of gene 17 of bacteriophage T4. *J Mol Biol* 1995, **247**:604-617.

57. Leffers G, Rao VB: Biochemical characterization of an ATPase activity associated with the large packaging subunit gp17 from bacteriophage T4. *J Biol Chem* 2000, **275**:37127-37136.

58. Baumann RG, Black LW: Isolation and characterization of T4 bacteriophage gp17 terminase, a large subunit multimer with enhanced ATPase activity. *J Biol Chem* 2003, **278**:4618-4627.

59. Rao VB, Black LW: Cloning, overexpression and purification of the terminase proteins gp16 and gp17 of bacteriophage T4. Construction of a defined in-vitro DNA packaging system using purified terminase proteins. *J Mol Biol* 1988, **200**:475-488.

60. Kondabagil KR, Zhang Z, Rao VB: The DNA translocating ATPase of bacteriophage T4 packaging motor. *J Mol Biol* 2006, **363**:786-799.

61. Alam TI, Rao VB: The ATPase domain of the large terminase protein, gp17, from bacteriophage T4 binds DNA: implications to the DNA packaging mechanism. *J Mol Biol* 2008, **376**:1272-1281.

62. Kanamaru S, Kondabagil K, Rossmann MG, Rao VB: The functional domains of bacteriophage t4 terminase. *J Biol Chem* 2004, **279**:40795-40801.

63. Rao VB, Mitchell MS: The N-terminal ATPase site in the large terminase protein gp17 is critically required for DNA packaging in bacteriophage T4. *J Mol Biol* 2001, **314**:401-411.

64. Mitchell MS, Rao VB: Functional analysis of the bacteriophage T4 DNA-packaging ATPase motor. *J Biol Chem* 2006, **281**:518-527.

65. Goetzinger KR, Rao VB: Defining the ATPase center of bacteriophage T4 DNA packaging machine: requirement for a catalytic glutamate residue in the large terminase protein gp17. *J Mol Biol* 2003, **331**:139-154.

66. Franklin JL, Haseltine D, Davenport L, Mosig G: The largest (70 kDa) product of the bacteriophage T4 DNA terminase gene 17 binds to single-stranded DNA segments and digests them towards junctions with double-stranded DNA. *J Mol Biol* 1998, **277**:541-557.

67. Mosig G: Recombination and recombination-dependent DNA replication in bacteriophage T4. *Annu Rev Genet* 1998, **32**:379-413.

68. Sun S, Kondabagil K, Gentz PM, Rossmann MG, Rao VB: The structure of the ATPase that powers DNA packaging into bacteriophage T4 procapsids. *Mol Cell* 2007, **25**:943-949.

69. Bhattacharyya SP, Rao VB: A novel terminase activity associated with the DNA packaging protein gp17 of bacteriophage T4. *Virology* 1993, **196**:34-44.

70. Bhattacharyya SP, Rao VB: Structural analysis of DNA cleaved in vivo by bacteriophage T4 terminase. *Gene* 1994, **146**:67-72.

71. Kuebler D, Rao VB: Functional analysis of the DNA-packaging/terminase protein gp17 from bacteriophage T4. *J Mol Biol* 1998, **281**:803-814.

72. Rentas FJ, Rao VB: Defining the bacteriophage T4 DNA packaging machine: evidence for a C-terminal DNA cleavage domain in the large terminase/packaging protein gp17. *J Mol Biol* 2003, **334**:37-52.

73. Alam TI, Draper B, Kondabagil K, Rentas FJ, Ghosh-Kumar M, Sun S, Rossmann MG, Rao VB: The headful packaging nuclease of bacteriophage T4. *Mol Microbiol* 2008, **69**:1180-1190.

74. Sun S, Kondabagil K, Draper B, Alam TI, Bowman VD, Zhang Z, Hegde S, Fokine A, Rossmann MG, Rao VB: The structure of the phage T4 DNA packaging motor suggests a mechanism dependent on electrostatic forces. *Cell* 2008, **135**:1251-1262.

75. Lin H, Rao VB, Black LW: Analysis of capsid portal protein and terminase functional domains: interaction sites required for DNA packaging in bacteriophage T4. *J Mol Biol* 1999, **289**:249-260.

76. Leffers G, Rao VB: A discontinuous headful packaging model for packaging less than headful length DNA molecules by bacteriophage T4. *J Mol Biol* 1996, **258**:839-850.

77. Sabanayagam CR, Oram M, Lakowicz JR, Black LW: Viral DNA packaging studied by fluorescence correlation spectroscopy. *Biophys J* 2007, **93**: L17-19.

78. Fuller DN, Raymer DM, Kottadiel VI, Rao VB, Smith DE: Single phage T4 DNA packaging motors exhibit large force generation, high velocity, and dynamic variability. *Proc Natl Acad Sci USA* 2007, **104**:16868-16873.

79. Ray K, Ma J, Oram M, Lakowicz JR, Black LW: Single Molecule- and Fluorescence Correlation Spectroscopy-FRET Analysis of Phage DNA Packaging: Co-localization of the Packaged Phage T4 DNA Ends within the Capsid. *J Mol Biol* 2010, **395**:1102-1113.

80. Hendrix RW: Symmetry mismatch and DNA packaging in large bacteriophages. *Proc Natl Acad Sci USA* 1978, **75**:4779-4783.

81. Simpson AA, Tao Y, Leiman PG, Badasso MO, He Y, Jardine PJ, Olson NH, Morais MC, Grimes S, Anderson DL, *et al*: Structure of the bacteriophage phi29 DNA packaging motor. *Nature* 2000, **408**:745-750.

82. Guasch A, Pous J, Ibarra B, Gomis-Ruth FX, Valpuesta JM, Sousa N, Carrascosa JL, Coll M: Detailed architecture of a DNA translocating machine: the high-resolution structure of the bacteriophage phi29 connector particle. *J Mol Biol* 2002, **315**:663-676.

83. Lebedev AA, Krause MH, Isidro AL, Vagin AA, Orlova EV, Turner J, Dodson EJ, Tavares P, Antson AA: Structural framework for DNA translocation via the viral portal protein. *EMBO J* 2007, **26**:1984-1994.

84. Ray K, Oram M, Ma J, Black LW: Portal control of viral prohead expansion and DNA packaging. *Virology* 2009, **391**:44-50.

85. Baumann RG, Mullaney J, Black LW: Portal fusion protein constraints on function in DNA packaging of bacteriophage T4. *Mol Microbiol* 2006, **61**:16-32.

86. Hugel T, Michaelis J, Hetherington CL, Jardine PJ, Grimes S, Walter JM, Falk W, Anderson DL, Bustamante C: Experimental test of connector rotation during DNA packaging into bacteriophage phi29 capsids. *PLoS Biol* 2007, **5**:e59.

87. Draper B, Rao VB: An ATP hydrolysis sensor in the DNA packaging motor from bacteriophage T4 suggests an inchworm-type translocation mechanism. *J Mol Biol* 2007, **369**:79-94.

88. Oram M, Sabanayagam C, Black LW: **Modulation of the packaging reaction of bacteriophage t4 terminase by DNA structure.** *J Mol Biol* 2008, **381**:61-72.

89. Ray K, Sabanayagam CR, Lakowicz JR, Black LW: **DNA crunching by a viral packaging motor: Compression of a procapsid-portal stalled Y-DNA substrate.** *Virology* 2010, **398**:224-232.

90. Aathavan K, Politzer AT, Kaplan A, Moffitt JR, Chemla YR, Grimes S, Jardine PJ, Anderson DL, Bustamante C: **Substrate interactions and promiscuity in a viral DNA packaging motor.** *Nature* 2009, **461**:669-673.

91. Earnshaw WC, King J, Harrison SC, Eiserling FA: **The structural organization of DNA packaged within the heads of T4 wild-type, isometric and giant bacteriophages.** *Cell* 1978, **14**:559-568.

doi:10.1186/1743-422X-7-356
Cite this article as: Rao and Black: **Structure and assembly of bacteriophage T4 head.** *Virology Journal* 2010 7:356.

Leiman *et al. Virology Journal* 2010, **7**:355
http://www.virologyj.com/content/7/1/355

VIROLOGY JOURNAL

Morphogenesis of the T4 tail and tail fibers

Petr G Leiman[1*], Fumio Arisaka[2], Mark J van Raaij[3], Victor A Kostyuchenko[4], Anastasia A Aksyuk[5], Shuji Kanamaru[2], Michael G Rossmann[5]

Abstract

Remarkable progress has been made during the past ten years in elucidating the structure of the bacteriophage T4 tail by a combination of three-dimensional image reconstruction from electron micrographs and X-ray crystallography of the components. Partial and complete structures of nine out of twenty tail structural proteins have been determined by X-ray crystallography and have been fitted into the 3D-reconstituted structure of the "extended" tail. The 3D structure of the "contracted" tail was also determined and interpreted in terms of component proteins. Given the pseudo-atomic tail structures both before and after contraction, it is now possible to understand the gross conformational change of the baseplate in terms of the change in the relative positions of the subunit proteins. These studies have explained how the conformational change of the baseplate and contraction of the tail are related to the tail's host cell recognition and membrane penetration function. On the other hand, the baseplate assembly process has been recently reexamined in detail in a precise system involving recombinant proteins (unlike the earlier studies with phage mutants). These experiments showed that the sequential association of the subunits of the baseplate wedge is based on the induced-fit upon association of each subunit. It was also found that, upon association of gp53 (gene product 53), the penultimate subunit of the wedge, six of the wedge intermediates spontaneously associate to form a baseplate-like structure in the absence of the central hub. Structure determination of the rest of the subunits and intermediate complexes and the assembly of the hub still require further study.

Introduction

The structures of bacteriophages are unique among viruses in that most of them have tails, the specialized host cell attachment organelles. Phages that possess a tail are collectively called "Caudovirales" [1]. The family Caudovirales is divided into three sub-families according to the tail morphology: Myoviridae (long contractile tail), Siphoviridae (long non-contractile tail), and Podoviridae (short non-contractile tail). Of these, Myoviridae phages have the most complex tail structures with the greatest number of proteins involved in the tail assembly and function. Bacteriophage T4 belongs to this sub-family and has a very high efficiency of infection, likely due to its complex tails and two sets of host-cell binding fibers (Figure 1). In laboratory conditions, virtually every phage particle can adsorb onto a bacterium and is successful in injecting the DNA into the cytosol [2].

Since the emergence of conditional lethal mutants in the 1960's [3], assembly of the phage as well as its molecular genetics have been extensively studied as reviewed in "Molecular biology of bacteriophage T4" [4]. During the past ten years, remarkable progress has been made in understanding the conformational transformation of the tail baseplate from a "hexagon" to a "star" shape, which occurs upon attachment of the phage to the host cell surface. Three-dimensional image reconstructions have been determined of the baseplate, both before [5] and after [6] tail contraction using cryo-electron microscopy and complete or partial atomic structures of eight out of 15 baseplate proteins have been solved [7-14]. The atomic structures of these proteins were fitted into the reconstructions [15]. The fact that the crystal structures of the constituent proteins could be unambiguously placed in both conformations of the baseplate indicated that the gross conformational change of the baseplate is caused by a rearrangement or relative movement of the subunit proteins, rather than associated with large structural changes of individual proteins. This has now provided a good understanding of the

* Correspondence: petr.leiman@epfl.ch
[1]Ecole Polytechnique Fédérale de Lausanne (EPFL), Institut de physique des systèmes biologiques, BSP-415, CH-1015 Lausanne, Switzerland
Full list of author information is available at the end of the article

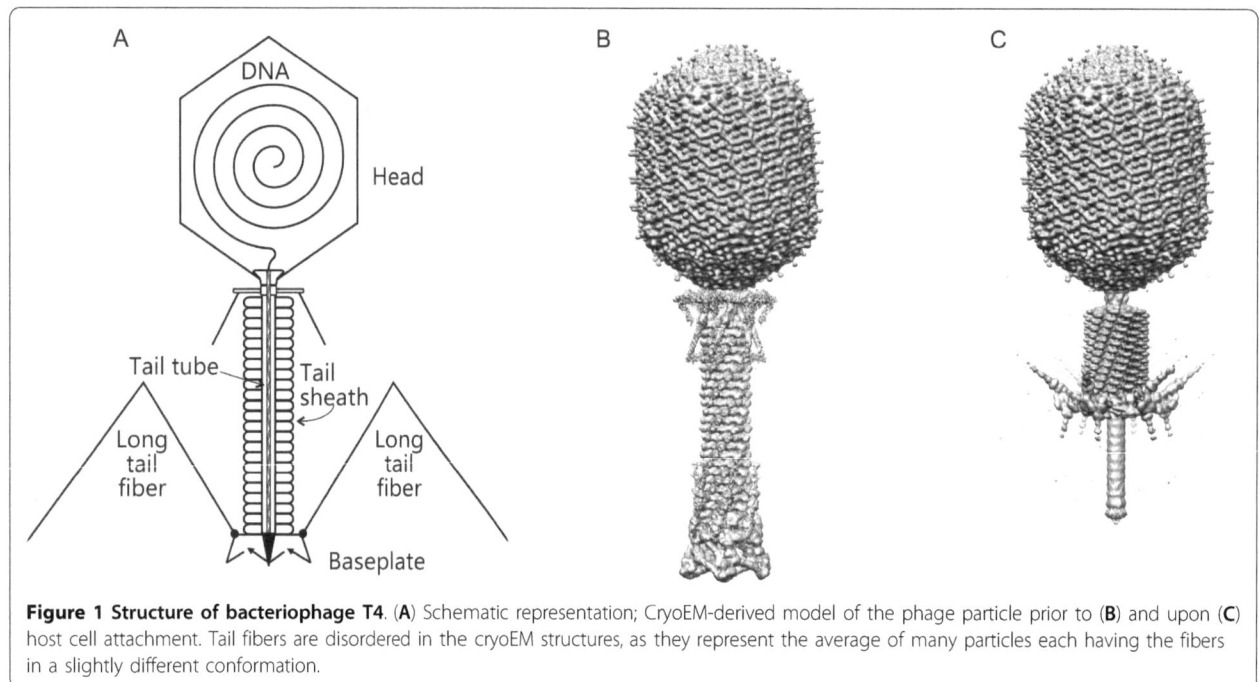

Figure 1 Structure of bacteriophage T4. (A) Schematic representation; CryoEM-derived model of the phage particle prior to **(B)** and upon **(C)** host cell attachment. Tail fibers are disordered in the cryoEM structures, as they represent the average of many particles each having the fibers in a slightly different conformation.

mechanics of the structural transformation of the base-plate, which will be discussed in this review.

Assembly Pathway of the Tail

The tail of bacteriophage T4 is a very large macromolecular complex, comprised of about 430 polypeptide chains with a molecular weight of approximately 2×10^7 (Tables 1, 2 and 3). Twenty two genes are involved in the assembly of the T4 tail (Tables 1, 2 and 3). The tail consists of a sheath, an internal tail tube and a baseplate, situated at the distal end of the tail. Two types of fibers (the long tail fibers and the short tail fibers), responsible for host cell recognition and binding, are attached to the baseplate.

The assembly pathway of the T4 tail has been extensively studied by a number of authors and has been reviewed earlier [16-20]. The main part of the assembly pathway has been elucidated by Kikuchi and King [21-23] with the help of elaborate complementation assays and electron microscopy. The lysates of various amber mutant phage-infected cells were fractionated on sucrose density gradients and complemented with each other *in vitro*. The assembly pathway is strictly ordered and consists of many steps (Figure 2). If one of the gene products is missing, the assembly proceeds to the point where the missing product would be required, leaving the remaining gene products in an "assembly naïve" soluble form, as is especially apparent in the baseplate wedge assembly. The assembly pathway has been confirmed by *in vivo* assembly experiments by Ferguson and Coombs (Table 1) [24] who performed pulse-chase

experiments using ^{35}S-labeled methionine and monitored the accumulation of the labeled gene products in the completed tail. They confirmed the previously proposed assembly pathway and showed that the order of appearance of the labeled gene products also depended on the pool size or the existing number of the protein in the cell. The tail genes are 'late' genes that are expressed almost simultaneously at 8 to 10 min after the infection, indicating that the order of the assembly is determined by the protein interactions, but not by the order of expression.

The fully assembled baseplate is a prerequisite for the assembly of the tail tube and the sheath both of which polymerize into the extended structure using the base-plate as the assembly nucleus (Figure 2). The baseplate is comprised of about 140 polypeptide chains of at least 16 proteins. Two gene products, gp51 and gp57A, are required for assembly, but are not present in the final particle. The baseplate has sixfold symmetry and is assembled from 6 wedges and the central hub. The only known enzyme associated with the phage particle, the T4 tail lysozyme, is a baseplate component. It is encoded by gene 5 (gp5).

The assembly of the wedge, consisting of seven gene products (gp11, gp10, gp7, gp8, gp6, gp53 and gp25), is strictly ordered. When one of the gene products is missing, the intermediate complex before the missing gene product is formed and the remaining gene products stay in a free form in solution. Gp11 is an exception, which can bind to gp10 at any step of the assembly. Recently, all the intermediate complexes and the complete wedge

Table 1 Tail proteins listed in the order of assembly into the complete tail 172425.

Protein	Monomer mass (kDa)	Oligomeric state in solution	Number of monomer copies in the tail	Location and remarks	Protein Data Bank accession code
gp11	23.7	Trimer	18	Wedge, STF[#] binding interface	1EL6
gp10	66.2	Trimer	18	Wedge, STF attachment	2FKK
gp7	119.2	Monomer	6	Wedge	
gp8	38.0	Dimer	12	Wedge	1N7Z
gp6	74.4	Dimer	12	Wedge	3H2T
gp53	23.0	ND*	6	Wedge	
gp25	15.1	Dimer[$]	6	Wedge	
gp5	63.7	Trimer	3	Hub	1K28
gp27	44.4	Trimer	3	Hub	1K28
gp29	64.4	ND	3	Hub, tail tube, Tape measure	
gp28	17.3	ND	1‡	Hub; the tip of gp5 needle?	
gp9	31.0	Trimer	18	Wedge, LTF[¶] attachment site	1S2E
gp12	55.3	Trimer	18	Baseplate outer rim, STF	1H6W, 1OCY
gp48	39.7	ND	6	Baseplate-tail tube junction	
gp54	35.0	ND	6	Baseplate-tail tube junction	
gp19	18.5	Polymer	138	Tail tube	
gp3	19.7	Hexamer	6	Tail tube terminator	
gp18	71.2	Polymer	138	Tail sheath	3FOA
gp15	31.4	Hexamer	6	Tail terminator	

[#] STF, short tail fiber.

* ND, not determined.

[$] P.G. Leiman, unpublished data.

[¶] LTF, long tail fiber.

‡ Copy number and presence in the tail are uncertain.

as well as all the individual gene products of the wedge were isolated, and the interactions among the gene products were examined [25]. An unexpected finding was that gp6, gp53 and gp25 interact with each other weakly. Gp53, however, binds strongly to the precursor wedge complex only after gp6 has bound. Similarly, gp53 is required for gp25 binding. These findings strongly indicated that the strict sequential order of the wedge assembly is due to a conformational change of the intermediate complex, which results in the creation of a new binding site rather than formation of a new binding site at the interface between the newly bound gene product and the precursor complex. Another unexpected finding was that the wedge precursor complexes spontaneously assemble into sixfold symmetrical star-shaped baseplate-like, 43S structure as soon as gp53 binds. The 43S

Table 2 Chaperones involved in the assembly of the tail, tail fibers and attachment of the fibers to the phage particle 7172343446274.

Protein	Monomer mass (kDa)	Oligomeric state in solution	Function	Protein Data Bank accession code
gp8	38.0	Dimer	Folding of gp6	1N7Z
gp26	23.9	ND*	Hub assembly chaperone	
gp51	29.3	ND	Hub assembly chaperone	
gp57A	5.7	Mixture: Trimer-Hexamer-Dodecamer	Folding of gp12, gp34, gp37	
gp38	22.3	ND	Folding of gp37	
gpwac	51.9	Trimer	LTF[¶] to baseplate attachment	1AA0
gp63	45.3	ND	LTF to baseplate attachment	

* ND, not determined.

[¶] LTF, long tail fiber.

Table 3 T4 fibers 17186265.

Fiber	Gene	Monomer mass (kDa)	No. of protein chains per fiber	Location
STF	12	55.3	3	Baseplate
LTF	34	140.0	3	Proximal part, connected to the baseplate
	35	30.0	1	Hinge region
	36	23.0	3	Distal part, hinge connection
	37	109.0	3	Distal part, receptor recognition tip
Head whisker	*wac*	51.9	3	Head-tail joining region

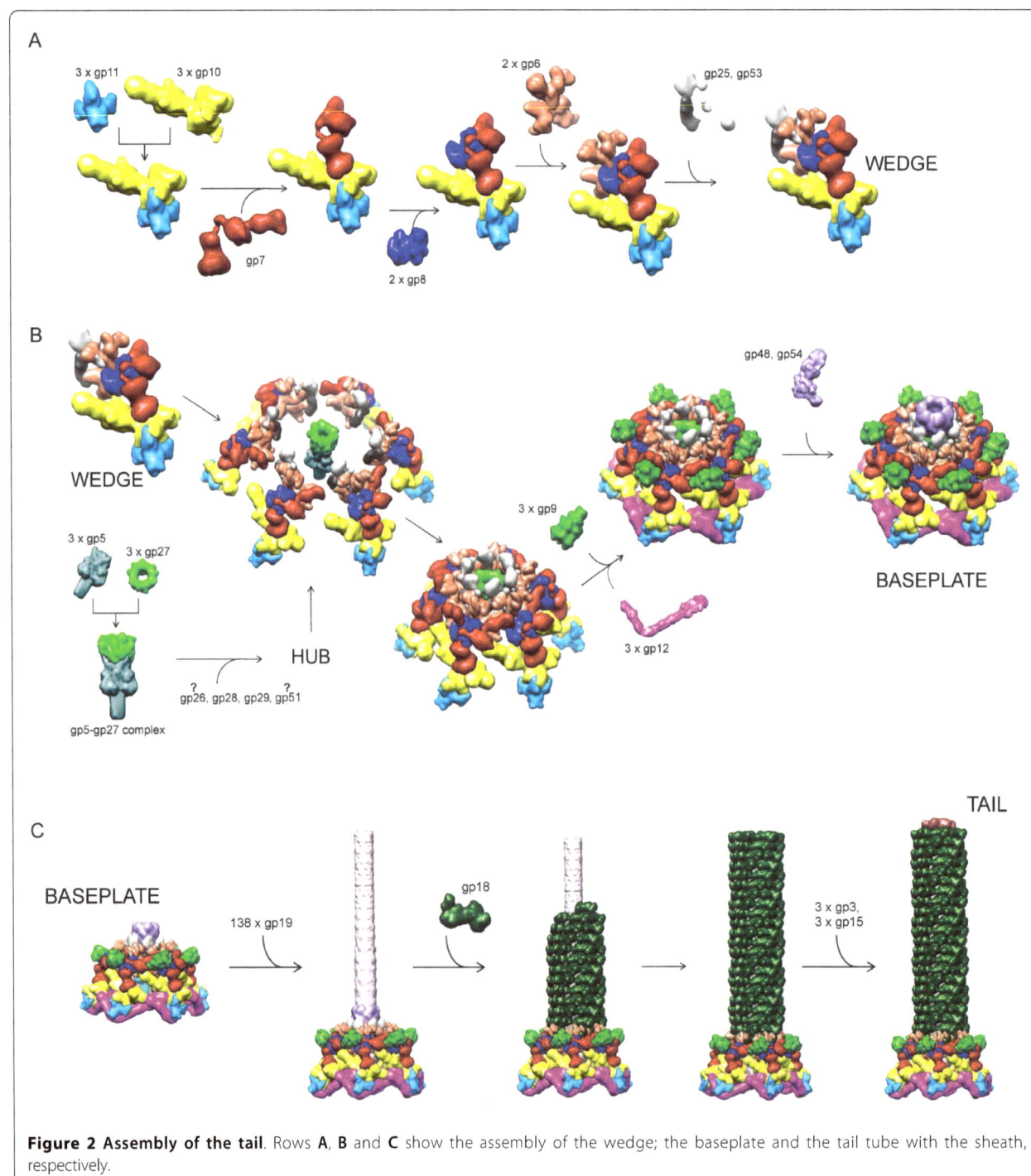

Figure 2 Assembly of the tail. Rows **A**, **B** and **C** show the assembly of the wedge; the baseplate and the tail tube with the sheath, respectively.

Leiman *et al. Virology Journal* 2010, **7**:355
http://www.virologyj.com/content/7/1/355

baseplate decreases its sedimentation coefficient to 40S after gp25 and gp11 binding, apparently due to a structural change in the baseplate [21-23]. Based on these findings, Yap *et al.* [25] have postulated that the 40S star-shaped particle is capable of binding the hub and the six short, gp12 tail fibers, to form the 70S dome-shaped baseplate, found in the extended tail.

Several groups studied the assembly and composition of the central part of the baseplate - the hub - and arrived at different, rather contradictory, conclusions [17]. The assembly of the hub is complicated by a branching pathway and by the presence of gp51, an essential protein of unknown function [26]. Structural studies suggest that the hub consists of at least four proteins: gp5, gp27, gp29 and another unidentified small protein, possibly, gp28 [5]. Recent genetic studies support some of the earlier findings that the hub contains gp26 and gp28 [27].

After the formation of the 70S dome-shaped baseplate containing the short tail fibers, six gp9 trimers (the "socket proteins" of the long tail fibers) bind to the baseplate. Gp48 and gp54 bind to the 'upper' part of the baseplate dome to form the platform for polymerization of gp19 for formation of the tube.

The detailed mechanism of the length determination of the tube is unknown, but the strongest current hypothesis suggests that gp29 is incorporated into the baseplate in an unfolded form. Gp29, the "tape-measure protein", extends as more and more copies of the tail tube protomer, gp19, are added to the growing tube[28]. At the end of the tube, the capping protein, gp3, binds to the last row of gp19 subunits (and, possibly, to gp29) to stabilize them. The tail sheath is built from gp18 subunits simultaneously as the tube, using the tube as a scaffold. When the sheath reaches the length of the tube, the tail terminator protein, gp15, binds to gp3 and the last row of gp18 subunits, completing the tail, which becomes competent for attachment to the head. Both gp15 and gp3 form hexameric rings [29].

The assembly pathway of the tail is a component of **Movie 1** (http://www.seyet.com/t4_virology.html), which describes the assembly of the entire phage particle.

Tail Structure

Structure of the baseplate and its constituent proteins

The tail consists of the sheath, the internal tail tube and the baseplate, situated at the distal end of the tail (Figures 1 and 2). During attachment to the host cell surface, the tail undergoes a large conformational change: The baseplate opens up like a flower, the sheath contracts, and the internal tube is pushed through the baseplate, penetrating the host envelope. The phage DNA is then released into the host cell cytoplasm through the tube. The tail can, therefore, be compared to a syringe,

which is powered by the extended spring, the sheath, making the term "macromolecular nanomachine" appropriate.

The baseplate conformation is coupled to that of the sheath: the "hexagonal" conformation is associated with the extended sheath, whereas the "star" conformation is associated with the contracted sheath that occurs in the T4 particle after attachment to the host cell. Before discussing more fully the baseplate and tail structures in their two conformations, the crystal structures of the baseplate constituent proteins as well as relevant biochemical and genetic data will be described.

Crystal structure of the cell-puncturing device, the gp5-gp27 complex

Gp5 was identified as the tail-associated lysozyme, required during infection but not for cell lysis [30]. The lysozyme domain of gp5 is the middle part of the gp5 polypeptide [31]. It has 43% sequence identity to the cytoplasmic T4 lysozyme, encoded by gene e and called T4L [32]. Gp5 was found to undergo post-translational proteolysis [31], which was believed to be required for activation. Kanamaru *et al.* [33] showed that the C-terminal domain of gp5, which they named gp5C, is a structural component of the phage particle. Furthermore, Kanamaru *et al.* [33] reported that 1) gp5C is an SDS- and urea-resistant trimer; 2) gp5C is responsible for trimerization of the entire gp5; 3) gp5C is rich in β-structure; 4) post-translational proteolysis occurs between Ser351 and Ala352; 5) gp5C dissociates from the N-terminal part, called gp5*, at elevated temperatures; and that 6) the lysozyme activity of the trimeric gp5 in the presence of gp5C is only 10% of that of the monomeric gp5*. The amino acid sequence of gp5C contains eleven **VXG**XXXXX repeats. Subsequent studies showed that gp5 forms a stable complex with gp27 in equimolar quantities and that this complex falls apart in low pH conditions (Figure 3). Upon cleavage of gp5, this complex consists of 9 polypeptide chains, represented as (gp27-gp5*-gp5C)$_3$.

The crystal structure of the gp5-gp27 complex was determined to a resolution of 2.9 Å [13]. The structure resembles a 190 Å long torch (or flashlight) (Figure 4) with the gp27 trimer forming the cylindrical "head" part of the structure. This hollow cylinder has internal and external diameters of about 30 Å and 80 Å, respectively, and is about 60 Å long. The cylinder encompasses three N-terminal domains of the trimeric gp5* to which the 'handle' of the torch is attached. The 'handle' is formed by three intertwined polypeptide chains constituting the gp5 C-terminal domain folded into a trimeric β-helix. The three gp5 lysozyme domains are adjacent to the β-helix. Two long peptide linkers run along the side of the β-helix, connecting the lysozyme domain with the gp5

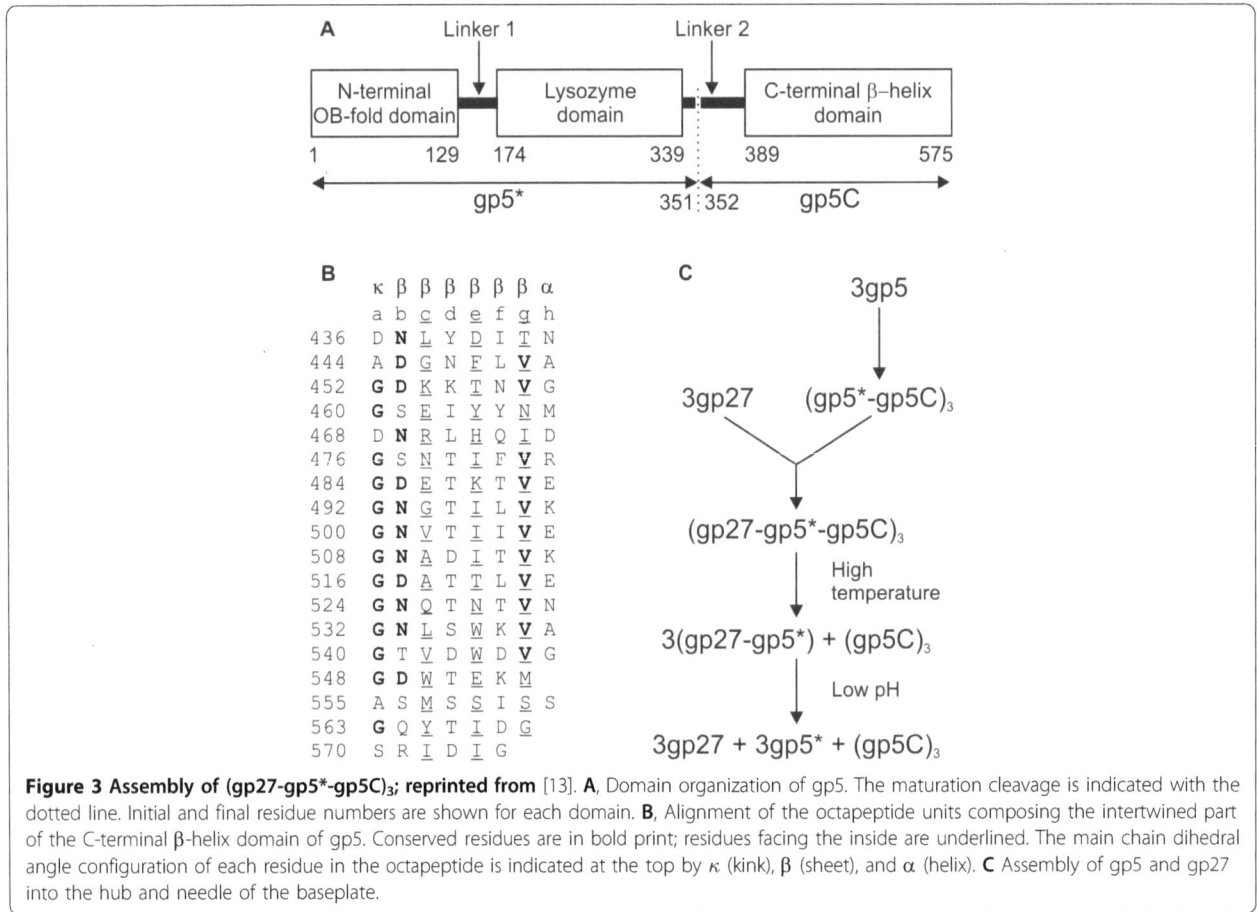

Figure 3 Assembly of (gp27-gp5*-gp5C)₃; reprinted from [13]. **A**, Domain organization of gp5. The maturation cleavage is indicated with the dotted line. Initial and final residue numbers are shown for each domain. **B**, Alignment of the octapeptide units composing the intertwined part of the C-terminal β-helix domain of gp5. Conserved residues are in bold print; residues facing the inside are underlined. The main chain dihedral angle configuration of each residue in the octapeptide is indicated at the top by κ (kink), β (sheet), and α (helix). **C** Assembly of gp5 and gp27 into the hub and needle of the baseplate.

N- and C-terminal domains. The linker joining the lysozyme domain to the β-helix contains the cleavage site between gp5* and gp5C.

Two domains of gp27 (residues 2 to 111 and residues 207-239 plus 307-368) are homologous (Figure 4). They have similar seven- or eight-stranded, antiparallel β-barrel structures, which can be superimposed on each other with the root mean square deviation (RMSD) of 2.4 Å between the 63 equivalent C_α atoms, representing 82% of all C_α atoms. The superposition transformation involves an approximately 60° rotation about the crystallographic threefold axis. Thus, these domains of gp27 form a pseudo-sixfold-symmetric torus in the trimer, which serves as the symmetry adjuster between the trimeric gp5-gp27 complex and the sixfold-symmetric baseplate. Notwithstanding the structural similarity of these two domains, there is only 4% sequence identity of the structurally equivalent amino acids in these two domains. Nevertheless, the electrostatic charge distribution and hydrophilic properties of the gp27 trimer are roughly sixfold symmetric.

Gp5* consists of the N-terminal OB-fold domain and the lysozyme domain. The OB-fold domain is a five-stranded antiparallel β-barrel with a Greek-key topology that was originally observed as being an oligosaccharide/oligonucleotide-binding domain [34]. It is clear now that this fold shows considerable variability of its binding specificity, although the substrate binding site location on the surfaces on most OB-folds has a common site [35]. It is unlikely that the gp5 N-terminal domain is involved in polysaccharide binding, as it lacks the polar residues required for binding sugars. Most probably, the OB-fold has adapted to serve as an adapter between the gp27 trimer and the C-terminal β-helical domain.

The structure of the gp5 lysozyme domain is similar to that of hen egg white lysozyme (HEWL) and T4L having 43% sequence identity with the latter. The two T4 lysozyme structures can be superimposed with an RSMD of 1.1 Å using all C_α atoms in the alignment. There are two small additional loops in gp5, constituting a total of 5 extra residues (Val211-Arg212 and Asn232-Pro233,-Gly234). The active site residues of HEWL, T4L and gp5 are conserved. The known catalytic residues of T4L, Glu11, Asp20, and Thr26, correspond to Glu184, Asp193, and Thr199 in gp5, respectively, establishing that the enzymatic mechanism is the same and that the

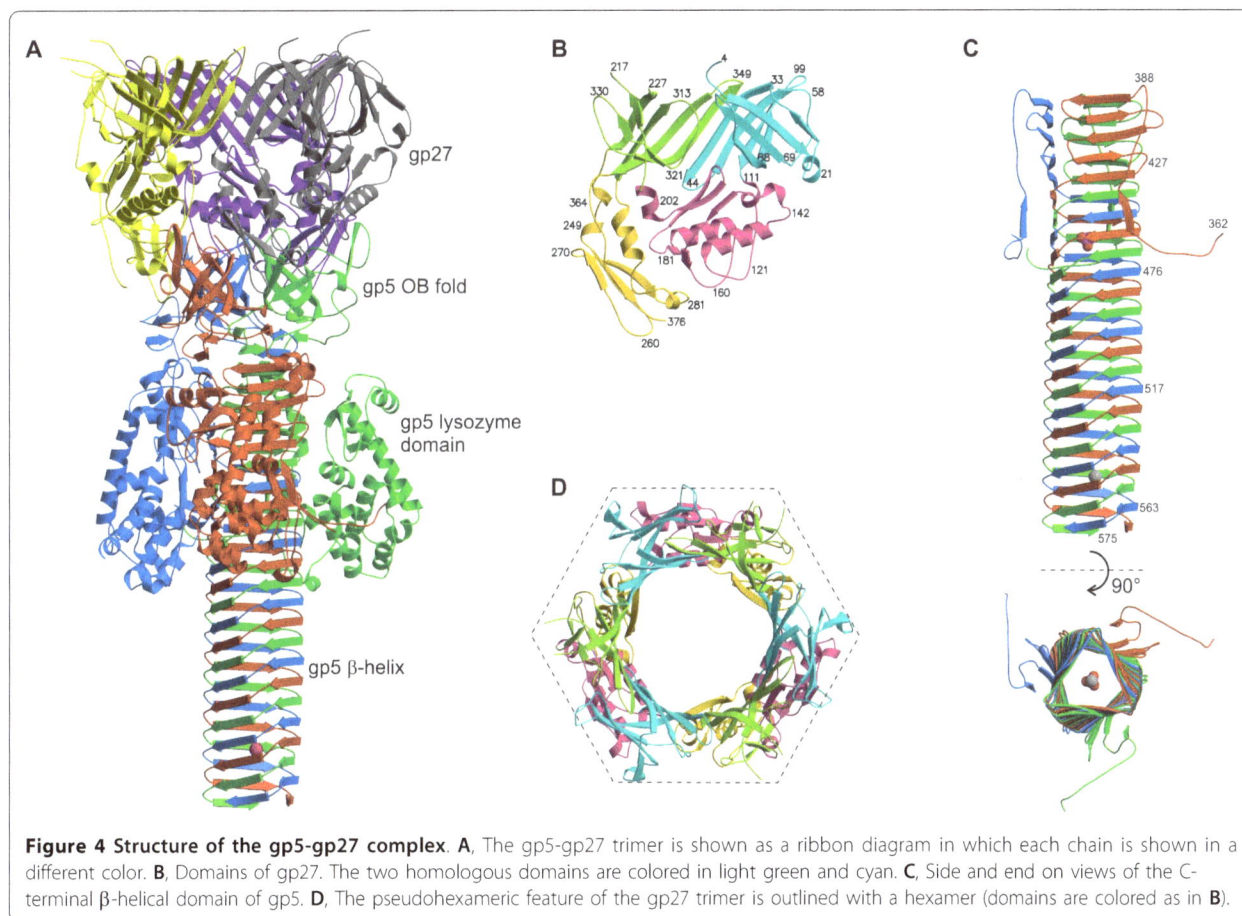

Figure 4 Structure of the gp5-gp27 complex. **A**, The gp5-gp27 trimer is shown as a ribbon diagram in which each chain is shown in a different color. **B**, Domains of gp27. The two homologous domains are colored in light green and cyan. **C**, Side and end on views of the C-terminal β-helical domain of gp5. **D**, The pseudohexameric feature of the gp27 trimer is outlined with a hexamer (domains are colored as in **B**).

gp5 lysozyme domain, T4L and HEWL have a common evolutionary origin.

By comparing the crystal structure of T4L with bound substrate [36] to gp5, the inhibition of gp5 lysozyme activity in the presence of the C-terminal β-helix can be explained. Both gp5 and T4L have the same natural substrate, namely *E. coli* periplasmic cell wall, the major component of which ((NAG-NAM)-LAla-D*iso*Glu-DAP-DAla [36]) contains sugar and peptide moieties. In the gp5 trimer, the linker connecting the lysozyme domain to the β-helix prevents binding of the peptide portion of the substrate to the lysozyme domain. At the same time, the polysaccharide binding cleft is sterically blocked by the gp5 β-helix. Dissociation of the β-helix removes both of these blockages and restores the full lysozyme activity of gp5*.

Gp5C, the C-terminal domain of gp5, is a triple-stranded β-helix (Figure 4). Three polypeptide chains wind around each other to create an equilateral triangular prism, which is 110 Å long and 28 Å in diameter. Each face has a slight left-handed twist (about 3° per β-strand), as is normally observed in β-sheets. The width of the prism face tapers gradually from 33 Å at the

amino end to 25 Å at the carboxy end of the β-helix, thus creating a pointed needle. This narrowing is caused by a decrease in size of the external side chains and by the internal methionines 554 and 557, which break the octapeptide repeat near the tip of the helix. The first 5 β-strands (residues 389-435) form an antiparallel β-sheet, which forms one of the three faces of the prism. The succeeding 18 β-strands comprise a 3-start intertwined β-helix together with the other two, threefold-related polypeptides. The intertwined C-terminal part of the β-helical prism (residues 436-575) is a remarkably smooth continuation of its three non-intertwined N-terminal parts (residues 389-435).

The octapeptide sequence of the helical intertwined part of the prism (residues **a** through **h**) has dominant glycines at position **a**, asparagines or aspartic acids at position **b**, valines at position **g**, and polar or charged residues at position **h**. Residues **b** through **g** form extended β-strands (Ramachandran angles $\phi \approx -129°$, $\psi \approx 128°$) that run at an angle of 75° with respect to the helix axis. The glycines at position **a** ($\phi = -85°$, $\psi = -143°$, an allowed region of the Ramachandran diagram) and residues at position **h** ($\phi = -70°$, $\psi = -30°$, typical for α-helices) kink the polypeptide chain by about 130° clockwise. The conserved valines at position **g** always point to the inside of the β-helix and form a "knob-into-holes" arrangement with the main chain atoms of the glycines at position **a** and the aliphatic part of the side chains of residues at position **c**. Asp436 replaces the normal glycine in position **a** and is at the start of the β-helix. This substitution may be required for folding of the β-helix, because the Asp436 O_δ atom makes a hydrogen bond with O_γ of Ser427 from the threefold-related polypeptide chain. The side chain oxygen atoms of Asp468, which also occupies position **a**, forms hydrogen bonds with residues in the lysozyme domain.

The interior of the β-helix is progressively more hydrophobic toward its C-terminal tip. The middle part of the helix has a pore, which is filled with water molecules bound to polar and charged side chains. The helix is stabilized by two ions situated on its symmetry axis: an anion (possibly, a phosphate) coordinated by three Lys454 residues and a hydrated Ca^{2+} cation (S. Buth, S. Budko, P. Leiman unpublished data) coordinated by three Glu552 residues. These features contribute to the chemical stability of the β-helix, which is resistant to 10% SDS and 2 M guanidine HCl. The surface of the β-helix is highly negatively charged. This charge may be necessary to repel the phosphates of the lipid bilayer when the β-helix penetrates through the outer cell membrane during infection.

Crystal structures of gp6, gp8, gp9, gp10, gp11 and gp12
Genes of all the T4 baseplate proteins were cloned into high level expression vectors individually and in various combinations. Proteins comprising the periphery of the baseplate showed better solubility and could be purified in amounts sufficient for crystallization. The activity was checked in complementation assays using a corresponding amber mutant phage. It was possible to crystallize and solve structures of the full-length gp8, gp9 and gp11 (Figure 5) [8-10]. The putative domain organization of gp10 was derived from the cryoEM map of the baseplate. This information was used to design a deletion mutant constituting the C-terminal domain, which was then crystallized [11]. A stable deletion mutant of gp6 suitable for crystallization was identified using limited proteolysis (Figure 5) [7]. Full-length gp12 showed a very high tendency to aggregation. Gp12 was subjected to limited proteolysis in various buffers and conditions. Two slightly different proteolysis products, which resulted from these experiments, were crystallized (Figure 5) [12,14]. Due to crystal disorder, it was possible to build an atomic model for less than half of the crystallized gp12 fragments [12,14].

Two proteins, gp6 and gp8, are dimers, whereas the rest of the crystallized proteins - gp9, gp10, gp11 and

gp12 - are trimers. None of the proteins had a structural homolog in the Protein Data Bank when these structures were determined. Neither previous studies nor new structural information suggested any enzymatic activity for these proteins. The overall fold of gp12 is the most remarkable of the six mentioned proteins. The topology of the C-terminal globular part is so complex that it creates an impression that the three polypeptide chains knot around each other [14]. This is not the case, however, because the polypeptide chains can be pulled apart from their ends without entanglement. Thus the fold has been characterized as being 'knitted', but not 'knotted' [14]. Gp12 was reported to be a Zn-containing protein [37] and X-ray fluorescent data supported this finding, although Zn was present in the purification buffer [14]. The Zn atom was found to be buried deep inside the C-terminal domain. It is positioned on the threefold axis of the protein and is coordinated by the side chains of His445 and His447 from each of the three chains, resulting in octahedral geometry that is unusual for Zn [12,14,38].

Although gp12, like gp5, contains a triple-stranded β-helix (Figure 5) these helices are quite different in their structural and biochemical properties. The gp12 β-helix is narrower than the gp5 β-helix because there are 6 residues (on average) per turn in the gp12 β-helix compared to 8 in gp5. The interior of the gp12 β-helix is hydrophobic, whereas only the interior of the C-terminal tip of the gp5 β-helix is hydrophobic, but the rest is quite hydrophilic, contains water, phosphate and lipid molecules (S. Buth, S. Budko, P. Leiman unpublished data). Furthermore, the gp12 β-helix lacks the well defined gp5-like repeat.

Many functional analogs of the T4 short tail fibers in other bacteriophages have enzymatic activity and are called tailspikes. The endosialidase from phage K1F and its close homologs from phages K1E, K1-5 and CUS3 contain a very similar β-helix that has several small loops, which create a secondary substrate-binding site [39-41]. The gp12-like β-helix can be found in tail fibers of many lactophages [42], and is a very common motif for proteins that participate in lipopolysaccharide (LPS) binding. However, most gp12-like β-helices do not possess LPS binding sites. Furthermore, unlike gp5, the gp12-like β-helix cannot fold on its own, requiring a chaperone, (e.g. T4 gp57A) for folding correctly [43,44]. Nevertheless, gp12-like β-helix might have enough flexibility and possesses other properties that render give it LPS binding proteins.

The T4 baseplate is significantly more complex than that of phage P2 or Mu, two other well studied contractile tail phages [45,46], and contains at least five extra proteins (gp7, gp8, gp9, gp10 and gp11), all positioned at the baseplate's periphery. T4 gp25 and gp6 have

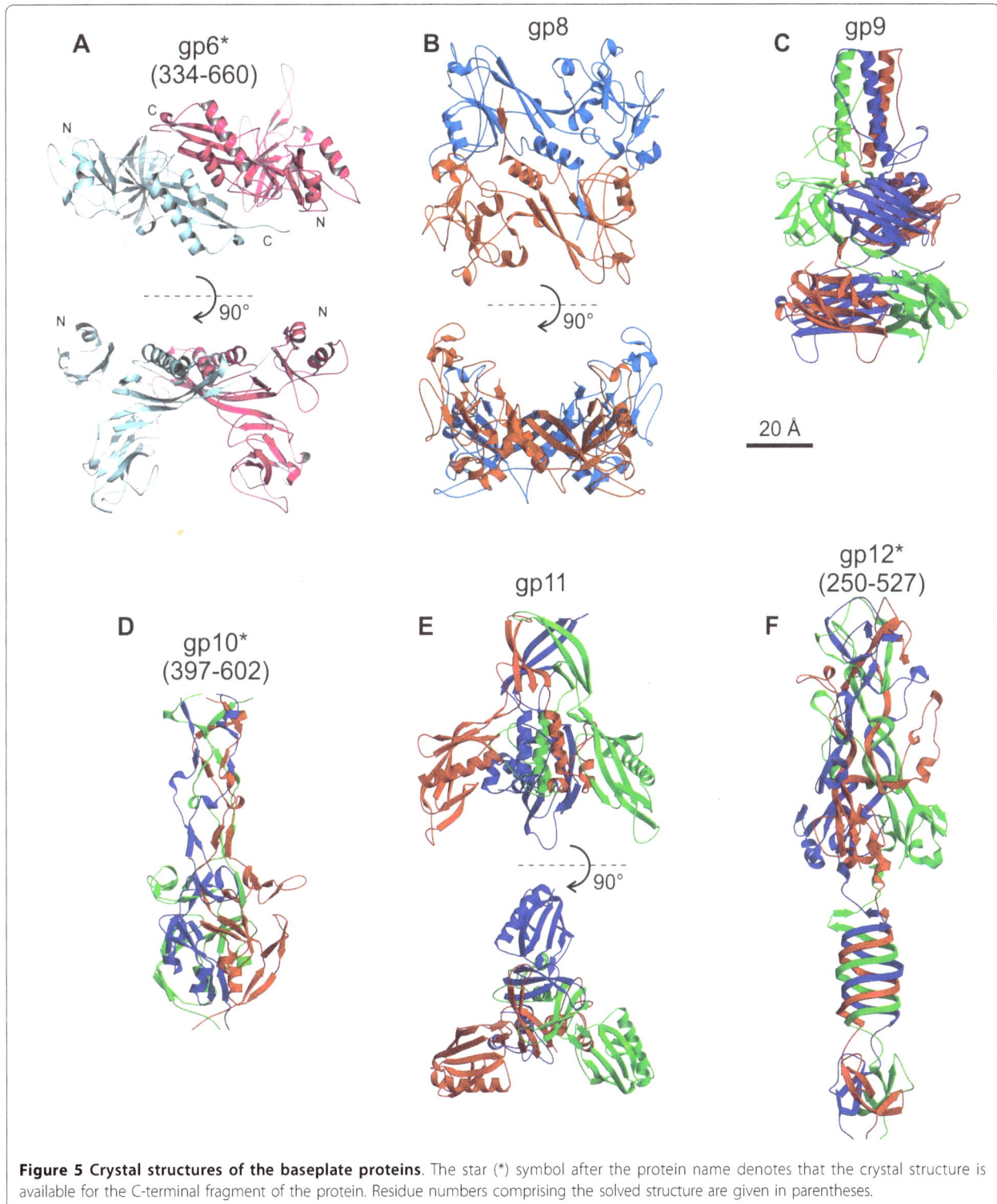

Figure 5 Crystal structures of the baseplate proteins. The star (*) symbol after the protein name denotes that the crystal structure is available for the C-terminal fragment of the protein. Residue numbers comprising the solved structure are given in parentheses.

genes W and J as homologs in P2, respectively ([45] and P. Leiman unpublished data). However, the origin and evolutionary relationships for the rest of the baseplate proteins cannot be detected at the amino acid level. The crystal structure of the C-terminal fragment (residues

397 - 602) of gp10 has provided some clues to understanding the evolution of T4 baseplate proteins [11].

The structures of gp10, gp11 and gp12 can be superimposed onto each other (Figure 5) suggesting that the three proteins have evolved from a common primordial

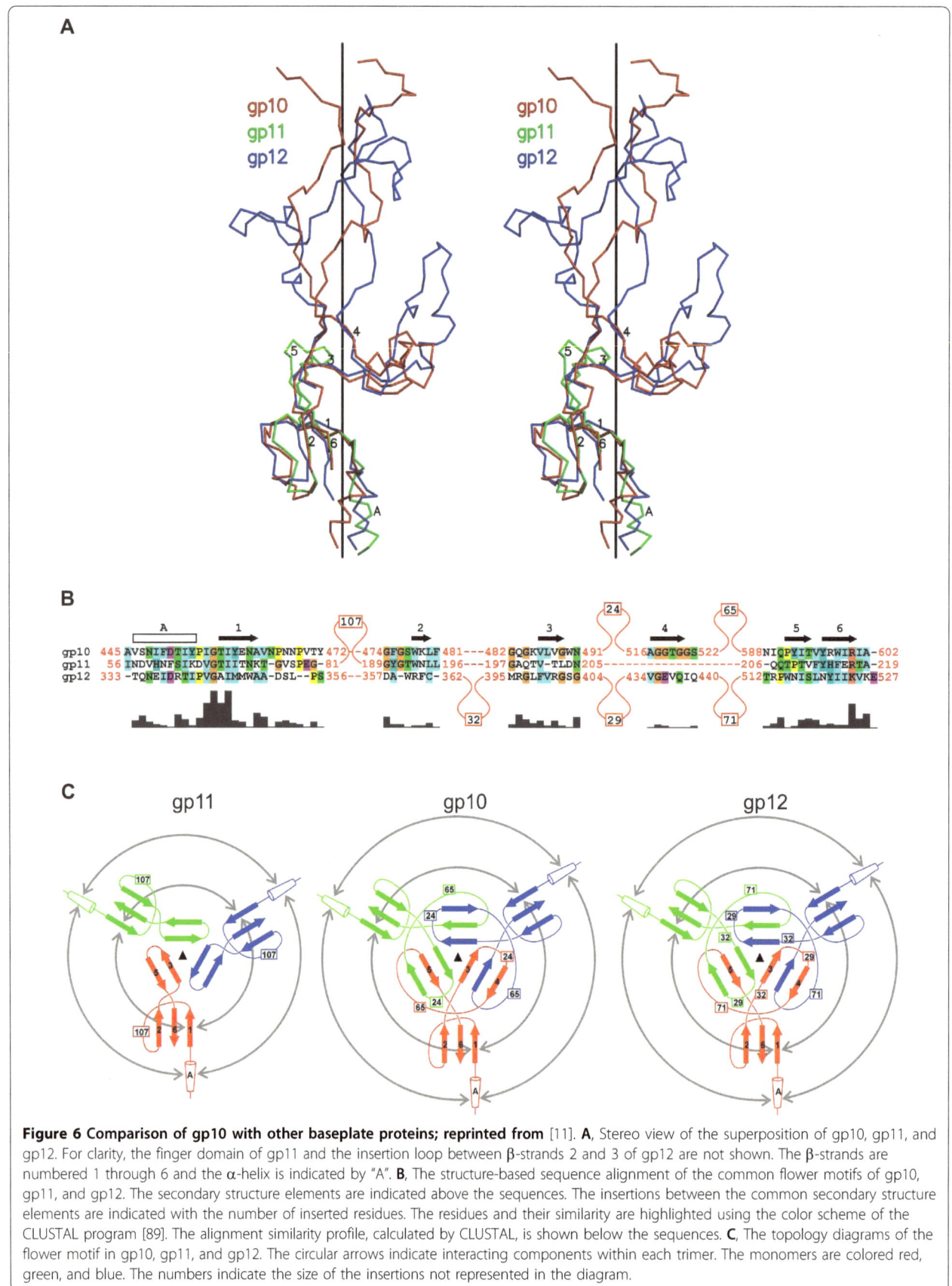

Figure 6 Comparison of gp10 with other baseplate proteins; reprinted from [11]. **A**, Stereo view of the superposition of gp10, gp11, and gp12. For clarity, the finger domain of gp11 and the insertion loop between β-strands 2 and 3 of gp12 are not shown. The β-strands are numbered 1 through 6 and the α-helix is indicated by "A". **B**, The structure-based sequence alignment of the common flower motifs of gp10, gp11, and gp12. The secondary structure elements are indicated above the sequences. The insertions between the common secondary structure elements are indicated with the number of inserted residues. The residues and their similarity are highlighted using the color scheme of the CLUSTAL program [89]. The alignment similarity profile, calculated by CLUSTAL, is shown below the sequences. **C**, The topology diagrams of the flower motif in gp10, gp11, and gp12. The circular arrows indicate interacting components within each trimer. The monomers are colored red, green, and blue. The numbers indicate the size of the insertions not represented in the diagram.

fold, consisting of an α-helix, a three-stranded β-sheet almost perpendicular to the helix, and an additional 2 or 3 stranded β-sheet further away from the helix (Figure 6). This structural motif is decorated by big loops inserted in various regions of the core fold, thus obscuring visual comparison. It is of significance that the three proteins are translated from the same polycistronic mRNA and are sequential in the genome. Furthermore, all three proteins are on the periphery of the baseplate and interact with each other. Apparently, over the course of the T4 evolution, these proteins have become more functionally specialized and have acquired or discarded subdomains that define the functions of the present proteins.

In addition to its structural role in the baseplate, gp8 functions as a chaperone for folding of gp6 (Table 2), which is insoluble unless co-expressed with gp8 [7]. Although wild type gp6 could not be crystallized, the structure of a gp6 mutant, constituting the C-terminal part of the protein (residues 334 - 660) has been determined [7]. The structure is a dimer, which fits well into the cryoEM map of both, the hexagonal and star-shaped baseplates [7].

Structure of the baseplate in the hexagonal conformation

The structure of the baseplate in the hexagonal conformation was studied both by using a phage mutant that produces the baseplate-tail tube complex (a g18⁻/g23⁻ double mutant), as well as by using wild type phage [5,47]. The star conformation was examined by treating the phage with 3 M urea in a neutral pH buffer [6] causing the tail to contract, but retaining the DNA in the head. This particle mimics the phage after it has attached to the host cell surface. Three-dimensional cryoEM maps of the baseplate and the entire tail in either conformation were calculated at resolutions of 12 Å and 17 Å, respectively (Figure 7). The available crystal structures were fitted into these maps.

The hexagonal baseplate is a dome-like structure with a diameter of about 520 Å around its base and about 270 Å in height. Overall, the structure resembles a pile of logs because its periphery is composed of fibrous proteins. The gp5-gp27 complex forms the central hub of the baseplate (Figure 7B). The complex serves as a coaxial continuation of the tail tube. Gp48 and/or gp54 are positioned between the gp27 trimer and the tail tube, comprised of gp19. The gp5 β-helix forms the central needle that runs along the dome's axis. A small protein with a MW of ~23 kDa is associated with the tip of the gp5 β-helix (Figure 7B). The identity of this protein is unclear, but the mass estimate suggests that it could be gp28. The tape measure protein, gp29, is almost completely disordered in the baseplate-tail tube structure. It is unclear whether gp29 degrades during the sample

preparation or its structure does not agree with the six-fold symmetry assumed in generating the cryoEM map.

The earlier cross-linking and immuno-staining analysis of interactions between the baseplate wedge proteins turned out to be in good agreement with the later cryoEM results [48-50]. This is impressive considering the limitations of the techniques employed in the earlier studies. In agreement with the earlier findings, the new high resolution data show that gp10, gp11 and gp12 (the short tail fibers) constitute a major part of the baseplate's periphery. Gp9, the long tail fiber attachment protein, is also on the periphery, but in the upper part of the baseplate dome. Gp8 is positioned slightly inwards in the upper part of the baseplate dome and interacts with gp10, gp7 and gp6. The excellent agreement between the crystallographic and EM data resulted in the unambiguous locating of most of the proteins in the baseplate.

Six short tail fibers comprise the outermost rim of the baseplate. They form a head-to-tail garland, running clockwise if viewed from the tail towards the head (Figure 8). The N-terminus of gp12 binds coaxially to the N-terminal domain of the gp10 trimer, and the C terminus of one gp12 molecules interacts with N terminus of the neighboring molecule. The fiber is kinked at about its center, changing its direction by about 90°, as it bends around gp11. The C-terminal receptor-binding domain of gp12 is 'tucked under' the baseplate and is protected from the environment. The garland arrangement controls the unraveling of the short tail fibers, which must occur on attachment to the host cell surface.

Gp10 and gp7 consist of three separate domains each, connected by linkers (Figure 8B). Gp7 is a monomer, and it is likely that each of its domains (labeled A, B and C in Figure 8B) is a compact structure formed by a single polypeptide chain. Gp10, however, is a trimer, in which the three chains are likely to run in parallel and each of the cryoEM densities assigned to gp10 domains is threefold symmetric. The angles between the threefold axes of these domains are close to 60°. This is confirmed by the fact that the trimeric gp10_397C crystal structure fits accurately into one of the three domains assigned to gp10. At the boundary of each domain, the three gp10 chains come close together thus creating a narrowing. Interestingly, the arrangement of gp10 domains is maintained in both conformations of the baseplate suggesting that these narrow junctions are not flexible. A total of 23% of the residues in the N-terminal 200 residues of gp10 are identical and 44% of the residues have conservative substitutions when compared to the N-terminal and middle domains of T4 gp9. A homology model of the N-terminal part of gp10 agrees reasonably well with the cryoEM density assigned to the gp10 N-terminal

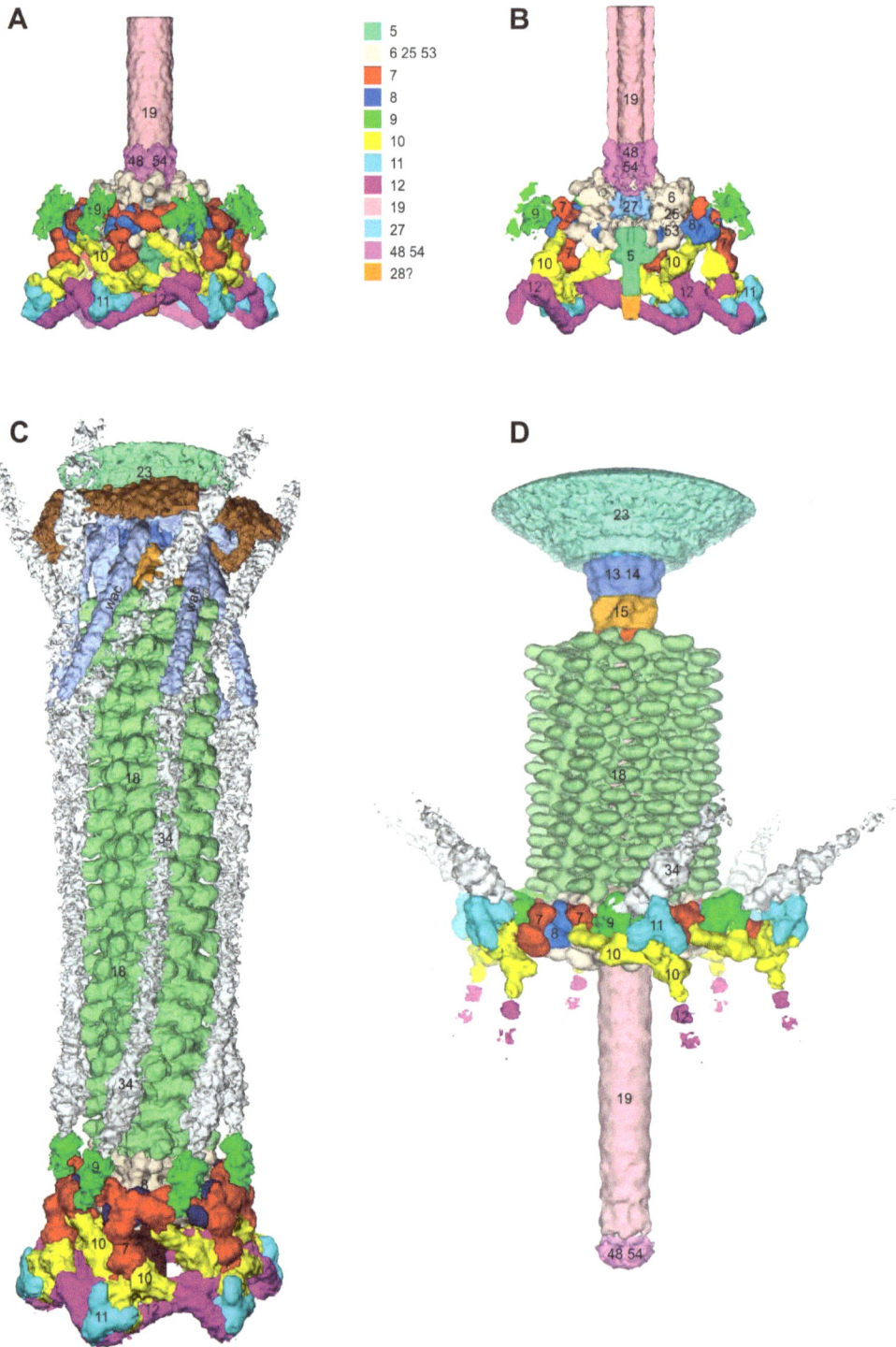

Figure 7 CryoEM reconstructions of the T4 tube-baseplate complex (A, B) and the tail in the extended (C) and contracted (D) conformation. Constituent proteins are shown in different colors and identified with the corresponding gene names. reprinted from [5,47] and [6].

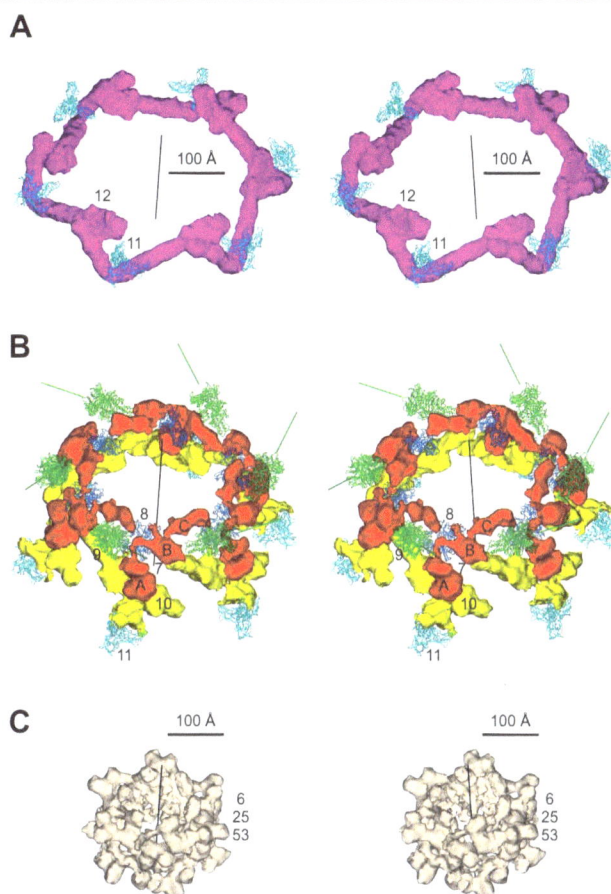

Figure 8 Details of the T4 baseplate structure; reprinted from [5]. Proteins are labeled with their respective gene numbers. **A**, The garland of short tail fibers gp12 (magenta) with gp11 structures (light blue C_α trace) at the kinks of the gp12 fibers. The six-fold axis of the baseplate is shown as a black line. **B**, The baseplate "pins", composed of gp7 (red), gp8 (dark blue C_α trace), gp10 (yellow), and gp11(light blue C_α trace). Shown also is gp9 (green C_α trace), the long tail fiber attachment protein, with a green line along its three-fold axis, representing the direction of the long tail fibers. **C**, Gp6, gp25, and gp53 density.

domain. The threefold axis of this domain in the cryoEM density coincides with that of the N-terminal part of gp12, which is attached to it. The middle domain of gp10 is clamped between the three finger domains of gp11.

Gp6, gp25 and gp53 form the upper part of the baseplate dome and surround the hub complex. The cryoEM map shows that the gp6 monomer is shaped like the letter S. Six gp6 dimers interdigitate and form a continuous ring constituting the backbone of the baseplate (Figures 8 and 9). Gp6 is the only protein in the baseplate, which forms a connected ring in both conformations of the baseplate. The N- and C-terminal domains of each gp6 monomer interact with two different neighboring gp6 molecules, i.e. the N terminal domain of chain 'k' interacts with the N terminal domain of chain 'k+1', whereas the C-terminal domain of chain 'k' interacts with the C terminal domain of chain 'k-1'. It is thus possible to distinguish two types of gp6 dimers,

depending on whether the N or C terminal domains of the two molecules are associated (Figure 9).

As there are only two molecules of gp6 per wedge, either the N-terminal or the C-terminal dimer has to assemble first (the intra-wedge dimer) and the other dimer is formed when the wedges associate into the ring structure (the inter-wedge dimer). Mutagenesis suggests that the Cys338 residue is critical for forming the N-terminal dimer, which therefore is likely to form the intra-wedge dimer [7]. The crystal structure represents the C-terminal inter-wedge dimer [7].

This finding is further supported by the baseplate assembly pathway. During assembly of the wedge, gp6 binds only after the attachment of gp8 [23,25]. Although a dimer of gp8 and a dimer of gp6 are present in each wedge [25], in the cryoEM baseplate map a single chain of the gp6 dimer interacts with a single chain of the gp8 dimer, whereas the other chain of the same gp6 dimer interacts with gp7. Together, gp8 and gp7 form a platform for binding of the N-terminal dimer of gp6, suggesting that the N-terminal dimer forms first during the assembly of the baseplate wedge, whereas C-terminal gp6 dimers form after six wedges associate around the hub.

The structures of the baseplate in the sheath-less tail tube assembly and in the complete tail are very similar, except for the position of gp9 (Figure 7) [5,47]. The N-terminal domain of gp9 binds to one of the gp7 domains, but the rest of the structure is exposed to the solution. The long tail fibers attach coaxially to the C-terminal domain of gp9. This arrangement allows gp9 to swivel, as a rigid body, around an axis running through the N-terminal domain, allowing the long tail fiber to move. In the extended tail structure, the long tail fibers are retracted and aligned along the tail (Figure 7c), whereas the tail tube-baseplates lack the long tail fibers. Thus, in the extended tail, the gp9 trimers point along the fibers, whereas in the tube-baseplate complexes, gp9 molecules are partially disordered due to their variable position and point sideways, on average. This variation in the positioning of gp9 is required to accommodate the full range of positions (and hence motion) observed for the long tail fibers [51].

Structure of the baseplate in the star conformation and its comparison with the hexagonal conformation

The star-shaped baseplate has a diameter of 610 Å and is 120 Å thick along its central sixfold axis. The central hub is missing because it is pushed through and replaced by the tail tube (Figure 10). Despite large changes in the overall baseplate structure, the crystal structures and the cryoEM densities of proteins from the hexagonal baseplate can be fitted into the star shaped baseplate. This indicates that the conformational changes occur as a result of rigid body movements of the constituent proteins and/or their domains.

The largest differences between the two conformations are found at the periphery of the baseplate. In the hexagonal conformation, the C-terminal domain of gp11 points away from the phage head, and its trimer axis makes a 144° angle with respect to the six-fold axis of the baseplate (Figure 10). In the star conformation, however, the gp11 C-terminal domain points towards the phage head, and the trimer axis makes a 48° angle with respect to the baseplate sixfold axis. Thus, upon completion of the baseplate's conformational change, each gp11 molecule will have rotated by almost 100° to associate with a long, instead of a short tail fiber. The long and short tail fibers compete for the same binding site on gp11. The interaction between gp10 and gp11 is unchanged in the two conformations. As a result, the entire gp10-gp11 unit rotates by ~100° causing the N-terminal domain of gp10 to change its orientation and

point towards the host cell surface (Figure 10). The short tail fiber, which is coaxially attached to the N-terminal domain of gp10, rotates and unfolds from under that baseplate and extends the C-terminal receptor-binding domain towards the potential host cell surface. In addition to the gp10-gp11 complex rotation and short tail fiber unraveling, domain A of gp7 swivels outwards by about 45° and alters its association with gp10, making the baseplate structure flat. This rearrangement brings the C-terminal domain of gp10 into the proximity of gp9 and allows the latter to interact with gp8. The structural information supports the hypothesis that the hexagonal-to-star conformational change of the baseplate is the result of a reorientation of the pins (gp7, gp10, gp11) [50] and additionally shows that the transformation also involves rearrangements of gp8, gp9, and gp12 situated around the periphery of the baseplate.

The association of gp10, gp11 and gp12 into a unit that can rotate by 100° is tight, but appears to be non-covalent. However, there could be at least one covalent bond that attaches this unit to the rest of the baseplate. Cys555, the only conserved cysteine in gp10 among all T4-like phages, is one of the residues that are involved in interactions between gp10 and domain B of gp7 in the baseplate. This cysteine might make a disulfide bond with one of eight cysteine residues in gp7, causing the gp10-gp11-gp12 complex and domain B of gp7 to act as a single rigid body during the conformational change of the baseplate. Unfortunately, residues 553-565 are disordered in the crystal structure of gp10_397C, and the exact structure of the region interacting with gp7 is uncertain. This is not surprising, as these residues might be prone to adopting various conformations, because the interaction with gp7 is not threefold symmetric.

The central part of the baseplate, which is comprised of gp6, gp25 and gp53, displays a small, but noticeable change between the two conformations of the baseplate. Both the N-terminal and C-terminal dimer contacts in the gp6 ring are maintained, but the angle between the gp6 domains changes by about 15°, accounting for the slight increase in the gp6 ring diameter (Figures 9 and 10). Therefore, the gp6 ring appears to have two functions. It is the inter-wedge 'glue', which ties the baseplate together and it is also required for maintaining the baseplate integrity during the change from hexagonal to star shaped conformations. At the same time, the gp6 ring is a framework to which the motions of other tail proteins are tied. The N-terminal domain of gp6 forms a platform onto which the first disk of the tail sheath subunits is added when the sheath it assembled. Therefore, the change in the gp6 domain orientations could be the signal that triggers the contraction of the sheath.

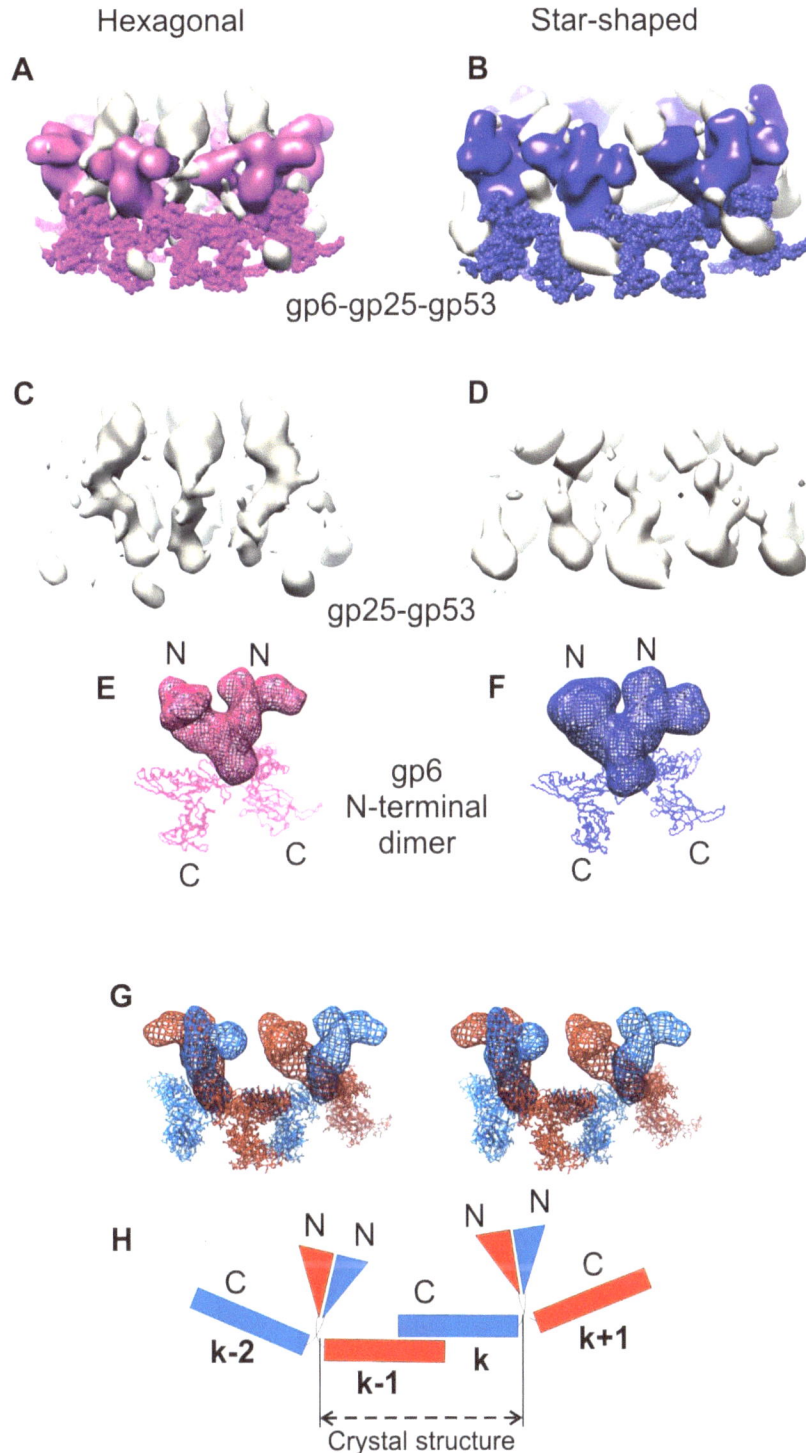

Figure 9 Arrangement of gp6, gp25 and gp53 in the baseplate; reprinted from [7]. **A, B**, Gp6 is shown in magenta for the "hexagonal" dome-shaped baseplate (left) and in blue for the star-shaped baseplate (right). The C-terminal part of gp6 corresponds to the crystal structure and is shown as a Cα trace with spheres representing each residue. The N-terminal part of gp6 was segmented from the cryo-EM map. The densities corresponding to gp53 and gp25 are shown in white. **C, D**, The densities of gp53 and gp25 after the density for the whole of gp6 was zeroed out. **E, F**, The N-terminal gp6 dimers as found in the baseplate wedge. The C-terminal domain is shown as a Cα trace, whereas the N-terminal domain, for which the structure remains unknown, is shown as a density mesh. **G**, A stereo view of the four neighboring gp6 molecules from the two neighboring wedges of the dome-shaped baseplate. The N-terminal part of gp6 is shown as a density mesh and the C-terminal part corresponds to the crystal structure. **H**, Schematic of the four gp6 monomers using the same colors as in **G**. The N-terminal part is shown as a triangle and the C-terminal part as a rectangle.

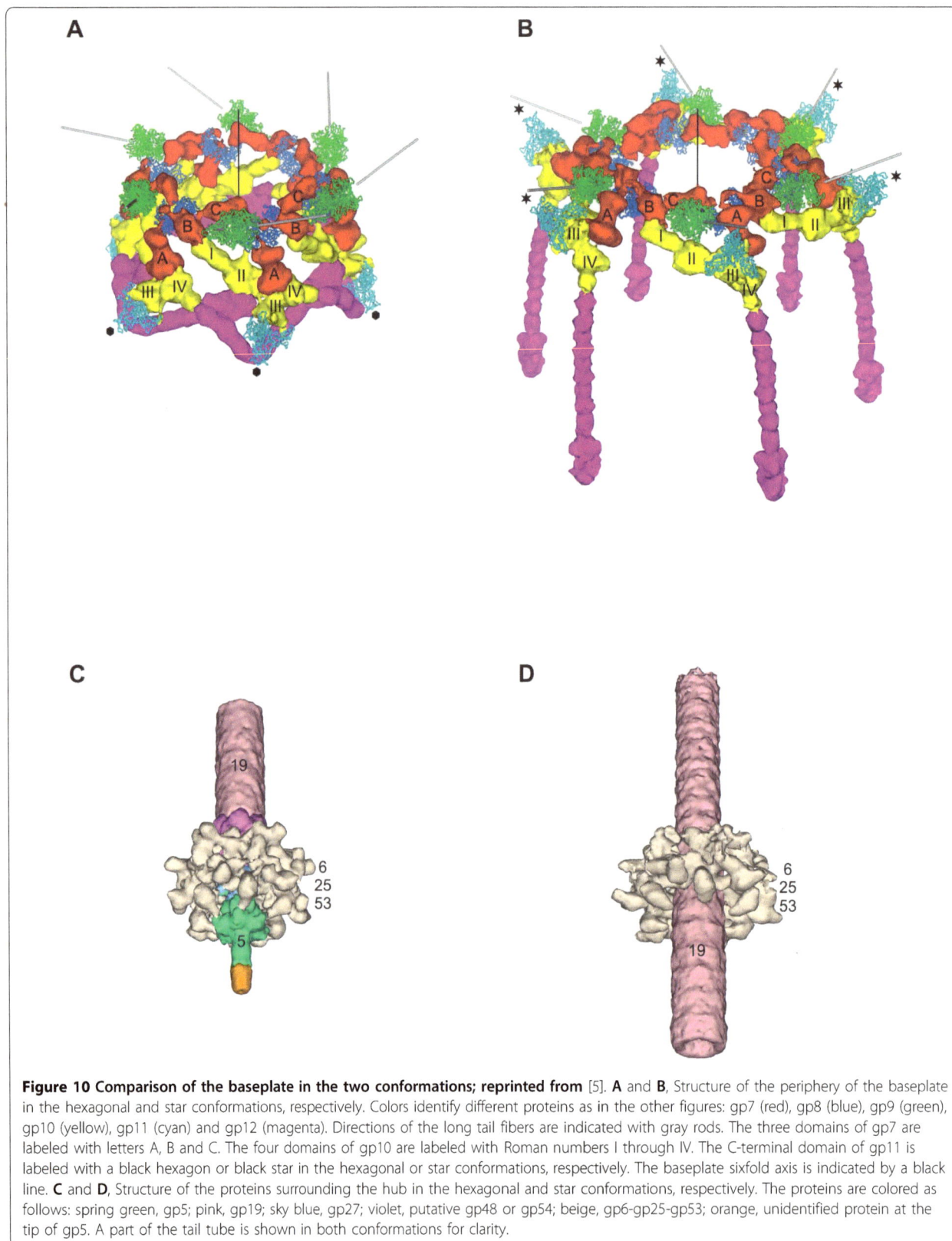

Figure 10 Comparison of the baseplate in the two conformations; reprinted from [5]. **A** and **B**, Structure of the periphery of the baseplate in the hexagonal and star conformations, respectively. Colors identify different proteins as in the other figures: gp7 (red), gp8 (blue), gp9 (green), gp10 (yellow), gp11 (cyan) and gp12 (magenta). Directions of the long tail fibers are indicated with gray rods. The three domains of gp7 are labeled with letters A, B and C. The four domains of gp10 are labeled with Roman numbers I through IV. The C-terminal domain of gp11 is labeled with a black hexagon or black star in the hexagonal or star conformations, respectively. The baseplate sixfold axis is indicated by a black line. **C** and **D**, Structure of the proteins surrounding the hub in the hexagonal and star conformations, respectively. The proteins are colored as follows: spring green, gp5; pink, gp19; sky blue, gp27; violet, putative gp48 or gp54; beige, gp6-gp25-gp53; orange, unidentified protein at the tip of gp5. A part of the tail tube is shown in both conformations for clarity.

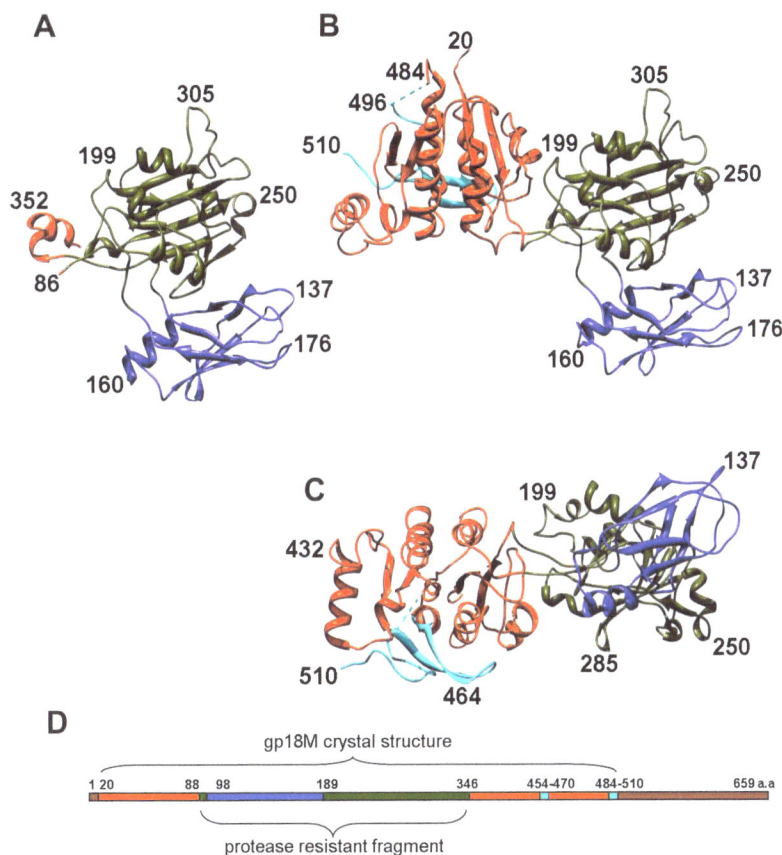

Figure 11 Structures of the gp18 deletion mutants reprinted from [53]. **A**, A ribbon diagram of the gp18PR mutant. The N terminus is shown in blue, the C terminus in red and the intermediate residues change color in spectral order. **B, C**, A ribbon diagram of the gp18M mutant (¾ of the total protein length). The three domains are shown in blue (domain I), olive green (domain II) and orange red (domain III); the β-hairpin (residues 454-470) and the last 14 C-terminal residues of gp18M are shown in cyan. **D**, Domain positions on the amino acid sequence, using the same color scheme as in (**B**) and (**C**). Brown indicates the part of gp18 for which the structure remains unknown.

Structure of the tail sheath in the extended and contracted conformation

Crystal structure of gp18

Recombinant, full-length gp18 (659 residues) assembles into tubular polymers of variable lengths called polysheaths, which makes crystallization and high resolution cryoEM studies difficult. However several deletion mutants that lack polymerization properties have been crystallized [52]. The crystal structures of two of these mutants have been determined. One of these is of a protease resistant fragment (gp18PR) consisting of residues 83-365. The other, called gp18M, is of residues 1-510 in which the C-terminal residue has been replaced by a proline (Figure 11). The crystal structure of the gp18PR fragment has been refined to 1.8 Å resolution and the structure of the larger gp18M fragment was determined to 3.5 Å resolution [53].

The structure of gp18M includes that of gp18PR and consists of Domains I, II and III (Figure 11). Domain I (residues 98-188) is a six-stranded β-barrel plus an α-helix. Domain II (residues 88-97 and 189-345) is a two layer β-sandwich, flanked by four small α-helices. Together, domains I and II form the protease resistant fragment gp18PR. Domain III (residues 24-87 and 346-510) consists of a β-sheet with five parallel and one anti-parallel β-strands plus six α-helices surrounding the β-sheet. The 24 N-terminal residues as well as residues 481 to 496 were not ordered in the gp18M crystal structure. The N and C termini of the structure are close in space, suggesting that the first 24 residues and residues 510-659 form an additional domain, Domain IV, which completes the structure of the full-length protein. The overall topology of the gp18 polypeptide chain is quite remarkable. Domain I of gp18 is an insertion into Domain II, which, in turn, is inserted into Domain III, which is inserted between the N and C termini comprising domain IV.

Fitting of the gp18M structure into the cryoEM map of the tail showed that the protease resistant part of gp18 is exposed to the solution, whereas the N and C

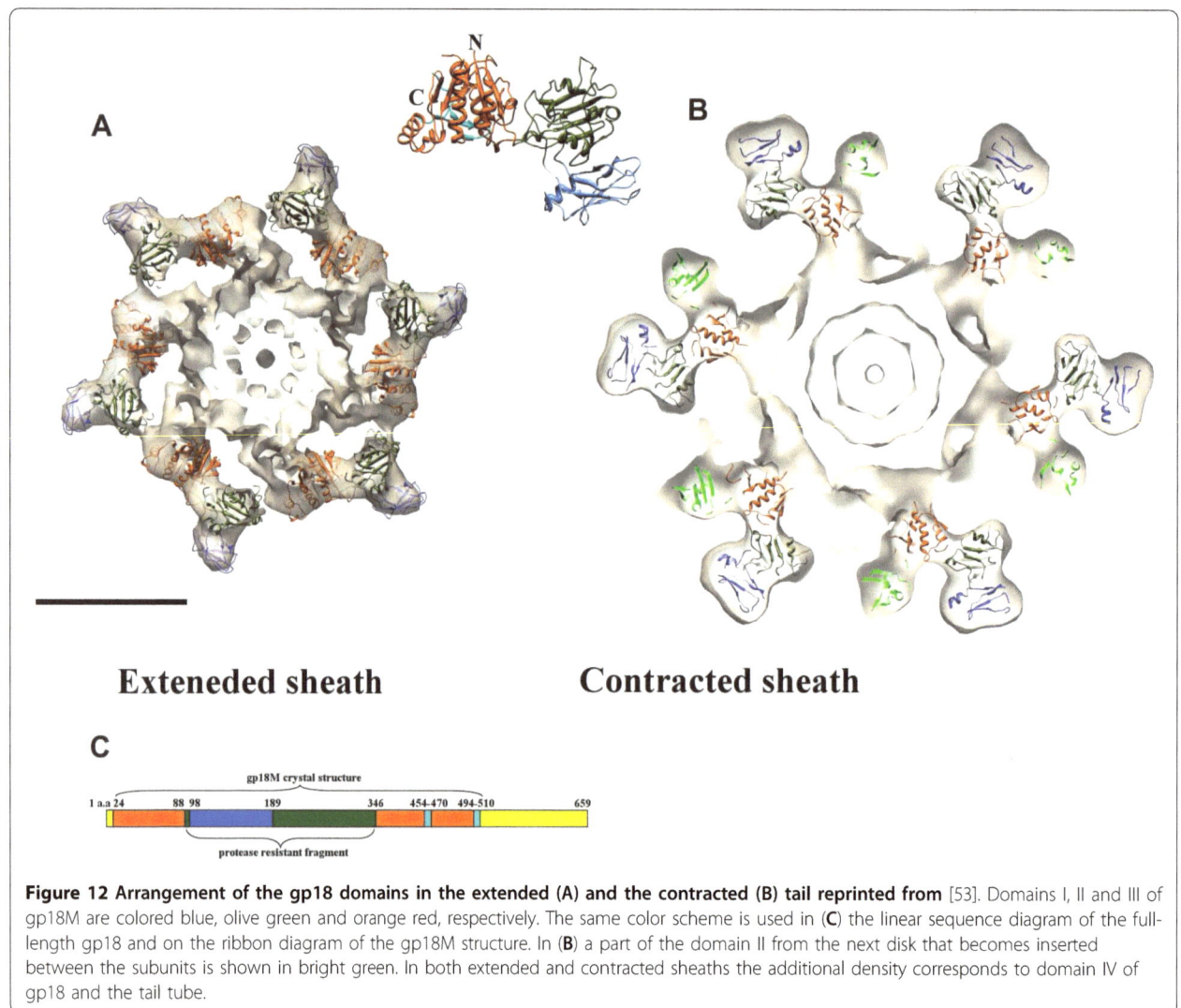

Figure 12 Arrangement of the gp18 domains in the extended (A) and the contracted (B) tail reprinted from [53]. Domains I, II and III of gp18M are colored blue, olive green and orange red, respectively. The same color scheme is used in (**C**) the linear sequence diagram of the full-length gp18 and on the ribbon diagram of the gp18M structure. In (**B**) a part of the domain II from the next disk that becomes inserted between the subunits is shown in bright green. In both extended and contracted sheaths the additional density corresponds to domain IV of gp18 and the tail tube.

termini, which form Domain IV, are positioned on the interior of the tail sheath (Figure 12). The exposed and buried residues in each conformation of the sheath are in agreement with previous immuno-labeling and chemical modification studies [54,55]. Domain I of gp18 is protruding outwards from the tail and is not involved in inter-subunit contacts. The other three domains form the core of the tail sheath with Domains III and IV being the most conserved parts of tail sheath proteins among T4-related bacteriophages (Figure 12). Despite the fact that Domain I has apparently no role in gp18-gp18 interactions, this domain binds to the baseplate in the extended tail sheath. Thus, one of the roles of Domain I may be to initiate sheath assembly and contraction. Domain I also binds the long tail fibers when they are retracted. It was previously shown that three mutations in Domain I (G106→S, S175→F, A178→V) inhibit fiber retraction [56]. These mutations map to

two loops close to the retracted tail fiber attachment site on the surface of the extended tail sheath, presumably abrogating binding of the tail fibers.

Structure of the extended sheath and the tube

The 240 Å-diameter and 925 Å-long sheath is assembled onto the baseplate and terminates with an elaborate 'neck' structure at the other end (Figures 13 and 14). The 138 copies of the sheath protein, gp18, form 23 rings of six subunits each stacked onto one another. Each ring is 40.6 Å thick and is rotated by 17.2° in a right-handed manner relative to the previous ring. The sheath surrounds the tail tube, which has external and internal diameters of 90 Å and 40 Å, respectively. The area of contact between the adjacent gp18 subunits with the neighboring gp18 subunit in the ring above is significantly greater than that between neighboring subunits within a ring (about 2,000 Å^2

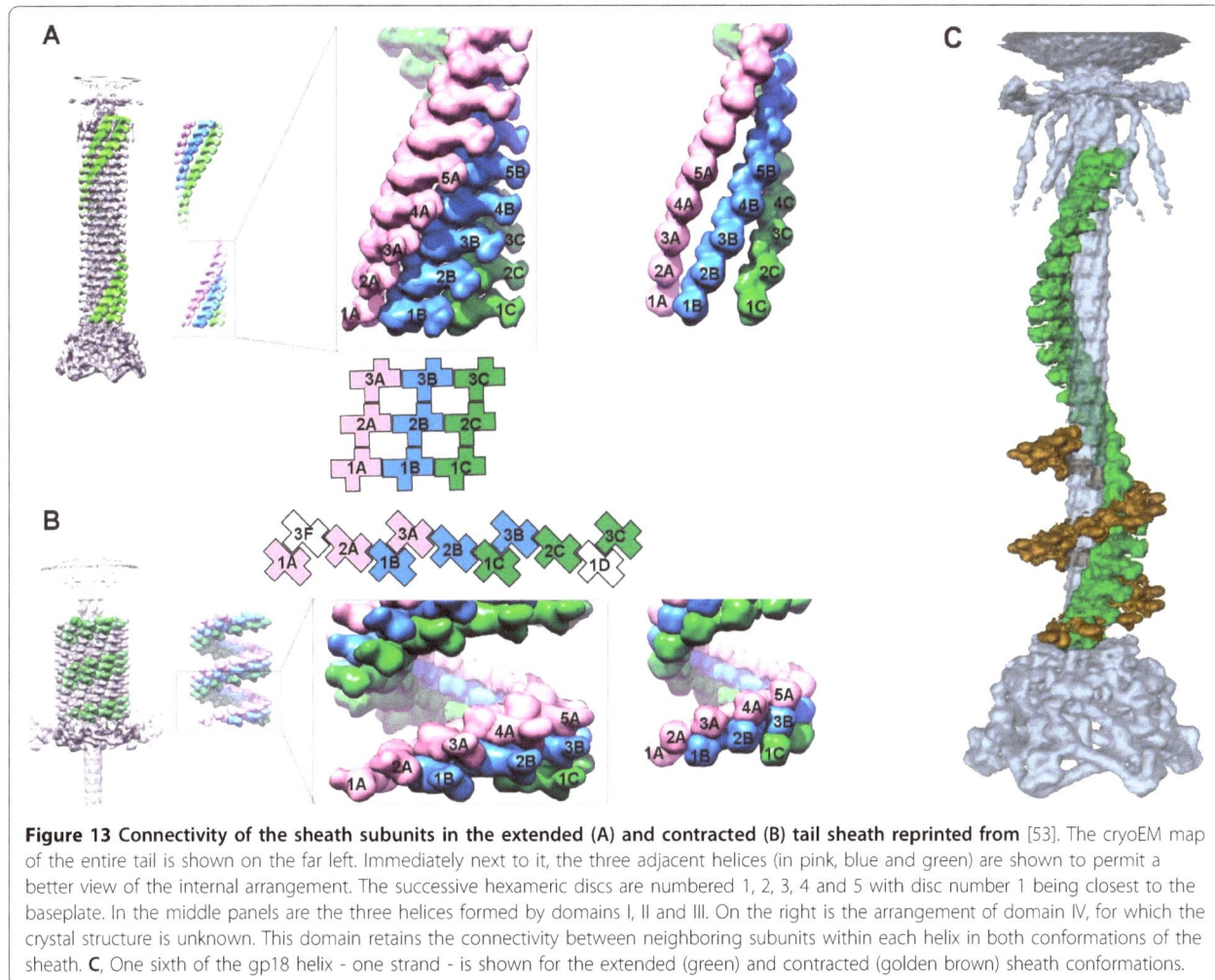

Figure 13 Connectivity of the sheath subunits in the extended (A) and contracted (B) tail sheath reprinted from [53]. The cryoEM map of the entire tail is shown on the far left. Immediately next to it, the three adjacent helices (in pink, blue and green) are shown to permit a better view of the internal arrangement. The successive hexameric discs are numbered 1, 2, 3, 4 and 5 with disc number 1 being closest to the baseplate. In the middle panels are the three helices formed by domains I, II and III. On the right is the arrangement of domain IV, for which the crystal structure is unknown. This domain retains the connectivity between neighboring subunits within each helix in both conformations of the sheath. **C**, One sixth of the gp18 helix - one strand - is shown for the extended (green) and contracted (golden brown) sheath conformations.

versus 400 Å^2). Thus, the sheath is a six-fold-symmetric, six-start helix (Figure 13).

The tail tube (also called the "core" in the literature) is a smooth cylinder, lacking easily discernable surface features. Nevertheless, it can be segmented into individual subunits of the tail tube protein gp19 at an elevated contour level. The subunits are arranged into a helix having the same helical parameters as those found for the gp18 helix.

Structure of the contracted sheath

The contracted sheath has a diameter of 330 Å and is 420 Å long (Figures 7 and 13). The gp18 subunits form a six-start right-handed helix with a pitch of 16.4 Å and a twist angle of 32.9° situated between radii of 60 Å and 165 Å. The sheath has an inner diameter of 120 Å and does not interact with the 90 Å-diameter tail tube, in agreement with previous observations [57]. Upon superimposing the midsection of the sheath onto itself using the helical transformation, the correlation coefficient

was found to be 0.98, showing that there is little variation in the structure of the gp18 subunits and that the sheath contracts uniformly.

The structure of gp18 subunit in the contracted tail is very similar to that in the extended tail. The internal part of the gp18 subunits retains its initial six-start helical connectivity, which is formed when the sheath is first assembled onto the tail tube. This helix has a smaller diameter in the extended conformation and interacts with the tail tube, thus stabilizing the sheath. This was further confirmed by fitting of the gp18M crystal structure into the cryoEM density maps of the tail sheath. The structure fits as a rigid body into both the extended and contracted conformations of the sheath, suggesting that contraction occurs by sliding of individual gp18 subunits over each other with minimal changes to the overall fold of the sheath protein (Figure 12). During contraction each subunit of gp18 moves outwards from the tail axis while slightly changing its orientation. The interactions between the C-terminal domains of gp18 subunits in the extended confirmation appear to be preserved in the contracted form, maintaining the integrity of the sheath structure. However, the outer domains of gp18 change interaction partners and form new contacts. As a result, the interaction area between the subunits increases about four times.

The helical symmetry of the sheath shows that the first and last layers in the extended and contracted conformations are related by a 378.4° (1.05 turns) rotation and 723.8° (2.01 turns) rotation, respectively. Assuming that the association of the sheath and tail tube subunits in the neck region is fixed, the tube will thus rotate by 345.4° - almost a full turn - upon tail contraction (Figure 13C).

Although the diameter of the tube is the same, the symmetry and gp19 subunit organization bear no resemblance to that of the extended or contracted sheath. The tail tube subunits in phage with a contracted tail appear to have an organization that is slightly different to that found in the virus with an extended sheath. However this might be an artifact of the image reconstruction procedure used to view the details of the tail tube, because the tail tube is internal to the sheath, which has a repetitive structure that might have influenced the reconstruction procedure.

The neck region lacks the fibritin and other proteins in the contracted tail map. This sample was prepared by diluting a concentrated phage specimen into 3 M urea. There is little doubt now, that this harsh treatment caused the observed artifacts. Recent experiments showed that the fibritin and other proteins remain associated with the phage particle if the latter is subjected to slow dialysis into 3 M urea. In this procedure, the tails uniformly contract and their structure is identical to that found in the earlier studies (A. Aksyuk, unpublished observations).

Structure of the neck region

The neck consists of a several sets of stacked hexameric rings consisting of gp3, gp15, and gp13 or gp14 (Figure 14). The gp3 terminates the tail tube, followed by gp15, and then by gp13 and/or gp14 closest to the head.

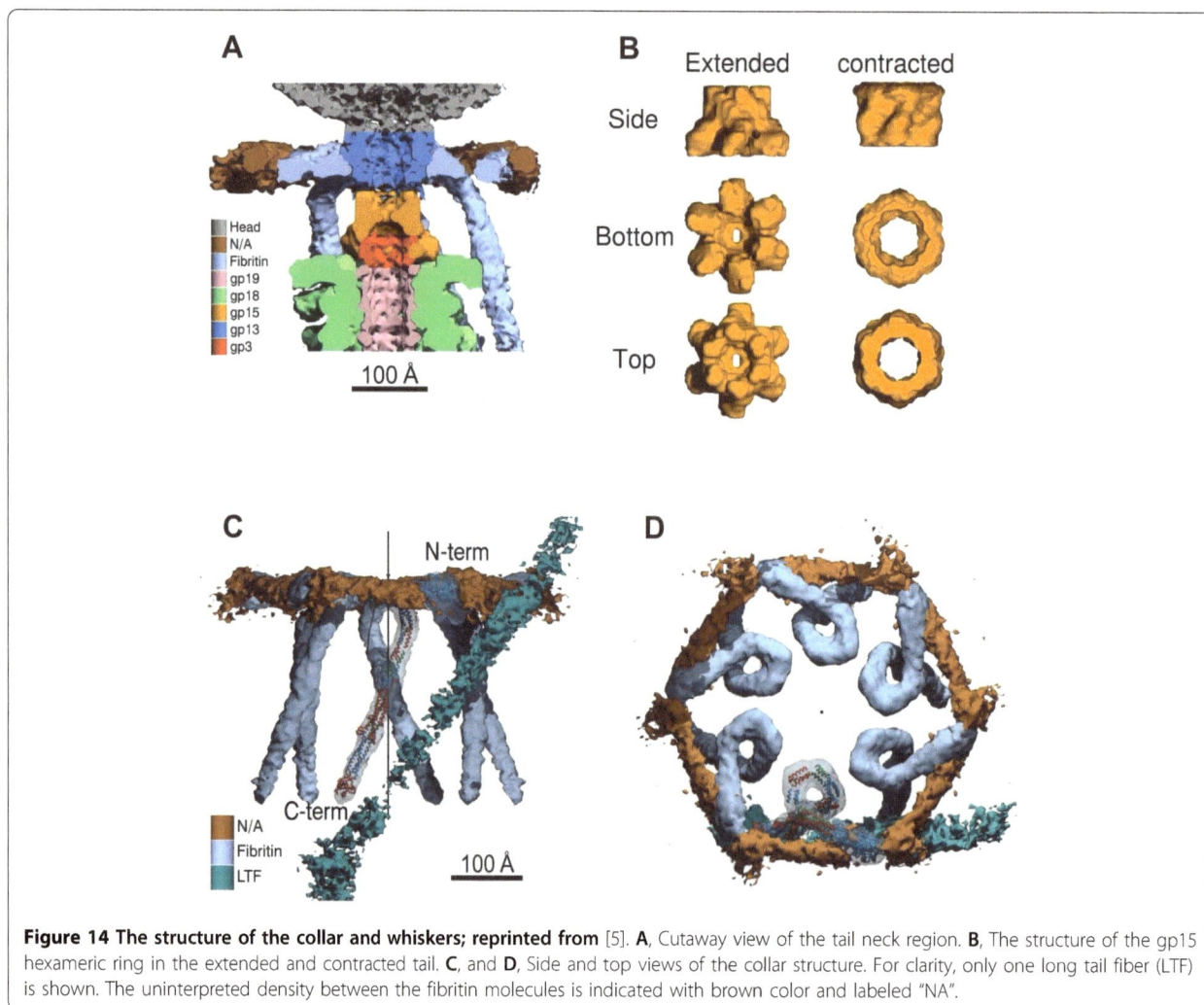

Figure 14 The structure of the collar and whiskers; reprinted from [5]. **A**, Cutaway view of the tail neck region. **B**, The structure of the gp15 hexameric ring in the extended and contracted tail. **C**, and **D**, Side and top views of the collar structure. For clarity, only one long tail fiber (LTF) is shown. The uninterpreted density between the fibritin molecules is indicated with brown color and labeled "NA".

In the cryoEM reconstruction of the wild type phage, the channel running through the length of the gp19 tube is filled with a roughly continuous density at an average diameter of ~20 Å. This might be the extended molecule(s) of gp29 tape measure protein or phage DNA. The former proposition is more likely, as the tail channel is blocked by the gp15 hexamer, which forms a closed iris with an opening of only 5-10 Å and should prevent the DNA from entering the tail.

The neck is surrounded by a 300 Å diameter and 40 Å thick collar, consisting at least in part of fibritin (gp *wac*) [58]. Fibritin is a 530 Å-long and 20 Å-diameter trimeric fiber [59]. The atomic structure of the N- and C-terminal fragments of fibritin is known [60,61]. The rest of this fiber has a segmented coiled coil structure and can be modeled using the known structure and the repetitive nature of its amino acid sequence [59-61]. The cryoEM map of wild type T4 could be interpreted with the help of this model.

Each of the six fibritin trimers forms a tight 360° loop, which together create the main part of the collar and the whiskers (Figure 14). Both the N and C termini of the fibritin protein attach to the long tail fiber. The C-terminal end binds to the 'kneecap' region of the long tail fiber, comprised of gp35, whereas the N terminus most probably binds to the junction region of gp36 and gp37. The fibritin's 360° loop interacts with gp15 and is in the N-terminal part of the protein. This is in agreement with earlier studies that found that the N terminus of fibritin is required for its attachment to the phage particle. The six fibritins and the long tail fibers are bridged together by six copies of an unknown fibrous protein to form a closed ring. This protein is about 160 Å long and 35 Å in diameter.

Tail Fiber Structure and Assembly
Overall organization and subunit composition
The long tail fibers of bacteriophage T4 are kinked structures of about 1440 Å long with a variable width of up to about 50 Å. They can be divided into proximal and distal half-fibers, attached at an angle of about 20° [62]. In adverse conditions for phage multiplication, the long tail fibers are in a retracted conformation, lying against the tail sheath and head of the bacteriophage. In the extended conformation, only the proximal end of the fiber is attached to the baseplate. The long tail fibers are responsible for initial interaction with receptor molecules [2]. The distal tip of the long tail fibers can recognize the outer membrane protein C (ompC) or the glucosyl-α-1,3-glucose terminus of rough LPS on *E. coli* [63]. Titration experiments showed that the phage particle has to carry at least three long tail fibers to be infectious [64].

The long tail fiber is composed of four different gene products: gp34, gp35, gp36 and gp37 (Figure 15) [65].

The proximal half-fiber, or the "thigh", is formed by a parallel homo-trimer of gp34 (1289 amino acids or 140 kDa per monomer). In the intact phage, the N-terminal end of gp34 is attached to the baseplate protein gp9 [8], while the C-terminal end interacts with the distal half-fiber, presumably with gp35 and/or gp36. Gp35 (372 residues; 40 kDa and present as a monomer) forms the "knee" and may be responsible for the angle between the proximal and distal half-fibers. The distal half-fiber is composed of gp35, trimeric gp36 (221 amino acids, 23 kDa) and gp37 (1026 amino acids; 109 kDa). The gp36 protein subunit is located at the proximal end of the distal half-fiber, forming the upper part of the "shin", while gp37 makes up the rest of the shin, including the very distal receptor-recognizing tip (or "foot"), which corresponds to the C-terminal region of gp37.

The four structural genes of the long tail fiber and the chaperone gp38 are located together in the T4 genome. Genes 34 and 35 are co-transcribed from a middle-mode promoter, gene 36 from a late promoter, while genes 37 and 38 are co-transcribed from another promoter [66]. The gp34 protein is the largest T4 protein, followed by the baseplate protein gp7 the second-largest protein and gp37 the third-largest protein in the baseplate.

Despite their extended dimensions, the long tail fibers appear to be stiff structures, because no kinked half-fibers have been observed in electron micrographs. Moreover, the angle between the half fibers in the complete fiber does not deviate very far from 20° on average. The stiffness may be necessary for transmitting the receptor recognition signal from the tip of the fiber to the baseplate and for bringing the phage particle closer to the cell surface as the baseplate changes its conformation. No atomic resolution structures for the long tail fibers, their components or their chaperones (see next section) have yet been published.

In the cryoEM reconstruction of the wild-type T4, the fibers are in the retracted configuration (Figure 7), likely caused by the unfavorable for infection conditions of the cryoEM imaging procedure (a very high phage concentration and a very low salt buffer). The density corresponding to the long tail fibers is quite poor (Figure 7). This is likely caused by the variability of the positions of the long tail fibers. The 700 Å-long proximal half-fiber and the about 2/3 of the 740 Å-long distal part are present in the cryoEM map. The proximal half-fiber is bent around the sheath, forming about a quarter of a right-handed helix.

Assembly: folding chaperones and attachment proteins
A phage-encoded molecular chaperone, gp57A, is required for the correct trimerization of long tail fiber proteins gp34 and gp37 [62]; and for the short tail fiber

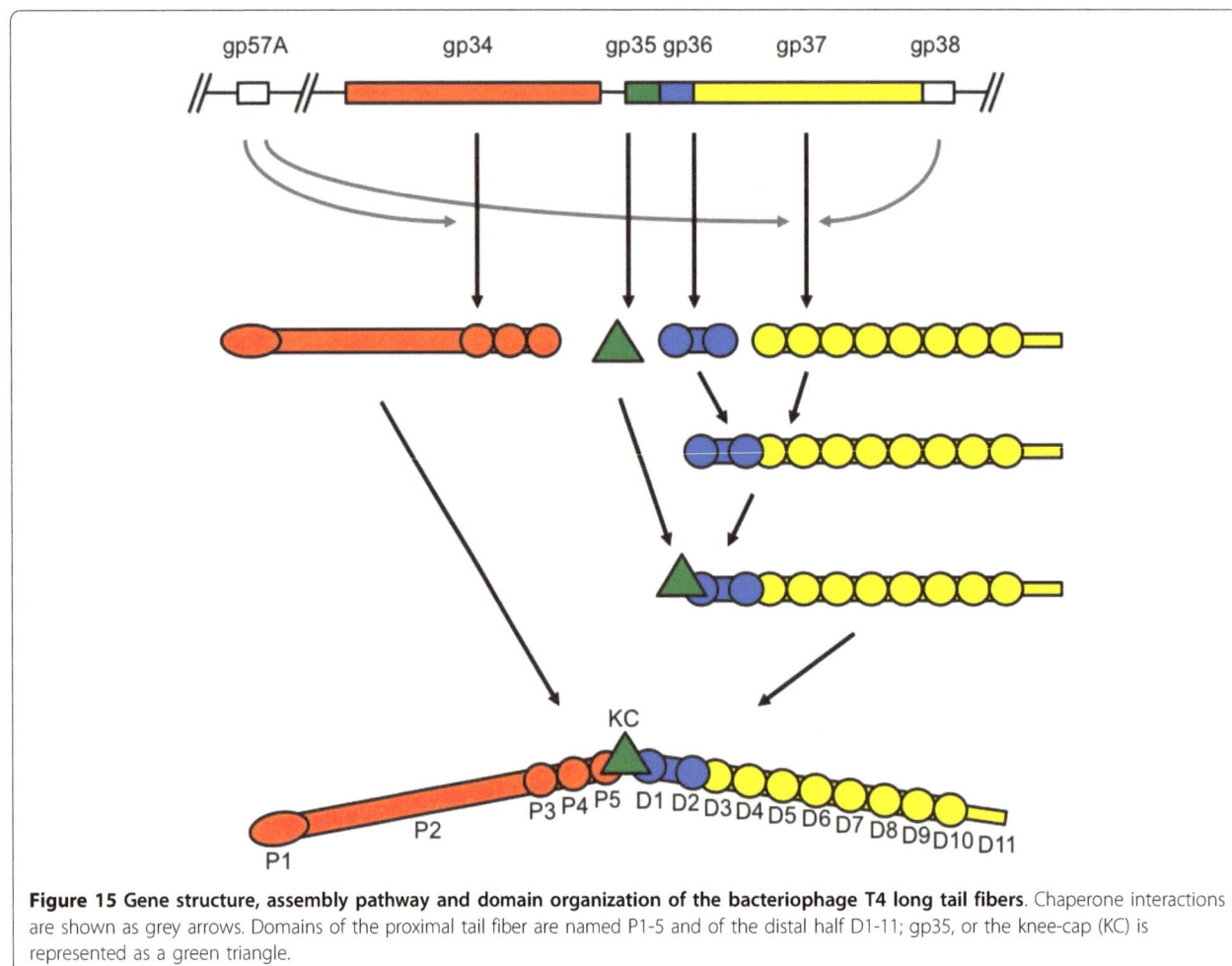

Figure 15 Gene structure, assembly pathway and domain organization of the bacteriophage T4 long tail fibers. Chaperone interactions are shown as grey arrows. Domains of the proximal tail fiber are named P1-5 and of the distal half D1-11; gp35, or the knee-cap (KC) is represented as a green triangle.

protein gp12 [67] (Table 2). Gp57A appears to be a rather general T4 tail fiber chaperone and is needed for the correct assembly of the trimeric short and long tail fiber proteins gp12, gp34, and gp37 [68]. Gp57A is a small protein of 79 residues (8,613 Da) that lacks aromatic amino acids, cysteines and prolines. *In vitro*, it adopts different oligomeric states [44]. The specific chaperone gp38 must be present [68] for the correct trimeric assembly of gp37. The molecular basis of the gp38 and gp57A chaperone activities are unclear, but it has been proposed that gp57A functions to keep fiber protein monomers from aggregating unspecifically, while gp38 may bring together the C-terminal ends of the monomers to start the folding process [62]. Qu *et al.* [69] noted that extension of a putative coiled-coil motif near the C-terminal end of gp37 bypasses the need for the gp38 chaperone. The extended coiled-coil may function as an intramolecular clamp, obviating the need for the intermolecular gp38 chaperone.

Two parts of the long tail fiber (the distal and proximal half-fibers) assemble independently. The three

proteins of the distal half-fiber interact in the following order. Initially trimeric gp36 binds to the N-terminal region of gp37, and then monomeric gp35 binds to gp36, completing the assembly of the distal half-fiber. Joining of the two half-fibers presumably takes place spontaneously.

Attachment of the assembled long tail fiber to the phage particle is promoted by gp63 and the fibritin (gp *wac*) [62], although neither of these proteins is absolutely essential (Table 2). Unlike gp63, the fibritin is a component of the complete phage particle and constitutes a major part of the neck complex (see above). In the absence of the fibritin, the long tail fibers attach to fiberless particles very slowly. The whiskers are also be involved in the retraction of the long tail fibers under unfavorable conditions. Gp63 has RNA ligase activity and may function as such in infected cells. However, the isolation of gene 63 mutants that affect RNA ligase activity, but not tail fiber attachment activity suggests that gp63 is a bifunctional protein that promotes two physiologically unrelated reaction [70].

Structural studies of the long tail fiber

Scanning transmission electron microscopy of stained and unstained particles has been used to study the structure of intact long tail fibers, proximal half-fibers and distal half-fibers [65]. The proximal half-fiber, gp34, consists of an N-terminal globular domain that interacts with the baseplate. It is followed by a rod-like shaft about 400 Å long that is connected to the globular domain by a hinge. The rod domain seen by EM correlates with a cluster of seven quasi-repeats (residues 438 to 797 [65]), which are also present six times in gp12 and once in gp37. One of these repeats is resolved in the crystal structure of gp12 (amino acids 246 to 290 [12]). This structural motif consists of an α-helix and a β-sheet. The proximal half-fiber terminates in three globular domains arranged like beads on a stick.

EM has shown that the proximal and distal half-fibers are connected at an angle of about 160°. A hinge is present between the proximal and distal half-fibers, forming the "knee". Density, associated with the presence of gp35, a monomer in the long tail fiber, bulges asymmetrically out on the side of the fiber forming the reflex angle (i.e. at the opposite side of the obtuse angle) [65].

The distal half-fiber, composed of gp36 and gp37, consists of ten globular domains of variable size and spacing, preceding a thin end domain or "needle" with dimensions of about 150 by 25 Å [65]. Based on its relative molecular mass (compared to that of the other long tail fiber components), gp36 should make up about one sixth of the distal half-fiber and thus likely composes at least the two relatively small proximal globules, the thin rod in between them, and perhaps the third globule. The remaining seven or eight globules and the needle or "foot" would then be gp37. A single repeat, similar to those also present in gp12 and gp34 is found in the N-terminal region of gp37, (amino acids 88-104). Residues 486 to 513 of gp37 show strong similarity to residues 971 to 998 of gp34 and are likely to form a homologous structural motif. Another sequence similarity has been observed between residues 814-860 and residues 342-397 of gp12 [65]. In gp12, these residues form the collar domain [12,14]. Gp34, gp36, and gp37 are predicted to mainly contain β-structure and little α-helical structure. However, their limited sequence similarity with each other, with the T4 short tail fiber protein gp12 and with other fiber proteins makes structure prediction difficult. *Streptococcus pyogenes* prophage tail fiber was shown to contain an extended triple β-helix in between α-helical triple coiled-coil regions [71], while the bacteriophage P22 tail needle gp26 has a very small triple β-helical domain and extensive stable α-helical triple coiled-coil regions [72]. A general principle may be that folding of the above mentioned fiber proteins starts near the C-terminus, as is the case for the adenovirus vertex fibers [73].

In general, trimeric fibrous proteins require a chaperone 'module' for folding. This module can be a small domain of the same polypeptide chain or a separate protein (or several proteins) [74]. Simultaneous co-expression of gp37, gp57A and gp38 has been used to obtain mg-amounts of soluble gp37 [75]. Correct folding of the trimeric protein was assessed by gel electrophoresis, cross-linking and transmission electron microscopy studies. The C-terminal fragments of gp37 appear to be folded correctly, showing that folding behavior of gp37 resembles that of gp12 [38].

The Infection Mechanism

Structural transformation of the tail during infection

The following observations suggest that the hexagonal conformation of the baseplate and the extended state of the sheath both represent high energy metastable assemblies. Purified baseplates have been shown to switch spontaneously into the star conformation [50]. In the absence of either the baseplate or the tail tube, the sheath assembles into a long tubular structure similar to that of the contracted sheath [57]. The tail sheath contraction is irreversible, and the contracted tail structure is resistant to 8 M urea [76]. These observations suggest that the baseplate in the hexagonal conformation together with its extended sheath can be compared to an extended spring ready to be triggered [77].

By combining all the available experimental information on T4 infection, it is possible to describe the process of attachment of the phage to the host cell in some detail (Figure 16, **Movie 2** http://www.seyet.com/t4_virology.html). The long tail fibers of the infectious phage in solution are extended, and most possibly move up and down due to the thermal motion [51,78,79]. Attachment of one of the fibers to the cell surface increases the probability for the other fibers to find cell surface receptors. The attachment of three or more of the long tail fibers to their host cell receptors is possible only when they point towards the host cell surface. This configuration of the tail fibers orients the phage particle perpendicular to the cell surface.

As the gp9 trimer is coaxial with the proximal part of the long tail fiber, gp9 proteins swivel up and down following the movements of the long tail fibers as the phage particle travels in search of a potential host cell. When the long tail fibers attach to the host cell surface and their proximal parts point down, several new protein-protein interactions at the periphery of the baseplate are initiated: 1) gp9 binds to the C-terminal domain of gp10; 2) the long tail fiber binds to a gp11 trimer. These interactions are likely causing gp11 to dissociate from gp12 leading to destabilization of the gp12 garland. The baseplate then unlocks from its high energy metastable hexagonal state. The A domain of

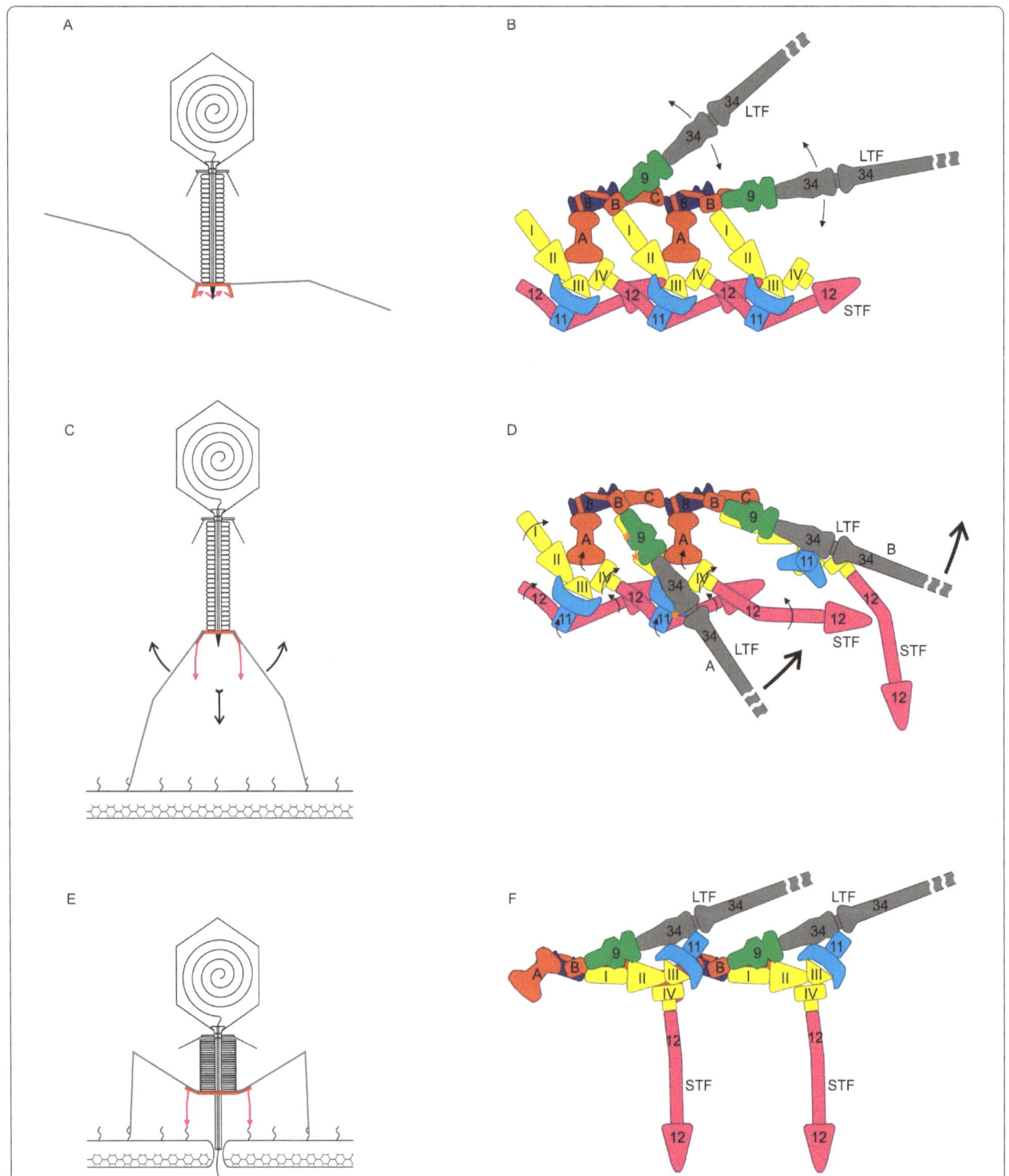

Figure 16 Baseplate conformational switch schematic reprinted from [6]. **A** and **B**, The phage is free in solution. The long tail fibers are extended and oscillate around their midpoint position. The movements of the fibers are indicated with black arrows. The proteins are labeled with their corresponding gene numbers and colored as in other figures. **C** and **D**, The long tail fibers attach to their surface receptors and adapt the "down" conformation. The fiber labeled "A" and its corresponding attachment protein gp9 interact with gp11 and with gp10, respectively. These interactions, labeled with orange stars, probably initiate the conformational switch of the baseplate. The black arrows indicate tentative domain movements and rotations, which have been derived from the comparison of the two terminal conformations. The fiber labeled "B" has advanced along the conformational switch pathway so that gp11 is now seen along its threefold axis and the short tail fiber is partially extended in preparation for binding to its receptor. The thick red arrows indicate the projected movements of the fibers and the baseplate. **E** and **F**, The conformational switch is complete; the short tail fibers have bound their receptors and the sheath has contracted. The phage has initiated DNA transfer into the cell.

gp7 swivels outwards and the entire gp10-gp11-gp12 module rotates, causing the C-terminal domains of the short tail fibers to point towards the host cell surface, thus preparing them for binding to the host cell receptors. Gp9 and the long tail fibers remain bound to the baseplate pins (the gp7-gp10-gp11 module), during this transformation.

During the conformational change of the baseplate, the long tail fibers are being used as levers to move the baseplate towards the cell surface by as much as 1000 Å. As the lengths of the two halves of the fiber are close to 700 Å each, such a large translation is accomplished by changing the angle between them by about 100°.

The conformational changes, which are initiated at the periphery of the baseplate, would then spread inwards into the center of the baseplate causing the central part of the baseplate (gp6, gp25 and gp53) to alter its conformation and thus initiating sheath contraction. The process of sheath contraction is accomplished by rotating and sliding the gp18 sheath subunits and progresses through the entire sheath starting at the baseplate (**Movie 3** http://www.seyet.com/t4_virology.html). The contracting sheath then drives the tail tube into the host membrane. The baseplate hub, which is positioned at the tip of the tube, will be the first to come in contact with the membrane. The membrane is then punctured with the help of the gp5 C-terminal β-helix and the yet unidentified protein (gp28?), which caps the tip of the gp5 β-helix. Subsequent tail contraction drives the tail tube further, and the entire gp5-gp27 complex is then translocated into the periplasmic space. The three lysozyme domains of the gp5 trimer start their digestion of peptidoglycan after the gp5 β-helix has dissociated due to the steric clashes with the peptidoglycan. This process results is a hole in the outer part of the cell envelope, allowing the tail tube to interact with the cytoplasmic membrane initiating phage DNA transfer. As mentioned above, the tail contraction involves rotation of the tail tube by an almost complete turn. Thus, the tail tube drills, rather than punctures, the outer membrane.

The fate and function of gp27 in the infection is unknown. Gp27 does not appear to form a trimer in the absence of gp5 [13], but it is possible that gp27 might be able to maintain its trimeric form upon its association with the tail tube because the gp27 trimer is a smooth coaxial continuation of the tail tube with a 25 Å diameter channel. Furthermore, the lysozyme-containing N-terminal part of gp5 (gp5*) might be able to dissociate from gp27 in the periplasm (due to the lower pH [13]) to open the gp27 channel. Gp27 may thus form the last terminal pore of the tube through which the phage DNA and proteins enter the host cell. Possibly, gp27 might interact with a receptor in or at the cytoplasmic membrane.

The above speculation that the gp27 trimer may serve as the terminal opening of the tail tube is supported by the crystal structure of a gp27 homolog called gp44 from bacteriophage Mu (a contractile tail phage) [80]. Although T4 gp27 and Mu gp44 have no detectible sequence similarity, the two structures have very similar folds [80]. Gp44, however, forms a stable trimer in solution and most probably serves as a centerpiece of the Mu baseplate. Gp45 is a glycine-rich protein from the Mu tail, making it a possible ortholog of gp5.

Conclusion
Contractile tail evolution and relation to other biological systems
There is building a body of evidence proving that all tailed phages have a common ancestor. The evolutionary relationship cannot be detected in their amino acid sequences, but structural studies show that capsid proteins of all tailed phages have a common fold (the HK97 fold) and that the portal proteins are homologous [81-83]. As the DNA packaging processes in all tailed phages are similar, their ATPases and many other structural proteins are also most probably homologous.

The recently discovered and incompletely characterized bacterial type VI secretion system (T6SS) appears to be related to a phage tail [84]. The T6SS is one of the most common secretion systems present in at least 25% of all Gram-negative bacteria, and is associated with an increased virulence of many pathogens [85]. Similar to other secretion systems, T6SS genes are clustered in pathogenicity islands containing 20 or more open reading frames. The hallmark of the T6SS expression is the presence of the conserved Hcp protein in the external medium [86]. VgrG proteins represent the other most common type of protein found secreted in a T6SS-dependent fashion. It was shown that in *Vibrio cholerae*, VgrG-1 is responsible for T6SS-dependent cytotoxic effects of *V. cholerae* on host cells including *Dictyostelium discoideum* amoebae and J774 macrophages [87]. The C terminus of VgrG-1 encodes a 548 residue-long actin cross-linking domain or ACD [87], which is also found embedded in a secreted toxin of *V. cholerae* called RtxA. VgrG orthologs in bacterial species other than *V. cholerae* carry a wide range of putative effector domains fused to their C termini [87].

The crystal structure of the N-terminal fragment the *Escherichia coli* CFT073 VgrG protein encoded by ORF c3393 shows a significant structural similarity to the gp5-gp27 complex, despite only 13% sequence identity [84]. The crystal structure of Hcp1 [88], the most abundant secreted protein in T6SS-expressing *Pseudomonas aeruginosa* strain PAO1, shows that it is homologous to the tandem 'tube' domain of gp27, which interacts with the T4 tail tube. Hcp1 is a donut-shaped hexamer with external and internal diameters of 85 Å and 40 Å, respectively. These hexamers stack on top of each other head-to-tail to form continuous tubes in the crystals. Some Hcp proteins can form tubes *in vitro* [84]. The homology of these two key proteins to the phage tail proteins and the fact that VgrG is translocated across a lipid membrane into a target cell suggest that the T6SS machine and phage tails might have a common ancestor.

Many evolutionary questions deal with the chicken and egg paradox. Whether the phage tail has evolved from the T6SS or vice versa is one of those questions. Clearly, the phage and its host benefit from coexistence and are capable of exchanging not only small proteins and protein domains, but also large and sophisticated supramolecular assemblies.

Abbreviations

Gp: gene product; HEWL: hen egg white lysozyme; LPS: lipopolysaccharide; NAG: N-acetylglucosamine; NAM: N-acetylmuramic acid; ORF: open reading frame; RMSD: root mean square deviation; T4L: lysozyme of T4 phage encoded by gene *e*; T6SS: bacterial type VI secretion system.

Acknowledgements

This article is part of the series of reviews on "Bacteriophage T4 and its Relatives". We are very grateful to Jim Karam for organizing this series, which is an essential update to the previously published but now somewhat outdated books and reviews on T4.
Movies 1 and 2 were made in close collaboration with the Seyet LLC, a scientific visualization company. We thank Seyet for their support.

Author details

[1]Ecole Polytechnique Fédérale de Lausanne (EPFL), Institut de physique des systèmes biologiques, BSP-415, CH-1015 Lausanne, Switzerland. [2]Tokyo Institute of Technology, Graduate School of Bioscience and Biotechnology, B-9 4259 Nagatsuta, Midori-ku, Yokohama, 226-8501, Japan. [3]Centro Nacional de Biotecnologia, Dpto de Estructura de Macromoleculas, Campus Cantoblanco c/Darwin 3, E-28049 Madrid, Spain. [4]Duke-NUS Graduate Medical School, Program in Emerging Infectious Diseases, 30 Medical Drive, Brenner Centre for Molecuar Medicine, NUS, 117609, Singapore. [5]Purdue University, Department of Biological Sciences, 915 W. State Street, West Lafayette, IN 47907-2054, USA.

Authors' contributions

PGL, FA and MJvR made a summary of the available data and wrote the text. MGR carried out a major revision of the text. PGL, VAK, AAA and SK prepared the figures. PGL and Seyet LLC created Movies 1 and 2. AAA created Movie 3. All authors read and approved the final manuscript.

Competing interests

The authors declare that they have no competing interests.

Received: 25 October 2010 Accepted: 3 December 2010
Published: 3 December 2010

References

1. Ackermann HW: **Bacteriophage observations and evolution.** *Res Microbiol* 2003, **154**:245-251.
2. Goldberg E, Grinius L, Letellier L: **Recognition, attachment, and injection.** In *Molecular Biology of Bacteriophage T4*. Edited by: Karam JD. Washington, D.C.: American Society for Microbiology; 1994:347-356.
3. Epstein RH, Bolle A, Steinberg C, Kellenberger E, Boy de la Tour E, Chevalley R, Edgar R, Susman M, Denghardt C, Lielausis I: **Physiological studies of conditional lethal mutants of bacteriophage T4D.** *Cold Spring Harbor Symposia on Quantitative Biology* 1963, **28**:375-392.
4. Eiserling FA, Black LW: **Pathways in T4 morphogenesis.** In *Molecular Biology of Bacteriophage T4*. Edited by: Karam JD. Washington, D.C.: American Society for Microbiology; 1994:209-212.
5. Kostyuchenko VA, Leiman PG, Chipman PR, Kanamaru S, van Raaij MJ, Arisaka F, Mesyanzhinov VV, Rossmann MG: **Three-dimensional structure of bacteriophage T4 baseplate.** *Nat Struct Biol* 2003, **10**:688-693.
6. Leiman PG, Chipman PR, Kostyuchenko VA, Mesyanzhinov VV, Rossmann MG: **Three-dimensional rearrangement of proteins in the tail of bacteriophage T4 on infection of its host.** *Cell* 2004, **118**:419-429.
7. Aksyuk AA, Leiman PG, Shneider MM, Mesyanzhinov VV, Rossmann MG: **The structure of gene product 6 of bacteriophage T4, the hinge-pin of the baseplate.** *Structure* 2009, **17**:800-808.
8. Kostyuchenko VA, Navruzbekov GA, Kurochkina LP, Strelkov SV, Mesyanzhinov VV, Rossmann MG: **The structure of bacteriophage T4 gene product 9: the trigger for tail contraction.** *Structure Fold Des* 1999, **7**:1213-1222.
9. Leiman PG, Kostyuchenko VA, Shneider MM, Kurochkina LP, Mesyanzhinov VV, Rossmann MG: **Structure of bacteriophage T4 gene product 11, the interface between the baseplate and short tail fibers.** *J Mol Biol* 2000, **301**:975-985.
10. Leiman PG, Shneider MM, Kostyuchenko VA, Chipman PR, Mesyanzhinov VV, Rossmann MG: **Structure and location of gene product 8 in the bacteriophage T4 baseplate.** *J Mol Biol* 2003, **328**:821-833.
11. Leiman PG, Shneider MM, Mesyanzhinov VV, Rossmann MG: **Evolution of bacteriophage tails: Structure of T4 gene product 10.** *J Mol Biol* 2006, **358**:912-921.
12. van Raaij MJ, Schoehn G, Burda MR, Miller S: **Crystal structure of a heat and protease-stable part of the bacteriophage T4 short tail fibre.** *J Mol Biol* 2001, **314**:1137-1146.
13. Kanamaru S, Leiman PG, Kostyuchenko VA, Chipman PR, Mesyanzhinov VV, Arisaka F, Rossmann MG: **Structure of the cell-puncturing device of bacteriophage T4.** *Nature* 2002, **415**:553-557.
14. Thomassen E, Gielen G, Schutz M, Schoehn G, Abrahams JP, Miller S, van Raaij MJ: **The structure of the receptor-binding domain of the bacteriophage T4 short tail fibre reveals a knitted trimeric metal-binding fold.** *J Mol Biol* 2003, **331**:361-373.
15. Rossmann MG, Arisaka F, Battisti AJ, Bowman VD, Chipman PR, Fokine A, Hafenstein S, Kanamaru S, Kostyuchenko VA, Mesyanzhinov VV, Shneider MM, Morais MC, Leiman PG, Palermo LM, Parrish CR, Xiao C: **From structure of the complex to understanding of the biology.** *Acta Crystallogr D Biol Crystallogr* 2007, **63**:9-16.
16. Berget PB, King J: **Isolation and characterization of precursors in T4 baseplate assembly. The complex of gene 10 and gene 11 products.** *J Mol Biol* 1978, **124**:469-486.
17. Coombs DH, Arisaka F: **T4 tail structure and function.** In *Molecular Biology of Bacteriophage T4*. Edited by: Karam JD. Washington, D.C.: American Society for Microbiology; 1994:259-281.
18. Leiman PG, Kanamaru S, Mesyanzhinov VV, Arisaka F, Rossmann MG: **Structure and morphogenesis of bacteriophage T4.** *Cell Mol Life Sci* 2003, **60**:2356-2370.
19. Mosig G, Eiserling F: **T4 and related phages: structure and development.** In *The Bacteriophages*. Edited by: R C, ST A. Oxford: Oxford University Press; 2006.
20. Kikuchi Y, King J: **Assembly of the tail of bacteriophage T4.** *J Supramol Struct* 1975, **3**:24-38.
21. Kikuchi Y, King J: **Genetic control of bacteriophage T4 baseplate morphogenesis. III. Formation of the central plug and overall assembly pathway.** *J Mol Biol* 1975, **99**:695-716.
22. Kikuchi Y, King J: **Genetic control of bacteriophage T4 baseplate morphogenesis. II. Mutants unable to form the central part of the baseplate.** *J Mol Biol* 1975, **99**:673-694.
23. Kikuchi Y, King J: **Genetic control of bacteriophage T4 baseplate morphogenesis. I. Sequential assembly of the major precursor, in vivo and in vitro.** *J Mol Biol* 1975, **99**:645-672.
24. Ferguson PL, Coombs DH: **Pulse-chase analysis of the in vivo assembly of the bacteriophage T4 tail.** *J Mol Biol* 2000, **297**:99-117.
25. Yap ML, Mio K, Leiman PG, Kanamaru S, Arisaka F: **The baseplate wedges of bacteriophage T4 spontaneously assemble into hubless baseplate-like structure in vitro.** *J Mol Biol* 2010, **395**:349-360.
26. Snustad DP: **Dominance interactions in Escherichia coli cells mixedly infected with bacteriophage T4D wild-type and amber mutants and**

their possible implications as to type of gene-product function: catalytic vs. stoichiometric. *Virology* 1968, **35**:550-563.

27. Nieradko J, Koszalka P: Evidence of interactions between Gp27 and Gp28 constituents of the central part of bacteriophage T4 baseplate. *Acta Microbiol Pol* 1999, **48**:233-242.

28. Abuladze NK, Gingery M, Tsai J, Eiserling FA: Tail length determination in bacteriophage T4. *Virology* 1994, **199**:301-310.

29. Akhter T, Zhao L, Kohda A, Mio K, Kanamaru S, Arisaka F: The neck of bacteriophage T4 is a ring-like structure formed by a hetero-oligomer of gp13 and gp14. *Biochim Biophys Acta* 2007, **1774**:1036-1043.

30. Kao SH, McClain WH: Baseplate protein of bacteriophage T4 with both structural and lytic functions. *J Virol* 1980, **34**:95-103.

31. Mosig G, Lin GW, Franklin J, Fan WH: Functional relationships and structural determinants of two bacteriophage T4 lysozymes: a soluble (gene e) and a baseplate-associated (gene 5) protein. *New Biol* 1989, **1**:171-179.

32. Matthews BW, Remington SJ: The three dimensional structure of the lysozyme from bacteriophage T4. *Proc Natl Acad Sci USA* 1974, **71**:4178-4182.

33. Kanamaru S, Gassner NC, Ye N, Takeda S, Arisaka F: The C-terminal fragment of the precursor tail lysozyme of bacteriophage T4 stays as a structural component of the baseplate after cleavage. *J Bacteriol* 1999, **181**:2739-2744.

34. Murzin AG: OB(oligonucleotide/oligosaccharide binding)-fold: common structural and functional solution for non-homologous sequences. *Embo J* 1993, **12**:861-867.

35. Arcus V: OB-fold domains: a snapshot of the evolution of sequence, structure and function. *Curr Opin Struct Biol* 2002, **12**:794-801.

36. Kuroki R, Weaver LH, Matthews BW: A covalent enzyme-substrate intermediate with saccharide distortion in a mutant T4 lysozyme. *Science* 1993, **262**:2030-2033.

37. Zorzopulos J, Kozloff LM: Identification of T4D bacteriophage gene product 12 as the baseplate zinc metalloprotein. *J Biol Chem* 1978, **253**:5543-5547.

38. van Raaij MJ, Schoehn G, Jaquinod M, Ashman K, Burda MR, Miller S: Identification and crystallisation of a heat- and protease-stable fragment of the bacteriophage T4 short tail fibre. *Biol Chem* 2001, **382**:1049-1055.

39. Leiman PG, Molineux IJ: Evolution of a new enzyme activity from the same motif fold. *Mol Microbiol* 2008, **69**:287-290.

40. Stummeyer K, Dickmanns A, Muhlenhoff M, Gerardy-Schahn R, Ficner R: Crystal structure of the polysialic acid-degrading endosialidase of bacteriophage K1F. *Nat Struct Mol Biol* 2005, **12**:90-96.

41. Stummeyer K, Schwarzer D, Claus H, Vogel U, Gerardy-Schahn R, Muhlenhoff M: Evolution of bacteriophages infecting encapsulated bacteria: lessons from Escherichia coli K1-specific phages. *Mol Microbiol* 2006, **60**:1123-1135.

42. Spinelli S, Desmyter A, Verrips CT, de Haard HJ, Moineau S, Cambillau C: Lactococcal bacteriophage p2 receptor-binding protein structure suggests a common ancestor gene with bacterial and mammalian viruses. *Nat Struct Mol Biol* 2006, **13**:85-89.

43. Matsui T, Griniuviene B, Goldberg E, Tsugita A, Tanaka N, Arisaka F: Isolation and characterization of a molecular chaperone, gp57A, of bacteriophage T4. *J Bacteriol* 1997, **179**:1846-1851.

44. Ali SA, Iwabuchi N, Matsui T, Hirota K, Kidokoro S, Arai M, Kuwajima K, Schuck P, Arisaka F: Reversible and fast association equilibria of a molecular chaperone, gp57A, of bacteriophage T4. *Biophys J* 2003, **85**:2606-2618.

45. Haggard-Ljungquist E, Jacobsen E, Rishovd S, Six EW, Nilssen O, Sunshine MG, Lindqvist BH, Kim KJ, Barreiro V, Koonin EV, Calendar R: Bacteriophage P2: genes involved in baseplate assembly. *Virology* 1995, **213**:109-121.

46. Morgan GJ, Hatfull GF, Casjens S, Hendrix RW: Bacteriophage Mu genome sequence: analysis and comparison with Mu-like prophages in Haemophilus, Neisseria and Deinococcus. *J Mol Biol* 2002, **317**:337-359.

47. Kostyuchenko VA, Chipman PR, Leiman PG, Arisaka F, Mesyanzhinov VV, Rossmann MG: The tail structure of bacteriophage T4 and its mechanism of contraction. *Nat Struct Mol Biol* 2005, **12**:810-813.

48. Watts NR, Coombs DH: Analysis of near-neighbor contacts in bacteriophage T4 wedges and hubless baseplates by using a cleavable chemical cross-linker. *J Virol* 1989, **63**:2427-2436.

49. Watts NR, Coombs DH: Structure of the bacteriophage T4 baseplate as determined by chemical cross-linking. *J Virol* 1990, **64**:143-154.

50. Watts NR, Hainfeld J, Coombs DH: Localization of the proteins gp7, gp8 and gp10 in the bacteriophage T4 baseplate with colloidal gold:F(ab)2 and undecagold:Fab' conjugates. *J Mol Biol* 1990, **216**:315-325.

51. Kellenberger E, Stauffer E, Haner M, Lustig A, Karamata D: Mechanism of the long tail-fiber deployment of bacteriophages T-even and its role in adsorption, infection and sedimentation. *Biophys Chem* 1996, **59**:41-59.

52. Efimov VP, Kurochkina LP, Mesyanzhinov VV: Engineering of bacteriophage T4 tail sheath protein. *Biochemistry (Mosc)* 2002, **67**:1366-1370.

53. Aksyuk AA, Leiman PG, Kurochkina LP, Shneider MM, Kostyuchenko VA, Mesyanzhinov VV, Rossmann MG: The tail sheath structure of bacteriophage T4: a molecular machine for infecting bacteria. *Embo J* 2009, **28**:821-829.

54. Arisaka F, Takeda S, Funane K, Nishijima N, Ishii S: Structural studies of the contractile tail sheath protein of bacteriophage T4. 2. Structural analyses of the tail sheath protein, gp18, by limited proteolysis, immunoblotting, and immunoelectron microscopy. *Biochemistry* 1990, **29**:5057-5062.

55. Takeda S, Arisaka F, Ishii S, Kyogoku Y: Structural studies of the contractile tail sheath protein of bacteriophage T4. 1. Conformational change of the tail sheath upon contraction as probed by differential chemical modification. *Biochemistry* 1990, **29**:5050-5056.

56. Takeda Y, Suzuki M, Yamada T, Kageyama F, Arisaka F: Mapping of functional sites on the primary structure of the contractile tail sheath protein of bacteriophage T4 by mutation analysis. *Biochim Biophys Acta (BBA) - Proteins & Proteomics* 2004, **1699**:163-171.

57. Amos LA, Klug A: Three-dimensional image reconstructions of the contractile tail of T4 bacteriophage. *J Mol Biol* 1975, **99**:51-64.

58. Dewey MJ, Wiberg JS, Frankel FR: Genetic control of whisker antigen of bacteriophage T4D. *J Mol Biol* 1974, **84**:625-634.

59. Efimov VP, Nepluev IV, Sobolev BN, Zurabishvili TG, Schulthess T, Lustig A, Engel J, Haener M, Aebi U, Venyaminov SY, Potekhin SA, Mesyanzhinov VV: Fibritin encoded by bacteriophage T4 gene wac has a parallel triple-stranded alpha-helical coiled-coil structure. *J Mol Biol* 1994, **242**:470-486.

60. Boudko SP, Londer YY, Letarov AV, Sernova NV, Engel J, Mesyanzhinov VV: Domain organization, folding and stability of bacteriophage T4 fibritin, a segmented coiled-coil protein. *Eur J Biochem* 2002, **269**:833-841.

61. Tao Y, Strelkov SV, Mesyanzhinov VV, Rossmann MG: Structure of bacteriophage T4 fibritin: a segmented coiled coil and the role of the C-terminal domain. *Structure* 1997, **5**:789-798.

62. Wood WB, Eiserling FA, Crowther RA: Long tail fibers: genes, proteins, structure, and assembly. In *Molecular Biology of Bacteriophage T4*. Edited by: Karam JD. Washington, D.C.: American Society for Microbiology; 1994:282-290.

63. Henning U, Hashemolhosseini S: Receptor Recognition by T-Even-Type Coliphages. In *Molecular Biology of Bacteriophage T4*. Edited by: Karam JD. Washiington, D.C.: American Society for Microbiology; 1994:291-298.

64. Wood WB, Henninger M: Attachment of tail fibers in bacteriophage T4 assembly: some properties of the reaction in vitro and its genetic control. *J Mol Biol* 1969, **39**:603-618.

65. Cerritelli ME, Wall JS, Simon MN, Conway JF, Steven AC: Stoichiometry and domainal organization of the long tail-fiber of bacteriophage T4: a hinged viral adhesin. *J Mol Biol* 1996, **260**:767-780.

66. Kutter E, Guttman B, Batts D, Peterson S, Djavakhishvili T, Stidham T, Arisaka F, Mesyanzhinov V, Ruger W, Mosig G: Genomic map of bacteriophage T4. In *Genomic Maps*. Edited by: O'Brien SJ. New York: Cold Spring Harbor Laboratory Press; 1993:1-27.

67. Burda MR, Miller S: Folding of coliphage T4 short tail fiber in vitro. Analysing the role of a bacteriophage-encoded chaperone. *Eur J Biochem* 1999, **265**:771-778.

68. Hashemolhosseini S, Stierhof YD, Hindennach I, Henning U: Characterization of the helper proteins for the assembly of tail fibers of coliphages T4 and lambda. *J Bacteriol* 1996, **178**:6258-6265.

69. Qu Y, Hyman P, Harrah T, Goldberg E: In vivo bypass of chaperone by extended coiled-coil motif in T4 tail fiber. *J Bacteriol* 2004, **186**:8363-8369.

70. Runnels JM, Soltis D, Hey T, Snyder L: Genetic and physiological studies of the role of the RNA ligase of bacteriophage T4. *J Mol Biol* 1982, **154**:273-286.

71. Smith NL, Taylor EJ, Lindsay AM, Charnock SJ, Turkenburg JP, Dodson EJ, Davies GJ, Black GW: Structure of a group A streptococcal phage-

encoded virulence factor reveals a catalytically active triple-stranded beta-helix. *Proc Natl Acad Sci USA* 2005, **102**:17652-17657.

72. Olia AS, Casjens S, Cingolani G: **Structure of phage P22 cell envelope-penetrating needle.** *Nat Struct Mol Biol* 2007, **14**:1221-1226.

73. Mitraki A, Barge A, Chroboczek J, Andrieu JP, Gagnon J, Ruigrok RW: **Unfolding studies of human adenovirus type 2 fibre trimers. Evidence for a stable domain.** *Eur J Biochem* 1999, **264**:599-606.

74. Marusich EI, Kurochkina LP, Mesyanzhinov W: **Chaperones in bacteriophage T4 assembly.** *Biochemistry (Mosc)* 1998, **63**:399-406.

75. Bartual SG, Garcia-Doval C, Alonso J, Schoehn G, van Raaij MJ: **Two-chaperone assisted soluble expression and purification of the bacteriophage T4 long tail fibre protein gp37.** *Protein Expr Purif* 2010, **70**:116-121.

76. Arisaka F, Engel J, Klump H: **Contraction and dissociation of the bacteriophage T4 tail sheath induced by heat and urea.** *Prog Clin Biol Res* 1981, **64**:365-379.

77. Caspar DL: **Movement and self-control in protein assemblies. Quasi-equivalence revisited.** *Biophys J* 1980, **32**:103-138.

78. Simon LD, Anderson TF: **The infection of Escherichia coli by T2 and T4 bacteriophages as seen in the electron microscope. II. Structure and function of the baseplate.** *Virology* 1967, **32**:298-305.

79. Simon LD, Anderson TF: **The infection of Escherichia coli by T2 and T4 bacteriophages as seen in the electron microscope. I. Attachment and penetration.** *Virology* 1967, **32**:279-297.

80. Kondou Y, Kitazawa D, Takeda S, Tsuchiya Y, Yamashita E, Mizuguchi M, Kawano K, Tsukihara T: **Structure of the central hub of bacteriophage Mu baseplate determined by X-ray crystallography of gp44.** *J Mol Biol* 2005, **352**:976-985.

81. Fokine A, Chipman PR, Leiman PG, Mesyanzhinov VV, Rao VB, Rossmann MG: **Molecular architecture of the prolate head of bacteriophage T4.** *Proc Natl Acad Sci USA* 2004, **101**:6003-6008.

82. Simpson AA, Tao Y, Leiman PG, Badasso MO, He Y, Jardine PJ, Olson NH, Morais MC, Grimes S, Anderson DL, Baker TS, Rossmann MG: **Structure of the bacteriophage phi29 DNA packaging motor.** *Nature* 2000, **408**:745-750.

83. Orlova EV, Gowen B, Droge A, Stiege A, Weise F, Lurz R, van Heel M, Tavares P: **Structure of a viral DNA gatekeeper at 10 A resolution by cryo-electron microscopy.** *Embo J* 2003, **22**:1255-1262.

84. Leiman PG, Basler M, Ramagopal UA, Bonanno JB, Sauder JM, Pukatzki S, Burley SK, Almo SC, Mekalanos JJ: **Type VI secretion apparatus and phage tail-associated protein complexes share a common evolutionary origin.** *Proc Natl Acad Sci USA* 2009, **106**:4154-4159.

85. Bingle LE, Bailey CM, Pallen MJ: **Type VI secretion: a beginner's guide.** *Curr Opin Microbiol* 2008, **11**:3-8.

86. Raskin DM, Seshadri R, Pukatzki SU, Mekalanos JJ: **Bacterial genomics and pathogen evolution.** *Cell* 2006, **124**:703-714.

87. Pukatzki S, Ma AT, Revel AT, Sturtevant D, Mekalanos JJ: **Type VI secretion system translocates a phage tail spike-like protein into target cells where it cross-links actin.** *Proc Natl Acad Sci USA* 2007, **104**:15508-15513.

88. Mougous JD, Cuff ME, Raunser S, Shen A, Zhou M, Gifford CA, Goodman AL, Joachimiak G, Ordonez CL, Lory S, Walz T, Joachimiak A, Mekalanos JJ: **A virulence locus of Pseudomonas aeruginosa encodes a protein secretion apparatus.** *Science* 2006, **312**:1526-1530.

89. Chenna R, Sugawara H, Koike T, Lopez R, Gibson TJ, Higgins DG, Thompson JD: **Multiple sequence alignment with the Clustal series of programs.** *Nucl Acids Res* 2003, **31**:3497-3500.

doi:10.1186/1743-422X-7-355
Cite this article as: Leiman *et al.*: **Morphogenesis of the T4 tail and tail fibers.** *Virology Journal* 2010 **7**:355.

www.ingramcontent.com/pod-product-compliance
Lightning Source LLC
Chambersburg PA
CBHW041700210326
41598CB00007B/482